THE HISTORY OF THE BAROMETER

Evangelista Torricelli. *(Courtesy of the Trustees of the British Museum)*

THE
HISTORY
OF THE
BAROMETER

BY

W. E. KNOWLES MIDDLETON

THE JOHNS HOPKINS PRESS
BALTIMORE

DEDICATION

*This book is dedicated to the memory of
Dr. E. W. R. Steacie, F.R.S., formerly
President of the National Research Council
of Canada, as an expression of my most
profound respect and admiration.*

PREFACE

WHEN IN 1941 I was writing *Meteorological Instruments,* I found myself constantly obliged to forego the pleasure of giving historical references, which would have been quite foreign to the practical purpose of that book, and in any event could not have given an adequate idea of the history of any meteorological instrument. But I did acquire a liking for the subject, which gradually ripened into a strong desire to write an extended account of the history of some of these instruments, especially the barometer. I realized, however, that a project of this sort would have to await the opportunity of spending a good deal of time in libraries and museums in Europe, as well as in North America, and it was not until twenty years later that this opportunity occurred.

The barometer, of course, is much more than a meteorological instrument, having been important in many researches in physics and chemistry, and indeed having been regarded for two centuries as an essential part of the equipment of most laboratories. Its invention also had an immediate effect on man's view of the world, as I shall try to show. It is therefore surprising that there have not been several books written about its history; but as far as I am aware, no work has hitherto been published on the history of the barometer as a scientific instrument, rather than as a decorative object.

This book is solely concerned with the barometer as an instrument, and its scientific background, and only occasional reference has been made to any of the countless barometers which have been designed and made purely as pieces of furniture to be hung in the hall at the bottom of the stairs. The exceptions are unusual types of barometer generally found in such situations. I believe that this is an essential limitation in view of the purpose of the book.

I am less happy about another limitation, this time geographical, imposed upon me by the division of the world which darkens the present decade. In my case this limitation has been not only political but also linguistic and financial, so that I can claim to have made a study of the history of the barometer only in Western Europe and North America. I have visited a number of museums and other institutions and have seen many barometers, the more interesting of which are mentioned in the Appendix. There are undoubtedly many interesting barometers that I have not examined, and I trust that their owners and curators will forgive me for having passed them by.

The concept "barometer" has been interpreted rather strictly as meaning any instrument, no matter how it works, used for measuring the pressure *of the atmosphere.* This has led to the omission of the many instruments designed primarily to measure pressures artificially produced, even though they work like barometers of various kinds. I have also stopped short of the specialized instruments used in aircraft for measuring height, feeling that these ought rather to form part of a history of aeronautics. I have not discussed the measurement of pressure by the observation of boiling points, which seems to belong to the story of temperature measurement, being indeed involved in the development of thermometric scales. Within the boundary imposed by my definition of a barometer, I have tried not to limit this history in any temporal sense, and I hope it is reasonably complete up to 1960.

There is one *caveat* which I think it important to put in, and this concerns instrument makers. In preparing a book of this sort, there is, I find, a temptation to spend a wholly unjustifiable amount of time in searching for biographical details of the clever workmen to whom we owe so much of the development of the art. Much more has been written about the instrument makers of some countries than of others, this being at least partly a reflection of the relative social status of the profession or trade. England, where a number of the leading instrument makers in the eighteenth and early nineteenth centuries became Fellows of the Royal Society, is fairly well served in this respect; France and Germany much less so, and other European countries scarcely at all. One can undoubtedly accumulate a great deal of information by searching directories and civic records, and indeed that is how the rare and invaluable books on instrument makers are compiled; but I have had to keep constantly in mind the fact that this is a book about the barometer, and that time is a non-renewable resource. I have therefore—with some reluctance—adopted the plan of incorporating in footnotes whatever information of sufficient interest I have been able to gather about the makers of particular barometers referred to in the text. It should be pointed out that while the great majority of barometer makers in the eighteenth and nineteenth centuries made *only* decorative instruments of no scientific interest whatever, thus complicating matters enormously, there were still a relatively large number who did contribute to the evolution of the barometer as an instrument. This contrasts with the history of the telescope, for example, which can profitably be based on a series of biographies of famous makers, as Rolf Riekher has recently demonstrated (*Fernrohre und ihre Meister*, Berlin, 1957). The reader will find much more information about English makers than about those of other countries. While I realize that this is a fault, I plead that it is because the English themselves have paid more attention to these valuable men.

As far as possible, this book has been written from primary sources. Every effort has been made to ensure the accuracy of citations, and I shall be grateful to any readers who will take the trouble to point out errors and, of course, omissions. In quoting from printed documents of the seventeenth and eighteenth centuries, the spelling has been preserved, but the unessential capitals and italics favored by the printers of that time have been done away with. On the other hand manuscripts have been quoted as exactly as possible. Translations from other languages, unless otherwise noted, are my own. Except for a few short passages in French, usually not longer than a single sentence, foreign-language quotations appear in English in the text.

A word about the illustrations: With a few obvious exceptions, these have been chosen, or prepared, not to decorate the book but to clarify the text. I believe it is highly desirable in a book of this kind to reproduce original diagrams as far as possible, but where the desired information could be presented more clearly by a new drawing, I have not hesitated to use one. In some instances this has been done simply because an otherwise desirable original could not be properly reproduced.

In the preparation of a work of this scope I am naturally under obligation to a great many people, not least to the Directors and staff of the many museums and libraries that I had the pleasure of visiting, and at which I invariably received unfailing kindness and indispensable help. I trust that these ladies and gentlemen will interpret the references to their institutions in the text as an expression of heartfelt gratitude. I ought, however, to make special reference to a few librarians upon whose time and patience I must have made serious inroads: Mr. I. Kaye, Librarian of the Royal Society, and Mr. N. H. Robinson, the Assistant Librarian; Mr. R. S. Read of the Royal Meteorological Society; Mr. B. C. Lack of the Meteorological Office, Bracknell; Dr. Friedrich Klemm of the Deutsches Museum, Munich, who kindly put a private study room at my disposal; Mme A. Gauja of the Académie des Sciences, Paris; and finally, Mr. Jack E. Brown and his staff at the National Research Council Library in Ottawa, with whom I have had the pleasantest personal and professional relations for a number of years.

I wish to record my deep obligation to the late Dr. E. W. R. Steacie, whose untimely death in 1962 deprived Canada of a very great man. As President of the National Research Council of Canada, he took a practical interest in this book, which made its completion possible at a much earlier date than I had been able to hope for. I also wish to acknowledge the co-operation of Dr. L. E. Howlett, Director of the Division of Applied Physics of the same institution.

Professor Derek J. de Solla Price of Yale University read Part I of the book in manuscript and made valuable suggestions, for which I am extremely grateful. Professor Harry Woolf of The Johns Hopkins University and Dr. Robert P. Multhauf of the Smithsonian Institution, Washington, have been most helpful. I am very much obliged to Mr. Stephen K. Marshall and Mr. Nicholas Goodison for valuable suggestions about passages translated from the Latin.

Thanks are due to the Académie des Sciences for permission to quote from the *Régistres;* to the Royal Society for permission to quote from several manuscripts in their archives, and to reproduce Figure 6.6; to the History of Science Society for permission to reproduce part of Chapters 1 and 2, which originally appeared in *Isis;* and to the Union Internationale d'Histoire et de Philosophie des Sciences for permission to reprint part of Chapter 6 from *Archives internationales d'Histoire des Sciences.* Credits for certain of the illustrations are given in their captions.

My wife went far beyond the general encouragement and support that husbands normally receive, by typing the entire draft copy of the manuscript as well as later alterations, and by reading the proofs.

Finally, I wish to record the willing co-operation of the staff of the Johns Hopkins Press.

November, 1963
London, England   W. E. KNOWLES MIDDLETON

xii

CONTENTS

xiv

List of Figures

Part I

THE SCIENTIFIC BACKGROUND

"La scoperta del barometro ... fece cambiare l'aspetto della fisica, come il telescopio quello dell'astronomia, la circolazione del sangue quello della medicina, la pila del Volta quello della fisica molecolare."

Vincenzio Antinori, in his *Notizie istoriche relative all'Accademia del Cimento*, Florence, 1841, p. 27.

Chapter 1

The Prehistory
of the Barometer

THE invention of six valuable scientific instruments within a few decades of the seventeenth century undoubtedly had an immense effect in accelerating the development of science. These instruments were the telescope, the microscope, the air pump, the pendulum clock, the thermometer, and the barometer. They all made experiments and measurements possible which had been unimaginable before; but two of them directly challenged some venerable doctrines about the constitution of the world. These were the barometer and the telescope. It is worth while to try to find out how such devices came into being.

We are often told that the barometer was "invented" by Evangelista Torricelli about 1643. In the next chapter we shall examine the sense in which this is true; but the barometer did not appear out of nothing, and a history of the instrument must begin with a review of what had happened in this field before 1643.

The ideas held in classical antiquity about space, substance, matter, and vacuum are of great interest, but for our present purpose I shall endeavor to set out in a simplified way the state of thought on these questions at the end of the sixteenth century. It is often stated that learning at this time was dominated by the ideas of Aristotle (384–322 B.C.) as interpreted in the light of Christian dogma. It would be more correct to say that it was based on the ideas of a long series of commentators on Aristotle, many of whom had found it difficult to reconcile some of his conclusions with the teachings of the Church.

One such idea was that the vacuum must be *logically* impossible. Aristotle defined space as the extension of an object filling it; then if the word "vacuum" means a space which contains nothing, but is capable of containing something, there can be no vacuum. The most fundamental property of a body is extension, so that everything which has extension must be material. Aristotle's principal argument against the vacuum was, in fact, that in a vacuum there can be no dimensions, up or down, north or south, east or west. He was somehow able to accept a limit to the universe, but without admitting that there can be a vacuum

3

even outside this. What is more, light cannot penetrate a vacuum, so that nothing can be seen through one, and thus there cannot be a vacuum even between the earth and the stars.

It is quite outside the province of this book to enter into the details of this question, or to provide any sort of guide to the millions of words devoted to it through eighteen hundred years.[1] We may note that Democritus (ca. 420–370 B.C.) and Lucretius (ca. 98–55 B.C.) denied the impossibility of a vacuum, being of the opinion that there must be a vacuum between the discrete particles (atoms) of which, they thought, all matter is composed. In general, however, the belief that a vacuum is impossible was almost universally held until the end of the sixteenth century, though in the thirteenth century theologians were obliged to deny that a vacuum was *logically* impossible, because that would mean that God could not produce one if he wished. But they insisted that no natural or finite force could produce one.[2]

This belief in the impossibility of a vacuum carried with it some interesting consequences. For example, it was held that no force could pull apart two perfectly flat and polished plates. If they could be pulled apart, then they were not flat. A bellows that was completely tight could not be opened by any force whatever; if it could be opened, it must have leaked. Similarly

the limited height to which water could be sucked by a pump was ascribed to the imperfections of pumps. Water was believed to shrink on freezing (as nearly all other substances do), and the rupture of closed bottles of water was thought to be nature's way of avoiding the production of a vacuum.

Another idea which developed gradually during the Middle Ages was that of a semi-material fluid, an *aether*, a *spiritus mundi*, a "subtle matter" that was finer than air and could pass through the pores of solid bodies. As we shall see, this concept was to give rise to a great deal of controversy in the seventeenth century.

The Peripatetic philosophers, as the followers of Aristotle are called, believed that every substance seeks its own proper position in the universe. Of the four "elements," earth and water are heavy, air and fire light; so that it was almost universally denied that air has weight. This in spite of the fact that Aristotle wrote, ". . . in its own place every body has weight except fire, even air. It is proof of this that an inflated bladder weighs more than an empty one."[3] Only toward the end of the sixteenth century was there a fairly general revolt (outside the Schools) against the doctrine that air in itself is really weightless,[4] but it was still felt that air is without weight when in its natural place, surrounded by other air.

---

[1] A general and easily available account is to be found in E. J. Dijksterhuis, *The Mechanization of the World Picture*, trans. C. Dikshoorn (Oxford, 1961), *passim*.

[2] *Ibid.*, pp. 173–74.

[3] Aristotle, *On the Heavens*, trans. W. K. C. Guthrie (Loeb Classical Library; London & Cambridge, Mass., 1939) p. 355. This must have been one of the "thought experiments," popular even into the seventeenth century, that were imagined but never carried out.

[4] De Waard, *L'expérience barométrique* (Thouars, 1936), p. 38.

Regarding the pressure of the air, the situation was somewhat different. It was not that the pressure of fluids was denied by the ancients; it was simply not understood. One of the first to have a clear idea of vertical and lateral pressure in a liquid was a thirteenth-century writer, a pupil of Jean de Némore,[5] whose book was published three centuries later. "Therefore," he wrote, "the deeper the liquid is, the more it will compress the deeper parts; . . . for they are compressed both by the upper parts and by those next to them." The figure which accompanies the argument leaves no doubt that he was aware of lateral pressure.

This unknown scholar was far ahead of his time. Even in 1612 Galileo denied forcibly that there was any pressure in air or water. ". . . and even if we then add a very large quantity of water above the level which is equal to the altitude of the solid, we shall not on that account increase the pressure or weight of the parts surrounding the said solid."[6] And in 1615 he saw no difference between air and water in this respect. "Note that all the air in itself and above the water weighs nothing. . . . Nor let anyone be surprised that all the air weighs nothing at all, because it is like water."[7] Galileo was one of the greatest of men, but this seems to have been one of his blind spots.

It ought to be kept clearly in mind that the weight and the pressure of the air are two separate ideas; and we shall find in what follows that this was frequently lost sight of in the seventeenth century. The elasticity of the air is a third concept.

Let us now review briefly the state of knowledge and feeling about the vacuum and about the weight and pressure of the air in the period 1610–1630. The time was certainly ripe for the revival of the belief in the possibility of a vacuum, but to the clerics the very name of the vacuum was anathema, being associated with the atomistic theories of Epicurus and Lucretius,[8] which were felt to be heretical.

Whatever he may have thought, however wrong he may have been about the weight and pressure of fluids, Galileo had no doubts about the possibility of a vacuum. In 1612 he wrote many annotations in his copy of a book entitled *De phaenomenis in orbe lunae* by one Julius Caesar de Galla. After the phrase (concerning the vacuum) *"Quod si non sensu, ergo non intellectu,"* Galileo wrote: "If the vacuum cannot be recognized either by the senses or by the intellect, how have you managed to find out that it does not exist?"[9]

Another natural philosopher who was quite clear about the vacuum was Isaac Beeckman (1588–1637), who was born at Middelbourg in Holland. As a young man he seems to have accepted the idea of the vacuum and become aware that air presses in every direc-

[5] Iordani, *Opusculum de ponderositate, Nicolai Tartaleae studio correctum novisque figuris auctum* (Venetiis, 1565). De Waard, *ibid.*, p. 69, has given the relevant Latin text.
[6] Galileo, *Le Opere*, ed. nat., IV (1894), 79.
[7] *Ibid.*, p. 167.

[8] Lucretius' *The Nature of the Universe* is available in an excellent translation by R. E. Latham (Penguin Classics, 1950).
[9] Galileo, *Le Opere*, ed. nat., IIIA (1892), 350.

tion. In his journal,[10] sometime in 1614, he wrote:

Now what is the reason that bodies are moved in any direction, so that a vacuum may not exist in nature? I answer: it happens that air, after the manner of water, presses upon things and compresses them according to the depth of the superincumbent air. But some things remain undisturbed, and are not perpetually moved about, because they are everywhere equally compressed by the air above them, just as our divers are pressed by the water. But things rush towards an empty space with great force, on account of the immense depth of the superincumbent air, and in this way the weight of the air arises. But air must not be said to be heavy, because we walk in it without any pain, as indeed the fishes move in water, suffering no compression.

We should probably interpret the "must not be said" (*non dicendus est*) as "people will not say," because in several other parts of his Journal, Beeckman makes it quite clear that *he* would say that the air has weight. He also compares the air to a sponge, which can be condensed, but because of its elasticity tries to return to its previous state.[11] He knew[12] that air was enormously more compressible than water, which could also be compressed, but imperceptibly. He correctly explained the action of a suction pump by an appeal to the pressure of the air,[13] and in 1618 he had the courage to defend his views on this subject

in his thesis for the M.D. degree.[14] We do not know how this was received at the University of Caen, but he need not have been so brave, for his thesis was on intermittent fever, and the subject of atmospheric pressure seems to have been brought in just because he could not keep his ideas to himself.

This concern with vacua and pumps and the pressure of the air might be called the practical part of Beeckman's work. He also established himself in a complicated philosophical position regarding the supposed subtle matter or *aether*, with which, he thought, even the vacuum was filled.[15] This brought him into relations, and finally into collision, with René Descartes (1596–1650), a very great philosopher, most of whose ideas about physics have turned out to be wrong. Descartes never admitted the existence of a vacuum; nevertheless he will appear importantly in the history of the barometer at a later date.[16]

Beeckman, whose work might never have been known except for the patient scholarship of Cornelis de Waard,[17] continued to write publicly and privately about the vacuum, and it was probably as well for him that he lived in the Netherlands. We may leave him with a characteristic statement in a letter that he wrote to the Reverend

[10] Isaac Beeckman, *Journal tenu par lui de 1604 à 1634 publié avec une introduction et des notes par Cornelis de Waard* (4 vols., La Haye, Martinus Nijhoff, 1939–1953), I, 36.

[11] *Ibid.*, p. 46.

[12] *Ibid.*, pp. 42, 47, 175.

[13] *Ibid.*, pp. 46–47, 79, 200.

[14] There is a unique but incomplete copy of this in the British Museum, 1179.d.9(3).

[15] This is a little remote from our subject. The interested reader should refer to De Waard, *L'expérience barométrique*, Chap. VI, for a detailed discussion of the similarities and differences between the ideas of Beeckman and Descartes on these questions.

[16] See p. 45 below.

[17] The MS of his journal was destroyed in the war of 1939–1945.

Father Marin Mersenne in Paris on October 1, 1629:[18]

You bring forth good arguments about the vacuum. Indeed, if a vacuum is said to exist in the pores of air, water, lead, etc., or if all the space between the outermost bound of our atmosphere and the stars is said to be empty, nothing follows that is absurd. Really, although the philosophers babble about the necessity of all things being united, of the propagation of accidents and visible appearances in the air, of the impossibility of motion in a vacuum, etc., these seem to me to be old wives' tales; for I admit nothing in philosophy, unless it is represented to the imagination as being perceptible to the senses.

On June 2, 1631, Descartes wrote a letter,[19] probably to his pupil Reneri, in which many of these ideas were developed, with the difference that wool, rather than a sponge, was introduced as an analog, and that the Cartesian hypothesis of vortices is made much of. It has been claimed by the historian E. Gerland that in this letter "Descartes . . . had already given out the idea of the barometer, still without its having been proven by trial."[20] In view of this strong statement, it seems desirable to present a translation of the relevant part of the letter in question:

[18] In *Correspondance du P. Marin Mersenne, religieux minime. Commencée par Mme Paul Tannery, publiée & annotée par Cornelis de Waard* (Paris, 1936), II, 282–83.

[19] René Descartes, *Oeuvres*, eds. Chas. Adam et Paul Tannery (Paris, 1897–1911), I (1897), 205–8.

[20] E. Gerland in *Report of the International Meteorological Congress Held at Chicago, Ill., August 21–24, 1893*. ed. O. L. Fassig (*U.S. Weather Bureau Bull.*, no. 11, pt. 3 [Washington, 1896], p. 690).

SIR,

To resolve your difficulties, imagine the air to be like wool, and the aether which is in its pores to be like eddies of wind which move hither and thither in this wool; and consider that this wind, which acts from all sides between the little fibers of the wool, prevents them from pressing as hard against one-another as they might do without it. For they all have weight, and press upon one-another as much as the agitation of this wind can let them, so that the wool which is upon the earth is compressed by all that which is above, even to beyond the clouds; and this makes a great weight. So that if we had to raise a part of this wool which is, for example, at the point marked O (see Figure 1.1), with all that above on the line OPQ, we should need a very considerable force. Now this weight is not commonly felt in the air when we push it upwards; because if we raise a part, for example that which is at the point E towards F, that which is at F goes in a circle towards GHI and comes back to E; and so its weight is not felt at all, no more than that of a wheel is felt in turning it, when it is perfectly balanced on its axis. But in the example that you bring forward of a tube DR closed at the end D, where it is attached to the plane AB, the quicksilver which you suppose to be in it cannot begin to descend all at once unless the wool which is near R goes towards O, and that which is at O goes towards P and Q, and so on, till all the wool in the line OPQ is raised, and this, taken all together, is very heavy. For, the tube being closed at the top, no wool—I mean air—can come in to replace the quicksilver when it descends. You will say that wind—I mean aether—can very well enter through the pores of the tube. I admit this; but consider, that the aether which will come in can come only from the sky above; for although there is some everywhere in the

8

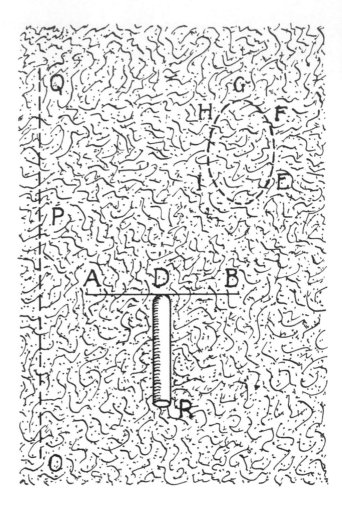

Fig. 1.1. Descartes' sketch in his letter to Reneri.

pores of the air, nevertheless there is no more than is necessary to fill them. Consequently if there is a new space to fill in the tube, aether which is in the sky above the air must come in, after the air has risen in its place.

And so that you will make no mistake, you must not think that this quicksilver cannot be separated from the top of the tube by any force, but only that it needs as much force as is required to raise all the air which there is from that point to above the clouds.

The remainder of the letter concerns the expansion of air by heat.

Surely it is unreasonable to see in this a prevision of the barometer. Apart from the fact that nothing is said about a cistern, there is no indication that Descartes realized that if the tube were long enough only part of the quicksilver would come out.

Others were thinking about the weight. and pressure of the air, for example Jean Rey (1583–1645)[21] and

[21] *Essai sur la recherche de la cause pour laquelle l'estain et le plomb augmentent de poids quand on les calcine* (Bazas, 1630). In this work Rey foreshadowed Lavoisier.

especially Giovanni Batista Baliani (1582–1666) of Genoa. On July 27, 1630, Baliani wrote very deferentially to the great Galileo about the explanation of an experiment he had made in which a siphon, led over a hill about 21 meters high, failed to work. Although the siphon had first been filled with water, when one end was opened the level in that limb dropped to about 10 meters above the reservoir, and when the other end was unstopped the level in the other limb dropped also.[22]

Galileo replied promptly enough, but somewhat condescendingly, on August 6, 1630: "The cause of such an effect rather worried me when I first looked into it; but it came to me at last that it should not be so recondite, but on the contrary quite manifest, as happens with true causes after they have been discovered."[23] He attributed the effect to an energy or attractive force belonging to the vacuum. It seems strange to us that a "nothing" should be thought to produce a force; but Govi[24] believes that Galileo was still affected to some extent by his scholastic education. At any rate, he persisted in his belief in the *"forza o resistenza del vuoto"* instead of accepting the correct explanation of the limited height to which water could be raised by a suction pump or a siphon. He thought that when a certain height was reached,

about 18 *braccia*,[25] the water column simply broke off, like a cord to which too large a weight had been attached.[26] He never seems to have understood the pressure of the air. It is quite astonishing that this very great man, who fought against so many of the old prejudices of the Peripatetic school, did not see the point in this instance. "But", to quote Bertelli,[27] "the fame of Galileo . . . nevertheless will remain forever luminous and undying, even though there escaped him some truths that in the sequel were well proved and that were also quite easy to deduce." We meet him again twice in the story of the barometer, the first time as the recipient of another letter from Baliani[28] dated October 24, 1630; a long letter in which it was made clear that Baliani had believed a vacuum could exist from the time he first knew that air had weight; and that "the higher we go in the air the less heavy it is."[29] Before his experiment with the siphon he would have believed that to produce a vacuum "a greater force would be required than that which can be exerted by the water in a pipe not over 80 feet long."[30] It will be seen that Baliani had come to the same conclusions as Beeckman; but Baliani was

[22] Galileo, *Le Opere*, ed. nat., XIV (1904), 124–25. This correspondence and its implications are discussed by Gilberto Govi, "Nota intorno al primo scopritore della pressione atmosferica," *Atti della R. Accad. delle scienze di Torino*, Vol. 2 (1867), pp. 562–81. Govi did not, of course, know about Beeckman.

[23] *Le Opere*, ed. nat., XIV (1904), 127–30.

[24] Govi, "Nota intorno, etc.," p. 567.

[25] The Florentine *braccio*, or ell, was about 0.54 meter according to De Waard, *L'expérience barométrique*, p. 187.

[26] Galileo, *Dialogues Concerning Two New Sciences* (1638). English trans. by F. Crew and A. de Salvio (New York, 1914), p. 16ff.

[27] P. Timoteo Bertelli, "Contributo alla storia del barometro," *Rivista Geogr. Ital.*, Vol. 13 (1906), pp. 169–81, 242–60. Quotation from p. 247.

[28] Galileo, *Le Opere*, ed. nat., XIV (1904), 157–60.

[29] *Ibid.*, p. 159.

[30] *Ibid.*, p. 160.

nearer the center of things, and more people knew of his views.

Among these were a group of people in Rome, friends of Rafael Magiotti (1597–1658), including Evangelista Torricelli, and probably Gasparo Berti, a young man who seems to have been born at Mantua, and who had a considerable reputation as a mathematician and astronomer. He also seems to have been exceedingly modest and to have left almost no writings. Galileo's *Discorsi*[31] arrived in Rome in December, 1638, and greatly excited the members of this circle. De Waard, who has examined all the relevant documents, is of the opinion that the *Discorsi* stimulated Magiotti to suggest a more convenient experiment than that with the siphon to study the production of a vacuum,[32] though I think that the idea might also have come directly from Galileo. It appears to have been Berti who actually carried out the experiment, but Magiotti was there, and so were Father Athanasius Kircher (1602–1680) and probably Father Nicolo Zucchi (1586–1670). Unfortunately the date of the experiment is very uncertain, as the only accounts of it were written some years later. It can scarcely have been earlier than 1639; on the other hand Berti certainly died before January 2, 1644, on which date his death is mentioned in a letter.[33] It may seem extraordinary that the circumstances of such an important experiment should be so uncertain, until we reflect that the mere word "vacuum" could raise very strong feelings in some quarters.

There are four accounts of Berti's experiment: by Magiotti, by Emmanuel Maignan, by Zucchi, and by Kircher. Of these that of Maignan is by far the most circumstantial; that of Magiotti is the account of an eyewitness written from memory five, six, or seven years later; the others were written by Jesuits who, though able scientists (Kircher in particular) had their minds made up in advance about the impossibility of a vacuum. Their accounts are also to some extent contradictory. The description of the experiment given by Maignan is of such importance to our subject that an English rendering will be given here.

Emmanuel Maignan or Magnanus (1601–1676) was born in Toulouse, and at the age of 35 was called to Rome, where he taught in the Convent on Monte Pincio. He undoubtedly had a very fine mind, and embraced the new scientific theories with great zeal. About 1650 he returned to Toulouse, where he wrote a very considerable work on Natural Philosophy in four volumes, which was greatly admired. It is on pages 1925 to 1936 of this work[34] that the description of Berti's experiment is given:

I refer to the notable experiment performed in Rome several years ago, when I myself dwelt there, by that most clever and erudite man Gaspar Berti, who was indeed my greatest friend. I relate faithfully how I was told by that same author of the entire affair. Seven or eight days later he told me of it and showed me

[31] See note 26 above.
[32] *L'experiénce barométrique*, pp. 101–10.
[33] *Ibid.*, p. 109.

[34] *Cursus philosophicus concinnatus ex notissimis cuique principiis, ac praesertim quod res physicas instauratas ex lege Naturae sensatis experimentis passim comprobata* (Toulouse, 1653). The four volumes are paged as one.

drawings, and indeed read to me two or three times what he had written down, carefully explaining what I had not been able to see with my own eyes, held by I know not which little distraction of my school. Thus I believe that I had nothing less than others had who saw these experiments, although I remember having read the writings of two of them who, though they wrote of this as if they had been eyewitnesses, do not wholly agree with the above-mentioned author, or indeed with each other, as I shall soon show. But if you do not believe, dear Reader, that I have eyes, you may believe that I have ears and that I heard from nearby. Because certainly in this affair I do not think I am going to injure anyone if with a good grace, and not disparaging his sincerity, I follow the author faithfully in fact and manner and circumstances.

Let me now come to the subject. This distinguished Gaspar, of whom I have spoken, erected a rather long leaden tube AB (Figure 1.2) out of doors on the wall of the tower of his house in which the stairs are, and made it secure by means of ropes fastened to iron clamps. I am obliged to say that I do not remember its length exactly, but I know that it must have been more (clearly very little more) than forty palms. The upper end A of the tube was opposite one of the windows of the tower, while the lower end B was not far from the ground, and was provided with a brass tap or cock R, this being within the cask EF, filled with water for the purpose. To the upper end A was fitted and carefully joined and cemented a glass vessel in the form of a flask, rather large but very solid, and having two necks and mouths, the wider one below, into which the end of the tube was inserted at A, as into a box; the narrower above at C, of lead or else tin, as is the custom, and well made so that it would fit the stem of the threaded brass screw (cochlea) D very closely, this being the most solid

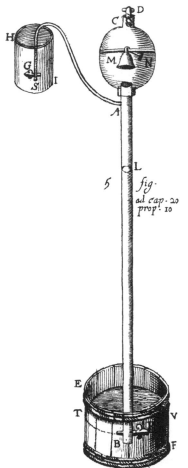

Fig. 1.2. Berti's experiment. (*Courtesy of the Trustees of the British Museum*)

kind of stopper and the most suitable for the business in hand.[35]

This being made ready, the tap R closed and the cask EF filled with water to about its middle, the entire tube, as well as the glass vessel, was filled from above through the opening C, right to the top. Then the opening C was closed with the screw D, in order to seal the entire apparatus.

[35] Probably an error in judgment, in view of what happened.

But by an unhappy chance the flask was broken when the screw, being turned many times and gradually thrust in, drove the water inwards from the leaden mouth of the flask, since, as I said, the water filled it to the top, even between the turns of the screw. For because, as I stated in Chap. 17, proposition 10, water struck or pressed in this way does not permit itself to be compressed, therefore being struck thus, it sought another place for itself and shattered the flask. So this excellent experimental work was discontinued, until on another day a new flask was cemented to the tube in a similar manner and very well closed by a screw plug put in from above, so that no air could get in anywhere.

At length, when the tap R was opened the water flowed (contrary to the hope of many)[36] out of the pipe into the cask EF, to an easily observable height; but not all of it flowed out, and it soon stood quite still. This was clear, because a mark was made in the cask at the surface of the water, and next day it was found that the water in it had remained exactly at the mark, although the tap R had been open all the time.[37] Then, when this tap R had been carefully closed again, the screw D was taken out above. And as soon as it was taken out, behold! the air rushed in with a loud noise, filling the space previously abandoned by the water. Then by lowering a sounding-line it was determined how much water was inside, or rather to what height the water stood in the tube, and it was found to stand about 18 cubits above the level of the water in the cask, at the mark L. I remember indeed that this was the number of cubits, or *braccia* as the Tuscans say, because it

was this number and this distance to which at that time, according to an observation made by Galileo, it was said that water could rise in a pump, as I have noted in prop. 8, no. 9.

On another day the experiment was tried in a different manner. The larger tube already mentioned was pierced near A at the side, and a curved lead tube AS, furnished with a tap G for convenience, was joined to it and soldered with tin. When everything was ready, water was poured in at C, as before. The plug D was put in, and the mouth S of the tube, with its tap G, was bent down into a vessel HI full of water, placed higher than the window,[38] so that obviously no external air could enter in any way through the tap G, when the water, as I am about to relate, receded and flowed out through the tap R. For when this was opened, the water flowed out as before to the mark L, as was apparent from the height of the mark previously made in the cask EF. Next the tap R was closed and the tap G opened. And then the small tube sucked up the water continually supplied to the vessel HI, until at last the entire space was filled, the tube above L, the flask, and the small tube, except that there appeared an air-bubble of a certain size under the mouth C.

Indeed the space occupied by this bubble, which some thought a vacuum, was really found to be full of air; because the tube AS and the vessel HI were raised a little so that the water in the latter was above C, so that by its weight it could run from above through A into the flask and fill it completely, as it seemed before that it could not be sucked in. But no water could be sent in farther, and so it was certain that the above-mentioned little space beneath C was not a vacuum,

[36] This may be an oblique reference to the Jesuits, especially Kircher and Zucchi.

[37] This way of conducting the experiment must have been imposed by the circumstance that the tube AB was made of lead, and therefore opaque.

[38] *Super fenestra.* The vessel HI must have been higher than shown in the figures; possibly on the roof or parapet.

but was indeed a plenum and a bubble of air, or as some preferred to think, of aether.

But opinions varied about where this air or aether came from, and whether or not it alone, rarefied, filled up the space which the water had left empty. For some said that this small portion of air had come in through the pores of the glass or of the lead, while others wanted it squeezed out of the water and then expanded to fill the whole flask and the tube down to L. And even that it could not be expanded any more by rarefaction, and that therefore the water was held suspended at this mark L by some cohesive force of the air, so that it neither descended nor flowed out any farther.

These things were reported by my good friend Gaspar. I add my own opinion. To be sure, when first asked I said that this air could not have come in through any pores or have been squeezed out of the water, so that it alone might occupy the space which was previously taken up not by it only, but by the water also. This for the reasons which I have given in similar circumstances in the preceding proposition.[39] I next added that this small quantity of air might well have come suddenly through the mouth of the tube S while the water was being swiftly sucked up by it and to that end continually being poured into the vessel HI, which was rather small. For in this filling and rapid emptying, or rather pouring and stirring up at the same time, it would be easy for some air to be mixed with the water, as happens in foam. Besides, as a result of this sudden fall and turbulence air springs up like a wind out of the water itself, as is well known in the fountains of Tivoli and of Tusculum. Whence I also added that it was at least possible that the air alone filled the space left by the water

when it went out through the tap R, since it had come in together with the new water through the tap G; certainly, because it was able to enter this way, we cannot prove directly, on the strength of this experiment, the thing that many people were loudly affirming, namely that the tube and flask, while they appeared empty after the water had flowed out, were full of this air, which had expanded.

I said above that the water would not have remained at the mark L unless it had been sustained by the balancing force of the surrounding air,[40] just as in the pump mentioned in prop. 8, nos. 8 and 9; especially as in the two cases the height of the water is the same, namely in the tube above TV (i.e., above the level of the water lying in the cask) to the mark L, and in the pipe of the pump from the level of the external water as far as I, where the piston is in the third figure.

Finally I said that I was not greatly troubled about the vacuum. First, because its impossibility cannot be proved without a circular argument; seeing that it cannot be proved except by experiments, while in these same experiments it is clearly assumed; all the more because the method of these experiments is given by false reasoning or none at all. Secondly, likewise (as you may see by this whole chapter), all these experiments are interpreted as favorably as possible, whether the fear of it is admitted or excluded. Thirdly, in particular, because (if I may omit other reasons) that concerning which there seems to be the greatest difficulty does not follow from the admission of the vacuum, namely that an accident,[41] that is to say

[39] Maignan, *Cursus*, etc., pp. 1885–1924. This deals with the Torricellian experiment, but he mentions (p. 1886) Valerianus Magnus (see Chap. 3), rather than Torricelli.

[40] This was published in 1653. It would be of the greatest interest to know whether Maignan had come to this conclusion at the time the experiment was made, thereby anticipating Torricelli to some extent (see p. 23 below).

[41] It was an old philosophical problem whether light was an "accident" or a "substance." Maignan was here being very "advanced."

light, would be without a subject or else would be generated from nothing inside that empty flask. No, I say, because light is a real body, even though extremely tenuous, as I have elsewhere stated fully enough in its proper place;[42] like the magnetic virtue, of which I have already shown in Chap. 14, prop. 33 that it is a substance and consists of magnetic spirits; and like other similar virtues.

These arguments were not entirely displeasing to a man equally well versed in physics and mathematics. He added that when he heard about the magnet there came into his mind what his great friend (and mine) Father Athanasius Kircher had declared; that in order to get more evidence on which the vacuum in the said flask might be admitted or rejected, the experiment should be tried again, with a bell such as M in the flask, and a suitable iron hammer N. This could be attracted by a magnet outside, and then the magnet could be removed so that the hammer could fall and strike the bell. If, he said, the sound of this blow should be heard, it would be all over with the vacuum, since sound cannot be produced in one. But though he had reflected on this, it still remained untried.[43]

To which, as I well remember, I at once responded that it is certainly impossible that sound should be produced in a vacuum or be propagated through it; but that the cause of the sound in the vacuum could very well be the bell and hammer in the empty flask; and from the percussion of the hammer and the bell there might arise a vibratory motion that would produce all the sound there was (as

you will see in Chap. 25). And besides, the vibrations of the bell in the vacuum could have been communicated by the support to the parts which hold it up in the interior of the flask, and therefore to the flask itself, which is in contact with the support.[44] Then by way of the flask to the external air; and by this the same motion is propagated to the ear, received by the ear-drum, and perceived. This is what we call sound, as I shall say in the above-mentioned Chap. 25, no. 23.

What is more, I added, we have also the example of the deaf man who, although his ears are open and well exposed to the sound carried by the air, yet hears nothing at all, but who joyfully and perfectly perceives musical sounds obviously not carried by the air, but only communicated by the musical vibrations of a harder body. Try this yourself: stop up your ears as well as possible, and grasp firmly in your teeth the peg of a lute played by yourself if you can, or by someone else; and you will hear all the music very well, even more loudly and agreeably than you will hear it on another occasion with open ears. We see from this how much better and more clearly the sound waves or vibrations of which sound consists are communicated to the internal organ of sensation (in which it is heard) when, the strings having been struck, the whole body of the lute, agitated to the same rhythms as the strings, first stimulates the teeth (for which the air can in no way suffice), and when from these the sound is carried by the jaws and by the solid bones of the head to the brain or to the above-mentioned organ, than they are communicated when they are con-

---

[42] Maignan described his corpuscular conception of light in his *Perspectiva horaria sive de Horographia gnomonica tum theoretica, tum practica libri quatuor*, (Romae, 1648) (quoted by de Waard, *L'expérience barométrique*, p. 189, note 6).

[43] But according to both Kircher and Zucchi, it must have been tried later.

[44] It is to be noted that when Boyle made acoustic experiments in a vacuum twenty years later he suspended his bell by a thin cord. *New Experiments Physico-Mechanicall touching the Spring of the Air, and its Effects*, etc. (Oxford, 1660), p. 205. (*Works*, 1772 ed., I, 62).

ducted in the usual manner from the strings to the surrounding fluid air, and thence to the ear, as I shall explain in the place above mentioned.

Consequently, I said, sound waves could be excited in the same way in those solid bodies the bell, the support, and the flask; and be conducted from these to the external air and so carried to the ear, so that sound would finally be heard. Therefore from the sound heard in this particular experiment it could scarcely be decided whether the flask was really empty or full.

Now these were our conversations at the time, and I do not know whether after this, before he departed this life,[45] he made the experiment with this little bell on some other occasion. I heard nothing whatever of the kind from him. I only read in Father Athanasius Kircher (in his remarkable book *Musurgia*)[46] and Father Nicolas Zucchi, who wrote a number of years later[47] that the experiment had in fact been carried out by Gaspar and that they had helped with it, and applied the magnet.

And I protest that I should never think of denying, in whole or in part, the good faith of these distinguished men. But it is not inappropriate if in my zeal for the truth, and in a friendly way, I propose some little difficulty of my own, which came to me in reading their writings on these questions.

There follows a criticism of the ac-counts which Fathers Zucchi and Kircher gave of the experiment.

It is more than probable that Father Maignan, one of the keenest minds of the seventeenth century, had formed opinions about the pressure of the atmosphere similar to those of Beeckman and Baliani, and this before the time of Berti's experiment. Whether or not he had at that time recognized, as did Torricelli some years later, that the weight of the column of liquid was balanced and sustained by that of a column of air extending to the confines of the atmosphere, we cannot know. We should like to think so; but Maignan wrote in 1653, after nearly all the experiments and discussions which we shall consider in the next two chapters; and he would have been more than human if, in relating his conversations with Berti, he could have been entirely unaffected by what he had since learned. However this may be, his reasoning was very sound, and demonstrates a praiseworthy hospitality to new ideas.

It seems impossible, at least at present, to establish the date of Berti's experiment more closely than indicated above (p. 10). It is not even clear that the idea was Berti's; and it is much more likely to have been due to Magiotti or even to Galileo, though if Galileo thought of the experiment, he was most probably concerned only to determine by another method how far the "attraction" of the vacuum would extend. In the Nationalbibliotek in Vienna there is an autograph letter[48]

---

[45] De Waard (*L'expérience barométrique*, p. 110) concludes that Berti probably died in the second half of 1643.

[46] *Musurgia universalis sive Ars magna consoni et dissoni in X libros digesta*, etc. (Rome, 1650).

[47] In 1648, in an anonymous pamphlet *Experimenta vulgata non vacuum probare, sed plenum et antiperistasim instabilire*, published at Rome. Republished in his *Nova de machinis philosophia*, etc. (Romae, 1649), pp. 102ff.

[48] MS 7049, letter CXXVII. De Waard, *L'expérience barométrique*, pp. 178–81, gives the Italian text, apparently somewhat edited.

from Raffaello Magiotti in Rome to Father Mersenne in Paris, dated *"il giorno di S. Gregorio* [25 May] 1648," which describes the experiment, obviously from memory. After some preliminary courtesies, there follows this paragraph:

As to the history of the quicksilver experiment, your Reverence may know that the many wells of Florence which are cleaned out every year by means of siphons "by attraction," gave Signor Galileo occasion to observe the height of such attraction, which was always the same, about 18 ells in that Tuscan measure; and so in every siphon or cylinder (as we like to say) however wide or narrow. From this had their origin his speculations about that matter, which were put into his work on the resistance of solids.

Afterwards Signor Gaspar Berti here in Rome, made a lead siphon (*syfone*), which rose about 22 ells from his courtyard to a room. . . .[49]

Magiotti then describes an experiment somewhat different from that shown by Maignan, and as Kircher and Zucchi both support Maignan as far as the main lines of the apparatus go, we can suppose that in six or seven years Magiotti had forgotten the details. Then he writes a very interesting phrase: "Berti believed he could convince Galileo with this experiment."[50] I think, with de Waard, that this adds to the strong probability that the experiment was performed before Galileo died on January 8, 1642. At any rate, everyone is agreed that it was done by Berti at his house, and there is little

doubt that it preceded the more famous experiment made by Viviani at the suggestion of Torricelli. This idea was violently attacked by Professor Mario Gliozzi thirty years ago, but he seems to have missed Maignan's book (which is not exceptionally rare), and, much more understandably, Magiotti's letter to Mersenne.[51]

Let us return to the experiment itself. It was certainly very well conceived, and Berti, who was said to be good at experiments,[52] may well have designed it. The possibility of using a sounding line after the tap at the bottom of the pipe was closed, and the one at the top opened, in order to find the height of the water in the opaque pipe, was an ingenious touch.

The correct explanation of the experiment was certainly not agreed upon at the time. If Galileo heard about it in his lifetime, he most certainly did not share Magiotti's views; and those of Kircher and Zucchi were made clear in the publications already referred to. Naturally the interest of most people was in the presence or absence of a vacuum. It is difficult for us in the twentieth century to realize how cer-

---

[49] *Ibid.*, recto.

[50] *Ibid.*, recto. "Il Sig' Berti credeva con questa esperienza convincere il Sig' Galileo."

[51] Mario Gliozzi, *Origini e sviluppi dell'esperienza Torricelliana* (Torino, 1931). It is interesting that G. Hellmann (*Abh. Preuss. Akad. d. Wiss., Phys.-Math. Kl.* [1920], pp. 1–38), while referring more than once to Berti, does not seem to realize that he preceded Torricelli or that his experiment was more than a repetition of the Torricellian experiment. On p. 37 of this paper, Berti is referred to as in the early 1650's.

[52] Zucchi, *Nova de machinis philosophia*, etc., p. 102, writes "Gasper Berti, Romanus, Vir nobilis, & in Physicis, Mathematicisque solidè doctus, singulari in experimentis capientis solertia."

tain many people were that there could not possibly be a vacuum above the column of water. It would have been as hard for a Jesuit to believe in the vacuum in 1640 as for a scientific man in 1960 to believe in the physical reality of the Indian rope trick. But when they were finally convinced that they could not have air without an offense to the intellect, they were happy, as we shall see, to settle for *aether*, or "subtle matter," or some quintessence or other; for anything, physically demonstrable or not, as long as it had the property of filling space, and the further property of transmitting light. This belief became good physics much later; for if a post-Newtonian scientist had been asked what filled the top of the barometer tube, might he not have replied, "the luminiferous aether?"

But at this early stage, before more decisive experiments had been made, there could indeed have been air; and the idea of using a little bell to find out was excellent, though not original.[53] The use of the lodestone to move the clapper of the bell was even more ingenious, and has many successors in twentieth-century vacuum technique.

[53] In a letter to Galileo dated April 11, 1615, Giovanfrancesco Sagredo explains how he had a vase made in the glassworks of Murano, and sealed up while it was very hot, so that when it cooled the "fiery spirit" (*lo spirito igneo*) escaped, and very little air was left within. This was proved in two ways: (1) he had had a little hawk's bell left inside, and when it was cold and the bell was moved, no sound was made "se non in quanto percoteva nel vetro et, per consequenza, faceva un suono esterno" (except in so far as it struck inside the glass and, in consequence, made an external sound); (2) the mouth of the cold vase was broken open under water, which entered and nearly filled it. (Galileo, *Le Opere*, ed. nat., XII [1902], 168.)

Father Maignan's criticism regarding the conduction of sound through the solid mounting was entirely just; but, nevertheless, the fact that sound was heard must have been a severe setback to the vacuists.

The transmission of light through the space raised most serious questions. The Peripatetics were bound to believe that a vacuum, if it could occur at all, would have to be completely opaque. Maignan simply denied this consequence, alleging that light is really an extremely tenuous *substance*. We shall see that this difficulty took half a century to fade out of men's minds.

Thus the immediate results of the experiment were more favorable to the plenists than to the vacuists. The plenists had various ideas as to how air could get into the flask—through the glass; between the water and the inside of the tube; or out of the water, widely believed to be full of air; or simply as the result of evaporation of the water, which was thought to turn to air on evaporating, the idea of a vapor not having occurred to anyone at the time.

The violent "boiling" of the water in the pipe was of course incomprehensible, except on the assumption that water was turning to air, because no one would believe that so much air could be *dissolved* in the water, as indeed it could not. In the letter from Magiotti to Mersenne already noticed, it is made clear that even in 1648 Magiotti had not decided about the bubbles in the tube.

Guericke repeated the experiment at Magdeburg, probably after 1654; he thought that the bubbles were effluvia coming from the walls of the tube as

well as from the water.[54] Boyle, in
1660, thought them vapor,[55] though we
may doubt the clarity of his ideas on
the subject. These were only two of a
long list of "water-barometers."[56]

Having come this far, let us not make

the mistake of saying that the barometer was "invented" at Rome in one of
the years 1640–1643. Berti's apparatus
was not a barometer; if words mean
anything, a barometer must be an instrument to measure pressure, and
apart from the verbal quibble that a
measuring instrument must have a
scale, Berti and his friends were interested, not in the measurement of pressure, but in producing a vacuum. What
we have been describing was a splendid physical experiment, somehow
almost forgotten for three hundred
years.

[54] Otto von Guericke, *Experimenta nova
(ut vocantur) Magdeburgica de Vacuo Spatio*,
etc. (Amstelodami, 1672), p. 95. These
bubbles would be nucleated on the walls of
the tube.

[55] Boyle, *New Experiments Physico-Mechanicall*, pp. 144–45. (*Works*, 1772 ed., I,
29.)

[56] See Chap. 14.

Chapter 2

The Torricellian Experiment

IF AN experiment as fundamental as Berti's had been performed three hundred years later, the scientific world would have echoed with it. In the twentieth century medals are awarded for work of less relative importance. In the 1640's it appears to have remained entirely unknown except to the people actually involved, until Magiotti wrote to Mersenne in 1648 and Zucchi published his anonymous pamphlet in the same year.

Who actually saw it, besides Berti? Certainly Kircher, Magiotti, and Zucchi; and as we have seen, Maignan knew all about it. De Waard[1] thinks Michelangelo Ricci (1619–1692) may have been there, but there is really no direct evidence for this assumption; nor is there any reason to believe that Torricelli witnessed the experiment. If Ricci was there, it would be strange if Torricelli did not hear of it, for on his deathbed in 1647, he dictated a statement to Lodovico Serenai in which he refers to

Ricci as "the best friend I had in Rome."[2] In his will he made Ricci his mathematical executor.[3] Nevertheless, there is no positive evidence that Ricci told him. But Magiotti, near the end of the letter from which we have already quoted, specifically states that he wrote to Torricelli about the experiment, with the suggestion that if the water had been sea water, it would have stood lower in the tube. "They" (presumably Torricelli and Viviani), he tells Mersenne, made experiments and finally came to use quicksilver.[4]

Nothing that we know about Magiotti would lead us to believe that after Torricelli's death he would lay claim to this small part of the credit if he had not actually made this suggestion, though neither his letter to Torricelli, nor any reply which its recipient may have made, has survived. It will be remem-

[1] *L'expérience barométrique* (Thouars, 1936), p. 105.

[2] Giovanni Ghinassi, *Lettere fin qui inedite di Ev. Torricelli, precedute dalla vita di Lui* (Faenza, 1864), p. lix.

[3] *Ibid.*, p. lxiii.

[4] Vienna, Nazionalbibliothek, ms. 7049, letter CXXVII, verso.

bered that Magiotti also wrote that Berti hoped to convince Galileo. The two statements together increase the probability that Berti's experiment was performed in the latter part of 1641, after Torricelli left Rome.

There is one piece of evidence which suggests that the idea of using quicksilver was, at least indirectly, attributable to Galileo himself. In his edition of Galileo's works,[5] E. Alberi notes that among the papers at Florence there is a copy[6] of the original Leiden edition of the *Discorsi* (1638), with manuscript corrections and additions in the hand of Viviani, "with the approval of Galileo himself, as is drawn to our attention more than once in those marginal notes."[7] One of these additions (Figure 2.1) comes after the passage in the *Discorsi* in which the limited height to which water can be drawn up by a pump is ascribed to the breaking of the water column, like a cord too heavily loaded; it reads:

And it is my belief that the same result will follow in other liquids, such as quicksilver, wine, oil, etc., in which the rupture will take place at a lesser or greater height than 18 braccia, according to the greater or lesser specific gravity of these liquids in relation to that of water, reciprocally; always measuring such heights perpendicularly, however.[8]

To Viviani, this must have constituted a very broad hint.

What sort of a man was this Torricelli?

He was born on October 15, 1608, the son of Gaspari Torricelli, but his place of birth is not certain, except that it was in the vicinity of Faenza. At the little Torricelli museum there, you may see a drawing of an old farmhouse which claimed the honor of being his birthplace. In later life he felt himself to be from Faenza. His family was not well off, and he seems to have been orphaned at an early age. By good fortune he had in Faenza a paternal uncle Alessandro, who had become a monk and taken the name of Jacopo. This uncle, a man of some learning, looked after the boy like a father and led him to the study of the liberal arts. Evangelista seems also to have studied mathematics with the local Jesuits. Then, probably in 1627, he went to study under Castelli at the University of Rome, then called *la Sapienza*. Castelli had never had a more apt pupil.

Torricelli very soon found himself under the influence of the writings of the great Galileo. In a letter written on September 11, 1632, he said that he "felt very fortunate to have been born in a century which was able to know and to write the praises of a Galileo, an oracle of nature."[9] He absorbed Galileo's work with avidity, but without in any way neglecting the mathematical studies which soon made him an accomplished and original mathematician. He also seems to have been a modest man of great personal charm. At the invitation of Galileo, Torricelli moved to Florence in 1641, but their closer

[5] *Le Opere. Prima edizione completa condotta sugli autentici manoscritti Palatini* (16 vols.; Florence, 1842–56).
[6] Gal., 79, Div. II, p. v, n. 9.
[7] *Le Opere*, ed. cit., XIII, xii.
[8] *Ibid.*, p. 21.

[9] Ghinassi, *Lettere fin qui inedite*, etc., p. xiii. Nearly all my information about the life of Torricelli has been taken from this work.

DEL GALILEO. 17

per la sustanza, ò porosità del vetro, ò del legno, aria, ò altra più tenue, e spiritosa materia, si vedrà radunare (cedendogli l'acqua) nell' eminenza v, *le quali cose, quando non si scorgano, verremo assicurati l'esperienza esser con le debite cautele stata tentata, e conosceremo l'acqua non esser distraibile, nè il vetro esser permeabile da veruna materia benche sottilissima, fuori che dal fuoco.*

Sagr. *Et io mercè di questi discorsi ritrouo la causa di vn' effetto, che lungo tempo m'hà tenuto la mente ingombrata di marauiglia, e vota d'intelligenza. Osseruai già vna Citerna, nella quale per trarre l'acqua fù fatta fare vna Tromba, da chi forse credeua, mà vanamente, di poterne cauar con minor fatica l'istessa, ò maggior quantità, che con le secchie ordinarie; & hà questa tromba il suo stantuffo, e animella sù alta, siche l'acqua si fà salire per attrazzione, e non per impulso, come fanno le Trombe, che fanno l'ordigno da basso. Questa sin che nella Citerna vi è acqua sino ad vna determinata altezza, la tira abbondantemente, mà quando l'acqua abbassa oltre à vn determinato segno, la Tromba non lauora più. Io credetti, la prima volta che osseruai tale accidente, che l'ordigno fusse guasto, e trouato il Maestro, acciò lo raccomodasse, mi disse che non vi era altrimente difetto alcuno fuor che nell' acqua, la quale essendosi abbassata troppo, non patiua d'esser' alzata à tanta altezza; e mi soggiunse nè con Trombe, nè con altra machina, che solleui l'acqua per attrazzione, esser possibile farla montare vp capello più di diciotto braccia, e sieno le Trombe larghe, ò strette, questa è la misura dell' altezza limitatissima. Et io sin hora sono stato così poco accorto, che intendendo, che vna corda, vna mazza di legno, e vna verga di ferro si può tanto, e tanto allungare, che finalmente il suo proprio peso la strappi, tenendola attaccata in alto, non mi è souuenuto, che l'istesso molto più ageuolmente accaderà di vna corda, ò verga di acqua. E che altro è quello, che si attrae nella Tromba, che vn Cilindro di acqua, il quale hauendo la sua attaccatura di sopra, allungato più, e più, finalmente arriua à quel' termine, oltre al quale tirato dal suo già fatto souerchio peso, non altrimente che se fusse vna corda, si strappa?* Salu.

collaboration ended almost as soon as it had begun, when Galileo died in January 8, 1642. Thus it is only in a figurative sense that Torricelli can be called the pupil of Galileo.

After the death of the latter, Torricelli thought of returning to Rome, but was prevented when the Grand Duke Ferdinand II made him his Philosopher and Mathematician, a post which he held for the rest of his life. In 1644 he published a collection of mathematical works, *De sphaera et solidis sphaeralibus libri duo*, etc., *De motu gravium*, etc., and *De dimensione parabolae solidique*, etc., in one quarto volume which seems to have been the only scientific book which he published.[10]

One of the matters dealt with in this book was the calculation of the area of the cycloid, and on its appearance he was promptly accused of plagiarism by Roberval and Pascal. It seems certain that while Roberval had the priority, Torricelli's work was entirely independent.

More germane to our subject is his work in hydraulics, in which he made important discoveries, extending the findings of Galileo and Castelli and leaving the science with a new look. He also seems to have been skilled in optical technique, his lenses being widely admired and sought after.

Torricelli died on October 24, 1647, having filled his thirty-nine years with diverse and memorable activity. This was the man who thought about the suggestion thrown out by Magiotti. It appears that the actual experiment was not made by Torricelli himself, but that he *imagined* what would happen and confided his thoughts to his great friend Viviani, who had the apparatus made, procured the mercury, and was the first to make the experiment and to see the effect *forecast* by Torricelli. So says Carlo Dati,[11] a pupil of Torricelli's who, nineteen years later, pseudonymously published the first real account of the experiment, and the letters in which it was first described. In view of the marginal notes found by Alberi, it is at least as probable that Viviani suggested the use of mercury.

The correspondence between Torricelli in Florence and Ricci in Rome in June, 1644, is the heart of this portion of our story, and besides describing the experiment, shows that Torricelli had thought about it hard enough to be able to explain the results with superlative clarity. I have attempted translations of the three letters below.

It is evident that at an earlier date Torricelli must have informed Ricci that experiments were in progress, for on June 11, 1644, the latter wrote to his friend, "I live in a great desire to know the success of those experiments that you indicated to me."[12] He had not long to wait, for on the same day Torricelli wrote him the famous letter describing the experiment.[13]

---

[10] He left many things in manuscript. See Ghinassi, *Lettere fin qui inedite*, etc.

[11] *Lettere a Filaleti di Timauro Antiate, Della vera storia della Cicloide, e della famosissima esperienza dell'argento vivo*, 24 Gennaio 1662 (Firenze, 1663), p. 20.
[12] Torricelli, *Opere*, eds. G. Loria & G. Vassura (Faenza, 1919), III, 189.
[13] *Ibid.*, pp. 186–88. The version given by Dati (*Lettere a Filaleti*, etc., pp. 20–21) should be preferred, as he specifically states that he has copied it from the original in the possession of Ricci.

Most Illustrious and Reverend Sir:

Some weeks ago I asked Signor Antonio Nardi for some of my demonstrations on the area of the cycloid, requesting him to direct them to you, or rather to Signor Magiotti, after he had seen them. I have already hinted to you that some sort of philosophical experiment was being done concerning the vacuum; not simply to produce a vacuum, but to make an instrument which might show the changes of the air, now heavier and coarser, now lighter and more subtle. Many have said that [the vacuum] cannot happen; others that it happens, but with the repugnance of nature, and with difficulty. I really do not remember that anyone has said that it may occur with no difficulty, and with no resistance from nature.[14] I reasoned thus: if I found a very obvious cause, from which resulted this resistance that is felt in trying to produce a vacuum, it would seem vain to try to attribute that resistance to the vacuum itself, as it would clearly derive from the other cause. On the contrary, making some very easy calculations, I find that the cause I adopted (*i.e.*, the weight of the air) ought by itself to produce a greater resistance than it does when we attempt to make a vacuum. I say this because some philosopher, seeing that he could not escape confessing that the gravity of the air is the cause of the resistance that is felt in producing a vacuum, would not say that he conceded the operation of the weight of the air, but would persist in his assertion that Nature also helps by her repugnance to the vacuum. We live submerged at the bottom of an ocean of elementary air, which is known by incontestible experi-

ments to have weight, and so much weight, that the heaviest part near the surface of the earth weighs about one four-hundredth as much as water. Then writers have observed regarding the twilight that the vaporous air is visible above us for about fifty or fifty-four miles. But I do not think it as much as this, because I should then admit that the vacuum ought to produce a much greater resistance than it does, even if there is this escape for these writers, that this weight, given by Galileo, refers to the lowest air, frequented by man and animals, but above the peaks of high mountains the air begins to be very pure, and of much less weight than the four-hundredth part of the weight of water.[15]

We have made many glass vessels such as those shown at A and B, wide, and with necks two ells long (Figure 2.2). When these were filled with quicksilver, their mouths stopped with the finger, and then turned upside-down in a vase C which had some quicksilver in it, they were seen to empty themselves, and nothing took the place of the quicksilver in the vase which was being emptied. Nevertheless the neck AD always remained full to the height of an ell and a quarter and a finger more. To show that the vessel was perfectly empty, the basin was filled up to D with water; and on raising the vessel little by little, it was observed that when the mouth of the vessel reached the water, the quicksilver in the neck came down, and the water rushed in with horrible violence and filled the vessel completely up to E.

While the vessel AE was empty, and the quicksilver, though very heavy, was sustained in the neck AC, we discussed this force that held up the quicksilver against its natural tendency to fall down.

---

[14] K. Schneider-Carius, *Wetterkunde, Wetterforschung. Geschicte ihrer Probleme*, etc. (Munich, 1955), pp. 62–64, has translated this letter into German. But I am convinced that he has mistaken the sense of these three sentences.

[15] Galileo's value was too great by a factor of more than two.

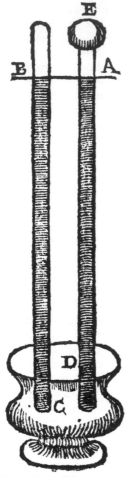

Fig. 2.2. Torricelli's experiment.

silver enters the glass *CE*, to being in which it has neither inclination nor repugnance, and rises there to the point at which it is in balance with the weight of the external air that is pushing it! Water, then, in a similar vessel but very much longer, will rise to about eighteen ells, that is to say, as much higher than the quicksilver as quicksilver is heavier than water, in order to come into equilibrium with the same cause, which pushes the one and the other.

This reasoning was confirmed by making the experiment at the same time with the vessel *A* and with the tube *B*, in which the quicksilver always stopped at the same level *AB*; an almost certain sign that the force was not within; because the vase *AE* would have had more force, there being more rarefied and attracting stuff, and this much more vigorous by virtue of its greater rarefaction than that in the very small space *B*.

With this principle I then tried to preserve all the kinds of repugnance which are felt to be in the various effects attributed to the vacuum, and up to this moment I have not met one of them that does not go well with it. I know that many objections will occur to you, but I hope also that thinking about the matter will appease them.

I have not been able to succeed in my chief intention, to find out with the instrument *EC* when the air is coarser and heavier and when more subtle and light; because the level *AB* changes from another cause (which I never thought of), that is, it is very sensitive to heat and cold, exactly as if the vase *AE* were full of air.

And humbly I pay my respects.

From your devoted and most deeply obliged servant,

V. TORRICELLI

Florence, 11 June, 1644

This is a remarkable letter. We should note that in the second sentence

It has been believed until now[16] that it was something inside the vessel *AE*, either from the vacuum, or from that extremely rarefied stuff; but I assert that it is external, and that the force comes from outside. On the surface of the liquid in the basin presses a height of fifty miles of air; yet what a marvel it is, if the quick-

[16] Torricelli must be referring to Galileo's belief in the attraction exerted by the vacuum. Neither Beeckman (of whom he had probably never heard) nor Baliani (of whom he certainly had) would have subscribed to this belief.

Torricelli shows that he has already made an enormous advance on his predecessors; he is trying to produce an instrument. Doubtless his keen mind saw little difficulty in the idea of a vacuum, and he must have heard "great argument" about it, until perhaps he was a little tired of the subject. But he has to show that he knows the cause of the resistance that we feel when we try to produce a vacuum. It is the weight of the air (il peso dell'aria). Note that he uses the word weight, not pressure; sometimes he speaks of the gravity (la gravità) of the air; and there must be at least enough of it, according to his calculations, to balance out the weight of the mercury.

Then comes that remarkable image: Noi viviamo sommersi nel fondo d'un pelago d'aria elementare (one suspects that the elementare was added to make a musical cadence to this fine phrase) —we live submerged at the bottom of an ocean of air. Galileo must have realized this, too; but thought the ocean exerted no pressure.

Torricelli's calculation necessarily used Galileo's experimental value, which was far from correct, for the ratio of the weights of equal volumes of air and water. His respect for Galileo's determination outweighed his confidence in the estimates of the height of the atmosphere derived from observations on the twilight. But he knew the upper air was lighter, as did Baliani before him;[17] only he ascribes its lightness to its greater purity, as Aristotle would have done.

Next follows the familiar experiment, including the brilliant idea of using

water to show that the space above the quicksilver was empty of air. This part of the experiment made a great impression on all who tried it in the next decade. The use of two vessels of greatly different volume was another excellent idea. We cannot know whether Torricelli or Viviani thought of these devices.

In the discussion of the experiment there is one very curious phrase: "either from the vacuum, or from that extremely rarefied stuff" (o di Vacuo, o di quella roba sommamente rarefatta). The quella sounds just slightly satirical, as if Torricelli had heard about the Cartesian "subtle matter" and did not think much of it.

His confession of failure at the end of the letter is interesting. There ought not to have been enough sensitiveness to heat and cold to mask the variations in the atmospheric pressure, and it is possible that the experiment was set up again with the tube that had been used for the experiment with the water, but without drying it out very well.

Knowing Ricci, who was not one to accept physics on faith, Torricelli expected objections. He got them, and almost by return of post.[18] On June 18, 1644, Ricci replied:[19]

. . . I admire your noble ardor in taking into consideration a thing not touched by anyone up to now, which has at the same time so much probability, except for two or three objections, which I have to present to you, and which I beg you to

[17] Galileo, Le Opere, p. 160.

[18] Yves Renouard, in L'histoire et ses méthodes (Paris, 1961), p. 115, says that the courier between Florence and Rome ordinarily took three or four days at that period.
[19] Torricelli, Opere, III, 193–95. Part of this letter is in Dati, Lettere a Filaleti, etc., pp. 21–22. I have translated only pertinent portions.

resolve for me, inasmuch as I know that you will be able to do so with ease. I think this is the truest and most reasonable way to deal with questions of this kind.

First of all, it seems to me that if one could exclude the action of the air in weighing upon the external surface of the [quick-]silver in the vessel, placing there a cover with only one hole through which the glass tube may pass, and then completely plugging it up everywhere in order that there might no longer be any communication whatever, then in such a case the air above the vessel would not succeed in weighing on the surface of the [quick-]silver, but would weigh on the cover; and if then the quicksilver remains up in the tube as at first, the effect can no longer be attributed to the weight of the air holding it up as if in equilibrium.

Secondly, taking a syringe, which seems to be used a good deal in this subject,[20] let us have its piston completely pushed in, so that it may exclude all other bodies by its volume. Then, plugging up the hole at its tip, and drawing the piston backwards by force, we feel a very great resistance, not only when we hold the syringe down and pull the piston, on the handle of which the air presses, upwards, but likewise when we hold it and pull in any other direction. In this case it really does not seem easy to understand how the weight of the air could do this.

Finally, a body immersed in water does not contend with all the water which stands above it, but only with that which is displaced by the immersed body, and which is not greater in volume than the body. Because I conceive that the same doctrine may apply to the balancing of the [quick-]silver, this should then contend with just as much air as its own mass,[21] and if so, how can the air ever preponderate? This is what has put up

my foolishness in opposition to your wisdom, but as an excuse I offer the desire I have to understand perfectly the solution of these objections, and to become instead the absolute defender, just as I am the sincere admirer, of this and of all your other discoveries, which are all so pleasing to me.

The rest of the letter deals with mathematical questions.

On June 28 Torricelli dealt at length with these "objections." The letter[22] begins with a long paragraph about his troubles in getting his book through the press (every author will smile sympathetically at his statement that this is really much more fatiguing than it was to write it!), and the impossibility of making Ricci the lenses he asked for, in such hot weather. It continues:

I think it superfluous to answer your three objections on the subject of my fantasy about the apparent resistance in making a vacuum, because I hope that the solution came to you after you had written your letter. As to the first, I reply: if you infer that the top, cemented so that it covers the surface of the basin, does it in such a way that it touches the quicksilver in the basin, then that which is raised in the neck of the vessel will remain raised as before, not by the weight of the atmosphere (*sfera aerea*), but because there will be no room for it in the basin.

But if you would have the cover put on so that it still takes some air inside, then I ask whether you will admit that the enclosed air will have the same degree of condensation as the external air? In this case the quicksilver will be held up as before (as in the example of the wool that I shall now give); but if the air that you

---

[20] The textbooks of hydrostatics were full of syringes.

[21] *mole*. But probably volume is meant.

[22] Dati, *Lettere a Filaleti*, etc., pp. 22–23; Torricelli, *Opere*, III, 198–201.

include is more rarefied than the external air, then the suspended metal will descend by the proper amount; if now it were infinitely rarefied, *i.e.*, a vacuum, then the metal would descend all the way, provided that the enclosed space were able to take it.

The vessel *ABCD* (Figure 2.3) is a cylinder filled with wool or other compressible material—let us say air. This vessel has two ends, *BC* fixed and *AD* movable, adapted to it. Let *AD* be loaded above with a lead weight *E* that weighs 10,000,000 lbs. I know you will understand how much force the bottom *BC* will feel. Now if we forcibly push in a plane or sharp knife *FG* so that it goes in and cuts the compressed wool, than I say that if the wool *BCFG* is compressed as before, the bottom *BC* will suffer the same pressure as it felt at first, even if it no longer feels any of the superimposed weight of the lead *E*.

You will beg that I should not continue to weary you. Now as to the second point. Once there was a philosopher who, seeing a cask being broached by a servant, dared to say to him that the wine would never come out, because the nature of gravity is to press downwards, not horizontally and in all directions. But the servant proved conclusively that even if liquids naturally gravitate downwards, they push out everywhere, and squirt out in every direction, even upwards, as long as they find places to go to, *i.e.*, places that resist with a force less than the force of these liquids. If you will entirely submerge a jug in water with its mouth down, and then make a hole in its bottom so that the air may come out, you will see with what violence the water moves from the bottom to the top to fill it up. You will certainly demand that I should weary you no further.

The third objection does not appear to me to be too pertinent; certainly it is less valid than the others, even if, being taken

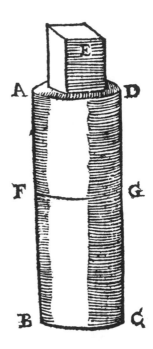

Fig. 2.3. To answer Ricci's first objection.

from geometry, it may appear the strongest of all. That a body immersed in water contends only with as much water as its own mass (*mole*) is true; but it does not seem to me that the metal held up in the neck of this vessel can be said to be immersed either in water or in air, or in glass, or in a vacuum. We can say only that it is a heavy fluid body, one surface of which is limited only by a vacuum, or almost a vacuum, which weighs nothing at all; the other surface is confined by air, pressed by so many miles of air piled up above it. On that account the surface that is not pressed upon at all ascends, driven away from the other; ascends until the weight of metal raised up comes to equal the weight of the air pressing in the other direction. You may imagine the vessel *A* joined to a tube *BCD*, open at *D*, as shown in the drawing (Figure 2.4), and that the vessel *A* is full of quicksilver.

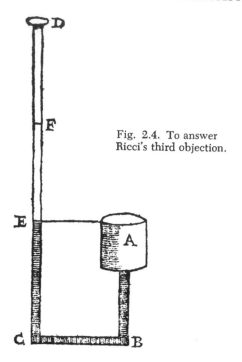

Fig. 2.4. To answer Ricci's third objection.

matical discussion about one of the conic sections.]

This most interesting letter discloses a good deal more of Torricelli's thought about the atmosphere. It will be seen that he was quite clear about the elasticity of the air, like Beeckman twenty-four years earlier. While Beeckman compared the atmosphere to an immense sponge,[23] Torricelli thought of wool, as Descartes had done before. In interpreting his model, we must of course assume that the knife which divides the wool is held fixed in relation to the bottom of the vessel. It is interesting that Boyle, making experiments with his air pump, also thought of the analogy of wool,[24] and this before Dati had published these letters.

But before describing his imaginary experiment with the wool, Torricelli predicts what will happen if the experiment with the mercury is made in more rarefied air, or even in a vacuum. We shall see in the next chapter how these predictions were exactly fulfilled in France shortly after his death.

Ricci's second objection seems frivolous, if we fail to remember that he was a mathematician rather than a physicist. It is here answered by a joke, introduced and concluded by a phrase which surely means, "now he *can't* be as stupid as that!" The third difficulty is more abstruse; we must remember that hydrostatics was only just coming into being as a unified science. At any rate, Torricelli's explanation by means of an experiment (probably imaginary) is superlatively clear.

It is certain that the metal will rise in the tube up to its level *E*; but if the instrument is immersed in water up to the mark *F*, the quicksilver will not rise to *F*, but only until the level in the tube is above the level *EA* in the vessel by about one-fourteenth part of the height to which the water *F* stands above the level of the vessel *A*. Of this you may be as certain as if you had made the experiment. Now here we see, for example, that the water might be fourteen ells deep, and the metal in the tube *ED* might rise only one ell; so this single ell of the metal does not balance the same amount of water, but the entire depth of water that is between *A* and *F*. In these cases, as you know, we pay no attention to the width and thickness of the volumes, but only to the vertical distances, and to the specific gravity (*grauità in specie*), not the absolute weight. But perhaps I have said too much; if I could talk to you, perhaps you might remain better satisfied.

[The letter ends with a short mathe-

[23] "Gelyk een sponse; die groot is." (Beeckman, *Journal*, II, 157.)

[24] Robert Boyle, *New Experiments Physico-Mechanicall touching the Spring of the Air, and its Effects*, etc., (Oxford, 1660), pp. 23–24. (*Works*, ed. cit., I, 11.)

Let us review Torricelli's position in these matters in June, 1644. He must have decided, on what evidence we do not know, that the weight of the column of air varies; for he announced in the second sentence of the first letter that he had wanted to make an instrument *(fare uno strumento)* to show this variation. *This is the real and sufficient reason for ascribing the invention of the barometer to Torricelli.* It is true that at the end of the letter he is obliged to confess failure; but there is little doubt that he and Viviani would have succeeded, had they pursued the matter further.

Torricelli was quite clear not only about the weight of the air but also about its elasticity; he realized the essential difference between aerostatics and hydrostatics. It is true that he could not see how so little mercury could balance so much air; but he had to contend with Galileo's very erroneous determination of the relative densities of air and water, which indeed he seems to have accepted rather uncritically. Nevertheless, his reverence for the great man who had inspired him did not prevent him from discarding the idea of the resistance or force of the vacuum.

In his introduction to the letters exchanged between Torricelli and Ricci, Dati[25] says that after repeating the experiment several times and "thinking of a good part of those observations that have been made since by others who have set up this beautiful experiment," Torricelli tried introducing fishes, large flies, and butterflies into the vacuum to observe whether they could live, make sounds, or fly; but they were too much affected by going up through the mer-

cury for the experiment to be satisfactory. Dati also says that Torricelli "began to write to his friends and tell them about it, and especially to Michaelangelo Ricci of Rome."[26] But no other letters are known, and even Dati does not mention the names of Torricelli's other correspondents. It must be noted that Dati's entire essay is expressed in terms of the most enthusiastic partisanship, which makes us suspect his reliability as to details.

We must now discuss two questions which can be answered only on the basis of probability. The first concerns the date of the experiment; the date, in fact, on which "the barometer was invented."

Almost every textbook of physics from that day to this gives the date as 1643. In this they are probably following the lead given by the famous *Accademia del Cimento,* or Academy of Experiment, founded in Florence in 1657 under the patronage of Prince Leopold, and dissolved ten years later under some ecclesiastical pressure. In 1667, before it came to its end, the Academy published a magnificent account of its experiments, which was reprinted in 1691, 1780, 1841, and 1957, and translated into English by Richard Waller in 1684. The 1841 edition is noteworthy in having a 133-page introduction by Vincenzio Antinori, and an appendix derived from manuscripts preserved at Florence, relating further experiments. In the first edition we read the following at the beginning of the chapter on "Experiments about the Natural Pressure of the Air":

That famous experiment with the quicksilver, which rose up before the great

---

[25] *Lettere a Filaleti,* etc., p. 20.

[26] *Ibid.*

intellect of Torricelli in the year 1643, is now known in every part of Europe, as is also the high and wonderful thought that he derived from it, when he began to speculate on the reason.[27]

Now all this really says about the date is that Torricelli thought about the experiment in 1643. That it was actually performed in 1643 seems extremely improbable. It should be noted that Ricci's letter asking for news of it crossed Torricelli's describing it, as they were both dated June 11, 1644, so that it cannot be argued that Torricelli was reminded by Ricci of something that he ought to have told him long ago. On the other hand he may have told Ricci quite a long time before, even in 1643 (in a letter now lost to us), that he was contemplating the now-famous experiment. But the letter of June 11 looks, to me at least, like an enthusiastic report of a result just obtained. It seems incredible that, even with his book going through the press, he would keep his best friend in the dark for five whole months about such a beautiful outcome. Even if the Academy had said that the experiment was *performed* in 1643, which they did not, we might well have doubts, especially in view of the fact that Dati, who was a member of the Academy, makes the theory precede the experiment (note 11 above), while it is the other way

round in the *Saggi*. A long twenty years had passed.

Once something gets into the textbooks it becomes almost immortal, and by 1743 one Georg Matthias Bose, a German professor, had improved on the Academy by picking May 2, 1643, as the date. He delivered a heroic oration, in Latin, almost unbelievably pompous, in celebration of the event.[28] I am afraid I was myself responsible for a small celebration of what I then believed, on the authority of the *Saggi*, to be the Tercentenary, at Toronto in 1943, though not on May 2.[29]

The second puzzle is why, as far as Italy is concerned, the experiment was kept almost secret; much more nearly secret, certainly, than the subject matter of recent Western military "security." For it was hidden as if it had been a mystic rite. The very next letter from Ricci to Torricelli that appears in the *Opere*[30] makes no mention whatever of the experiment, nor the next, nor the one after that; with the exception of one letter from Du Verdus in Rome, dated 9 [probably July] 1644,[31] nothing whatever about the experiment appears, as far as I can determine, in the volumes containing Torricelli's very extensive correspondence—not for all the three years that remained to him. The letter from Du Verdus, a Frenchman, was written after Ricci had shown

---

[27] *Saggi di naturale esperienze fatte nell'-Accademia del Cimento* (Firenze, 1667), p. 35. "E nota oramai par ogni parte d'Europa quella famosa esperienza dell'argentovivo, che l'anno 1643 si parò davanti al grande intelletto del Torricelli; e noto parimente è l'alto e maraviglioso pensiero ch'egli formò di essa, quand' ei ne prese a specular la ragione." (In the 1841 edition, which set out to "improve the spelling and make some important corrections," "specular" becomes "specificar"; which seems less likely.)

[28] in *Raccolta d'Opuscoli Scientifici e Filologici, tomo* XXXII (Venezia, 1745), pp. 33–58.
[29] The four papers read at this meeting by L. C. Karpinsky, G. S. Brett, John Satterly, and W. E. K. Middleton, are to be found in the *Journal of the Royal Astronomical Society of Canada* between December 1943 and February 1944.
[30] III, 206–7. It is dated July 2, 1644.
[31] *Ibid.*, pp. 210–11.

him Torricelli's letter of June 11, and shows its writer much impressed by the experiment, but in difficulties with his scholastic education.

Now while it is clear that he was busy with his mathematical researches, it is surely impossible to believe that Torricelli (not to mention Viviani) immediately lost all interest in an experiment which he must have felt to be important, and which must have suggested many others, as it did to many people all over Europe in the next two decades. Nor is it at all likely that Ricci would have refrained from discussing it further with his many friends in Rome,[32] except for one consideration which outweighed all others.

This consideration was prudence. We have to remember that all those involved were in a sense disciples of the great Galileo; they would be expected to have clearly in mind what had happened to him at the hands of the Holy Office in 1633; more, their lively imaginations could foresee that what happened to their famous master would be mild (he was, in fact, treated with surprising consideration) compared to what might happen to them. The idea of a vacuum was anathema to the Church, and in Italy the Church was, at the moment, almost omnipotent. Better let the whole thing drop. Better still, tell Father Marin Mersenne in Paris. This is what Du Verdus did, with the results we shall describe in the next chapter. And in the decade following Torricelli's death, his experiment seems to have been mentioned in only two works published in Italy.[33]

There is more evidence for this caution. In the years between 1645 and 1664, a globe-trotting Frenchman called Balthasar de Monconys (1611–1665) traveled over a good deal of Europe, making a point of visiting "philosophers," but not forgetting to record other things in his copious notebooks. From the account of his travels,[34] it is evident that his curiosity was immense.

Now in November, 1646, he was in Italy, and on November 3, he went to see Torricelli. He went again on the 4th and apparently bought some glasses from him; then they went to Mass together. On the 6th he went again to see Torricelli, "qui ajusta mes Lunettes, & me donna des pierres de Boloigne, & des auoines sauuages pour voir les Temps secs ou humides."[35] Next Torricelli gave him a lesson in astronomy; then after dinner he saw Nardi and Viviani. But not a word about the famous experiment. On the 7th Torricelli showed him the Grand Duke's thermometers and described how they were made; but still nothing about the

[32] G. Hellmann (Abh. Preuss. Akad. d. Wiss., Phys.-Math. Kl., no. 1 [1920], pp. 30–32) notes that Casati, although he lived in Rome, learned of the experiment only when, three years later, Magni's work (see p. 42 below) was published; and that Fantuzzi in Bologna had not heard by 1648 that it had been done in Florence. Even Baliani seems not to have learned about it until 1647, and then only from Mersenne in Paris; this has been documented by De Waard, L'expérience barométrique, p. 128.

[33] See De Waard, ibid., p. 130. According to M. Del Gaizo, Atti d. Soc. Ital. di storia crit. d. sci. med. e. nat. (Faenza, 1909), p. 128, Borelli obtained copies of Torricelli's two letters from Ricci in 1658. That it was of interest to do so is further evidence of the way the experiment was hidden.

[34] Journal des voyages de M. de Monconys, etc. (3 vols., 4°; Lyon, 1655–66; 2d ed., 2 vols. 4°; Lyon, 1677).

[35] Ibid., I (2d ed.) 115. Incidentally, this was well before the supposed date of the invention of the oat-beard hygroscope by Hooke.

tube and the mercury.[36] The day after, de Monconys left for Pisa.

It is quite unlikely that if de Monconys had heard anything about the experiment he would have failed to mention it in his journal. We may take it that it was simply not being talked about. This impression is strengthened by the fact that in June, 1664, de Monconys was in Florence again, and this time Viviani described the experiment to him. ". . . le soir Monsieur Viuian me vint voir, qui me dit comment il montroit la pression de l'air par cet instrument."[37] By this time the *Accademia del Cimento* were doing their experiments, and the whole matter was in the open.

I do not think that this interpretation of the events is contradicted by the fact that the Grand Duke Ferdinand II decreed a "triumph" for his mathematician,[38] which must have been a recognition of his other work as well, and in any case was probably embarrassing to Torricelli. Ferdinand was never very successful in his struggles against the authority of the Pope, and the very fact of this advertisement may well have determined Torricelli to play the matter down. At any rate, there seems to be no indication that he wrote any more about it, although a few months later he did repeat the experiment for Mersenne on being asked to do so.[39]

Nor can much be made of the undoubted fact that in February, 1645, the experiment was performed in Rome by the Cardinal Giovanni Carlo de' Medici,

the Grand Duke's brother. There is nothing to indicate that this was done in public; indeed Kircher,[40] who was there with Zucchi,[41] indicates that the experiment was performed for him as a special mark of favor. Emmanuel Maignan in his *Cursus philosophicus* refers to having witnessed the experiment "in Rome six years ago."[42] There is no evidence that Mersenne saw the experiment in Rome, although he was there at the time; and this is a further indication that it was a private affair. It was not the sort of occasion that Mersenne would gladly have missed.

What is Torricelli's real position in the history of the barometer? Beeckman and Baliani had preceded him in understanding the omni-directional pressure of the air, and Torricelli nowhere claimed to have originated this hypothesis. In his letters to Ricci he confirmed and explained it. Berti had made a similar but much less convenient experiment, for the improvement of which Magiotti had provided a valuable suggestion. Viviani, with the benefit of a broad hint from Galileo on the use of quicksilver, did the experimental work. Torricelli's contribution was to imagine that an *instrument* could be devised "which might show the changes of the air." This instrument became the barometer; and it is in this sense alone that Torricelli was its inventor.

[36] *Ibid.*, p. 131.

[37] *Ibid.*, II, 475.

[38] See Ghinassi, *Lettere fin qui inedite*, etc., p. xxvi–xxvii.

[39] Mersenne, *Novae observationes physico-mathematicae* (Parisiis, 1647), p. 216.

[40] *Musurgia* (Romae, 1650), p. 11.

[41] *Experimenta vulgata non vacuum probare*, etc. (Romae, 1648), p. 4, quoted by de Waard, *L'expérience barométrique*, pp. 177–78.

[42] P. 1897. The *Cursus* was published in 1653; but the "approbation" is dated 28 Feb. 1650 (fol. 12ᵛ), so that this may well have been Giovanni de' Medici's experiment. We shall come back to it in Chapter 3.

The "Extraordinary Effervescence"

" A PEINE Torricelli a-t-il communiqué sa découverte à son ami Michel-Ange Ricci," wrote the great historian Pierre Duhem in 1912, "à peine celui-ci en a-t-il fait part à quelques italiens et français curieux de Physique, qu'un extraordinaire effervescence agite l'Europe savante; chacun veut voir 'l'expérience du vif-argent;' chacun s'efforce de la repétér, de la modifier, de l'interpreter."[1]

It is an exaggeration, however, to say that Europe was excited about the matter. Italy was certainly not, nor England immediately, nor the Germanic countries. In fact for about four years this effervescence was confined to France and to a few Frenchmen abroad, except for the Roman experiment referred to in Chapter 2, and for an event which took place in Poland, which we shall relate in due course. But as far as France was concerned, Duhem's metaphor cannot be improved upon. "For

who, indeed," wrote Pierius in 1648, "could have imagined the controversies this experiment has excited in this city for the last two years?"[2]

The experiment done in February, 1645, by Giovanni Carlo de' Medici had no immediate effect on the intellectual climate, because it was kept secret for several years; so we had better deal with it first, especially as in some aspects it was identical with experiments performed elsewhere more than a decade afterward.

There are three witnesses to the fact that the experiment was performed in Rome, two of them unequivocal about the date.[3] Of their accounts, that of Emmanuel Maignan is by far the most circumstantial. He definitely states, for example, that the experiment was made in tubes of widely different shapes and sizes, and that he saw this in Rome.[4]

---

[1] In A. Maire, *L'oeuvre scientifique de Blaise Pascal. Bibliographie critique et analyse de tous les travaux qui s'y rapportent* (Paris, 1912), Preface by P. Duhem, p. vi. This is a most valuable work.

[2] Jacobus Pierius, *Ad experientiam nuper circa vacuum R. P. Valeriani Magni demonstrationem ocularem*, etc. (Paris, 1648). Dedication to Petit, n.p.

[3] See footnotes 40, 41, and 42 in Chap. 2.

[4] *Cursus*, p. 1888. ". . . cum enim ego Romae haec omnia sim expertus."

Our Figure 3.1 illustrated his description. Supposing that he is worthy to be believed, it follows that the Cardinal or someone in his *entourage* had done a good deal of thinking about the experiment. Could it have been Kircher, who had a reputation (perhaps exaggerated) for scientific acumen and had witnessed the experiment of Berti? We do not know. But they went further. The five tubes, presumably as in the figure, were

kept under observation; and it was found that the changes in the level of the mercury could not be correlated with the observed changes in temperature and humidity, so that "the height of the mercury is not changed by the alteration of the air around it, but by that of all the air in a broad region."[5] This experiment has a remarkable resemblance to one performed some twelve years later in Florence, which will be discussed later on.

Maignan also made mention of several other experiments, using the results to prove to his own satisfaction that it is indeed the pressure of the air which holds the mercury up. But as in the course of the discussion he uses the results of other researches carried out elsewhere some years later, it is unsafe to assume that he also saw all these experiments in Rome in 1645.

[5] *Ibid.*, p. 1898.

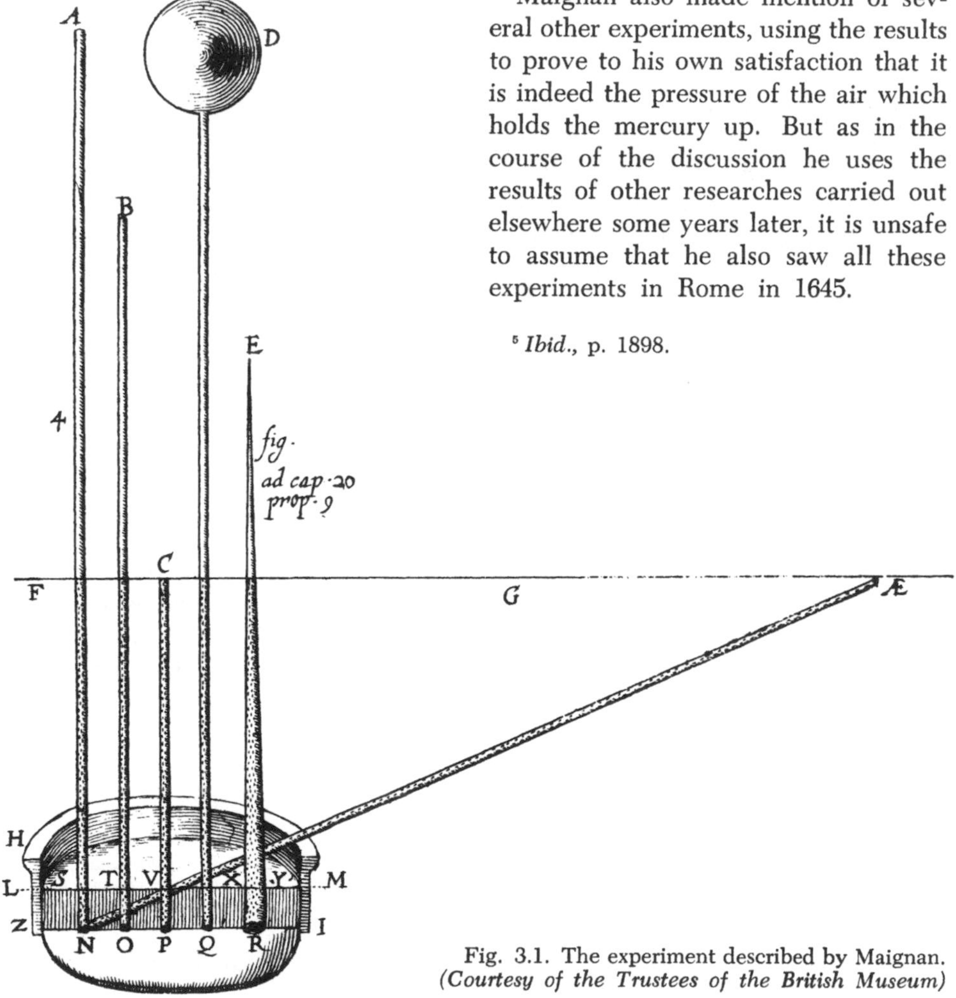

Fig. 3.1. The experiment described by Maignan.
(*Courtesy of the Trustees of the British Museum*)

Let us get back to the letter written by Du Verdus. Marin Mersenne (1588–1648), to whom Du Verdus wrote, was an indefatigable cleric who performed for mid-seventeenth-century science the functions assumed a few years after his death by the *Philosophical Transactions* and the *Journal des Sçavans*. His enormous correspondence served as a means of disseminating the latest results of experiment and speculation over a large part of Europe, and there were few scholars of any prominence who did not use his study as a clearinghouse for scientific information and gossip.[6]

The document we are considering can scarcely be called a letter at all. It is written on the two sides of the first half of a folded sheet of paper,[7] the third page[8] being blank, and the back page[9] containing only the address. From the folds and the evidence of soiling, it seems to have been simply folded in eight for transmission. I go into these details because of the contents. It consists simply of two extracts from the letters of Torricelli to Ricci which I have already reproduced in translation. The recto of folio 45 (Figure 3.2) is headed "*Extraict d'une Lettre du Sieur Torricelli,*" and in the margin "*Al. Sʳ. Ricci.*" The verso has the caption "*Extraict d'une autre Lettre du mesme au mesme*

sur le mesme sujet." The extracts are copied in Italian and are in agreement with Dati[10] except for small differences. There is no salutation, no signature, no comment, although there was an entire blank page on which they might have been written. It is not impossible that a second sheet containing a covering letter was inserted, though I should judge from the original folds that this is unlikely.

This is not very important to us. What is of great interest is whether the extracts on folio 45 were all Du Verdus sent, especially of the all-important letter of June 11, 1644. I think they were. Du Verdus ends each extract with an "&c." Furthermore he draws a short line under the first one at the left-hand margin. There is a mystery, however, and that is what the lower sketch on fol. 45ʳ is doing there at all. It refers, as the reader will remember, to the third "objection" dealt with in the letter of June 28, and this part of the letter is not copied by Du Verdus. I think this sketch is a later addition by someone else; the style of the sketch is entirely different from the other one on fol. 45ʳ and the one on 45ᵛ, both of which have a great deal of shading; and the capital letter A is of a form that Du Verdus does not use elsewhere.

The portion of the first letter that Mersenne received was, if these conclusions are correct, that beginning "We have made many glass vessels . . ." and ending ". . . by virtue of its greater rarefaction than that in the very small space *B*." Du Verdus made one important mistake in his copying; instead of "50 miles of air" he had "500 miles of

---

[6] The publication of his correspondence was begun by Mme Paul Tannery and continued with the co-operation of Cornelis de Waard, *Correspondance du P. Marin Mersenne, religieux minime,* (4 vols.; Paris, 1934–1955). The fourth volume took it only through 1634, so that the correspondence of the period in which we are interested has not been collected.

[7] Bibliothèque Nationale, *fonds. lat. nouv. acq.* 2338, fol. 45 recto and verso.

[8] *Ibid.,* fol. 46ʳ.

[9] *Ibid.,* fol. 46ᵛ.

[10] See page 22 above.

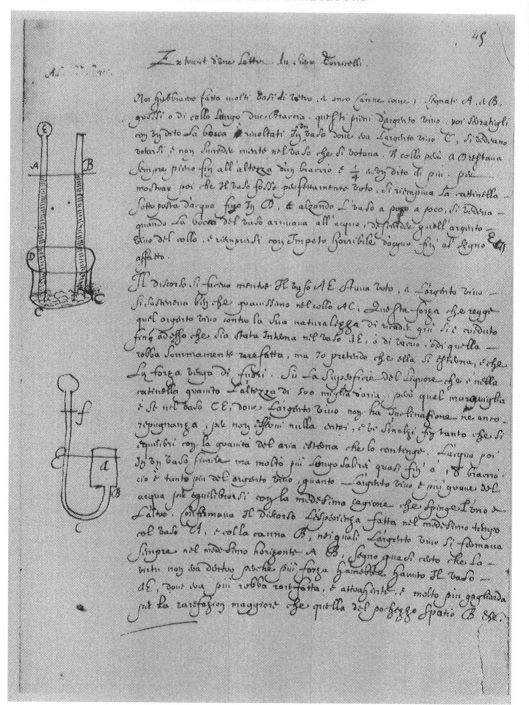

Fig. 3.2. The copy of part of Torricelli's first letter. (*Bibliothèque Nationale*)

air," and Mersenne had not the earlier part of the letter, which would have shown him at once that this was wrong.

Mersenne got enough to know how to perform the experiment, and also to realize that Torricelli explained it by means of the pressure of the air, but there were two things he did not get; Torricelli's thoughts about the facility with which a vacuum can be produced, and the statement of his "chief intention," to make an instrument to show the changes in the pressure of the air.

From the second letter Du Verdus extracted the passage beginning "As to the first [objection], I reply . . ." and ending "You will beg that I should not continue to weary you." This gave Mersenne the answer to Ricci's first "objection," and that is all. Since Mersenne had not read Ricci's letter, this could not have been very informative. But this extract contains one sentence the importance of which will become evident as we proceed:

In this case the quicksilver will be held up as before . . . but if the air that you include is more rarefied than the external air, then the suspended metal will descend by the proper amount; if now it were infinitely rarefied, i.e., a vacuum, then the metal would descend all the way, provided that the enclosed space were able to take it.

Mersenne left Paris for Rome in October, 1644. At the end of December he was in Florence, where Torricelli showed him the experiment. Three years later Mersenne wrote:

. . . it is certain that the vacuum was observed with a glass tube in Italy before it was in France; by, I believe, the illustrious geometer Evangelista Torricelli,

who showed me the tube in the year 1644 in the admirable schools of the Grand Duke of Tuscany. Moreover, we were first informed about this observation at Rome by his particular friend Michelangelo Ricci, the distinguished ornament of the whole Academy of Geometry.[11]

There follows a very brief description of the experiment, but it is clear that at least in 1647 Mersenne did not believe that it is really the pressure of the air which balances the column of mercury.[12]

In July, 1645, Mersenne returned to Paris, and in the autumn of that year he and Pierre Chanut, the Ambassador to Sweden, endeavored without success to repeat the experiment.[13] They could not get adequate glass tubes.

The first man to make the experiment outside of Italy was Pierre Petit, Intendant of Fortifications, who was at Rouen, where there was a good glass works. This was probably in October, 1646. At any rate, on November 19, 1646, Petit wrote from Paris to Chanut at Stockholm a long letter which came to twenty-four printed pages when it was published in the succeeding year.[14]

---

[11] Novarum observationum physico-mathematicarum (Paris, 1647), III, 216.

[12] Ibid., p. 217.

[13] J. Thirion, "Pascal, L'horreur du vide et la pression atmosphérique," Rev. des questions scientifiques (Brussels), 3e sér., Vol. 12 (1907), pp. 394–95. This important work, which extends over three years (12 [1907], 383–450; 13 [1908], 149–251; 15 [1909], 149–201) will be referred to as Thirion.

[14] Pierre Petit, Observation touchant le vuide faite pour la première fois en France, contenue en une lettre écrite à Monsieur Chanut . . . par Monsieur Petit . . . le 10 novembre 1646. Avec le discours qui en a esté imprimé Pologne sur le mesme sujet, en Ieuillet 1647 (Paris, 1647). The true date of the letter may be November 10, 19, or 26. (See Brunschvicg in his edition of Pascal's Oeuvres, I, 328n.)

"C'est de l'experience du Toricelli [*sic*] touchant le vuide, dont ie vous veux entretenir," writes Petit, "si vos affaires plus serieuses vous en peuuent donner le loisir."[15] (We should note that he knew the name of Torricelli.) He then tells how he had tried it at home with a very small tube, but had not had mercury enough; so he told Pascal about it, "qui fut rauy d'ouyr parler d'vne telle experience, tant pour sa nouueauté que par ce que vous sçauez qu'il y a longtemps qu'il admet le Vuide."[16] So on his return from a trip to Dieppe they did the experiment together with a larger tube (indeed quite a large tube), and enough mercury.

Petit's account of the experiment is so detailed and so full of naïve enthusiasm that we shall let him tell it in his own words:

When I came back to Rouen, we went together to the glass works, where I had a tube (*sarbatane*[17]) made, four feet long, and as big as the little finger inside, and had it stopped up at one end, or as they say in the trade, hermetically sealed. We next went to a shop (*chez un espicier*[18]) to get as much mercury as we should need, about forty or fifty pounds. Using a little funnel of folded paper, as we had forgotten to have one made of glass, and because those of tinned iron are not as suitable on account of the tin, we filled

our tube very gently, with its lower, closed end resting in a very spacious bowl or basin of wood. An earthenware pot which was not as wide at the bottom, and deeper, would have been still better. The tube being thus quite filled with mercury, I put some in the pot[19] to the depth of three fingers, and on top of this we put as much water. This done, and having put a finger on the tube, which was so full of mercury that the finger caused some to spirt out, we lifted the pipe gently, supporting it by the bottom and the middle, for fear that the weight might break it, or that I might have too much difficulty in lifting it. Then, always holding my hand firmly against the pipe, I plunged it through the water and the mercury, until my finger was at the bottom of the vase. At this point we took some time to consider whether there was some small quantity of air at the top of the pipe, but we saw none whatever, but on the contrary it was apparently entirely full of mercury. After this, when I took my finger gently away from underneath, and let the tube touch the bottom of the vase, we saw the mercury descend and leave the top of the tube, not all at once, nor in an instant, nor too slowly either, but like water being poured from a water-pot [*esguière*]; and what is very wonderful, it sank more than 18 inches, which is an extraordinary distance. . . .[20]

Then they began to philosophize, Pascal saying (probably as devil's advocate) that the space might be full of air which had passed through the pores of the glass. Why, then, said Petit, did not air continue to enter, and the mercury fall? And the success of air thermometers showed that air did not pass through glass. And if it had

---

[15] *Ibid.*, p. 3.
[16] *Ibid.*, p. 4.
[17] E. Littré. *Dictionnaire de la langue française*, says *sarbacane* means a blowpipe, or a speaking tube. He remarks that *sarbatane* is the earlier form, and that the word has been corrupted by confusion with *canne*.
[18] No dictionary that I have been able to consult has given me any help with this word in the present context. An *épicier* was, first, a dealer in spices, later by extension, a grocer, and why he should deal in quicksilver I do not know.

[19] *La terrine* this time, which is what would have been better just before.
[20] Petit, *Observation touchant*, etc., pp. 5–7.

passed through the water and the mercury (which it could not), why only enough to fill 18 inches of the tube?

But here is what seems decisive: after we had looked for a long time, and with astonishment, on this apparent or veritable vacuum, and had measured it and made a mark on the tube, I raised it up gently, and—a strange thing—the apparent vacuum got longer by as much as the depth of mercury in the bottom of the vase, without the level or height of that which was in the tube changing in any way, or rising, as I should have thought it would.[21]

It is to be noted that Petit was clearly entranced with the demonstration of the vacuum and was not thinking at this time about the pressure of the air.

After raising and lowering the tube several times, he finally brought the end up into the water, whereupon the mercury all fell out and was replaced by water, which entirely filled the tube "sans qu'il y en parust vn seul grain de reste par le haut. . . ."[22] Here was something to think about! This was one more proof that air could not have passed through pores in the tube.

Pascal was very enthusiastic and felt that this proved that there had been a vacuum in the tube, but Petit warned him that those who believed in the *horror vacui* would insist that the space was full of rarefied air, even so little that when condensed again it could not be seen.

Petit also warned Chanut that if one tries to repeat the experiment, one must "essuyer ou chauffer ladite Sarbatane en sorte que toute l'eau & humidité qui est dedans s'euapore, & qu'elle demeure

seiche."[23] And the mercury must be dried too, or little drops of water will be mixed with it in the tube, and destroy the continuity of the mercury. "Ce qui pourroit seruir de pretexte aux Peripateticiens, de dire que c'est de là que vient leur rarefaction pretendue, & conuersion d'vn Element en l'autre."[24]

These precautions show Petit in a very favorable light as an experimenter. They should have been taken by Torricelli and Viviani, but the failure of the latter to make a working barometer strongly suggests that they were not. Otherwise the experiment was conducted exactly as Du Verdus had described it to Mersenne. Unless we can believe in a most extraordinary coincidence in the trick of putting the water over the mercury, Mersenne must have passed the description of the experiment on to Petit in its entirety. But neither Petit or Pascal seem to have even considered at this time that the elevation of the mercury might be due to the pressure of the air, and Petit was obliged to take refuge in the following extremely vague explanation: ". . . it seems to me that we might say that the vacuum, if there is one, or the rarefaction, is in no way limited, but follows the force of the agent which causes it, in such a way that the weight of the mercury [has] only the force to descend to a certain height and consequently to produce vacuity or rarefaction in all the remainder of the tube. . . ."[25] It appears certain from this that Mersenne, who did not himself believe in the pressure of the air, cannot have

---

[21] *Ibid.*, pp. 8–9.
[22] *Ibid.*, p. 10.

[23] *Ibid.*, p. 14.
[24] *Ibid.*, p. 15.
[25] *Ibid.*, p. 17.

transmitted Torricelli's brief reference to it. The ensuing history of the subject might have been very different if Du Verdus had copied all of Torricelli's first letter. Mersenne might have been convinced and have sent it all to Roberval or Petit. Pascal would surely have seen at once that it was the correct explanation.[26]

But I am too bold in predicting what Blaise Pascal might have done. The life and work of this extraordinary man seem to arouse extremes of partisanship, especially in France. A particularly violent controversy over Pascal's scientific work broke out in the first decade of the twentieth century, and this was almost exclusively concerned with his part in the history of the barometer. It was started by an extremely erudite but highly tendentious series of articles by Félix Matthieu, who accused Pascal of plagiarism and deliberate lying.[27] This raised the intellectual temperature a good deal, as may be imagined; a number of people took a hand in the quarrel, but it was finally calmed down to a great extent by J. Thirion in the very scholarly and, I think, entirely fair series of papers already referred to.[28] While this work was coming out, the first three volumes of a magnificent new

edition of Pascal's works[29] came to Thirion's attention and caused him to modify his first conclusions a little, but Pascal's reputation nevertheless emerged rather tarnished.

Thirion's and Matthieu's papers together come to about 420 pages. My readers will understand from this that a full and adequate treatment of the history of the barometer in the years 1646 to 1648 would throw the present book entirely out of balance. I shall do my best to summarize what I believe to be the most probable interpretation of the documents; I may as well state in advance that I am in general agreement with the position taken by Thirion.[30]

One example of the sort of accusation brought against Pascal arose from a consideration of the published letter from Petit to Chanut from which we have just quoted. This is introduced by a dedication signed "Dominicy," to Seguier, the Chancellor of France, and by an *Au lecteur* which is not signed at all.[31] In this preface occurs the following passage: ". . . on verra que la gloire de l'inuention appartient à l'Italie, &

---

[26] E. J. Dijksterhuis, *The Mechanization of the World Picture*, Trans. C. Dikshoorn (Oxford, 1961), p. 448, thinks that even at this time Pascal "was fully aware that the experiments . . . could be explained by atmospheric pressure." But Dijksterhuis' very interesting discussion of Pascal's work gives no hint of any of the controversies that have raged around it.

[27] Félix Matthieu, "Pascal et l'expérience du Puy-de-Dôme," *La Revue de Paris*, Vol. 13, part II (1906), pp. 565–89, 772–94; Vol. 13, part III (1906), pp. 179–206; Vol. 14, part II (1907), pp. 176–224, 347–78, 835–76.

[28] Note 13 above.

[29] *Oeuvres de Blaise Pascal, publiées suivant l'ordre chronologique, avec documents complémentaires, introductions et notes, par Léon Brunschvicg, Pierre Boutroux, et Félix Gazier* (14 vols.; Paris, Hachette, 1904–1914). Reference will be made to this edition as Brunschvicg.

[30] It is unfortunate that both Matthieu's and Thirion's papers are relatively inaccessible. I have not found all of the former anywhere in London, for example.

[31] Marc Antoine Dominicy was a historian who was concerned with the history of France. It is remarkable that both Brunschvicg (I, 326) and Thirion ("Pascal, etc." [1907], pp. 428–31) seem to assume that Dominicy wrote the *au lecteur* as well as the dedication, or that we are supposed to think that he did.

selon mon aduis à cet admirable Philosophe & Mathematicien Galilee non pas à Toricelli: aussi celle de l'auoir obseruée le premier en France ne peut estre disputée a Monsieur Petit, . . ."[32]

Now Thirion came to the conclusion that the editor of this little book was none other than Pascal himself, and that the latter (who, of course, really knew about Torricelli all the time) had the ambition at that moment to be considered the successor of Galileo. Pascal might well have written the unsigned preface, and Thirion's account of his supposed ambitions is supported by various passages in the *Expériences nouvelles*.[33]

The experiment created a sensation in Rouen. In the winter of 1646–1647 Pascal set it up at the glass factory in great tubes filled with water and wine. Among the spectators were Périer (Pascal's brother-in-law), Guiffart, Pierius, and Auzout, as well as several Jesuits.[34] Jacobus Pierius was moved to write a short dissertation on the vacuum,[35] and Guiffart a long and rambling examination of the possibility of their being a vacuum above the mercury in the tube.[36] The *privilège du Roy* for this was given on April 20, 1647. It appears that Pascal liked conferences and public demonstrations; his editor, Brunschvicg, admits that in them "il . . . goûte d'une façon plus directe la joie de la gloire; en même temps aussi, par l'assentiment

ou par la contradiction qu'il rencontre, il mesure mieux la portée de sa découverte."[37]

All this had nothing to do with the pressure of the air, but was intended only to show the fragility of the scholastic thesis about the *horror vacui*. Pascal's double experiment with the water and the wine was designed to confute those who asserted that the space was filled with volatile "spirits" from the liquid. He asked such people to predict what would take place, and of course they knew that wine is more "spirituous" than water, so they forecast that the water would stand higher than the wine. The experiment was made, and the reverse happened, because, of course, the wine was lighter. The young Pascal was a superb *metteur en scène*, and was obviously enjoying himself. He also made the experiment of inclining the tube with the mercury, so that the mercury filled it, which led Guiffart[38] to speculate on the enormous amount of mercury that might be held up if a long tube were set tangent to the surface of the earth.

Dijksterhuis[39] has emphasized that it was possible at the time to hold any one of four different views on the explanation of the experiment, since one could believe or not believe in a vacuum and could (quite independently of one's views about the vacuum) believe that the column was held up by the pressure of the air, or by some attraction or other within the tube. For those who refused to believe in a vacuum, there was a good choice of substances

---

[32] Petit, *Observation touchant*, etc., fol. a5[v].
[33] See p. 43 below.
[34] Thirion, "Pascal, etc.," (1907), p. 397. Brunschvicg, II, 8.
[35] No title or date or place of impression. 14 pp.
[36] *Discours du vuide sur les expériences de M. Paschal et le traicté de M. Pierius*, etc. (Rouen, 1647).

[37] Brunschvicg, II, 5.
[38] *Discours du vuide*, etc., p. 239 ff.
[39] *The Mechanization*, pp. 444–45.

with which to fill the tube: rarefied air, vapor from the liquid, small particles of fire *(igniculi)*, or some "subtle matter" or "ether" which could have whatever properties were needed in order to explain the observed phenomena, and which was therefore invaluable to both the Peripatetics and the followers of Descartes.

In the spring of 1647 Pascal moved to Paris. Then on July 24, 1647, Des Noyers, a Frenchman attached to the court at Warsaw, wrote to Mersenne about an experiment performed in the presence of King Wenceslas VII and the Queen of Poland by Valeriano Magni (1587–1661), a Capucin. This was, quite simply, the Torricellian experiment, water and all. Des Noyers was plainly much impressed, and asked Mersenne what he thought.[40] He also enclosed a printed account[41] of the experiment. Again the interest is entirely in the demonstration of the possibility of a vacuum. As Thirion says of this book,[42] "Son titre est une déclaration de guerre aux péripatéticiens. . . . Pour Magni, comme pour Pascal, ce qui en fait l'intérêt, c'est le vide que l'expérience montre aux yeux, le vide où le mouvement se produit, que la lumière traverse, le vide qui convainc d'erreur Aristote et ses disciples; . . . il brandit

son tuyau comme une arme de guerre."[43]

Des Noyers' letter was answered in Latin by Roberval,[44] who easily demolished Magni's claim to priority and reported some experiments of his own, but said nothing about the pressure of the air. It is interesting that Roberval claimed to have a letter about the experiment from Torricelli to Ricci:

I have a letter that the most celebrated Evang. Torricelli, the Grand Duke of Tuscany's mathematician, sent to his friend the very learned Angelo Ricci in Rome at the end of 1643,[45] written in Italian; which contains nothing but a controversy between those two distinguished men who, as happens to nearly all, felt differently about such an experiment.[46]

The use of the singular means that Torricelli's second letter was all he had; he goes on to say that "this letter and some others" were sent by Ricci to Mersenne. It is therefore still an open question how much of Torricelli's views Mersenne and Roberval had come to know. It is certain that at this time Roberval would not have agreed with those views, supposing he had known them. It would be strange, if he knew them, not to mention them in his long letter.

Roberval did point out that Magni had been in Italy at the beginning of

---

[40] Brunschvicg, II, 15–18.

[41] *Demonstratio ocularis loci sine locato: corporis successiuè moti in vacuo: luminis nulli corpori inhaerentis. A Valeriano Magno Fratre Capvccino exhibita,* etc. (Warsaw, [1647]). The permission to print is dated July 12. For the rather complicated bibliography of this work see G. Hellmann, *Abh. Preuss. Akad. d. Wiss., Phys.-Math. Kl.* (1920), no. 1, pp. 33–34.

[42] "Pascal, etc.," (1907), pp. 402–3.

[43] His anti-Aristotelianism later cost him a cardinal's hat, and even got him into prison for a time in Vienna. (De Waard, *L'expérience barométrique,* p. 127.)

[44] Brunschvicg, II, 20–35. Gilles Personne de Roberval (1602–1675) was a mathematician of considerable celebrity.

[45] The letters we know of are dated June, 1644.

[46] Brunschvicg, I, 21–22.

1645 and had met Mersenne there. As soon as Magni was shown Roberval's letter he wrote a reply,[47] in which, while acknowledging the priority of Torricelli, he stoutly maintained that at no time until then had he even heard of Ricci and Torricelli, "not that these men's names were not distinguished, but that compared to them, I was obscure." Nor had he at any time, while he was in Florence, heard the word "vacuum."[48] Mersenne had said nothing to him about it in Rome, and in short, he had heard nothing from friends and seen nothing in print or in manuscript about the experiment. He even attached two fragments of letters from ecclesiastics in Rome, both dated September, 1647, and both indicating that the *Demonstratio ocularis* had been generally received in Rome as a complete novelty which at first seemed ridiculous, as they had heard nothing of the discussion between Torricelli and Ricci.

In a further treatise,[49] Magni, with great frankness, published not only this reply but also Roberval's letter. There seems to me to be only one cause for doubt that he had, in fact, done the experiment entirely independently, and that is the circumstance that he also thought of putting water above the mercury, even though he put a small vase of mercury in a larger vase of water. This use of water is not, or so it seems to me, an obvious extension of the fundamental experiment. Whether

or not he was an independent discoverer, one fact is clear: he was the first to publish a complete account of the experiment, although Pascal pretended in 1651 that he had published his account first.[50] Thirion[51] believes that Pascal could not abide competition at this period, and must have been very much annoyed by the appearance of Magni's book when he was working at a book about his own experiments.[52]

This little work is claimed to be merely an abstract of a treatise that he is preparing (which never appeared); he is writing it to obtain priority. He says that in 1644 Mersenne told Perier, Perier told him, about the experiment made in Italy "il y a environ quatre ans," and that they made the same experiment at Rouen, "et trouvasmes de poinct en poinct ce qui avoit esté mandé de ce pays là, sans y avoir pour lors rien remarqué de nouveau."[53]

But he has made numerous other experiments since, which have convinced him that "l'espace vuide en apparence, qui a paru dans les experiences, est vuide en effet de toutes les matières qui tombent sous les sens, et qui sont connuës dans la Nature."[54] Reading the *Experiences nouvelles*, one senses that at this period Pascal, when he wrote the word "vacuum," really meant an empty space. After describing some experiments (possibly imaginary) with siphons filled with water and with mercury, he gives seven "maxims," eight "proposi-

---

[47] Brunschvicg, II, 503–6.
[48] *Ibid.*, p. 504. ". . . nec tamen aliquando sonuit mihi in illa Urbe vox ista, Vacuum."
[49] *Admirando de vacuo scilicet, Valeriani Magni Demonstratio ocularis de possibilitate vacuo,* etc. (Warsaw, n.d., but evidently November, 1647). Quoted in Hellmann, *Abh. Preuss. Akad. d. Wiss.,* etc., p. 34.

[50] Brunschvicg, II, 490.
[51] Thirion, "Pascal, etc.," (1907), p. 404.
[52] *Expériences nouvelles touchant le vuide,* (Paris, 1647). (Brunschvicg, II, 55–76.) The *permission* is dated October 8, 1647.
[53] Brunschvicg, II, 59.
[54] *Ibid.,* p. 61.

tions," and seven more "maxims" which are the conclusions about the vacuum which he has drawn from his researches. The "maxims" may be summarized as follows: there is a certain force which tends to prevent the occurrence of an empty space between bodies, but this force is limited, and equal to that with which water about 30 feet deep tends to flow downward. A force only slightly greater may produce a vacuum as large as desired. The "propositions" declare that the space which appears empty is not filled with air that has come from outside, nor with air which has been between the pores of the surrounding bodies; it is not filled with an imperceptible morsel of air, extraordinarily dilated ("que quelques-uns soutiendront se pouvoir rarefier assez pour remplir tout le monde, plustost que d'admettre du vuide"[55]). Nor it is full of a small amount of quicksilver or water, "qui . . . se rarefie et se convertit en vapeurs,"[56] nor of a "more subtle air" which is mixed with the external air and goes through the pores of the glass. In fact, the space which appears empty is just that. As his "conclusion," he states that

After having demonstrated that none of the substances that can be perceived by the senses, or of which we have knowledge, fill this apparently empty space, it will be my opinion, until someone has demonstrated to me the existence of some matter which fills it, that it is truly empty, and destitute of all matter.[57]

In view of this clear statement, it is amusing to read in Perier's introduction to the *Traitez de l'equilibre des liqueurs, et de la pesanteur de la masse de l'air,* etc.,[58] the following disclaimer:

Throughout, where the word *Vacuum* is found, it must not be imagined that Mr. Pascal had the intention of proving that there could be an absolutely empty space, but only that by this word vacuum he always means a space empty of all substances which can be perceived by the senses, as he notes in several places.

Finally Pascal lists five "objections" that can be raised to contradict his conclusion: that it is against common sense; that it accuses nature of impotence; that many everyday experiences show that nature cannot suffer a vacuum; that an imperceptible form of matter fills the space; and that light could not get through a vacuum. He does not comment on any of these "objections."

The mention of the fourth one of course annoyed Réné Descartes, who chose to consider it a personal attack on his "subtle matter."[59]

It will be seen that in 1647, at least, Pascal was taking an extreme "vacuist" position, no matter what Perier may have wished people to believe after his death. It was a much more extreme position than that taken by any other seventeenth-century writer, including Boyle and Hooke, as we shall see in Chapter 4. And to Pascal the only thing that mattered at this time was the existence of the vacuum. Nowhere in the book is there any indication that Pascal knew or cared about the pressure of the air, or that he thought it was the weight of the air which balanced the weight of the column of mercury.

[55] Brunschvicg, II, 72. This, of course, is what actually occurs.
[56] *Ibid.*, p. 73.
[57] *Ibid.*, p. 73.

[58] Paris, 1663, fol. 2.
[59] Brunschvicg, II, 165 and 408.

Nor, while he refers to the experiment as having been made in Italy, does he mention the name of Torricelli, although it seems impossible that Petit, who knew it, should not have mentioned it during their experiments at Rouen. And there was Petit's letter to Chanut.

Let us, for the moment, come back to René Descartes, who, as is well known, categorically denied the possibility of a void, filling all apparently empty space with two kinds of "subtle matter" having wonderful properties by means of which almost everything could be explained. Now Descartes visited Pascal on September 23, 1647, and again on the 24th, as we learn from a letter written by Pascal's sister Jacqueline to his other sister, Madam Perier.[60] The first meeting was somewhat spoiled by the presence of Roberval, for whom Descartes felt a well-returned dislike, but the second was quieter. Jacqueline reported that Descartes, contrary to what Pascal had thought, was a believer in the pressure of the air, "mais par une raison que mon frère n'approuve pas."[61] Roberval, on the other hand, denied the pressure of the air at this time.

Unfortunately, Jacqueline Pascal's letter says nothing about the celebrated experiment known ever since as the experiment on the Puy-de-Dôme, in which for the first time a barometer was set up at various elevations on a mountain; and this is very unfortunate, because

on December 13, 1647, in a letter to Mersenne, Descartes claimed to have suggested the famous mountain experiment:

> I'auois auerti M. Pascal d'experimenter si le vif-argent montoit aussi haut, lorsqu'on est au-dessus d'vne montagne, que lorsqu'on est tout au bas; ie ne sçay s'il l'aura fait.[62]

We shall probably never know who first suggested the experiment, and the opinions of historians are various. Thirion[63] concluded, even after reading what Brunschvicg had to say, that it was Descartes; this opinion is shared by J. O. Fleckenstein,[64] who thinks it likely that at their famous meeting, Pascal scarcely heard what Descartes had said about it, because of his strong feeling that the latter's philosophy was just Aristotelianism in another form. The most charitable view is that while Descartes did make the original suggestion, it penetrated into Pascal's subconscious and came out later; but our acceptance of this hypothesis must depend on our judgment of the letter which Pascal wrote to his brother-in-law Florin Perier, and which we shall discuss below.[65] Meanwhile it must be recorded that Mersenne, after receiving Descartes' letter of December 13, wrote about the proposed experiment, first to Huygens, then to Le Tenneur, who was,

---

[60] *Ibid.*, pp. 43–48.
[61] *Ibid.*, p. 46. This was, of course, the introduction of vortices of "subtle matter" for which Pascal had no use. Descartes had believed in the pressure of the air since at least 1631 (see p. 7 above), but he would have nothing to do with the idea of a vacuum.

[62] *Oeuvres*, eds. Chas. Adam et Paul Tannery, Vol. 5 (Paris, 1903), p. 99. Brunschvicg, II, 165.
[63] "Pascal, etc.," (1909).
[64] *Experientia*, (Basel, Vol. 4, 1948), p. 163.
[65] P. Duhem (*Rev. Gén. des Sci.*, Vol. 17 [1906], pp. 814–17) believed that Mersenne conceived the experiment quite independently. F. Strowski, in his edition of Pascal's works (1923, I, 69, note), also thought that Mersenne had the first idea of the experiment.

as far as he knew, at Clermont. He invited Le Tenneur to make the experiment on the Puy-de-Dôme.[66] On January 16, 1648, Le Tenneur, who had moved to Tours, replied that he was no longer in the Auvergne and therefore could not do it. Furthermore, he added, ". . . je vous diray que je pense, avec Roberval, que cela seroit entierement inutile et que le mesme chose se trouveroit en haut qu'en bas."[67] He then confessed that he had been quite unable to understand the phenomenon of the Torricellian tube as described by Mersenne.[68] The idea of a cylinder of air balancing the mercury was a stumbling-block. ". . . je vous prie, ou est ce cylindre d'air imaginaire, et comme quoy disent ils qu'il agit au travers d'un tuyau vuide d'air?"[69] Few people at the time understood hydrostatics. The rest of the letter is very interesting, showing as it does an intelligent man who sees that the Peripatetic arguments are "chansons et railleries, et en un mot, de signalés moyens pour faire voir la futilité de leur resonement,"[70] but is puzzled by the "subtle matter" of Descartes, whose authority he respects. "La matiere subtile de M. Descartes ne me contente pas. Que ce soit celle qui estoit meslée avec le mercure, qui s'en soit separée, il me semble que cela ne ce peut," etc.[71] One gets the feeling that the phenomenon might have been understood much earlier and more

generally in France except for the complications introduced by Descartes. In England, where he had much less influence, the Torricellian experiment seems to have been accepted with greater ease; but there were also fewer theological difficulties.

Nevertheless in the history of the barometer, Descartes has one positive contribution to his credit. If a meter must be an instrument with a scale, then by furnishing the Torricellian tube with a paper scale he was the first to make a barometer. In the letter to Mersenne referred to above[72] occurs the following passage:

. . . so that we may also know if changes of weather and of location make any difference to it, I am sending you a paper scale two and a half feet long, in which the third and fourth inches above two feet are divided into lines; and I am keeping an exactly similar one here, so that we may see whether our observations agree.[73]

We do not know whether the scale was glued to the tube or not, or indeed how either the tube or the scale was mounted; but here, probably for the first time, was an elementary instrument for the routine measurement of atmospheric pressure.

Let us come back to Pascal and discuss the famous letter that he is supposed to have written to his brother-in-law Perier on November 15, 1647.[74]

[66] Descartes, *Oeuvres*, ed. cit., Vol. 5, p. 102.
[67] *Ibid.*, p. 103.
[68] *Novarum observationum physico-mathematicarum F. Marini Mersenni*, etc. (Paris, 1647), preface. Quoted by Brunschvicg, II, 150–51.
[69] Descartes, *Oeuvres*, ed. cit., Vol. 5, p. 103.
[70] *Ibid.*, p. 104.
[71] *Ibid.*, p. 105.

[72] See p. 45 above.
[73] *Oeuvres*, ed. cit., Vol. 5, p. 99.
[74] Brunschvicg, II, 153–62. Some readers will note that I have passed over the controversy between the Pascals, father and son, and the Jesuit Father Noël. This exchange shows the almost automatic reaction of a convinced and yet extremely courteous Aristotelian to the idea of a vacuum, and is very interesting

Florin Perier was "general conseiller en la Cour des Aides" at Clermont.

Almost at the beginning of this letter, Pascal says he is inclined to attribute to the pressure of the air the effects which the philosophers impute to the *horror vacui*. He then declares that we must be very careful in rejecting opinions held for a long time, but after the experiment, which he performed in Perier's presence "ces jours passez" with two tubes, one inside the other, he is convinced of the pressure of the air. But it would also be possible to explain this experiment by a [limited] *horror vacui*. There is therefore only one decisive experiment:

C'est de faire l'experience ordinaire du Vuide plusieurs fois en mesme jour, dans un mesme tuyau, avec le mesme vif-argent, tantost au bas et tantost au sommet d'une montagne, eslevée pour le moins de cinq ou six cens toises,[75] pour esprouver si la hauteur du vif-argent suspendu dans le tuyau, se trouvera pareille ou differente dans ces deux scituations.[76]

Perier lives at Clermont, which is at the foot of a high mountain, the Puy-de-Dôme, and Pascal hopes he will make this experiment. On this assurance he has told "tous nos curieux de Paris" of the idea, and especially Father Mersenne, who has written about it to

Poland, Sweden, Holland, etc. Will he please do it as soon as possible, to satisfy the impatience of so many people who are waiting to hear the result?

This is the letter which Matthieu roundly declared to be a falsification, written for publication, rather than for Perier, and in the summer of 1648 rather than in November, 1647.[77] The textual difficulties have been dealt with at length by Brunschvicg and by Thirion; I shall confine myself to the main points which are felt to cast suspicion on the authenticity of this letter, or rather of its date and destination:

1. No manuscript is known, nor any of an answer from Perier.

2. It was published[78] together with Perier's account of the experiment which he made in September, 1648. The letter was published without the usual *Privilège* and by a small bookbinder, Charles Savreux, rather than by Pascal's ordinary publishers.

3. It is difficult to believe that Pascal could have been so suddenly convinced of the role of the pressure of the air.

4. None of the letters that Mersenne is supposed to have already written to various countries have survived. On the other hand, as we have already noted, Mersenne did write later to Le Tenneur, who was, he thought, at Clermont. It is inconceivable that he would have done this if he had known that Pascal had made a similar request to Perier.

5. The letter was supposedly writ-

---

for the light it throws on Pascal's rigorous scientific method, but is of minor importance in the history of the barometer. For an excellent analysis see Dijksterhuis, *The Mechanization*, pp. 448–51.

[75] 1 *toise* or fathom = 6 feet.

[76] Brunschvicg, II, 159–60. Father Cherubin, a Capucin of Orleans, in his *Effets de la force et de la contiguité des corps*, etc. (Paris, 1679), pp. 113–14, even managed to explain away the mountain experiment while denying the pressure of the air.

[77] Matthieu, "Pascal et l'expérience, etc.," (1906), p. 206.

[78] Blaise Pascal, *Recit de la grande expérience de l'equilibre des liqueurs*, etc. (Paris, 1648).

ten in November, and yet Pascal seemed to expect that the Puy-de-Dôme could be climbed almost at once, i.e., in winter.

I think that we may agree with Thirion in doubting that the letter was actually written in 1647. What then was the motive for such a piece of falsification?

It has been thought that the key to the matter is to be found in the experiment of the "vacuum within the vacuum," referred to in Pascal's letter to Perier. This is an experiment in which a barometer is set up in another barometer tube, i.e., in a vacuum. The mercury in the inner tube falls to the level of that in its cistern. Then, when air is let into the outer tube, so that its pressure acts on the cistern of the inner barometer, the mercury in the inner tube rises to its usual height. It is not necessary that all the air should be let in at once; a partial vacuum in the outer tube would lead to the mercury in the inner tube remaining part of the way up.

It is agreed that an experiment of similar import, but more convenient in design than that described by Pascal, was done in the summer of 1648, either by Adrien Auzout or by Roberval.[79] It can be maintained with at least some probability that Pascal also did the experiment he described. But as to whether he did it for Perier at some time[80] in October or November, 1647, I think the historian may be pardoned

for desiring something more than Pascal's own word for the date in view of the difficulties mentioned above, especially Mersenne's letter to Le Tenneur. The controversy about this rumbles on, and may never reach a satisfactory conclusion. Louis Lafuma, for example, stoutly defends the memory of Pascal in an essay[81] which seems to me to be just as partisan as that of Matthieu on the other side. Let us now leave this matter to the *pascaliens,* and get on with the history of the barometer.

It should be noted again that the part of Torricelli's second letter to Ricci that Mersenne received from Du Verdus contained a clear forecast of the results that would be obtained from the experiment of the vacuum within a vacuum; but there is no certainty that Pascal ever saw this letter. We have Perier's word for it, in his *Preface* to the posthumous *Traictez de l'equilibre des liqueurs et de la pesanteur de la masse de l'air,*[82] that Pascal heard of Torricelli's "thought" only in 1647, but considered it only a simple conjecture, needing experimental proof. This last phrase rings true in view of what we know of Pascal's scientific skepticism.

The history of the barometer between January and June, 1648, was somewhat blank until Leon Brunschvicg discovered a long manuscript by Roberval which was begun on May 15 of that year and finished, Brunschvicg believed, in October.[83] It is probable that

[79] Brunschvicg, II, 291–93.

[80] I have not gone into the subject of Perier's movements in the autumn of 1647 (see Brunschvicg, II, 155). Exhaustive investigation has shown, as far as I understand it, that it would have been possible for him to have paid a short visit to Paris during the period in question, but that it is not known for certain that he did so.

[81] *Controverses pascaliennes,* (Paris, 1952), pp. 62–74.

[82] Paris, 1663. (Translated into English by I. H. B. and A. G. H. Spiers, [New York, 1937]).

[83] *E. P. de Roberval de Vacuo Narratio ad nobilem virum dominum des Noyers,* etc. B.N., ms. f. lat. 11197 fol. 26 sqq. In Brunschvicg, II, 310–40 and 359–61.

Roberval never sent it to its destination, feeling, Brunschvicg thinks, that the Puy-de-Dôme experiment had completely changed the state of affairs.

This manuscript is interesting in many ways. In one of many experiments, Roberval heated the top part of the tube, finding (as Torricelli had) that the mercury went down somewhat. But whatever "body" was in the tube, Roberval did not think it was air, for he had removed the air very carefully. (It may have been a little water.) He found that a small amount of air let into the tube depressed the mercury more than an equal volume of water, and that the longer the tube (i.e., the bigger the vacuum space) the less effect a given volume of air had. He found the correct answer in the spontaneous dilatation of the air, and it was at this point that he realized, to his own surprise, that the force that kept the mercury up must be external to the tube, namely the pressure of the air. To prove this, he devised the experiment of the vacuum within the vacuum.

But he did not describe the apparatus, and we must go to a later account,[84] which ascribes the experiment to Adrien Auzout (1622–1691). "In order," says Pecquet, "that you should not continue to hold the opinion of the Ancients against the argument according to which the weight of the external air is balanced by the mercury inside, I am glad to show you the experiment of the vacuum within the vacuum, first successfully tried by the sagacity of that very clever man, Auzout."[85] There follows a detailed description of the ex-

periment, illustrated by the figure reproduced as Figure 3.3.

A flask AB, with a very long neck, has a small side tube G, open at the end, which can be sealed with a piece of bladder. After a small rectangular cistern has been installed, another tube CF is sealed into the flask at B. The cistern D is partly filled with mercury, and A stopped up (with the finger, he says) while the whole thing is filled with mercury through F.[86] Next F was sealed up. Then when the finger was removed from A under the surface of the mercury in D, all the mercury ran out except what was in AE and in the little cistern C. The height AE was found to be 27 inches. Then when the bladder at G was pierced with a pin, the mercury fell out of AB to the level in D and rose 27 inches in CF.

"And what, O reader, must we conclude from this? The external air balanced the mercury in the cylinder AE; and therefore, air is heavy even in its own sphere."[87] He goes on to say that air, in its elastic nature, imitates sponge or wool.

This somewhat awkward experiment was later improved, as we shall see in Chapter 4, but nevertheless it was very impressive at the time.

Another experiment by which Roberval hoped to convince the skeptics was to tie up a tiny bladder (of a carp), leaving as little air in it as possible. When this was introduced into the vacuum space, it became inflated. This

---

[84] Jean Pecquet, *Experimenta nova anatomica*, etc. (Paris, 1651), pp. 56–58.

[85] *Ibid.*, p. 56.

[86] According to Pecquet; perhaps G was left open as long as possible. I should have thought it better to seal F and do most of the filling through A, but the small cistern may not have been well enough anchored.

[87] *Ibid.*, p. 58.

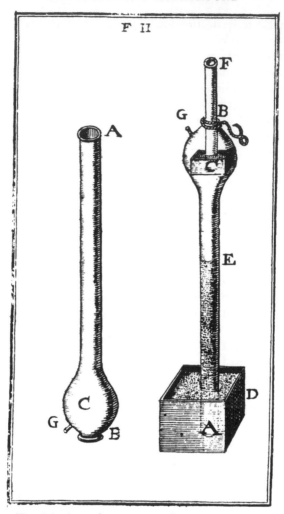

Fig. 3.3. Auzout's apparatus, after Pecquet, 1651.

convinced many people, but not, of course, Descartes, who had made up his mind that nature had to follow his speculative ideas, and insisted that the bladder must have been inflated by his "subtle matter." He was convinced that the Torricellian experiment could not be made in a completely closed chamber; the mercury could not fall, because it would have nowhere to go.[88]

The position that Roberval found himself in after all this was, first, that it was the pressure of the air that sustained the mercury; and second, that there was not a vacuum at the top of the tube, but very greatly rarefied air. This interpretation of the observations was confirmed for him by experiments made by Mersenne on the sound of a bell in the space.[89] As in Kircher's ex-

[88] See Brunschvicg, II, 296–301.

[89] Brunschvicg, II, 308.

periments,[90] not enough care was taken to prevent vibrations reaching the glass through the supports. Mersenne died on September 1, 1648, still of the opinion that the space was not a vacuum.

Three weeks later, on September 19, Perier took the Torricellian tube up the Puy-de-Dôme.[91] He got up at five that morning and saw that the weather looked promising. Then he sent messages to several people, locally distinguished, whom he had persuaded to witness the experiment: a reverend Father Superior, a Canon of the Cathedral (Mosnier), two legal gentlemen, and a doctor. They all assembled at eight o'clock in the garden of the monastery, which is in the lowest part of Clermont.

Perier had re-distilled (rectifié) 16 pounds of mercury. He took two similar glass tubes, hermetically sealed at one end, and made the Torricellian experiment in each, bringing them close together and dipping in the same vessel (careful man!), to be sure that the mercury in the two tubes stood at the same height, as indeed it did, 26 inches, 3½ lines. He emptied the tubes and did the experiment again, and then once more, and obtained the same result. We are not told how the height was measured, only that it was "au dessus de la superficie de celuy du vaisseau."

Leaving one tube set up at the monastery in charge of one of the monks, who was to observe it frequently throughout the day, the party took the other tube and the remainder of the mercury up the Puy-de-Dôme, which was estimated to be 500 fathoms high.

When they set the barometer up, the mercury stood at only 23 inches, 2 lines. It is hard for us to feel the excitement of that moment, and we had better let Perier tell it in his own words:

. . . thus between the heights of the quicksilver in these two experiments, there was a difference of three inches and one-and-a-half lines; which ravished us all with admiration and astonishment, and surprised us so much that for our own satisfaction we wished to repeat it. That is why I did it, very exactly, five times more at various places on the summit of the mountain, either under cover in the little chapel which is there, or out-of-doors, once[92] in a sheltered place, once in the wind, once while the weather was fine, and once during the fog and rain which came up now and then, having purged the tube of air with great care each time; and in all these experiments the height of the quicksilver always turned out to be the same . . . which fully satisfied us.[93]

They were witnessing one of the great moments in the history of ideas, and Perier, at least, was aware of it. I cannot get rid of the impression that he did not really expect the experiment to turn out the way it did.

Rather more than half way down the mountain, at a place called Lason de l'Arbre, they found a height of 25 inches in the tube; they were delighted to see "la hauteur du vif argent se diminuer, suivant la hauteur des lieux."[94] When they got back to the monastery, the good Father Chastin was able to assure them that the height of the mercury had been constant all day; and as a final precaution, they again twice filled

[90] See p. 14 above.
[91] Pascal, Recit de la grande expérience, etc. (Brunschvicg, II, 350–58.)

[92] No more literal translation will make sense of "five times."
[93] Brunschvicg, II, 354.
[94] Brunschvicg, II, 355.

the tube that they had taken up the mountain, always getting the same result, 26 inches, 3½ lines.

The next day Perier went up the highest tower of the cathedral and to one or two other places in Clermont, obtaining small differences in the height of the mercury.

At the end of his letter Perier regrets that while the mercury column was measured as carefully as possible, the elevations on the mountain were determined much less exactly. They seem to have been guessed at. Thus the value of the experiment lay in its certain demonstration that the air has weight, and its refutation of the doctrine of the *horror vacui;* but it furnished no good rule for the determination of heights with the barometer.

In 1943 P. Humbert[95] drew attention to the fact that there is another account of the Puy-de-Dôme experiment in the works of Gassendi,[96] who got his information from Canon Mosnier, one of those who made the ascent with Perier. The numbers given by Mosnier differ in a systematic way from those reported by Perier, and Humbert thinks it probable that the measurements were actually made with a scale divided according to the *pied de Macon* (33.57 cm.), and reported in these terms by Mosnier, while Perier, in communicating with Pascal, converted the results to the *pied de Roi,* or *de Paris* (32.5 cm.).[97] It seems more probable that he converted the readings than that he used a different scale, for Gassendi tells us that a

scale was fixed to the glass tube, four feet long, which was taken up the mountain.

Perier's description of the ascent, together with the controversial letter from Pascal, was published in the autumn of 1648, in the famous and excessively rare 20-page pamphlet already referred to.[98] It is hard to believe that Pascal intended a wide distribution of this little work. It is evident that Descartes had not even heard of it by June 11, 1649, when he wrote to Carcavi:

. . . I beg you to let me know the success of an experiment which I have heard that Mr. Pascal has made or caused to be made on the mountains of Auvergne, in order to find out whether the quicksilver rises higher in the tube when it is at the foot of the mountain, and how much higher it rises than at the top. I should have a right to expect this from him rather than from you, since it was I who advised him, two years ago, to make this experiment, and who assured him that I did not doubt of its success even though I had not made it.[99]

Descartes thought that Pascal was still under the influence of Roberval, his *bête noir.*

It is even harder to see why Pascal did not seem to want the knowledge of this experiment more widely distributed; but I have no desire to re-embark on a subject which is of more importance to the criticism of Pascal than to the history of the barometer. What is very interesting in this connection is that Pascal himself carried a partly-

---

[95] *La Météorologie,* Paris (1943), pp. 222–24.
[96] *Opera omnia* (Lyon, 1658), I, 211.
[97] See Chapter 8 regarding the confusion in units.

[98] *Recit de la grande expérience de l'équilibre des liqueurs,* etc. For the bibliography of this see Maire, *L'oeuvre scientifique,* pp. 76–77 and Hellmann, *Abh. Preuss. Akad. d. Wiss.,* etc., p. 36.
[99] Descartes, *Oeuvres,* ed. cit., Vol. 5, p. 365.

filled "balon" up the Puy-de-Dôme in 1649 or 1650.[100]

Pascal's most extensive work on physical science[101] was not published until after his death, and by this time the pneumatic experiments of Guericke and of Robert Boyle had been given to the world, so that it had little influence on the subsequent history of the barometer. The book was, however, written long before, perhaps even by 1650, according to the editor, Perier.[102] But Pascal, who had had an overwhelming religious experience, and now thought science unworthy of the consideration of serious people, would not hear of its publication. It is largely a treatise on hydrostatics, containing the celebrated generalization known to every schoolboy as Pascal's Principle. Much of it could have been derived from the work of Simon Stevin (1548–1620), but it was probably original, and at any rate the systematization of hydrostatics and aerostatics was an achievement of major importance. The part of the work dealing with the pressure of the air describes a simple form of the experiment of the vacuum within a vacuum,[103] but, more important for our purpose, also describes with great clarity the siphon barometer, the invention of which is certainly due to Pascal:

The most suitable instrument for observing all these variations is a glass tube three or four feet high, sealed at the top and curved round at the bottom, to which is glued a band of paper, divided in inches and lines. For if it is filled with quicksilver, we shall see that this will fall part of the way, and will remain partly suspended, and we shall be able to note exactly the degree to which it is suspended; and it will be easy to observe the variations in the weight of the air which happen because of the changes of weather, and those which take place in carrying it to a more elevated place.[104]

It is possible that Pascal had this idea during the experiments at Rouen in 1646; for in Roberval's first letter to Des Noyers, written in September, 1647, there is the following passage: "I must certainly not fail to mention that the observation of the mercury was made by the same Mr. Pascal in the recurved tubes that we commonly call siphons."[105] On the other hand this may refer merely to the demonstration that mercury will not flow through a siphon which is too high.

It is time to review a little. We have seen how Mersenne, who was a sort of one-man *Philosophical Transactions*, heard of the Torricellian experiment in the summer of 1644; how he found out more about it in Italy during the following winter, and in 1646 told Petit how to make the experiment, although, for want of adequate materials, he had not himself succeeded in doing so. Petit told Pascal, and in October, 1646, they made a long series of experiments at Rouen, with mercury, water, and wine. In July, 1647, before an account of these had been published, news arrived

---

[100] Brunschvicg, III, 200, note.
[101] *Traictez de l'équilibre des liqueurs et de la pesanteur de la masse de l'air, etc.* (Paris, 1663). This was translated into English by I. H. B. and A. G. H. Spiers (New York, 1937). (Brunschvicg, III, 156–292.)
[102] *Ibid.*, preface.
[103] Brunschvicg, III, 236. As noted by Perier in the preface, the engraving which illustrates this is quite wrong.
[104] *Traictez de l'equilibre*, etc., p. 100. (Brunschvicg, III, 233.)
[105] Brunschvicg, II, 32.

from Warsaw that Valeriano Magni had made the Torricellian experiment there, claiming it as his own idea. Roberval showed Magni that Torricelli had the priority, but Magni maintained that he had thought of the experiment entirely independently.

Meanwhile in September, 1647, both Mersenne and Descartes, independently or not, seem to have suggested the experiment of carrying a Torricellian tube up a mountain, an idea which Pascal later claimed for his own. Some time before July, 1648, the experiment of the vacuum within the vacuum was performed, either by Auzout or by Roberval or both, and also possibly by Pascal. Finally on September 19, 1648, Perier and his Clermont friends took the tube up the Puy-de-Dôme.

Opinions may differ as to whether the mountain experiment or the experiment of the vacuum within the vacuum was the crucial one, or indeed whether either was crucial. The defenders of Aristotle seem to have had trouble with both, but perhaps more with the mountain experiment. These die-hards, of whom there were a great many, used gallons of ink during the remainder of the seventeenth century in their vain attempt to maintain the impossibility of a vacuum. Indeed, there were isolated examples in the eighteenth and even in the nineteenth century, though these later ones should be classified as eccentrics. For the most part the seventeenth-century Aristotelians were honest and learned men whose greatest fault was an exaggerated respect for authority. In general their writings had no consequences whatever, but there were one or two who goaded greater men into making experiments to refute them. Some of these experiments were of very great importance, and we shall discuss them in the next chapter.

Chapter 4

Seventeenth-Century Experiments and Speculations

IN THE year 1663, the young Royal Society was planning to entertain Charles II, the monarch who had given it its charter (and little else). The President, Lord Brouncker, was looking for ideas. On July 30 Christopher Wren sent him a letter from Oxford which must have been welcome, as it contained numerous suggestions for "experiments" that might interest the king. Near the beginning of this letter occurs the following passage: "It is not every yeare, will produce such a Master-Expt as the Torricellian, and so fruitful as it is of New Expts, and therefore the Society hath deservedly spent much time upon it, and its offspring . . ."[1] It had indeed, and was still to do so. What is more, some of its most distinguished Fellows had made and were still to make experiments, the account of which will constitute a good deal of this chapter, for in the half century after 1650 more attention was paid to the barometer in England than in any other country.

The first barometric experiments in England were probably made by members of a group of men who began in the 1640's to hold "philosophical" meetings in London. This group gradually developed into the Royal Society.[2] The recollections of John Wallis (1616–1703) about those early days are of direct interest to us:

I take its [the Royal Society's] first Ground and Foundation to have been in London, about the year 1645. (If not sooner) when the same Dr. Wilkins (then Chaplain to the Prince Elector Palatine, in London) Dr. Jonathan Goddard, Dr. Ent, (now Sir George Ent) Dr. Glisson, Dr. Scarborough, (now Sir Charles Scarborough) Dr. Merrit, with my self and some others, met weekly . . . confining our selves to Philosophical Inquiries . . . as Physick, Anatomy, Geometry . . . [a long list] . . . the weight of the Air, the

---

[1] Royal Society, *Letter book* W.3, item 3 (ms. copy).

[2] Harold Hartley, *The Royal Society, Its Origins and Founders* (London, 1960).

Possibility or Impossibility of Vacuities, and Natures abhorrence thereof, the Torricellian Experiment in Quicksilver. . . .[3]

One of the "others" was Theodore Haak (1605–1690), a German who had moved to England in 1625, and was in correspondence with Marin Mersenne, to whom he wrote in clumsy French sometime in April, 1648:

We still remain very much obliged to you, Sir, for the communication of your experiment with the tubes and the quicksilver; for we have tried it two or three times in the company of men of letters and of quality, with much pleasure and astonishment . . . however, they still won't decide that there is a true vacuum in the glass behind [derrière] the quicksilver. . . .[4]

In July of the same year, he wrote again:

We have also tried mixing water with the quicksilver in the tube, and find notable differences in it, which oblige us to be more careful of exactness in our observations from now on. I should very much like to learn how you regulate your experiments so as not to spoil and lose a lot of mercury; and whether you use accurate glasses, or some picked at random. Also, I still do not properly understand the way of doing your last experiment with one tube inside the other, which ought to empty everything;[5] our attempt to do this has not yet succeeded.[6]

[3] John Wallis, A Defence of the Royal Society, etc. (London, 1678), p. 7.
[4] Quoted by Miss R. H. Syfret, Notes & Records Roy. Soc., Vol. 5 (1948), p. 135.
[5] I.e., obtain a complete vacuum in the tube, making the mercury descend all the way in the inner tube. See p. 48 above.
[6] Quoted in Oeuvres de Blaise Pascal, etc., eds. Léon Brunschvicg, Pierre Boutroux, et Félix Gazier (14 vols., Paris, 1904–1914), II, 307.

Note that this letter was written at a time when nothing had yet been published about the experiment of the vacuum within a vacuum, and when Roberval's second letter to Des Noyers was still, Brunschvicg believed, being composed.

Another letter dating from 1648, which may have been Robert Boyle's introduction to the Torricellian experiment, was written to him from London on May 9, 1648, by Samuel Hartlib, a Pole living in England, who is noted for having been a partisan of the educational ideas of J. A. Comenius.

There is an experiment how to shew, as they suppose, that there is or may be vacuum. It were too long to write all the particulars, but in brief thus: they prepare a long tube, like a weather glass, which is filled with quicksilver; and being stopped as close as may be with one's finger, the tube is inverted, and plunged in a vessel half or more full of quicksilver. The quicksilver in the tube will force the quicksilver in the vessel to rise, by adding more quicksilver to it, and so leaves a space in the top of the tube vacuum, as is supposed. But a bladder being hung in that vacuum was as perfectly seen as could be, so that there must be some body there to convey the action of sight to the eye, as I suppose, and divers others here. That bladder was made as flat as they could, when they put it in; and when the quicksilver left it, it swelled in that supposed vacuum like a little foot-ball. Thus far Sir Charles Cavendish[7] to Mr. Petty,[8] who honours and loves you. . . .[9]

[7] Charles Cavendish (1591–1654) was a noted mathematician who traveled on the continent from 1644 to 1651.
[8] Later Sir William Petty (1623–1687), a political economist, and one of the original Fellows of the Royal Society.
[9] Boyle, Works, (2d. ed.; London, 1772), VI, 77–78.

The date of this is again interesting; the experiment with the bladder was supposedly done by Roberval not long before. Note that Hartlib and his friends, like almost everyone else in the seventeenth century, had difficulty in accepting the vacuity of the tube because one could see through it perfectly well. The Aristotelian view was that light is an "accident," or quality of the transparent substance, and therefore in the absence of any substance there could be no light. Maignan[10] was in the minority in believing that light is an extremely tenuous substance, so that it could pass more easily through the void than through even the most transparent body. It was characteristic of the time to ignore the fact that nobody knew anything whatever of the real nature of light. In the seventeenth century only a few people of the caliber of Pascal, Newton, and Huygens had the strength of mind to insist on the necessity of testing by experiment assumptions such as these.

For a few years after 1648 we lose sight of whatever experiments were going on in England. In Italy Nicolo Zucchi, S.J. (1586–1670), one of those who witnessed Berti's experiment, and apparently one of the few Italians who knew what Viviani and Torricelli had done, had been making experiments with the tube and the quicksilver, and in 1648 published an anonymous pamphlet about it,[11] reprinted in 1649 under his name.[12] The titles of these works

express the conclusions he reached, and yet it is evident that Zucchi had made all the experiments he could think of, and was really trying to keep an open mind. He notes the great difficulty of filling the tube without bubbles. He is certain that what replaces the mercury in the tube is not air, because if any air could get in through the glass, more could get in. The fact that when the tube is inclined the mercury fills it again precludes the possibility that it is air; nor, for the same reason, can it be water. But there cannot be a vacuum, because if the space at the top of the tube is cooled, the mercury rises; if warmed, it falls. (It is entirely likely that he had a rather poor vacuum.) He concludes that the apparently empty part of the tube is filled with a tenuous substance or spirit "breathed out" by the mercury, and that an attraction between this and the mercury sustains the latter in the tube.

Zucchi's views are typical of those of an intelligent Aristotelian who had made or witnessed several experiments with the Torricellian tube, except that as time passed, more and more such people became willing to renounce any demonstrable material substance in the space, and to settle for an "ether" or "subtle matter" which was admittedly quite unobservable, but so fine that it could pass through glass or even mercury in both directions. Although this fell in nicely with the ideas of Descartes, not all those who adopted this view were Cartesians. It was entirely possible to demand an ether while denying the doctrine of vortices, and,

---

[10] See p. 13 above.

[11] *Experimenta vulgata non vacuum probare, sed plenum & antiperistasim instabilire,* etc. (Rome, 1648). There is a copy at Paris, BN, R.25606.

[12] "Exclusio vacui contra nova experimenta,"

in *Nova de machinis philosophia,* etc. (Rome, 1649), pp. 101–15.

indeed, by about 1690 the idea of an ether as a necessary medium for the propagation of light was becoming generally accepted, and has survived until recent times in spite of fundamental changes in ideas about the nature of light itself.[13] What is surprising is not that the vacuists should have been satisfied with anything as imponderable as an ether—this was simply a last resort—but that, as we shall see, first class scientists should have postulated the action of such an unobservable entity to explain some of the phenomena of the Torricellian tube.

With one or two exceptions, nothing is to be gained by dwelling on the effusions of the Peripatetics on the subject. The interested reader may refer to Vol. VII of Lynn Thorndike's monumental work.[14]

Let us now go to Toulouse and Emmanuel Maignan, who returned there from Rome in 1650. Maignan, it will be remembered, had been in Rome when Berti's experiment was made.[15] He had a good deal to say in his enormous textbook of physics[16] about the experiment with the mercury as well. His conclusions are epitomized in what must surely be one of the longest chapter headings any physics text has ever contained. I have tried to retain the sentence structure in translating it:

To relate and explain the experiment, revealed in recent times, in which a glass tube somewhat more than three palms long, hermetically sealed at one end and at first full of mercury, then at once inverted with its mouth down in a basin having four, five, or six inches of mercury put in it, is emptied as to its upper part, but retains mercury in its lower part to the height of three palms above the level of the mercury placed outside in the basin; and at the same time to explain not only this but also other phenomena of the same tube by the pressure of the external air only, without any fear of the vacuum, although perchance not without the vacuum itself.[17]

Maignan, after describing the experiment, demolishes to his own satisfaction the arguments of those who believe that (1) the mercury in descending leaves behind air or ether which had been mixed with it; (2) that such air or ether comes through the pores of the glass; (3) that it enters through the mouth of the tube when the finger is removed; (4) that a "spirit" is squeezed out *(expressus)* from the receding mercury and, rarefied, can fill the space by itself. He has heard, too, that gold, mercury, lye, and almost everything else can get through the "pores" of glass; but why then does the glass suddenly cease to be porous "at the precise moment of time at which the mercury . . . reaches the limit of its descent"?[18] Later, after arguing excellently in favor of the pressure of the air, he describes and discusses the Puy-de-Dôme experiment, drawing the correct conclusions.

He then describes at great length the Roman experiments which we have dis-

---

[13] See Sir Edmund Whittaker, *A History of the Theories of Aether & Electricity* (London, 1910), Chapter I.

[14] *A History of Magic and Experimental Science*, (8 vols., New York, 1923–1958).

[15] See p. 10 above.

[16] *Cursus philosophicus concinnatus ex notissimis cuique principiis, ac praesertim quod res physicas instauratas ex lege naturae sensatis experimentis passim comprobata* (4 vols., 8°; Toulouse, 1653), paged as one.

[17] *Ibid.*, p. 1885.

[18] *Ibid.*, p. 1892.

cussed in Chapter 3.[19] It would be most interesting to be able to date these experiments exactly. They were of the same sort, as far as the tube of mercury was concerned, as those Petit and Pascal made at Rouen in the autumn of 1646, and in some respects more comprehensive. He refers to the experiments as having been made in Rome six years before.[20] As the approval for printing the *Cursus* was given in February, 1650, this would still put the Roman experiments back to some time in 1644, which suggests that Maignan had miscounted his years or that his text was added to after being approved.

There is one aspect of his description of these experiments which is illuminating. In the tubes then used, the mercury fell a little when they were warmed, rose when they were cooled. His opponents maintained (reasonably enough, one might think) that this showed that there was something in the alleged vacuum. But no, says Maignan, this was due to the "spirits of fire" *(spiritus ignei)*, which can pass through the walls of the tube; these spirits, rushing around in vortices *(vorticatim)*, press not only on the walls of the tube but also on the surface of the mercury, which, being an easily movable barrier, descends a little.[21] The vortices are probably Cartesian; but is it not remarkable that even Maignan, so very clear-headed in many ways, and unwilling to let *aether* penetrate the glass tube, had to resort to the "spirits of fire" to get himself out of a real difficulty?

In the year of Maignan's book, Henry Power (1623–1668), later one of the first Fellows of the Royal Society, but at the time a physician at Halifax, in Yorkshire, made, or caused to be made, a number of experiments with the Torricellian tube. His account of these was not published until 1664,[22] but a manuscript[23] by John Spong refers them to 1653. Not surprisingly, the experiment is referred to as Pascal's, though Power shows[24] that he has read or at least heard about Valeriano Magni. Spong's notes are of a charming directness, beginning with a list of requirements[25] including "a quart of ☿ or thereabouts," [!] "ten or 12 Cylindricall glass trunks or tubes," a funnel, wooden dishes, and "a Blanket or Coverlet" to spread on the floor or ground "when you try the Experiment so none of the ☿ may be lost." There is a note that "the ☿ must be taken up with wooden spoones."

He did not find it easy. We read that ". . . this expm^t. cannot so cautiously be performed in any tube whatsoever but some little aire will be at the top of it when fil'd wth ☿ wch is exactly like the little cap of aire at the obtuse end of a new-lay^d egg, So y^t if you incline the tube to what angle soever . . . the ascending ☿ will never exactly fill the tube but the forementioned little cap of aire will be at its top."[26] Folio 6 begins:

May the sixt 1653 I went to the top of

[19] See p. 33 above.
[20] *Ibid.*, p. 1897. ". . . experimenta quae ego . . . celebraui Romae iam ante sex annos." See also note 42 in Chap. 2.
[21] *Ibid.*, p. 1910.
[22] Henry Power, *Experimental Philosophy in Three Books* (London, 1664).
[23] John Spong, *Dr. Pascall's Rare Experiments of ☿ Tryed and Augmented by Henry Power 1653 May 3^d* (Oxford, Ashmolean Ms. 1400A. XII). Quoted by permission.
[24] *Experimental Philosophy*, p. 94.
[25] *Dr. Pascall's Rare Experiments*, fol. 2^r.
[26] *Ibid.*, fol. 3^v.

the Hill at Hallifax to the Beacon and there I filled the tube of 45 inches w^th ☿ and imerst into a vessell of ☿ . . . and there it descended within ⅕ inch to the 26 inch viz ⅘ inch lower than the usual stint, it descended also just so much into a tube of 35½ inches though of a larger bore, And that the variation of the weather might be no hindrance to the exactness of the expm^t I tryed it both at the top & bottome of the hill within three houres tyme the weather not varying.

Let us hope that he had a fine May morning for the first such experiment, as far as I know, ever made in England.

While Henry Power's *Experimental Philosophy* is not a very important work, it may be worth-while to quote a passage from it as evidence of the sort of difficulty with the interpretation of the Torricellian experiment that was likely to occur to a reasonably intelligent man with a seventeenth-century university education. (He was a B.A. of Christ's College, Cambridge.) We must note that before he sent his book to the printer he must have heard a good deal of discussion of this very subject at the Royal Society; and he had been convinced that the pressure of the air determined the height of the mercury in the tube, but he could not admit a vacuum.

Valerianus Magnus, and some others are so fond to believe this deserted cylinder to be an absolute Vacuity, which is not only non-philosophical, but very ridiculous.
1. For, the space deserted hath both longitude, latitude, and profundity, therefore a body; for the very nature of a body consists onely in extension, which is the essential and unseparable property of all bodies whatsoever.
2. Again we have the sensible eviction of our own eyes to confute this suppositional vacuity; for we see the whole space to be luminous (as by *Obser*). Now light must either be a substance or else how should it subsist (if a bare quality) in a vacuity where there is nothing to support it?
3. Again, the magnetical effluxions of the earth are diffused through that seeming vacuity, as *per experiment*.
4. There is some air also interspersed in that seeming vacuity, which cannot be expelled upon any inclination of the tube whatsoever, as by *Obser*. is manifest.
5. The most full evidence against this pretended vacuity is from the returgenscency [*sic*] of the empty bladder suspended in this vacuity, for, how should it be so full blown from nothing? as is by *Exp*. most incomparably evinced.[27]

Later on (the book is full of inconsistencies, and seems to have been written at various times) he gives the correct explanation of the experiment with the bladder: ". . . the spontaneous dilatation and elastick rarefaction of that little remnant of ayr, skulking in the rugosities thereof."[28]

It is a little difficult to be certain what Power thought was in the apparently empty space, but by reference to experiments with water, for which he quotes Pascal, and also "Kircher and Birthius," he concludes "that it is not the effluviums of mercury that fill up that seeming vacuity."[29] And by making the experiment in a gun barrel, he shows that it cannot be filled only with light.

We may turn from Henry Power to a much more important man, Otto Guericke (1602–1686) of Magdeburg,

[27] Power, *Experimental Philosophy*, pp. 94–95.
[28] *Ibid.*, p. 117.
[29] *Ibid.*, p. 98.

who, according to Hellmann,[30] was the first to make the experiment in Germany, at Regensburg in 1654.[31] Otto Guericke made an immense contribution to seventeenth-century science by inventing the air pump, which, as Dijksterhuis rightly says, was one of the four greatest technical inventions of the century, the others being the telescope, the microscope, and the pendulum clock.[32] He was shown the experiment, he says, by Valeriano Magni, who gave him a copy of his *Demonstratio ocularis;*[33] but he later learned from other authors that the inventor was really Torricelli. He did not doubt that there was a vacuum at the top of the mercury; but not a complete vacuum, "because when mercury is poured in, some little bubbles remain close to the inner wall of the tube, and can be perceived to stick there; nor does it seem possible to put the mercury in without these. So when afterwards, at home, I compared such glass tubes, and often tried the experiment, in order to produce a true vacuum I arranged matters so that I could connect the tube with the air-pump described in chapter 4."[34]

Thus, almost as soon as it was invented, the air pump was used to assist the filling of a barometer.[35] Guericke describes how bubbles came out into the vacuum and the apparent volume of the mercury became less. He filled

the tube only a third full at first and pumped it; then another third, and finally after three fillings and three pumpings, he felt that when the tube was inverted he had a pretty good vacuum.

It is evident from his book that Guericke realized, just as clearly as Torricelli had, that the weight or pressure of the air was responsible for the raising of the liquid. His demonstrations of the pressure of the air by the use of evacuated vessels are famous.

At about this time, Ferdinand II, Grand Duke of Tuscany, who had been interested in the atmosphere for ten years, organized the first meteorological observing network. Observations of pressure, temperature, humidity, wind direction, and the state of the sky were made at Florence, Pisa, Vallombrosa, Curtigliano, Bologna, Parma, and Milan, and later well outside the Duke's dominions at Innsbruck, Osnabrück, Paris, and Warsaw. Except for the Florentine series, all the records seem to have been lost. In Florence the rising and falling of the "Torricellian tube" excited the greatest interest.[36] According to Vincenzio Antinori,[37] the form of barometer used in this network was that shown in Figure 4.1, taken from the *Saggi.*[38] A scale was made by fusing tiny glass beads to the tube. By this time the tube was clearly considered as being a scientific instrument, though it had not

---

[30] *Meteorol. Zeits,* Vol. 11 (1894), p. 449.
[31] *Ottonis de Guericke experimenta nova (ut vocantur) Magdeburgica de vacuo spatio,* etc. (Amsterdam, 1672), pp. 117–18.
[32] E. J. Dijksterhuis, *The Mechanization of the World Picture,* trans. C. Dikshoorn, (Oxford, 1961), p. 455.
[33] See p. 42 above.
[34] Guericke, *Experimenta nova,* etc., p. 118.
[35] See also Chapter 9.

[36] G. Hellmann, *Veröff k. Preuss. meteorol. Insts.,* Nr. 273 (Berlin, 1914), p. 139–40.
[37] *Saggi di naturali esperienze fatte nell'-Accademia del Cimento* (3d Florentine edition; Florence, 1841); "Notizie storiche," p. 44.
[38] *Saggi,* ed. cit., plate VII, fig. 25. I have always wondered how the tube was inverted into the cistern in this elegant arrangement.

Fig. 4.1. A barometer of the
Accademia del Cimento, 1657.

yet received a name. On December 13, 1657, Tommaso Bellucci wrote to an unknown correspondent, "I will tell you something about the new instrument. [He describes the setting up of the tube.] We say that the space inside the tube, abandoned by the quicksilver, is the vacuum,"[39] and also that the pressure of the air keeps the mercury up. The Grand Duke, who was active in these experiments, had found the mercury to fall "six degrees" *(sei gradi)* between evening and the next morning, "in which night the air was very full of water, and in consequence heavier, and so should . . . make the quicksilver rise. . . . It is added that fire can enter the tube, and so there is not a vacuum there."[40]   On the 28th of the same month, the famous Giovanni Alfonso Borelli (1608–1679) wrote to Viviani, again referring to the Grand Duke's "new and curious observation" that the mercury falls in humid weather, and incidentally complained that obstinate philosophers still refuse to believe in the pressure of the air.[41] The puzzle about the mercury going down in rainy weather remained entirely unsolved for another century and a half.

It was in 1657 that the short-lived *Accademia del Cimento,* or Academy of Experiments, was founded by the Duke. In the *Saggi,* published by this Academy in 1667 as it was about to drop out of sight, a large place is given to

[39] Quoted by V. Antinori in *Archivio meteorologico centrale Italiano nell' I. e R. Museo di Fisica e Storia naturale.* Prima publicazione, (Firenze, 1858), p. xxxiv. This interesting publication contains items of meteorological interest extracted from the ms. in what is now the National Library at Florence.

[40] *Ibid.*

[41] *Ibid.,* p. xxxv.

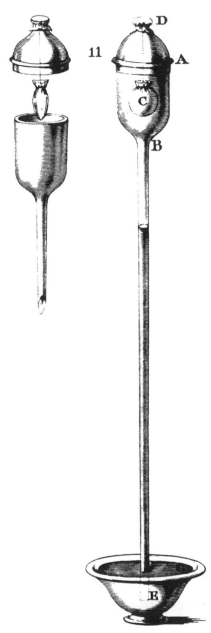

Fig. 4.2. The experiment with the little bladder.

"experiments belonging to the natural pressure of the air."[42]

[42] *Saggi*, ed. cit., pp. 23–47, and figs. 9–29.

The Academy's report, because of its corporate character, has a curiously impersonal flavor, much more like a modern scientific paper than most works of the period. After describing the Torricellian experiment and mentioning that in their "very long series of observations" they have found the height of the mercury to vary rather more according to the season and the weather than according to the variations of heat and cold, they show how the vertical height of the mercury is independent of the inclination of the tube.

They next repeated the experiment of the vacuum within the vacuum, which they called the *esperienza del Roberval*, in essentially the form described by Pecquet and referred to in Chapter 3. As an addition they left a small deflated bladder in the flask, which swelled up when the mercury left the flask in the first stage of the experiment. They also made a beautiful glass apparatus, shown in Figure 4.2, specially for this latter demonstration. They sealed a bell jar over the Torricellian tube and observed no change in the level. They sealed up the cistern with just a little air left in, as suggested in Torricelli's second letter to Ricci. They added a side tube and a syringe and found that they could raise or lower the level of the mercury in the tube at will. They devised a simpler form of the vacuum within the vacuum experiment, as shown in Figure 4.3, in which the tubes *AB* and *DE* are first both filled with mercury, *DE* being put inside *AB*, and the two inverted into the bowl *FG*, when the mercury sinks to its usual level *H* in both tubes. The end *B* of the larger tube is then closed

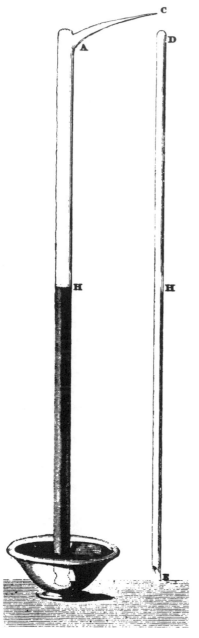

Fig. 4.3. Simpler form of Auzout's experiment.

with the finger, and the fine side tube broken off at *C*, letting the air in. The

mercury then rises in the inner tube, nearly to *D*.

They noted the variation of pressure as they went up and down the highest tower available and some of the hills around Florence and built several special air thermometers to investigate such small pressure differences.

According to Antinori, these pneumatic experiments were begun in August, 1657, and were the first that were tried by the Academy.[43] They were followed by experiments on the properties of the vacuum itself, which of course were more limited in scope than would have been the case if the air pump had been used.

Writing in 1670, Giovanni Alfonso Borelli (1608–1679), who with Rinaldini and Oliva was chiefly responsible for conducting the experiments, recalled[44] the observations made on the tower and stated that they found a one-tenth inch depression. When they did it on a hill near Florence, they measured the temperature with "the most perfect thermometers,"[45] and chose a time when it was the same above as it had been below. And "lest a suspicion should creep in that by the shaking of the mercury in the tube while it was being carried up, some very small portions of air might have been set free, which would have depressed the mercury a short time afterwards, we took care that the opening at the bottom of the tube was stopped up, so that no shaking of the mercury might be produced."[46]

[43] *Saggi*, ed. cit., "Notizie storiche," p. 95.
[44] Borelli, *De motionibus naturalibus, a gravitate pendentibus.* Rogio Julio 1670 (2d. ed., Leiden, 1686).
[45] *Ibid.*, p. 150. These were spirit-in-glass thermometers.
[46] *Ibid.*, p. 151.

This was one way of doing it, though not a way that would commend itself nowadays.

For a long time after this, not much of interest to our story was to be accomplished in Italy. We must now return to Oxford in England, where in 1654 the Honorable Robert Boyle (1627–1691) had settled after his Grand Tour on the Continent, and established a private laboratory. Boyle wrote an almost endless series of books and pamphlets, which were collected into five large volumes by Thomas Birch in 1744, and reprinted in an edition of six volumes in 1772.[47] It is to this edition that we shall refer. Boyle's style has been well described by Thorndike as "diffuse and rambling, apologetic and deprecatory, and without adequate terminal facilities,"[48] but although this is just, which makes his books boring to read, yet his works on physics and chemistry are of high importance for the history of science in the seventeenth century.

In 1660 Boyle published at Oxford the first of his physical treatises, *New Experiments Physico-Mechanicall touching the Spring of the Air, and its Effects, etc.*[49] This is the account of his extremely important series of experiments with that new "engine" the air pump, which was, as he acknowledges, invented by Otto Guericke; but it was greatly improved by Robert Hooke, then assisting Boyle, almost immediately they knew about it. The air pump offered the possibility of producing a near-vacuum in much larger spaces than was practicable with the Torricellian tube. Boyle used the air pump in order to establish many properties of the vacuum and of things placed in it. It will make the application of his experiments easier if we keep clearly in mind what Boyle meant when he wrote about "the air":

By the air I commonly understand that thin, fluid, diaphanous, compressible and dilatable body in which we breath, and wherein we move, which envelops the earth on all sides to a great height above the highest mountains; but yet is so different from the Aether [or Vacuum] in the intermundane or interplanetary spaces, that it refracts the rays of the moon or other remoter luminaries.[50]

The square bracket and its contents are in the original, and it is noteworthy that Boyle never was willing to state that there was *nothing* in the space above the mercury in the tube, especially because light and the effects of a magnet were transmitted through it.[51] The question about the existence of a vacuum seemed to him to be "rather a metaphysical than a physiological [i.e., physical] question."[52]

After describing the improved pump and the manner of using it, Boyle explains what he means by the spring, or elasticity, of the air, which "may perhaps be somewhat further explain'd, by conceiving the air near the earth to be such a heap of little bodies, lying one upon another, as may be resembled to a fleece of wooll."[53] It is probable that

---

[47] *The Works of the Honourable Robert Boyle.* 6 vols. *To Which Is Prefixed the Life of the Author. A New Edition. Edited by Thos. Birch* (London, 1772). (Fulton no. 241)

[48] Thorndyke, *A History of Magic*, VIII, 172.

[49] *Works* (1772), I, 1–117. (Fulton no. 13).

[50] *Ibid.*, V, 612–13.

[51] *Ibid.*, I, 36–37.

[52] *Ibid.*, I, 38.

[53] *Ibid.*, I, 11.

Boyle felt that this was more than a mere analogy, for we read that

This power of self-dilation, is somewhat more conspicuous in a dry spunge compress'd, than in a fleece of wooll. But yet we rather chose to imploy the latter on this occasion, because it is not like a spunge, an entire body, but a number of slender and flexible bodies, loosely complicated, as the air it self seems to be.[54]

On the same page we are asked to remember that air has weight, and that the upper parts compress the lower. The whole atmosphere can be compared to a tall pile of fleeces of wool. At this point Boyle briefly discusses Descartes' hypothesis of vortices and wisely refuses to decide for or against it.

Next, by an analogy strikingly similar to that used by Torricelli in his second letter to Ricci—which it is extremely unlikely that Boyle had ever seen[55]— he explains how the air in a bell jar on a plate can be at atmospheric pressure, even though the glass keeps the air above from pressing on that within.[56] The conclusion that the pressure inside and outside can be the same is emphasized by the consideration of a soap bubble.

Of the many experiments with the air pump described in his first book, the most important one from our point of view was the seventeenth, in which he set up the Torricellian experiment and then put the cistern and lower part

of the tube in his "receiver," sealing the stopper carefully around the tube with sealing wax.

. . . upon which closure, there appeared no change in the height of the mercurial cylinder: whence the air seems to bear upon the mercury rather by virtue of its spring, than of its weight; since its weight could not be supposed to amount to above two or three ounces, in comparison of such a cylinder of mercury as it would sustain.[57]

He then exhausted the receiver and got the mercury to fall nearly all the way, and when he let the air in again it went *nearly* all the way back to its original position, and Boyle perspicaciously ascribed the difference to air bubbles trapped in the mercury. He varied the experiment by using a tube only two feet long and found that after some pumping the mercury left the top of the tube and began to descend as before.

We may agree with James Bryant Conant that this seventeenth experiment was the critical one.[58] Boyle himself realized this. He repeated it in the presence of Thomas Willis (1621–1675), Seth Ward (1617–1689), and Christopher Wren (1632–1723); and it is pleasant to imagine these four friends, all propagandists of the new experimental science, discussing the results. He described it as "that experiment, whereof the satisfactory trial was the principal fruit I promised myself from our engine."[59]

He realized, of course, that the barometer made a splendid vacuum gauge,

[54] *Ibid.*, I, 12.

[55] It was published only in 1663. See Chapter 2.

[56] The bell jar on a plate, surrounding the tube from the air pump, was called the *receiver* by Boyle, and this somewhat astonishing nomenclature has continued through 300 years of the teaching of physics.

[57] *Ibid.*, I, 33.

[58] J. B. Conant, *Robert Boyle's Experiments in Pneumatics* (Cambridge, Mass., 1950), p. 25.

[59] *Works*, ed. cit., I, 33.

and he obtained plenty of experience in filling barometers. Like almost everyone else in his century, he found difficulty in getting rid of small bubbles and described the trick of letting a large bubble traverse the tube and gather them up. Only on one occasion, he tells us, was he able to fill a tube so that when it was inclined there would be no visible bubble at the sealed end. Further

. . . we have found, by experience, that in short tubes a little air is more prejudicial to the experiment than in long ones, where the air having more room to expand it self, does less potently press upon the subjacent mercury.[60]

One wonders how anyone who had discovered as much about the properties of air as Boyle had could need to find this particular effect "by experience"; but perhaps there is a danger of reading too much into Boyle's sentences, or at least of imagining that every word is significant.

It appears[61] that Christopher Wren had suggested to Boyle a long series of barometric observations, with the idea of testing a hypothesis of Descartes that the pressure should vary in synchronism with the tides, but Boyle doubted whether the "accidental mutation in the air" would permit this small effect to be discovered. This doubt was well founded; we know now that careful analysis of long series of observations of a much higher precision than Boyle could have attained is required for the investigation of tides in the atmosphere.

It is outside our province to discuss Boyle's many experiments with the air

pump, some of which were no doubt suggested by Hooke. But we must note that Boyle keeps coming back to the question of what the vacuum really is. At one point he quotes statements by some of the more violent plenists to show that they do not merely mean that an apparently empty space must have some "subtle matter" in it, thereby avoiding the necessity of a vacuum, but that they are actually denying that a space can be empty of ordinary air. This is an important distinction. Boyle admits that a vacuum entirely free from "all corporeal substance" is probably impossible. The meaning of the expression "nature abhors a vacuum" seems to him to be "that by the wise Author of nature . . . the universe, and the parts of it, are so contriv'd, that it is as hard to make a vacuum in it, as if they studiously conspired to prevent it."[62]

At about the time Boyle's book appeared, the Royal Society, just formed, was finding in the Torricellian tube a subject for speculation and experiment. On January 16, 1660/61, John Evelyn recorded in his famous *Diary* that he "went to the Philosophic Club, where was examined the Torricellian experiment." The *Journal Book* of the Society is more explicit, recording on that date that "My Lord Brounker, Mr Boyle, Sr Robert Moray, Sr Paul Neile, Mr Rooke, Mr Wren, Mr Ball, Dr Clarke, Mr Hill; Appointed a Committee to bring in the History of the Quicksilver Experiment."[63]

Even before this, they had taken up

---

[60] *Ibid.*, I, 39.
[61] *Ibid.*, I, 41.

[62] *Ibid.*, I, 75.
[63] Royal Society, *Journal Book*, Jan. 16, 1660/1.

the suggestion (first made in 1647 by Mersenne, though they probably did not know this) that the experiment should be made on the Peak of Teneriffe in the Canary Islands, of whose height exaggerated tales were common in the seventeenth century. On December 5, 1660, a committee had been appointed to "prepare some questions, in order to the Tryal of the Quicksilver Experim^t. upon Tenariff."[64] The results of the deliberations of this committee have the honor of being at the very beginning of the Society's first *Register Book:*

Questions propounded by L^d. Brounker and Mr Boyle and Agreed upon to be sent to Tenariff.

1. Try the Quicksilver Experiment at the top and at severall other Ascents of the Mountain, and at the end of the Experiment upon the top of the Hill, lift out the Tube from the Restagnant Quicksilver somewhat hastily, and observe if the remaining Mercury be impell'd with the usual force, or not. And take by Instrument (with what exactness may be) the true Altitude of every place where the Experiment is made, and observe at the same time, the Temperature of the Air, as to heat and cold, by a Weather Glass; and as to moisture and dryness[65] with a Hydroscope [*sic*], and note that sense the Experimenters have of the Air at those times respectively.

2. Carry up Bladders, some very little blown, some more, and others full blown; and observe how they alter upon the severall Ascents.[66]

There were also twenty other experiments and observations, but nothing seems to have come of these grandiose plans. Presumably the "Canary Merchants" were busy with their own affairs.

Soon after the *New Experiments* appeared in print, it came under attack from such diverse opponents as Thomas Hobbes (1588–1679), the nominalist philosopher, and the Jesuit Francis Line, alias Hall (1595–1675). However brilliant Hobbes may have been as a philosopher, he dealt in absurdities when he came to consider mathematics and natural science, and although Boyle devoted a great deal of space[67] to a careful refutation of his arguments,[68] they are really on a par with Hobbes's attempts to square the circle. Hobbes had got into the habit of feeling that he could not be wrong on any subject. He could. It is interesting that Boyle takes yet another opportunity to state that he has not declared for or against the possibility of a vacuum; nor, he says, has the Royal Society done so.

Francis Line was another circle-squarer,[69] but his book[70] was entirely different. He had made experiments with the tube, and was obviously worried by what he found; to explain the phenomena as he saw them, he succeeded in imagining a mechanism that no one else had thought of, a *funiculus*, an invisible elastic cord extending from

---

[64] *Ibid.*, Dec. 5, 1660.

[65] Note that the word "temperature" was applied also to humidity.

[66] Royal Society, *Register Book*, I, 1.

[67] Boyle, *An examen of Mr. T. Hobbes his dialogus physicus de naturà aëris*, etc., (London, 1662). (*Works*, ed. cit., I, 186–242) (Fulton no. 14).

[68] Hobbes, *Dialogus physicus sive de natura aëris*, etc. (London, 1661). (Fulton no. 312)

[69] In a letter dated Sept. 16, 1661, to Sir Robert Moray, Christian Huygens wrote "Je me trompe s'il est bon mathematicien." (Huygens, *Oeuvres*, III, 320.)

[70] Franciscus Linus, *Tractatus de corporum inseparabilitate, in quo experimenta de vacua . . . examinantur*, etc. (London, 1661). (Fulton no. 315)

the top of the mercury to the top of the tube, and holding the mercury up. He seems to have thought of this after making the experiment with a tube open at both ends, stopping the normally closed end with a finger as well as the other end. When this was inverted into the vessel of mercury and he removed the finger from the lower end, he felt a strong suction on the finger at the top as the mercury receded, the fleshy part being, of course, pushed hard against the end of the tube. He made up his mind that there was a pull beneath his finger, rather than a pressure on top, and imagined a *funiculus* to do the pulling.

In itself this fantastic idea was of little importance. But it led Boyle, who was ready to defend the doctrine of the pressure and elasticity of the air against all comers, to make further experiments in order to answer particular arguments by Linus, and one set of these led to the generalization known as Boyle's law. Boyle's answer to Linus[71] appeared in 1662.

However, before we deal with the discovery of Boyle's law, we should note the passage in Linus' book in which he was able to "explain," to his own satisfaction, the experiment of the vacuum within the vacuum.[72] It is very ingenious. To explain the falling of the mercury in the lower tube, he says that the letting in of the air relaxes the funiculus there. As to the rising of the mercury in the upper tube, Linus says that before the air was let in the forces

of the funiculi in the flask and in the upper tube were in balance, but that when the air was let in, the one in the flask was relaxed, enabling the funiculus in the upper tube to pull up the mercury.

Linus admitted that the air had weight, and also a little "spring," but denied that it had enough elasticity to sustain the mercury. This objection proved very fruitful, for it led Boyle to perform further experiments to try to find out just how much "spring" the air has; and although he did not succeed in this, he did find that it had much more than enough to support the quicksilver in the barometer tube.

These experiments were made some time before September 11, 1661, for on that date, at the Royal Society:

Mr Croon produced two Experiments one of Compression of Air with Quicksilver in a crooked Tube of Glass, the nipt end of which brok. (Mr Boyle gave an Account of his same Experiment made by compressing twelve Inches of Air to three Inches, with about a hundred Inches of Quicksilver) and the other Experiment with a cork. . . .[73]

Boyle took a U-tube with a short closed limb and a very long open limb. Along the short limb, he fixed a scale in quarter-inch divisions and poured in some mercury, managing to get a column of air exactly 48 divisions long when the mercury in the two arms was at exactly the same level, so that the small enclosed volume of air was at atmospheric pressure, on that occasion 29⅛ inches of mercury. He then had someone pour mercury into the long

[71] *A Defence of the Doctrine touching the Spring and Weight of the Air . . . Against the Objections of Franciscus Linus. (Works,* ed. cit., I, 123–85) (Fulton no. 14)
[72] Linus, *Tractatus,* pp. 60–64.

[73] Royal Society, *Journal Book,* Sept. 11, 1661.

A Table of the Condensation of the Air.

A	A	B	C	D	E
48	12	00	Added to 29 1/8 makes	29 2/16	29 2/16
46	11 1/2	01 7/16		30 9/16	30 6/16
44	11	02 13/16		31 15/16	31 12/16
42	10 1/2	04 6/16		33 8/16	33 1/7
40	10	06 3/16		35 5/16	35 --
38	9 1/2	07 14/16		37 --	36 15/19
36	9	10 2/16		39 5/16	38 7/8
34	8 1/2	12 8/16		41 10/16	41 2/17
32	8	15 1/16		44 3/16	43 11/16
30	7 1/2	17 15/16		47 1/16	46 3/5
28	7	21 3/16		50 5/16	50 --
26	6 1/2	25 3/16		54 5/16	53 10/13
24	6	29 11/16		58 13/16	58 2/8
23	5 3/4	32 3/16		61 5/16	60 18/23
22	5 1/2	34 15/16		64 1/16	63 6/11
21	5 1/4	37 15/16		67 1/16	66 4/7
20	5	41 9/16		70 11/16	70 --
19	4 3/4	45 --		74 2/16	73 11/19
18	4 1/2	48 12/16		77 14/16	77 2/3
17	4 1/4	53 11/16		82 12/16	82 4/17
16	4	58 2/16		87 14/16	87 3/8
15	3 3/4	63 15/16		93 1/16	93 1/5
14	3 1/2	71 5/16		100 7/16	99 6/7
13	3 1/4	78 11/16		107 13/16	107 7/13
12	3	88 7/16		117 9/16	116 4/8

AA. The number of equal spaces in the shorter leg, that contained the same parcel of Air diversly extended.
B. The height of the Mercurial Cylinder in the longer leg, that compress'd the Air into those dimensions.
C. The height of a Mercurial Cylinder that counterbalanc'd the pressure of the Atmosphere.
D. The Aggregate of the two last Columns *B* and *C*, exhibiting the pressure sustained by the included Air.
E. What that pressure should be according to the ***Hypothesis***, that supposes the pressures and expansions to be in reciprocal proportion.

Fig. 4.4. The experiment which established Boyle's Law. *(Courtesy of the Trustees of the British Museum)*

open limb until the column of air was 46 divisions long, and measured the difference in height of the two mercury columns. He did the same thing at 44 divisions, and so on down to 12, obtaining the results shown in Figure 4.4.[74] He recognized that the pressure on the enclosed volume of air was the sum of the barometer reading and the height difference in the U-tube. The last two columns in the table show a comparison of the observed results with those calculated on "the Hypothesis, that supposes the pressures and expansions to be in reciprocal proportion." There were further experiments, made by an obvious inversion of the method, which showed that air at less than atmospheric pressure obeys the same law, and there is another table[75] similar to that which we have reproduced, but referring to "dilated" air.

And now we must consider a passage which, I must confess, bothers me.

I shall readily acknowledge (writes Boyle) that I had not reduced the tryals I had made about measuring the expansion of the air to any certain hypothesis, when that ingenious gentleman Mr Richard Townely was pleased to inform me, that

[74] From *A Defence of the Doctrine,* etc., p. 60. *(Works,* ed. cit., I, 158). Professor Price of Yale University has suggested to me that this may be the very first time that completely "uncooked" experimental data were presented in support of a theoretical induction.

[75] *Ibid.,* p. 64 *(Works,* I, 160).

having by the perusal of my *Physico-mechanical experiments* been satisfied that the spring of the air was the cause of it, he had endeavoured . . . to supply what I had omitted concerning the reducing to a precise estimate how much air dilated of it self loses of its elastical force, according to the measures of its dilation.[76]

Richard Townley (or Towneley) of Townley Hall, Lancashire, was a self-effacing gentleman who seems to have gone about doing scientific good works. For example, he brought to light[77] the invention of the micrometer about 1641 by William Gascoigne, who had perished in the Civil War. It had been reinvented by Auzout.[78] Now a literal interpretation of the above passage simply means that Townley had shown Boyle that the law which the latter had already discovered for "condensed" air also applied to "dilated" air. This was the opinion of E. Gerland.[79] But still I wonder whether this is the correct interpretation. It seems incredible that Boyle, after computing such a table as the one we have reproduced, and finding a really remarkable agreement between "hypothesis" and experiment, should not have applied the same process to his experiments with expansion. Furthermore, it is obvious from his writings that Boyle's mind was not in the least a mathematical one. Nevertheless, it is much too late to suggest that the famous relation should be called "Townley's law," even if the matter were entirely clear.

The diversity of possible views about the Torricellian tube is further illustrated by a book written in the form of a letter to Linus by Gilbert Clerke (1626–1697?), a Presbyterian minister.[80] Clerke dealt one by one with Linus' objections to Boyle's work, but at the same time managed to deny that the mercury is held up by the pressure of the air. In this he seemed to think that he was supporting the philosophy of Descartes.

It was at about this time that the word "barometer" was first used, and almost certainly by Boyle. As far as I know, its first appearance in print is on page 27 of his *New Experiments and Observations touching Cold*, which was printed in London early in 1665.[81] It was pointed out by Shedd[82] that this work contains a note by Oldenburg, dated March 10, 1664/5, to the effect that it had been "near two years" since the author's papers had been given him to be copied, and on the strength of this Shedd wishes to assign the word to March 1662/3. In view of the phrase "near two years," perhaps we may say sometime in 1663. The context in which the word "barometer" appears is as follows:

. . . consulting the barometer (if to avoid circumlocutions I may so call the whole instrument wherein a mercurial cylinder of 29 or 30 inches is kept suspended after

---

[76] *Ibid.*, p. 63 (*Works*, I, 160).

[77] *Phil. Trans.*, Vol. 1 (1667), pp. 457–58.

[78] *Traité du micromètre* (Paris, 1667).

[79] *Geschichte der Physik* (Munich, 1913), p. 501.

[80] *Tractatus de restitutione corporum, in quo experimenta Torricelliana & Boyliana explicantur & rarefactio Cartesiana defenditur,* etc. (London, 1662).

[81] On Jan. 4, 1664/5, Boyle mentioned that it was being printed. (Royal Society, *Journal Book*)

[82] John C. Shedd, *Science*, Vol. 19 (1904), p. 109.

the manner of the Torricellian experiment) I found . . .[83]

A few pages later the word "baroscope" is used twice. This word seems to have been current in the Royal Society well before Boyle's book appeared, for there exists a copy of a letter, headed "Mr. Hook to Mr. Beale, jun 24, 1664," and beginning:

S[r], concerning what you say of yr baroscope, I can add nothing else, but yt unless if it be so carefully filled and inverted, yt nothing of air be left above ye ☿ it signifies nothing. . . .[84]

I have not been able to find the letter from Beale to which this is a reply.

By the end of 1665 Oldenburg seems to have decided that both words should be made more widely known; for we read in the *Philosophical Transactions* that "modern philosophers, to avoid circumlocutions, call that instrument, wherein a cylinder of quicksilver, of between 28. and 31. inches in altitude, is kept suspended after the manner of the Torricellian experiment, a *barometer* or *baroscope* . . ."[85]

The word *baromètre*, which for a long time was written without an accent, appeared, probably for the first time, in 1667, though not in print. In the early part of that year the *Académie royale des Sciences* was arranging an expedition to Madagascar. Adrien Auzout made out an extensive list[86] of

instruments and supplies for it, one item of which is "15 ou 20 liures de vif argent tant pour faire des barometres que pour faire les autres Experiences, et uoir comment il monte plus ou moins qu'icy, et mesme dans toux les lieux par lesquels on passera."[87]

Returning to the Royal Society and the years 1662 to 1664, there was a good deal of barometric experimentation. The decrease of pressure with height was a favorite topic. There exists a memorandum in Hooke's hand, undated, but from the writing, and the absence of the words "barometer" and "baroscope," probably written in 1662 or 1663. It is interesting because Hooke was speculating about the influence of temperature as well as that of height:

It will be worth wyall to make the Torricelliã Exp[t]. at the bottom and in some middle stations & at the top of the hills if possiby [sic] in the same Day in the same tube & wth the same Diligence & circumstancy in all particulars, vnless there be two or more correspondents, wch may make it in severall places at the same time, wch would be better because y[e] pressure of y[e] atmosphere is sometimes observed to alter very much in a short time.

Twill be worth trying wth a convenient weather glass how much the Included air will expand it self the neerer y[e] weather glass is carried to the top of the Hill. And to observe those Degrees & as near as may be to give the perpendicular hight of the several Stations, and the precise quantity of the airs extension. That is how many inches & quarters of room Soe many cubick inches of air doe in the severall Stations fill.

Twill be worth trying to observe by a seald up thermometer w[t] the degrees of

[83] *New Experiments and Observations touching Cold* (London, 1665), p. 27. *Works*, ed. cit., II, 487. (Fulton no. 70)

[84] Royal Society, *Classified Papers*, XX, no. 30.

[85] *Phil. Trans.*, Vol. 1 (Feb. 12, 1665/6) p. 153.

[86] Acad. r. des Sci., *Registres de Mathématiques 1667–Apr. 1668*, pp. 37–40. (Ms.) This was read on Jan. 11, 1667.

[87] *Ibid.*, p. 39.

heat and cold are in the severall Stations and to set them downe.

Twill be very instructive for the finding w^t of ye airs expansion is to be ascribed to the Removall of ye air's pressure & w^t to ye ambient heat. If before the tryall are [sic] made on ye mountain, or at least afterwards, both the commõ weatherglass and Thermoscope be together heated by the same degrees and the risings in ye one & other punctually observd, and markd both in the one & the other wth 1. 2. 3. &c. for by this meanes it will be very evident that ye stretching of ye air does not altogether proceed from the heat at ye top of ye mountain.[88]

Depths as well as heights were of interest. On December 3, 1662, it was recommended that Dr. Henry Power, who had just read a paper on coal mining, should make the Torricellian experiment underground at various depths, and also determine whether a closed bladder became more or less distended down a mine.

Some experiments of a new kind were made by Dr. Jonathan Goddard at the meeting on August 13, 1662.

There were taken severall Glass-Canes of different lengths and bores, the longest about four foot, and the shortest under the hight of the standing of the Quicksilver; all close and intire at the one end, with small lines firmly fixed to the close ends to hang them by; each of them was filled with Quicksilver, stopped with the fingers and, inverted and so immersed in a Vessell of Quicksilver, as is usuall: and then each was fastned by its line to one Scale of a Ballance; and so much weight was put into the other Scale, as brought all to exact Equilibration, the lower ends of the Canes playing at Liberty, between

the bottome of the vessell, and the Surface of the Quicksilver conteyned in it.

The event was, that the weight of each of the Canes, with the Quicksilver hanging in this State, was so much less than the weight of each of the canes, and of the Quicksilver taken up in them putt together, by so much as the ends of the same Canes with the Quicksilver conteyned in them, immersed within the stagnant Quicksilver, at the stopping and taking up, might be estimated to weigh; that is, was equall to the weight of all the parts of the same Cane and Quicksilver, that were above the surface of the Stagnant Quicksilver in the Vessell.[89]

The experiment was varied by putting the cistern on a balance, with mercury in it, and weighing it, then filling the tube with some of this mercury, inverting it, and hanging it up so that the open end was in the cistern but not touching the bottom. The original weight was now diminished by that of the mercury in the tube above the surface of the quicksilver in the cistern.

The interpretation of these experiments was far from easy. It was not obvious how the result could be obtained, because it looked as if the quicksilver must be in some way attached to the tube that was hung from the balance, so that its weight would be added to that of the tube, minus a small buoyancy correction. (The weighings were probably not sensitve enough to show up the much smaller error due to neglecting the difference in density of mercury and of the small immersed portion of the glass.) Hydrostatics was a very young science, and few people could be expected to see that the closed end of the barometer tube represented

[88] Royal Society, *Classified Papers*, XX, no. 14. I have quoted the entire document.

[89] Royal Society, *Register Book*, I, 185–86.

a horizontal projected area with the atmosphere pressing downward on top of it, and nothing whatever pressing upward underneath. Even the circumstance that the end of the tube was not flat but rounded presented a serious difficulty to all but the very few who were entirely familiar with the idea of the resolution of forces.

One of those who found it impossible to understand Goddard's experiments was Sir Matthew Hale, an eminent Judge who had a very much misplaced confidence in his abilities as a natural philosopher, and wrote widely on all sorts of subjects. He comes into our story with two books,[90] in the second of which he used this very experiment, among others, to give what he believed to be a watertight demonstration that it is not the pressure of the air that sustains the mercury in the Torricellian tube.[91] He thought, in fact, that the result could very well be explained by Linus' *funiculus*. Hale's effusions had an indirect effect of some interest which will be discussed in Chapter 6, page 100, as well as the more immediate result of engendering *A Discourse of Gravity and Gravitation, Grounded on Experimental Observations,* given by that sound mathematician John Wallis (1616–1703), who had a sharp and unsympathetic eye for nonsense, at least in natural philosophy. It was read to the Royal Society on November 12, 1674, and printed the following year as a pamphlet.

In 1663 the Royal Society set up a water barometer, and in the discussion of the best way of doing it, on June 22, 1663, they directed "that Mr. Boyle's engine should be applied to it, to see, whether the water could be raised higher by suction."[92] This must certainly be an example of the application of the Royal Society's motto, *Nullius in verba*! Observations of the water barometer, by Hooke, covering the period July 23 to August 29, 1663, have been preserved.[93]

During the ensuing winter, Hooke seems to have invented his wheel barometer,[94] and on August 17, 1664, the Torricellian (or Bertian) experiment was made with rather dilute spirits of wine, water, and brine, with the results we should expect.[95] And the next month, at Boyle's request, Hooke[96] twice made observations with a special barometer both at the top of St. Paul's steeple and at the base of its tower.

After about this time, the Royal Society as a body seems to have taken less interest in the barometer. Boyle continued his experiments. Freezing water slowly from the bottom, he showed that it expands when it turns to ice, "and therefore in these cases, we need not, nor cannot fly to I know not what *fuga vacui* for an account of the *Phaenomenon* [of closed vessels breaking] . . ."[97] One more nail in the coffin of Aristotelianism. It is noteworthy that Chris-

[90] *An Essay touching the Gravitation or Non-gravitation of Fluid Bodies,* etc. (London, 1673). *Difficiles nugae: or, Observations touching the Torricellian Experiment, and the Various Solutions of the Same, Especially touching the Weight and Elasticity of the Air* (London, 1674). (Anonymous)

[91] *Difficiles nugae,* pp. 160–62.

[92] Birch, *The History of the Royal Society,* (4 vols.; London, 1756–57), I, 266.

[93] Royal Society, *Classified Papers,* IV (1), no. 8.

[94] See p. 94 below.

[95] Birch, *The History,* I, 459.

[96] *Ibid.,* I, 465 and 467.

[97] Boyle, *New Experiments and Observations touching Cold,* p. 227. Works, ed. cit., II, 537. (Fulton no. 70)

tian Huygens (1629–1695) also thought it of interest to make experiments on the freezing of water, according to a memorandum at the *Académie des Sciences* in Paris.[98]

In 1669 Boyle published further experiments, which he had been making for some time.[99] The tenor of this book is shown by a passage in the preface:

. . . I shall have obtained a great part of what I aimed at, if I have shewn, that those very *phaenomena*, which the school-philosophers, and their party urge, and sometimes triumph in, as a clear proof of Natures abhorrency of a vacuum, may be not onely explicated, but actually exhibited, some by the gravity, and some also by the bare spring of the air. Which latter I now mention as a distinct thing from the other, not that I think it is actually separated in these tryals, (since the weight of the upper parts of the air does, if I may so speak, bend the springs of the lower,) but because that having in the already published experiments, and even in some of these, manifested the efficacy of the airs gravitation on bodies, I thought fit to make it my task in many of these, to shew, that most of the same things that are done by the pressure of all the superincumbent atmosphere acting as a *weight*, may be likewise performed by the pressure of a small portion of air, included indeed (but without any new compression) acting as a *spring*.[100]

Many of these experiments are highly ingenious, but the most direct is the first, in which a glass tube, open at both ends, stands up in a small vial filled half

full of mercury and sealed. When the whole thing is put under the receiver of the air pump, and this exhausted, the mercury rises as much as 27 inches. Boyle then shows that if, instead of a small vial, a quart bottle is used, the mercury could be raised nearly as high as the barometric column, but never any higher. Yet as far as the vacuum is concerned, Boyle was still determined to hedge. ". . . I have formerly made, and now renew a solemn profession, that I do not in this treatise intend to declare either for or against the being of a vacuum. . . ."[101]

I am inclined to think that the reason Boyle never made up his mind about this was the question posed by the transmission of light, about the nature of which no one at that period had any real knowledge whatever. This impression is confirmed by a consideration of the famous *Traité de physique* by the eminent Cartesian Jacques Rohault (1620–1675),[102] especially Chapter XII of Part I, which is entitled "Des mouvemens que l'on a coûtume d'attribuer à la crainte du vuide." This chapter is written with a complete appreciation of the facts of the pressure and elasticity of the air, but it is denied that there is a vacuum in the space above the mercury in the Torricellian tube. We are told why:

. . . the space left vacant by the mercury is filled with some sort of matter, because visible objects behind it still act on our eyes to make us see, exactly as they did before, which they could not do if there were a vacuum, because their action would be interrupted.[103]

---

[98] *Registres de physique* (ms.,) Jan. 8, 1667.
[99] *A Continuation of New Experiments Physico-Mechanical touching the Spring and Weight of the Air, and Their Effects* (Oxford, 1669). *Works*, ed. cit., III, 173–276. (Fulton no. 16)
[100] *Ibid.*, preface, sig. °, fol. 3r&v (Works ed. cit., III, 176–77).

[101] *Ibid.*, p. 29 (Works, III, 198).
[102] Paris, 1671.
[103] Rohault, p. 82.

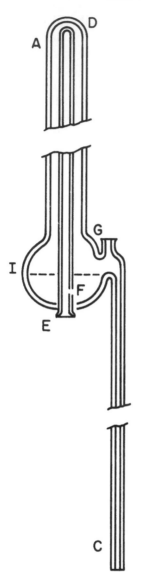

Fig. 4.5. Rohault's apparatus.

travers." It must be a more subtle matter.

Rohault gives a considerably improved version of Roberval's (or Auzout's) apparatus for the demonstration of the vacuum within a vacuum. This arrangement, which would be likely to appeal to a glass blower, is shown in Figure 4.5. To use it, the opening G was closed with a piece of bladder, then the whole thing was inverted and filled through E as far as possible. E was then closed, and the remainder of the apparatus filled through C before it was inverted again, this time into a dish of mercury. Note that there is a small hole in AE at F. When the finger was removed from C under the mercury, the level in CG fell to its proper height, that in DI and AE to the level shown by the dotted line. Finally the bladder over G was pierced, and the level in CG fell to that in the dish, while that in AE rose to its appropriate height.

Returning to Boyle's *Continuation of New Experiments*, we should note an extremely interesting passage in which he describes the first complete portable barometer that can be considered an instrument rather than a laboratory experiment.[104] This, a siphon barometer, is shown in Figure 2 of the plate which we have reproduced as Figure 4.6. Boyle also describes methods of filling it, one of them by means of a funnel with a thin, bent stem. The alleged portability of this instrument seems to have greatly impressed the Royal Society, for on June 4, 1668:

However, he believes that what is in the space is not ordinary air, because if one stops the tube with a finger under the mercury and inverts the tube, the mercury "tombe tout d'une piece, comme si c'estoit un corps dur, & on ne s'apperçoit point que rien monte au

[104] Boyle, *Continuation of New Experiments*, (1669), pp. 68–73. (*Works*, ed. cit., III, 219–23.)

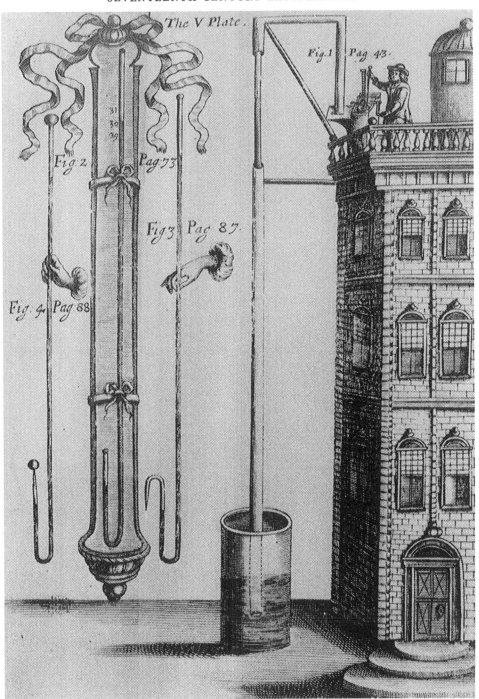

Fig. 4.6. Boyle's siphon barometer and water barometer.

The society being put in mind to give order for the making of portable baroscopes, contrived by Mr Boyle, to be sent into several parts of the world, the operator was ordered to attend Mr Boyle, to receive his directions for filling them aright; and that being done, to make some of them forthwith, to be sent not only into the most distant places of England, but likewise by sea into the East and West Indies, and other parts, particularly to the English plantations, as Bermudas, Jamaica, Barbados, Virginia, and New England; and to Tangier, Moscow, St. Helena, the Cape of Good Hope, and Scanderoon; . . .[105]

One wonders how many of these barometers would have arrived intact had this grand scheme been carried out.

At about this period there was a revival of interest in a phenomenon which had first been investigated in 1661 by Huygens, and received wide publicity in 1672.[106] Inspired by one of Boyle's experiments, he had taken a tube 9 inches long with a bulb at one end, filled it with water, inverted it in a dish of water, and put it under the bell jar of his improved air pump. When the bell jar was exhausted, the water fell out of the bulb and tube. But then he left the water in the vacuum for 24 hours, purging it of air; and when he repeated the experiment with the same water, the tube and bulb remained full. Then he let in a very small bubble of air, and did it again. The water descended.

When he was in England in 1663, he continues, he had repeated these experiments before the Royal Society; and Boyle had thought of trying them with mercury. By leaving some mercury under the exhausted bell jar for a time, and then filling a barometer with it, Boyle had succeeded in suspending the mercury to a height of 34, then 52, 55, and finally 75 inches. But the least bubble of air or shaking the tube would bring it down at once.[107]

This phenomenon is now known as cohesion and ascribed to intermolecular forces. But Huygens had an entirely different explanation:

Besides the pressure of the air, which keeps the mercury suspended at the height of 27 inches[108] in the Torricellian experiment, and of which we are convinced by an infinity of other effects that we see, I conceive of an additional pressure, stronger than this, coming from a substance more subtle than air, and which, without difficulty, penetrates glass, water, mercury, and all the other bodies which are clearly impenetrable to air. This pressure, added to that of the air, is capable of sustaining 75 inches of mercury and perhaps even more, inasmuch as it acts only against the lower surface, that of the mercury into which the open end of the tube dips. But as soon as it can act in the other direction as well, as happens when, by tapping the tube or by introducing a small bubble of air, we give it a way of beginning its action, then its pressure becomes equal in the two directions, so that there remains only the pressure of the air, which holds

---

[105] Birch, *The History*, II, 292.

[106] *J. de Sçavans*, Vol. 3 (1672, July 25), pp. 60–66. This was reproduced, and is more readily accessible, in *Mém. de l'Acad. r. des Sci. depuis 1666 jusqu'à 1699* (Paris, 1733), X, 529–36. It appears in translation in *Phil. Trans.*, Vol. 7 (1672), pp. 5027–30.

[107] These experiments are reported in Birch, *The History*, I, 274–75, 287, 295, 305, 310, and 320.

[108] Paris inches. Boyle in his *New Experiments* (1660) warned his readers that in reading the accounts of experiments with the Torricellian tube, the difference between French and English inches should be kept in mind. (*Works*, ed. cit., I, 38.)

the mercury up at the ordinary height of 27 inches.[109]

But why, one may ask, if the subtle particles can pass through glass, do they not press upon the liquid at the top of the tube even when it is quite full? Huygens thought that this is certainly a great difficulty, but believed that the pores in the glass are at some distance apart, so that only part of the pressure of the subtle matter would act at the top of the tube. But he was not really happy about this:

I confess that the solution I have just given does not satisfy me so fully that there remain no scruples whatever about it, but that does not prevent me from remaining very sure of the new pressure which I have supposed to exist in addition to that of the air. . . .[110]

He supported this notion by recalling that two optical flats wrung together do not separate in a vacuum, and that a siphon will work in a vacuum, presumably if the liquid used is free from air.

Such a hypothesis could scarcely go unchallenged. Reading about it in the *Philosophical Transactions,* John Wallis sat down and wrote a letter to the editor[111] which was published,[112] somewhat edited, a short time afterward. Taking a common-sense view, Wallis does not see the force of Huygens' argument; for why should such a "subtle matter" not act just as ordinary air does? And if it can pass through glass, why does the mercury ever stay up; and when it falls, why suddenly rather than slowly?

It is probable that almost nobody believed that the space above the mercury was really empty. Wallis, like Boyle, is careful to state exactly where he stands in the matter:

. . . where I speake of Vacuity, caus'd by the Torricellian Experiment, or such other way, I do Expressly Caution . . . not to be understood as affirming absolute Vacuity (which whether or no there bee, or can be in Nature, I list not to dispute) but at least an Absence of that Heterogeneous mixture, which we call Air, such as that is wherein wee breath. . . .[113]

But an *aether* to carry light and magnetic force was one thing, an *aether* with a strong pressure was another. Wallis proposed as an explanation the supposition that matter, being at rest, will remain so until it is set in motion by the application of some force; thus the tube can be shaken down.

People worried about this phenomenon for a long time. On May 28, 1684, Robert Hooke read a paper to the Royal Society which was, however, only published after his death.[114] In it he referred to the experiment made by Boyle in 1663 and adduced it as a conclusive proof of the existence of a "subtle fluid," but not of one such fluid only:

And I do believe, from that little insight I have had of the operations of Nature, that all the sensible part of the World is almost infinitely the least part of the body thereof, and but, as it were, the *cuticula* or outward filme of things; whereas that which fills up and compleats the space

---

[109] *Mém. de l'Acad.,* X, 533–34.
[110] *Ibid.,* p. 535.
[111] Royal Society, *Letter Book,* Vol. 5, p. 411–18 (ms. copy).
[112] *Phil. Trans.,* Vol. 7 (1672), pp. 5160–70.

[113] Royal Society, *Letter Book,* Vol. 5, p. 412.
[114] *The Posthumous Works . . . Published by Richard Waller* (London, 1705), pp. 365–68.

incompass'd by that filme consists of a multitude of insensible bodies, each of them as distinct in their natures and operations, as air and quicksilver, or any other two sensible bodies we can name. . . .[115]

This is a most astonishing statement to come from an experimental scientist of Hooke's ability. It seems to lead past Descartes, who after all wanted only two kinds of "subtle matter," and right back to the Middle Ages. But let us follow him farther. He explains the phenomenon in question by supposing that one of these fluids penetrates glass, but does not penetrate mercury,

. . . though quicksilver may be penetrated by a great number of other more subtle fluids, such as those which cause [!] gravity, magnetism, fluidity, &c. if at least it shall be found necessary by future experiments to ascribe those three properties to more than one fluid . . . this fluid hath a pressure every way analogous to the pressure of the air, and . . . this pressure is much greater than that of the air . . . there is no need of supposing it to have a springy nature like that of the air, since all the phaenomena may be solved without it.[116]

Hooke, truer to form than in these speculations, tried to make the "explanation" more intelligible by an experiment: drawing out a glass tube in a lamp until it was "almost as small as a hair," he broke it into short pieces, bound these "faggot-wise" with thread, and melted wax about them in such a way as not to stop their ends. He then used the bundle as a porous stopper for a larger glass tube. When this was pushed into a jar of water, open end down, the air in the tube passed freely through the porous plug and the tube filled with water; but when the tube was raised again, the water did not run out until the stopper was "some inches" above the surface of the water in the jar.

He suggested that experiments should be made to find out ". . . whether some glasses are not more porous than others, and consequently whether the mercury will not stand to a much greater height [presumably before being shaken] in tubes made of glass of a more opacous or more refracting substance than in tubes of a more transparent or less refracting substance."[117]

The matter was also taken up by Edme Mariotte (?1620–1684), in his De la nature de l'air.[118] Mariotte was in a minority among the Academicians in being an anti-Cartesian; but while denying the existence of the "subtle matter" of Descartes, he claimed that the top of the barometer tube was full of a matière aérienne, exhaled by the mercury as it descends. But if the mercury has been kept in a vacuum there is none of this that will easily come out, unless some is knocked out by a shock. In this way he proposed to explain the sticking of the mercury in the top of the tube and the fact that it could be dislodged

---

[115] Ibid., p. 366. It ought to be noted that the existence of a complex ether to explain gravity, the properties of light, magnetism, electricity, etc., was adopted as a hypothesis by Newton in a famous letter to Oldenburg dated Dec. 7, 1675 (The Correspondence of Isaac Newton, H. W. Turnbull, ed., Vol. I, [Cambridge, 1959], pp. 362–86), though he clearly stated that he would not defend this hypothesis and was making it with some reluctance. Hooke may have seen this letter.
[116] Ibid., p. 366.

[117] Ibid., p. 368.
[118] Paris, 1679. Also in Oeuvres (La Haye, 1740), I, 148–82. There is a paraphrase in Hist. Acad. r. des Sci. (Paris, 1733), I, 270–78 (Histoire for 1679), which is considerably clearer than the original.

by a smart blow. "Thus, said Mr. Mariotte, is demonstrated the law of nature by which all bodies, as soon as they are in contact, resist being separated, if some other body does not come between them."[119] This particular phenomenon seems to have baffled a good many people, and to have engendered more than its share of *ad hoc* hypotheses. Mariotte's "law of nature" would have had some validity if it could have been referred to molecular structure, but this was, of course, quite impossible at the time.

Mariotte's essay on the nature of the air contains the restatement of Boyle's Law which has caused every Frenchman since that time to think of it as "Mariotte's law," with no justification whatever. Boyle, as we have seen, probably with the aid of Townley, established the pressure-volume relation for air both above and below atmospheric pressure and published his results, with actual tables, in 1662. Furthermore, the experimental method finally adopted by Mariotte and the enameller Hubin, who was helping him, was essentially the same as that of Boyle. There is no doubt whatever that Boyle has the priority.[120]

As we near the end of the seventeenth century we find that the Torricellian experiment continued to have a great importance for physical science. One of the most eminent seventeenth-century physicists, Christian Huygens, felt that it was decisive proof of the existence of a luminiferous ether, necessary on his theory for the propagation of light.[121] This ether would obviously pass through substances, like glass, which are impermeable to air. Why should it not be present in the empty space between us and the stars?

An account of the experiment found a place in the fashionable courses of lectures on physics which were a feature of life in London and Paris at this time. For example, in 1712 William Whiston read a course of lectures illustrated with elaborate experiments performed by Francis Hauksbee. In the extremely rare pamphlet describing and illustrating this series, we read:

16th day. The several Phaenomena of the Torricellian Experiment exhibited and explained. Other experiments of the like nature, with fluids variously combined. Several sorts of Barometers, Thermometers, and Hygroscopes. The pressure of the Air shewn by Experiment to be different at different Altitudes from the Surface of the Earth.[122]

One of the "other experiments" was that of the vacuum within the vacuum, unequivocally described as "Monsieur Auzout's noble Experiment, to deter-

[119] *Hist. Acad. r. des Sci.*, I, 275.

[120] P. G. Tait (*Properties of matter*, [2d ed.; Edinburgh, 1890], Appendix 4) examines this question in some detail. He goes so far as to say that Mariotte "transcribes, or adapts, into his writings (without any attempt at acknowledgment) whatever suits him in those of other people. He seems to have been a splendidly successful and very early example of the highest class of what we now call the *Paper-Scientists*." (Page 321. The italics are Tait's.)

[121] Huygens, *Traité de la lumière* (Leiden, 1690), pp. 10–11. Quoted in *Oeuvres de Pascal*, ed. cit., II, 12, note 2.

[122] *A Course of Mechanical, Optical, Hydrostatical, and Pneumatical Experiments. To Be Performed by Francis Hauksbee; and the Explanatory Lectures Read by William Whiston, M.A.* (n.p. n.d., but to the copy in the Harvard University Library is attached a ms. note stating that this course of lectures was advertised in *The Spectator*, no. 275, for Tuesday, Jan. 15, 1712.) The passage quoted is on p. 3.

mine, that 'tis certainly the Air's Pressure that raises the Quicksilver in the Barometer."[123]

The barometer was also becoming an article of commerce in London and Paris, as gentlemen began to make a hobby of observing the weather. In his biography of his brother Francis, Baron Guilford, Roger North credits him with getting the trade started in London, probably not much after 1670:

His Lordship was much affected by the discoveries, which fell in the consequences of the Torricellian Experiment; whereby a new world of air, compressing every thing it touches, is reveal'd. He could not but observe a manifest connection between the alterations of the mercurial station, and the course of the winds and weather; but could not fix in his mind any certain rules of indication, but rather the contrary, viz. that events failed as often as corresponded with the ordinary expectation. But yet he would not give it over for desperate, and hoped that a more general observation might generate a better prognostic of the weather from it, than was yet known. And that must be expected from a more diffused, if not an universal, use of it, which could not then be thought of; because the instruments were rare, and confined to the cabinets of the virtuosi; and one was not to be had but by means of some of them. Therefore his Lordship thought fit to put some ordinary tradesmen upon making and selling them in their shops; and accordingly he sent for Jones,[124] the clock-maker in the Inner-Temple Lane; and, having shewn him the fabrick, and given him proper cautions in the erecting of them, recom-

mended the setting them forth for sale in his shop; and, it being a new thing, he would certainly find customers. He did so, and was the first person that exposed the instrument to sale publickly in London. But his Lordship, perceiving that his business lay in other operations he was more used to, and that he began to slight these, sent for Mr. Winn,[125] a famous instrument-maker over-against his house in Chancery-Lane, and did the like to him, who pursued the manufacture to great perfection, and his own no small advantage; and then others took it up, and few clock-makers, instrument-makers, cabinet-makers, and diverse other trades, were without them always in their shops, ready for sale: And all moving from the first essays, as I related, set on foot by his Lordship.[126]

Not long after this, continental Europe received a virtual invasion of barometer makers from Italy. Their products, often decorative but seldom of any scientific importance, can be seen in every museum, but have no place in a book such as this, with occasional interesting exceptions which will be noted in the following chapters.

By about 1690 the period of random search and philosophical dispute about the experiment was nearing its end, though there were questions left over, such as those about the luminescence of the barometer and the relation between height and barometric pressure, which took a long time to answer. The development of the barometer as a measuring instrument was well under way.

---

[123] *Ibid.*, p. 15.

[124] This was Henry Jones, who was apprenticed in 1654, and in 1691 and 1692 was Master of the Clockmakers' Company. *See* S. E. Atkins & W. H. Overall, *Some Account of the Worshipful Company of the Clockmakers of the City of London* (London, 1881), p. 169.

[125] This was Henry Wynn or Wynne (d. 1709). *See* E. G. R. Taylor, *The Mathematical Practitioners of Tudor and Stuart England* (Cambridge, 1954), pp. 242–43.

[126] Roger North, *Life of the Rt. Hon. Francis North, Baron Guilford,* etc. (London, 1742), p. 295.

Part II

THE MERCURY BAROMETER
AS AN INSTRUMENT

". . . a barometer, strictly consider'd, is, in truth, nothing else but a philosophical pair of scales, wherein (by the artful contrivance of a vacuum, and the restless endeavours of nature to restore and preserve an aequilibrium) a column of air is continually weighing against a column of quick-silver."—EDWARD SAUL, *An Historical and Philosophical Account of the Barometer, or Weather Glass* (2d ed.; London, 1735), p. 64.

Chapter 5

A General Survey

IN THE first four chapters we have discussed the seventeenth-century experiments which led to the invention of the mercury barometer and to the correct explanation of its action. We must now trace the long development of the barometer from the simple tube and vase employed by Torricelli to the neat and efficient instrument in use today.

If such a survey is to be anything more than a list, it cannot be a continuous progression in order of dates. Almost from the first, several lines of evolution are discernible, and these further divide themselves according to the means adopted to attain the various ends in view, so that a diagram of the development of the barometer would look much more like an elm tree than a railway line.

Torricelli himself knew that the pressure of the air could vary; and indeed, as we have seen, he intended to make an instrument to follow these changes. He seems not to have persisted in this purpose. The first instrument that can really be called a barometer was probably the tube provided with a paper scale by Descartes in 1648.[1]

Now one of the first things that must

have been noticed about the behavior of the Torricellian tube when it was left set up at a fixed station was the smallness and slowness of the movements of the upper mercury surface. In the seventeenth century people were not used to making fine measurements. It is therefore natural that they should have sought for methods of expanding the scale of the barometer so that the movements of the mercury could easily be seen. The history of these efforts will be the subject of Chapter 6. It is interesting to note that even in the twentieth century isolated attempts to expand the scale of the barometer have been made, although these are usually buried in the less busy files of the Patent Offices of the world.

Perier's ascent of the Puy-de-Dôme was the precursor of a long list of attempts to use the barometer in the measurement of height. If one is going to take a barometer up a mountain, something more easily portable and much more convenient than a glass tube, a bowl, and a flask of mercury is desirable, and for over two centuries the development of mountain barometers was the spur to the improvement of barometers in general. As they became more portable, they became more precise; but because many barometers

[1] See p. 46 above.

which have to be precise need not be very easily portable, and because the mechanical requirements for the attainment of the two *desiderata* are somewhat different, the search for portability (which began earlier) will be treated in Chapter 7, and the improvement of accuracy deferred until Chapter 9, after a chapter devoted to the history of the corrections to the barometer.

In the nineteenth century, as networks of meteorological stations began to develop on a national scale, it became desirable to have a special barometer at the central station of each network; a barometer which could be relied on for stability and precision of reading to a higher order of accuracy than seemed warranted at the dependent stations. Such special barometers became known as *Normalbarometer* (in Germany, where they were first developed), *Baromètres étalons,* or Standard barometers. Finally the increasing precision of physical and physicochemical measurements brought with it a requirement for the measurement of pressure with an accuracy quite unnecessary for meteorological purposes; and beginning with the *Bureau International des Poids et Mesures* at Sèvres, the great standardizing laboratories began to construct barometers (and

manometers) with which atmospheric and other pressures can be measured in absolute units with an overall accuracy which sometimes reaches one part in 2×10^5. Such barometers are often known as *Primary* barometers, and strictly speaking only such instruments should be called primary or normal; but in Chapter 10 we shall include an account of the earlier and simpler barometers developed for the chief stations of the national networks.

Man's laziness is often the motive power behind elaborate efforts to devise instruments or other devices to do his work for him. Chapter 11 will be devoted to an account of the development of recording mercury barometers, including devices for transmitting to a distance the value of the atmospheric pressure. It is interesting that attempts to record automatically some kinds of meteorological observations began almost as soon as the indicating instruments themselves had been invented.

Chapter 12 will recount the shorter but significant history of the barometer in North America. Finally this part of the book closes with an account of the explanation of the strange light which appears in the vacuum space of many mercury barometers when they are moved or shaken in the dark.

Chapter 6

The Expansion
of the Scale

1. GENERAL REMARKS

IN THE century after Torricelli, the effort to magnify the scale of the barometer resulted in six distinct forms, which will be treated here approximately in the order of their first appearance. They were:

1. Barometers with one or more liquids in addition to mercury.
2. The "wheel barometer."
3. The balance barometer.
4. The diagonal barometer.
5. The "square" or L-shaped barometer.
6. The conical barometer and its derivatives.

2. BAROMETERS WITH ADDITIONAL LIQUIDS

On September 24, 1650, Chanut wrote from his embassy at Stockholm to his friend Perier ". . . feu Monsieur Descartes s'estoit proposé de continuer cette même observation dans un tuyau de verre, vers le milieu duquel il y eust une retraite & un gros ventre, environ à la hauteur où monte à peu près le vif argent, au-dessus duquel vif argent mettant de l'eau jusqu'au milieu environ de la hauteur qui reste au dessus du vif argent; il auroit vû plus exactement les changemens."[1] Such an instrument would look something like Figure 6.1, and the surface of the water would move several times as far as that of the mercury in the Torricellian tube, ideally almost 13½ times as far if the bulb were very large and the upper tube very small. Chanut goes on to say that he has tried to get such an instrument made, but it was too much for the Stockholm glass blowers.

Of course it would not work in this form. Even if the water had been carefully boiled to get all the dissolved air out of it, the water vapor in the tube would make the instrument as much a thermometer as a barometer. Water was the wrong choice for a second liquid. This instrument was reinvented by C. O. Bartrum in 1892, suggesting

[1] In Pascal, *Traitez de l'equilibre des liqueurs et de la pesanteur de la masse de l'air* (2d ed.; Paris, 1698), pp. 207–8. (Brunschvicg, II, 437–38.)

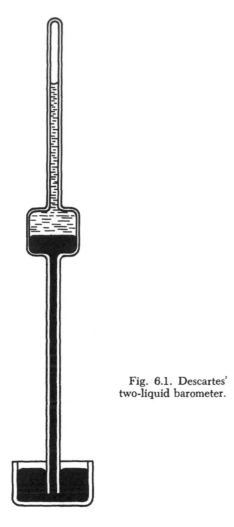

Fig. 6.1. Descartes' two-liquid barometer.

"a liquid of small specific gravity compared with that of mercury, such for instance as glycerine, aniline, methyl salycilate, or other suitable liquid"[2] to fill the upper tube. Bartrum's barometer, which has a magnification of 7.8, is in the Science Museum, London[3] and appears to be in working order. Looking at it, I reflected that the use of

modern organic liquids of extremely low vapor pressure, such as are used in vacuum technique, makes barometers of this type entirely practicable for special display purposes. Later I found that "low-vapour-tension hydrocarbons" had been suggested for this barometer as long ago as 1877.[4]

A somewhat better way of using two liquids was found in 1668 by Robert Hooke (1635–1703), then Curator of Experiments to the Royal Society, to which he demonstrated this "double barometer" on June 18.[5] Its general arrangement is as shown in Figure 6.2; the cylindrical bulbs A and B are long enough to accommodate the motion of the mercury for all likely pressures, and the indication is derived from the position of the water surface at D. The tube BC can be quite small, and the magnification can approach, but not reach, half the ratio of the specific gravity of the liquids. It is extremely unlikely that Hooke knew about Descartes' letter to Chanut at this time.

In 1672 Christian Huygens (1629–1695) reinvented both Descartes' "double barometer" and Hooke's, writing to Jean-Baptiste Du Hamel (1623–1706), secretary of the *Académie royale des Sciences*,[6] and to Henry Oldenburg (1616–1678) the secretary of the Royal Society.[7] Huygens realized that the form shown in Figure 6.2 was the bet-

[2] British Patent 12,092 (1892).
[3] Inventory no. 1950–252.

[4] F. Guthrie, *Phil. Mag.*, Vol. 3 (1877), pp. 139–41.
[5] Royal Society, *Journal* (ms.), June 18, 1668. Thomas Birch, *History of the Royal Society* (London, 1756–57), II, 298.
[6] Acad. r. des Sci., Paris, *Régistres* (ms., 1672), fol. 152. Printed in *Mém. Acad. r. des Sci.*, X (Paris, 1733), 540–44.
[7] Huygens, *Oeuvres Complètes.*, publ. par la *Société Hollandaise des Sciences* (22 vols., La Haye, 1888–1950), VII (1919), 242.

ter, because the water was not in the vacuum; but he recognized that there would still be a fairly large temperature effect, the water in the bulb B and the tube BC acting as a thermometer. Nevertheless his barometer was taken seriously by the *Académie,* and in 1677 Edme Mariotte (1620(?)–1684) thought it worth-while to calculate the magnifi-

cation, given the diameters of the tubes at B and at D and the ratio of the specific gravities of the two liquids.[8] It is rather surprising that even at this time the specific gravity of mercury was taken as 14, which differs from the correct value by more than 3 per cent.

In 1704 Guillaume Amontons (1663–1705) suggested that the temperature correction might be made zero by suitably proportioning the tubes;[9] but his arguments were unsound.

Hooke's two-liquid barometer, in spite of its imperfections, enjoyed an immense popularity. The only existing barometer that certainly belonged to the *Académie royale des Sciences* is of this type, and is preserved in the Archives of the Academy. It is on a decorative wooden panel, and bears the inscription "Fait par Jean Baptiste Prieur rue Ste Marguerite Faubourg St. Antoine." There is another, better preserved, at the *Conservatoire National des Arts et Métiers* (C.N.A.M.) in Paris, bearing the inscription "Baromètre composé par Mossy constructeur des instruments de phisique [*sic*] en verre de l'Academie des Sciences et de la société Roiale del' Medicine Quay Pelletier à la croix d'or à paris 1780." The magnification is about 10 and the scale is graduated from 27 to 29 "inches."[10] Mossy was a very well-known instrument maker who is represented by no less than seven barometers in the C.N.A.M., five of them of more conventional construction. At the end of a 24-page pamphlet describing some new barometers of his invention,

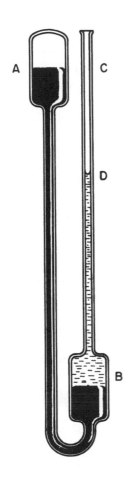

Fig. 6.2. Hooke's two-liquid barometer.

[8] Acad. r. des Sci., *Régistres* (1677), fol. 115r–116v.
[9] *Mém. Acad. r. des Sci.* (1704), pp. 164-72.
[10] C.N.A.M. no. 1578.

Changeux[11] recommended "Le sieur Mossi, marchand d'instrumens météorologiques & constructeur en toutes sortes de machines de physique, . . . j'ai cru qu'il convenoit de l'indiquer aux amateurs de la physique."

It is clear that all through the eighteenth century, even after great advances had been made in the accuracy of mercury barometers, there was still a large and continuing demand for instruments that would make the changes in atmospheric pressure evident at a glance. This is strikingly illustrated by the common practice of mounting an ordinary barometer and a two-liquid barometer on the same board; it was recognized that the two-liquid barometer could not be trusted to maintain a calibration. There are numerous examples of these instruments in European museums, especially in Holland, where they seem to have been a standard product of the Italian workmen who emigrated in large numbers all through the eighteenth century and well into the nineteenth. They are of too little scientific interest to justify a detailed treatment here,[12] but a word may be in order about the name which is usually inscribed on these instruments. It is frequently "Barometer en Contraroleur," but in examples at Leiden the last word is also spelled "Contraleur," "Controleur," and "Contrarolleur." Dr. Maria Rooseboom, director of the Leiden Museum, has recently discovered that *contraroller* is an early spelling of the French verb *contrôler*, and suggests that "contrarolleur" refers to the pointer with which these instruments are often provided, so that one may mark *(contrôler)* the oil level in order to compare the reading of the instrument with one at some later time.

The two-liquid barometer has been revived in modern times in somewhat different forms. A notable attempt to make a scientific instrument of it was made in 1850 by Debrun,[13] who called his device a *baromètre amplificateur*. It is shown in Figure 6.3. A mercury barometer has a fiducial point P in the vacuum space, and a closed cistern of adjustable volume and large area. Over the mercury there is water. Two tubes rise from the top of the cistern: a short tube A which extends down into the mercury, and a longer tube B in which the water can rise. The volume of the cistern being adjusted until the mercury just touches P, the height of the mercury column is AP. But the change in the level of the water in B will be to that of the mercury in A very nearly as the ratio of the densities of mercury and water. Debrun used an electric current to determine when the mercury touched the point P.

Two years later O. Braun of Berlin patented a similar device.[14] Anton

---

[11] *Description de Nouveaux Baromètres à appendices*, etc. (Paris, 1783). (Extract from Rozier's *J. de Phys.*, May, 1783), p. 24.

[12] Dr. Maria Rooseboom, the dedicated and enthusiastic director of the Museum voor de Geschiedenis der Natuurwetenschappen at Leiden, informs me that in preparing her *Bijdrage tot de Geschiedenis der Instrumentmakerskunst in de noordelijke Nederlanden* (Leiden, 1950), she felt obliged to neglect all instrument makers who were represented only by barometers. There were just too many of them. She generously allowed me to study her large file of cards, with the result that I also decided to neglect these makers, unless barometers of greater scientific importance had come down from them, which seldom seems to have happened.

[13] *J. de Phys.*, Vol. 9 (1880), pp. 387–89.

[14] D. R. Patent 20,451, 1882. See also *Zeits. für Instrum.*, III (1883), 151.

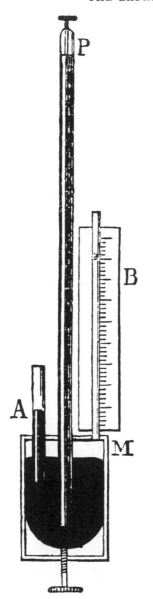

Fig. 6.3. Debrun's *baromètre amplificateur*.

water through a funnel, or removing it through a drain, surely not as good a method as that of Debrun. Steinhauser, however, did give an elaborate theory of the temperature errors of such an instrument.

In 1877 N. Goutkowski of St. Petersburg made a barometer in which a column of mercury of invariable height balanced almost the whole of the atmospheric pressure, the remainder being balanced by mineral oil only one-fifteenth as dense as mercury. A flexible diaphragm in a bulb divided the two liquids, and was stated to offer a resistance of only 1/60 mm. of mercury.[16] A different construction was patented in 1891[17] and described in 1892[18] by J. Joly, in whose barometer a column of mercury was suspended *over* a column of glycerine, and kept in place by a piston which nearly fitted the tube and was attached to a spherical float which held it up against the bottom of the mercury column. Most of the pressure of the air was balanced by 27 inches of mercury, and the magnification would, of course, be large. This tricky arrangement apparently worked, but one might have fears for its stability. Another variant of Hooke's barometer, with a cistern replacing the siphon tube, was patented in 1907 by B. Junquera;[19] it would have a very large temperature coefficient.

For scientific purposes barometers on Hooke's pattern are now of no interest; for display, they suffer from the damn-

Steinhauser's two-liquid barometer[15] also used water and mercury, but in this instrument the height of the column of water was altered by adding

[15] *Repert. der Phys.*, Vol. 23 (1887), pp. 277–96.

[16] *J. de Phys.*, sér. I, Vol. 6 (1877), pp. 195–96.
[17] British Patent 21,309 (1891).
[18] *Royal Dublin Soc., Proc.*, Vol. 7 (1892), pp. 547–51.
[19] British Patent 28,626 (1907); U.S. Patent 888,015 (1908).

ing defect that when the pressure goes up, the barometer goes down. But the instruments devised by Bartrum and by Goutkowski register right way up, and if constructed of the best available modern materials they would seem to offer the possibility of making good display barometers in a reasonable compass.

One of the worst defects of Hooke's two-liquid barometer is that the tube containing the lighter liquid becomes dirty in the region through which the surface of the liquid moves. Hooke and Huygens were both aware of this, and on February 3, 1685/6, the former described to the Royal Society an ingenious solution of the problem.[20] This is shown in Figure 6.4, and resembles the two-liquid barometer with the addition at the top of the open tube of a reservoir C, having the same diameter as the bulbs A and B. The tube BC now contains two immiscible liquids, such as alcohol and turpentine. The common surface D of these two liquids is the indicator of the atmospheric pressure.

From the standpoint of the time, this construction had three important advantages. Firstly, the combined height of the two liquids, pressing on the mercury in addition to the pressure of the air, would be almost constant; indeed, if the two immiscible liquids were chosen to have the same density, it could be exactly constant; therefore the level of the mercury in A would vary in the same way as in an ordinary siphon barometer. Secondly, the friction of the liquids against the narrow tube was more nearly uniform. Thirdly, there was the theoretical possibility of

making the magnification as large as desired, simply by making the tube BC smaller or the bulbs larger. This appealed greatly to the seventeenth-century mind.

The liquid BD still acted as a thermometer, and Hooke, realizing this,

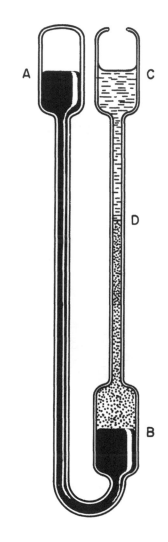

Fig. 6.4. Hooke's three-liquid barometer.

[20] *Phil. Trans.*, Vol. 16 (1686), pp. 241–44.

THE EXPANSION OF THE SCALE

proposed constructing a thermometer with a tube of the same diameter as BC and having the same volume of alcohol in it, and hanging it with the barometer.[21]

It is rather surprising, in view of its publication in the Philosophical Transactions, that two French Academicians should have imagined at different dates that this invention was theirs. The first was Guillaume Amontons (1663–1705),[22] who wrote in 1695 that about 10 years before, he had suggested such a construction to "Sʳ Hubin Émailleur[23] ordinaire du Roy." He was astonished to find that Hubin had already made such an instrument but "ce ne fut qu'à son retour d'Angleterre[24] où les mêmes choses luy [furent] proposées par Mʳ Hovckt de la Société Royale de Londres."[25] In view of this acknowledgment, it is all the more surprising that Philippe de la Hire (1640–1718) should have reinvented this barometer in 1708, describing it as an improvement on the double barometer of Huygens.[26] Not a word, of course, of Hooke. As this was printed in the Memoirs of the Academy, we can only suppose that no one took the trouble to disillusion its author.

The manifold disadvantages of this construction have been set out at some length by Jean André de Luc[27] (also spelled Deluc) in his Recherches sur les Modifications de l'Atmosphère;[28] they need not be emphasized here.

Three liquids were not to be the ultimate, for in 1710 Nicolas Gauger described a four-liquid barometer, which not only produced a large magnification of the scale, but was also shorter than a standard barometer, after the manner of the "folded" barometer which we shall come to in Chapter 7. Gauger was solving a set problem.[29] Referring to Figure 6.5, it is supposed that the cross-sectional area of the bulbs is 50 times that of the tubes, and that the vertical distance MQ, for example, is half the minimum barometric height. BDEF and MOQ contain mercury, MI contains oil of tartar, IH, spirits of wine, and HF, mineral oil. The amount of spirits of wine in IH is exactly equal to the internal volume of

[21] Ibid., p. 244.

[22] G. Amontons, Remarques et expériences phisiques sur la Construction d'une nouvelle Clepsidre, sur les Barometres, Termometres, & Hygrometres (Paris, 1695), pp. 145 ff. Reviewed in J. des Sçavans (1695), p. 181.

[23] "From the end of the 17th century, when barometers and thermometers began to be produced by craftsmen, they were made by enamellers because the graduated scale was inscribed on an enamelled metal plate. . . . The workmen whose specialty was the construction of barometers and thermometers soon formed a separate group in the guild of enamellers. They took the title of physicist and began to manufacture . . . a number of physical instruments. . . ." (Maurice Daumas, Les instruments scientifiques aux XVIIᵉ et XVIIIᵉ siècles, [Paris, Presses Universitaires de France, 1953], p. 130.)

[24] Daumas, ibid., p. 81, says that Hubin was an Englishman who had gone to Paris.

[25] Amontons, Remarques, etc., pp. 147-48.

[26] Mém. Acad. r. des Sci. (1708), pp. 154–67.

[27] Jean André de Luc (1727–1817), born in Geneva, was perhaps the most profound thinker about meteorological instruments in the eighteenth century. His Recherches sur les Modifications de l'Atmosphère (2 vols.; Geneva, 1772) will be frequently cited; his Idées sur la météorologie (London, 1786) is important in the history of that subject.

[28] Part I, Chap. 2.

[29] Résolution du problème posé dans le Journal de Trévoux du mois de mars dernier, pour la construction de nouveaux thermometres et de nouveaux barometres de toutes sortes de grandeurs. Par Mr G**** (Paris, 1710). Pamphlet. (BN, Rz. 3224, 8°. It is ascribed to Gauger in the BN catalog.)

the tube *IL*, and the filling is arranged so that when the pressure is at the bottom of its range the boundary between the oil of tartar and the alcohol is at *I*. The alcohol is colored. As the pressure rises, this boundary goes down to *L*. When it reaches *L*, the other boundary which was originally at *H* has now reached *I*, and this is used as an indicator of further changes in pressure. With the dimensions given, it

Fig. 6.5. Gauger's four-liquid barometer.

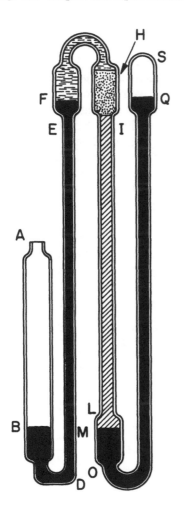

seems to be assumed that the maximum change in pressure would be about one inch of mercury. A "folded" three-liquid barometer with about half the magnification was also described.

3. THE "WHEEL BAROMETER"

The means of expanding the scale of the barometer that we have described so far may be considered as applications of hydrostatics. The expansion of the scale by mechanical means seemed natural enough to Robert Hooke, certainly one of the greatest mechanical talents of that age, and perhaps of any age.[30] Some time before 1664 he devised the "Wheel-barometer," for more than two centuries a familiar instrument in thousands of homes throughout Europe. The first published description of this is in his famous *Micrographia*,[31] which on November 24, 1664, was ordered to be printed by the printers to the Royal Society; but in the archives of that Society there exists a sketch,[32] probably antedating the *Micrographia*, which is reproduced in Figure 6.6. In the upper left-hand corner of the sheet is the indication "★2," creating the possibility that there may have been an even earlier version.

[30] And unjustly neglected until recently. *See* Margaret d'Espinasse, *Robert Hooke* (Berkeley & Los Angeles, 1956).

[31] *Micrographia, or some Physiological Descriptions of Minute Bodies, Made by Magnifying Glasses, with Observations and Inquiries Thereupon* (London, 1665). Preface, fol. c2–c3.

[32] Royal Society, *Classified Papers* XX, item 32. In the Society's *Journal Book* it is also recorded that on December 30, 1663, "Mr. Hook produced a little engine, to make the ascent and descent of quicksilver in glassecanes more discernable." This was probably the wheel barometer.

✳ 2

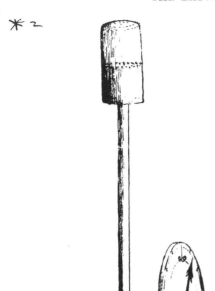

Fig. 6.6. Hooke's sketch of the wheel barometer. *(By permission of the Royal Society)*

There is also a later version in 1666.[33] The three versions have interesting differences, and I have thought it worth while to reproduce them all; Figure 6.7 is from the *Micrographia*, (Scheme I, Figure I), and Figure 6.8 from the *Philosophical Transactions*.

The early sketch (Figure 6.6) shows a simple bent-up tube with a cylindrical enlargement at the top, of a shape that would be unlikely to appeal

to a glassworker. A float resting on the mercury in the lower limb is partly counterweighted, the thread joining float and counterweight passing over, or perhaps round, a small pulley. On the axle of the pulley is a pointer, which indicates the atmospheric pressure on a suitable scale. We should note that the upper level of the mercury is about halfway up the cylindrical enlargement, so that most of the change in the height of the column is reflected in the motion of the float.

The figure in the *Micrographia* (Figure 6.7) is similar except that the upper enlargement has become spherical, probably at the suggestion of a glassblower. The mercury level is still in the middle of the enlargement. But there is another part of this figure, not always taken note of; the peculiar vessel *DEF* shown enlarged at the lower left. It is clear from Hooke's description[34] that he had found the filling of an ordinary siphon barometer somewhat difficult, and that this was an improvement:

The instrument is this. I prepare a pretty capacious bolt-head *AB*, with a small stem about two foot and a half long *DC*; upon the end of this *D*, I put on a small bended glass, or brazen[35] syphon *DEF* (open at *D*, *E*, and *F*, but to be closed with cement at *F* and *E*, as occasion serves) whose stem *F* should be about six or eight inches long, but the bore of it not above half an inch diameter, and very even; these I fix very strongly together by the help of very hard cement . . .

There follows a detailed description of the mounting of the tube and "siphon" on the board, and of the filling, which

[33] "A new Contrivance of *Wheel-Barometer*, much more easy to be prepared, than that, which is described in the *Micrography*; imparted by the Author of that Book." *Phil. Trans.*, Vol. 1 (1666), pp. 218-19. With a figure.

[34] *Micrographia*, preface, fol. c2–c3.
[35] Did Hooke not know that mercury attacks brass?

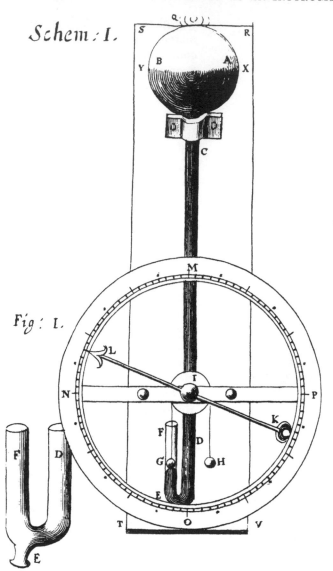

Fig. 6.7. The wheel barometer, from the *Micrographia*.

is done through the orifice E after temporarily stopping up the end F; E is then stopped with "pretty hard cement," with a piece of leather tied over it as an extra precaution. Then

Having thus fastned it, I gently erect again the glass after this manner: I first let the frame down edge-wayes, till the edge RV touch the floor, or ly horizontal; and then in that edging posture raise the end RS; this I do, that if there chance to be any air hidden in the small pipe E, it may ascend into the pipe F, and not into the pipe DC.

After opening *F*, Hooke next takes mercury out until the upper level is at the diametral plane of the "bolt-head" *AB*. In this way nearly all the movement is at the lower surface.

But because there really is some small change of the upper surface also, I find by several observations how much it rises in the ball, and falls in the pipe *F*, to make the distance between the two surfaces an inch greater than it was before; and the measure that it falls in the pipe is the length of the inch by which I am to mark the parts of the tube *F*, or the board on which it lyes, into inches and decimals.

Here is the original of the contracted barometer scale, which became popular two centuries later, as we shall see in Chapter 9. He goes on:

Having thus justned and divided it, I have a large wheel *MNOP*, whose outmost limb is divided into two hundred equal parts; this . . . is fixt on the frame RT. . . . In the middle of this . . . is placed a small cylinder, whose circumference is equal to twice the length of one of these divisions, which I find answer to an inch of ascent, or descent, of mercury: This cylinder *I*, is movable on a very small needle, on the end of which is fixt a very light index *KL*, all which are so pois'd on the axis, or needle, that no part is heavier than another: Then about this cylinder is wound a small clew of silk, with two small steel bullets at each end of it *GH*; one of these, which is somewhat the heavier, ought to be so big, as freely to move to and fro in the pipe *F*; by means of which contrivance, every the least variation of the height of the mercury will be made exceeding visible by the motion to and fro of the small index *KL*.

It appears that one revolution of the index was equivalent to two inches of mercury, so that each division of the scale came to a hundredth of an inch.

When now we consider the wheel barometer later described and illustrated in the *Philosophical Transactions* (Figure 6.8), it becomes evident that the title of the note cannot have been written by Hooke. For it transpires that the *raison d'être* of the modification is not to make it "more easy to be prepared."

. . . *ABC* represents the tube, which may be either blunt, or with a head, as *ABC* (by which latter shape, more room is allowed for any remainder of air, to expand the better). This is to be filled with quicksilver, and inverted as commonly; but into a vessel of stagnant mercury, made after the fashion of *IK*, that is, having its sides about 3 or 4 inches high, and the cavity of it equally big both above and below; and if it can be (besides that part, which is fill'd by the end of the mercuriall tube, that stands in it) of equal capacity with the hollow of the cane about *B*: For then the quicksilver rising as much in the hollow of *I*, as it descends at *B*, the difference of the height in the receiver *I*, will be just half the usual difference. . . . But, whether the difference be . . . bigger or less, 'tis no great matter, since by the contrivance of the wheel and index . . . the least variation may be made as sensible as desired. . . .[36]

It is clear that in the two years that had elapsed, Hooke had changed his ideas about the design of the instrument. Instead of trying to throw as much as possible of the motion of the mercury into the lower limb, he had now decided that it is better to lengthen the tube so that the "bolt-head" will act as a large reservoir to take any small bubbles of air which manage to get into the vacuum space. That this is the main change in the in-

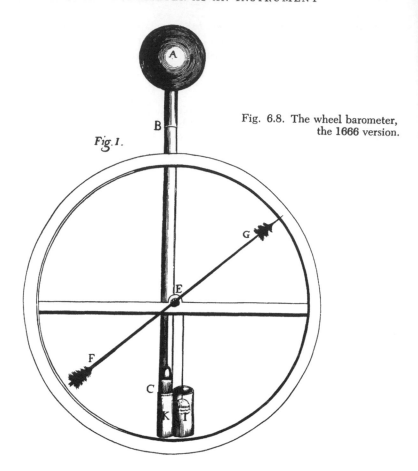

Fig. 1.

Fig. 6.8. The wheel barometer,
the 1666 version.

strument is evident both from the figure and the text. Nevertheless one may wonder how the tube was inverted into a vessel of the shape of *KI*, and whether, in fact, this was not similar to the earlier vessel *DEF*, the draughtsman having failed to notice the orifice *E* at the bottom. In that event it must surely have been Oldenburg who wrote the title of the later paper.

Except for its application to the recording barometer, which will be dealt with in Chapter 11, the wheel-barometer could, in the nature of things, be altered only in mechanical detail. Friction was the difficulty, and much in-

genuity was devoted to its reduction. In 1770, such a barometer was made for Keane Fitzgerald, with two dials, one having a pointer revolving once for 3 inches of mercury, the other once for an inch. "When a strong gust of wind rises, one may very plainly perceive the index of this barometer to sink several divisions, and rise again as it abates."[37] By making the moving parts extremely light (the index of straw, for example), and the tube large, William Snow-Harris seems to have been able to construct a wheel-barometer which would actually move with and indicate a pressure change of 0.001 inch of mercury.[38] An air pump was used in filling the tube, which contained 15 lb. of quicksilver. However, one suspects that this remarkable sensitivity must have decreased somewhat drastically as the mercury surface in the open limb became dirty.

The wheel barometer has had much excellent instrument making devoted to it. Mechanically (i.e., apart from decoration), the finest instrument of this sort that I have seen is in the Linceo Giulio Beccaria at Milan: a barometer with a dial about 35 cm. in diameter. The pulley in this barometer is about 12 mm. in diameter and is on a shaft about 2.5 mm. in diameter, supported by four anti-friction wheels each about 40 mm. across. The instrument bears the inscription "Barometro Multiplicatore di Antonio Frascoli Milano 1866 N 2," and was in working order in 1961.

In the Central Library at Florence there is a manuscript by the astronomer Leonardo Ximenes (1716–1786),[39] in which he suggested a simple balanced lever operated directly from the float by means of a chain. He seemed to feel that the reduction in friction would more than compensate for the nonlinearity of the scale and the awkward shape of the instrument.[40]

The shape of the wheel barometer lent itself extremely well to decorative treatment, and many beautiful examples by the great clockmakers are to be seen. It is entirely out of our province to discuss these, and the reader must be referred to special works dealing with barometers considered as decorative objects.[41] We shall note only that as early as 1688 a clockmaker called John Smith was complaining that wheel barometers were too expensive.[42]

4. THE BALANCE BAROMETER

In textbooks of physics from the early eighteenth century until the present time, the balance barometer and the diagonal barometer are almost universally ascribed to Sir Samuel Morland; but entirely without documentation.[43] The absence of evidence has clearly worried many authors. For example Halliwell, who in 1838 published a brief biography of Morland,

---

[38] British Association, *Report* for 1833, pp. 414-17.

[39] *Teoria de' difetti, e di rimedj di alcuni Barometri, a cui si aggiunge il progetto di un nuovo Barometro.* mss. Galileani, Torricelli, Vol. 24, fol. 70–95. (Autograph, n.d.; probably about 1780.)

[40] *Ibid.,* fol. 90r–91v.

[41] For England, G. H. Bell and E. F. Bell, *Old English Barometers* (Winchester, 1952).

[42] John Smith, *A Compleat Discourse of the Nature, Use and Right Managing of that Wonderful Instrument the Baroscope or quick-silver Weather Glass* (London 1688), p. 2.

[43] For example J. I. Desaguliers, *A Course of Experimental Philosophy* (London, 1744), p. 263; A. P. Deschanel, *Elementary Treatise on Natural Philosophy,* trans. J. D. Everett (5th ed.; London, 1880), p. 155; P. Musschenbroek, *Essai de Physique* (Leiden, 1751), p. 628. Professor E. G. R. Taylor (*The Mathematical Practitioners of Tudor and Stuart England* [London, 1954], p. 229) says that Sir Samuel made "a barometer which the king admired, but which Hooke declared to be a copy of one of his own which Morland had got hold of at Thomas Tompion's shop." I have been unable to trace the origin of this, and Professor Taylor cannot now (1961) remember where it came from. It is, of course, a most tantalizing statement.

writes, "Morland is said to have written a treatise on the barometer, which was answered by Lord North in another tract on the same subject; I have seen neither of them."[44] In a discussion of a balance barograph, J. D. Forbes stated that he had looked very hard for any record by Sir Samuel of the invention of the "statical barometer," but with no success. "It is perhaps likely," he wrote, "that the ascription of it to Morland, and the story of its presentation to Charles II, was a tradition among the London instrument makers.[45] The late H. W. Dickinson left in manuscript a biography of Sir Samuel,[46] in which he confesses that he has not been able to trace any publication of Morland's about the balance barometer. The question is therefore of some interest.

The principle of the balance barometer seems to have been investigated some years before anything of the sort was constructed. Mersenne[47] weighed a Torricellian tube hanging in its cistern; and the possibility was implicit in the experiments by Dr. Goddard described in Chapter 4, where it was noted that Sir Matthew Hale had entirely failed to understand them. Now it happened that Sir Matthew's *Essay touching the Gravitation or Non-gravi-*

tation of Fluid Bodies[48] came to the attention of Francis North, first Baron Guilford, a man of great intelligence who was invited into the Royal Society, but declined the invitation on the ground that he was too busy with affairs of state to give the Society the attention it deserved—an answer that must have been unique in the seventeenth century.[49] But he did find time to study hydrostatics and so come to the correct conclusions about the worthlessness of Sir Matthew's arguments. His studies may well have brought him into relations with Sir Samuel Morland, who in the 1670's was busy designing waterworks for Charles II. Let us allow Roger North to tell us the sequel, first noting that his account begins "about this time," and that his book is not arranged in such a way that a nearer estimate of the date can be arrived at.

About this time, the philosophical world was entertain'd about settling the grand affair of the mercurial barometer, and its indications. Among the rest, Sir Samuel Moreland publish'd a piece, containing a device to prolong the indicatory space from three inches, as in common tubes, to a foot, or more, as you please; and he defied all the virtuosi to resolve it. This he call'd a statick barometer; for it was contrived by suspending a common tube at one end of a plain ballance, and the other arm to be duly counterpoised, and drawn to a point directed to play against an arch of about a sextant, divided into three parts; and that was to correspond with the three inches on the

[44] J. O. Halliwell, *Brief Account of the Life Writings, and Inventions of Sir Samuel Morland, Master of Mechanics to Charles the Second* (Cambridge, 1838), p. 18.

[45] J. D. Forbes, *Roy. Soc. Edinburgh, Proc.* Vol. 3 (1857), pp. 481.

[46] I have been able to examine this through the courtesy of Mr. Arthur Stowers of the Science Museum, London. It was written about 1950, but is still unpublished.

[47] *Novarum observationum physico-mathematicarum Tomus III* (Paris, 1647), first preface.

[48] See note 90, p. 74.

[49] Roger North, *Life of the Rt. Hon. Francis North, Baron Guilford*, etc. (London, 1742), p. 285.

plate of an upright tube. The cistern was a cylindrical glass, of more than the double diameter of the tube; and, in that, charged with mercury, the tube, erected according to art, was immersed; and the moving of the mercury in the tube, higher and lower, was of no regard, but the index only. His Lordship wrote a paper in answer to the Knight's challenge, and consider'd this experiment according to the laws of hydrostaticks, and concluded that the mystery lay in the difference of specifick gravity between mercury and glass, which may be nearly as one to twenty.[50] The standing of the mercury, in the tube, is always taken upon the distance of the upper from the lower superficies; and, whatever happens, the mercury will find that distance as the pressure of the atmosphere requires. He consider'd also that the quantity of mercury, and the quantity of the glass tube, not immersed, taken together, were the sum of the whole weight above the stagnum, supposed to make an equilibrium against the counterpoids. This standing level, and the index pointing (for example) to 29½ inches, if the variation of the pressure comes to require a 30 inch column, then ½ inch mercury in weight is added on that side. This must draw down the tube into the stagnum, till so much of the glass tube is immersed, as shall answer that encrease of weight; and then the index riseth, because the tube and the mercury tend down into the stagnum. But as the glass goes down, the mercury seems to rise in the tube; for the column will always, as I said, answer the pressure, whether the tube goes up or down. His Lordship consider'd also that the specific weight of glass is so much less than that of mercury, that the glass tube must lose two or three inches to countervail one half, or perhaps one quarter of an inch of mercury, whether sinking into the stag-

num, or emerging from it, and so in proportion, as it shall happen: Which makes the opposite arm, with the index, make larger sweeps than the rising and falling, in common tubes, shew. His Lordship consider'd farther, that the stagnum not being very wide, as the tube sunk, the mercury there rose and swallow'd the glass faster than, if wider, it would do; and that it ought to be so adjusted, for quantities of mercury and glass, that the arm shall not play much above or below the level, which, otherwise, would create some impediment, if not inequality, in the motion; and lastly, that the arch must be graduated mechanically; for the measures must be taken as they happen, and will not be adjusted by calculates. It is obvious how, by this means, the beam moves, and stands in continual balance, and the index shews the barometrical action, by the arched and graduated plate, with advantage. But, in practice, the many frictions, as of the mercury in the tube, and of the glass in the stagnum, corrupt the nicety of the instrument (and in time exaggerating) so much, that it is not made use of but for shew.[51]

Thus there seems to be no doubt that Sir Samuel Morland did invent the balance barometer. More precisely, one type of balance barometer; for several other types were developed later, and the theory of these differs in essential respects from the type Sir Samuel hit upon. We shall defer these matters until we come to discuss recording barometers, only noting here that an essential requirement for the stability of the instrument described above is that the outer cross-section of the part of the tube that dips into the "stagnum" should be greater than the inner cross-section of the tube at the level of the

---

[50] One to six would be more nearly correct.

[51] *Ibid.*, pp. 293–94.

upper surface of the mercury. Using a plain straight tube as he did, Morland could not help fulfilling this requirement, although it would perhaps be unwise to assume that he was conscious of it.

As to the date, it was probably about 1675, or a year or two later. I have found no trace of either North's or Morland's manuscript, but Roger North's account is so circumstantial that he almost certainly had Francis North's paper in front of him when he was writing. As to Morland, his personal affairs, always rather complicated, became more so near the end of his life, and few of his autographs survive.[52]

It must be recorded that Robert P. Multhauf of the Smithsonian Institution has ascribed the invention of the balance barometer to Sir Christopher Wren,[53] on the basis of a single passage in Sprat's *History of the Royal Society*. This occurs in a lengthy eulogy of Wren, and I propose to quote rather more of it than the single phrase referred to by Multhauf; the passage reads

He has devis'd many subtil wayes for the easier finding the gravity of the atmosphere, the degrees of drought and moysture, and many of its other accidents. Amongst these instruments there are balances which are usefull to other purposes, that shew the weight of the air by their spontaneous inclination.[54]

It is rather surprising that the Journal Books of the Royal Society quite fail to ascribe such a barometer to Wren; this is remarked by Cope and Jones in their note on the above passage. I think it likely that Sprat was confusing the "weather-clock" which Wren had made at about this time[55] and the "statical baroscope" independently invented by Boyle and by Guericke.[56] The tone of the pages in which the passage occurs is one of immoderate enthusiasm—however justifiable—and it is unlikely that Sprat checked all his references. Furthermore there is the clause "which are usefull to other purposes,"[57] a description which would fit the "statical baroscope" but would scarcely apply to an instrument that included a long mercury column. Boyle, indeed, emphasized the use of his device as a demonstration instrument, in connection with the air pump.

Nevertheless Wren did have an idea about a "poysed weather glass" in 1678, as we shall see; but this was of a different sort.

The subsequent history of the balance barometer as an indicating instrument[58] is not very extensive. J. H. de Magellan, a transplanted Portuguese with more than a sufficiency of vanity, recorded[59] that while he had never

---

[52] See the *Dictionary of National Biography*, art. "Morland."

[53] U.S. National Museum, *Bulletin 228* (Washington, 1961), p. 108, note.

[54] Thomas Sprat, *The History of the Royal Society of London, for the Improving of Natural Knowledge* (London, 1667). Critical edition by J. I. Cope and H. W. Jones (St. Louis, Mo. 1958), p. 313.

[55] See H. E. Hoff and L. A. Geddes, *Isis*, Vol. 53 (1962), pp. 287–310. This meteorograph does not seem to have included a barometer.

[56] See Chapter 15 below.

[57] This is omitted from the phrase as quoted by Multhauf, U.S. National Museum, *Bulletin 228*.

[58] For balance barographs, see Chapter 11.

[59] *Description et usages de nouveaux baromètres*, etc. (Londres, 1779) (Paged from 87 to 164), pp. 155–58.

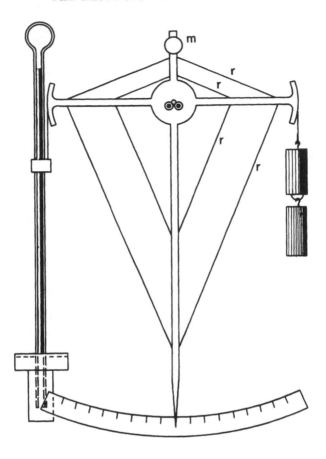

Fig. 6.9. Magellan's balance barometer.

seen a printed description, he had seen two such instruments: one by Adams,[60] made in 1760 for George III; the other commenced by Jonathan Sisson (1680–1760). He had acquired the latter and had it in his house; he had made some improvements in it. From the very rough figure given by Magellan we have constructed Figure 6.9, which

[60] Presumably George Adams the elder (c. 1704–1773), a famous maker of all sorts of physical and astronomical instruments.

shows a more advanced instrument than that described by Francis North, but operating on the same general principles. The scale of this barometer could be linear because the tube and counterweight were supported by chains from sectors. The stiffening of the beam by the taut wires rr and the use of friction rollers instead of knife edges to furnish an axis of rotation are interesting features. The adjustable weight m was probably to bring back the center of gravity near the axis in spite of the rather large pointer. On the evidence of the rest of his book, it seems doubtful whether he was really responsible for much improvement, and (p. 157) he does not seem to have understood the action of the instrument.

I have found no trace of the barometer that Adams is supposed to have made for George III. It does not figure in the present collection of the King's instruments.[61] Nor is it mentioned in the catalogue prepared for Queen Charlotte, presumably by Stephen Demainbray, the Court physicist of the time.[62]

In a later paper,[63] Magellan proposed to "improve" the balance barometer by enlarging the vacuum chamber with the intention of overcoming friction by having a greater force available. The fallacy in this was pointed out in 1867 by Secchi, of whom we shall hear much more in Chapter 11. "Il ne se doute pas," he writes, "de l'effet que cette modification aurait eu, savoir de rendre son instrument impossible."[64] Secchi's intention was to

combat an opinion which then prevailed that Magellan was the inventor of the "modern" weight barometer. Reading Magellan's papers today, one gets the impression that any ideas of value were poached from the London instrument makers whose shops he frequented, such as Nairne and Blunt.

The last notable application of this principle to an indicating barometer was probably the large instrument which was put into operation in Florence on November 8, 1860. Toward the end of 1859 the Marchese Cosimo Ridolfi, then Minister of Education in Tuscany, asked Filippo Cecchi (1822–1887) and his colleague Giovanni Antonelli (1818–1872) to build a barometer and a thermometer with very large scales, to be installed in the famous *Loggia dei Lanzi* on the south side of the *Piazza della Signoria*. The decorative requirements called for dials and pointers, which made the balance barometer a natural solution for this half of the project. A large iron tube *AB* (Figure 6.10), of 52 mm. internal diameter near the top, had its narrower lower section surrounded by an iron cup *C*, open at the top but welded mercury-tight at the bottom, in order to make the cross section at the surface of the mercury in the cistern slightly greater than that of the vacuum chamber, a condition which, as we have noted already, is essential for stability. When the tube had been filled and inverted in the cistern, it was suspended by a group of gold tinsel ribbons (which are very flexible) from one end of a light iron balance beam terminating in a circular arc, and counterpoised by a 16-kg. weight supported in a similar way from the other end of the beam. A second metallic ribbon

[61] London, Science Museum, *H. M. King George III Collection. Descriptive Catalogue.* (London, 1951), 92 pp.

[62] *A Catalogue of the Apparatus of Philosophical Instruments, Her Majesty has deposited in the Royal Observatory at Richmond, with an Account of the Presents, by Sundry Persons Made to Her Majesty's Collection.* ms., n.d., iii+23+ii leaves (Science Museum, London). Discussed by Robert S. Whipple, *Proc. Optical Convention, 1926.* (London, n.d.), pp. 502–28. Two other manuscript catalogues which were stated by Whipple to be in the Library of King's College (London) cannot now be found.

[63] *Observations sur la Phys.*, Vol. 19 (1782), pp. 108–25, 194–212, 257–73; 341–56.

[64] Angelo Secchi, *Comptes Rendus*, Vol. 65 (1867), p. 445.

from the top of the tube passed round a small pulley with two flat grooves, and was fastened to it. In the other groove was a similar ribbon supporting a 330-gm. weight. This pulley was on the axis of the pointer. The bearings were of a very ingenious anti-friction design. As for the beam, its cylindrical axis rolled on the bottom of rectangular holes F in a counterbalanced steel fork EF, which could itself oscillate on knife edges at x. The horizontal motion of the axis was limited by further rollers m and n, supported on the frame. An ingenious brake was provided to prevent movement of the

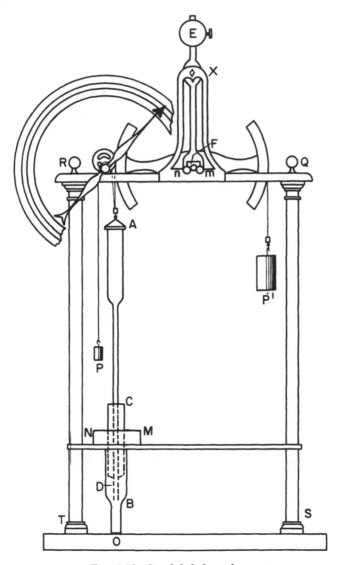

Fig. 6.10. Cecchi's balance barometer.

Fig. 6.11. The dial of Cecchi's barometer in the *Loggia dei Lanzi*. (*Mansell Collection*)

index due to gusts of wind, the index being out of doors.[65]

Never can scientific instruments have had more gorgeous surroundings, as may be guessed from Figure 6.11, which shows the dial of the barometer as a background to the marble group "Il ratto della sabina" by Jean Bologne, in the west arch of the *Loggia*. The dial of the thermometer, behind the eastern arch, had an even more distinguished neighbor, the "Perseus" of Cellini. The dials and pointers were complete, but not operative, in 1934; in 1961 there was no sign on the back wall of the *Loggia* that they had ever been there; and a middle-aged commissionaire at the door of the *Palazzo Vecchio* could not remember even having heard of them.

Antonelli worked out the theory of this barometer,[66] showing that if the tube floated freely the level of the mercury in the cistern would not change with changes of pressure. This is true only for this type of balance barometer; i.e., the type in which the beam is supported at its center of gravity and the counterweight is constant. It was apparently also realized that by proportioning the tube and the cistern in an appropriate way, the instrument could be compensated exactly for temperature changes at some mean pressure; but Cecchi and Antonelli do not seem to have developed this idea.

This is the place to mention a curious instrument, a combination of a siphon barometer and a balance, devised by C. Kraiewitsch in 1882[67] for the purpose of indicating small fluctuations in atmospheric pressure. A cup of mercury on one pan of a sensitive balance was connected by a siphon full of mercury to the open limb of a siphon barometer. The siphon continually transferred mercury backward and forward between the barometer and the cup on the balance, to equalize the two levels, and the resulting changes in the weight on the pan of the balance deflected its pointer. This could obviously be made with more sensitivity than there would be any use for.

A unique design for a weight barometer was patented in 1908 by the firm of Siemens and Shuckert.[68] This instrument had a fixed tube and a movable cistern hung on a beam and supported partly by a counterweight and partly by a cylinder floating in a fixed. vessel of mercury.

The balance barometer is really a fundamentally sound idea, especially if you do not mind the inconvenience of actually weighing the tube each time an observation has to be made. In 1749 G. W. Richmann[69] of St. Petersburg devised two forms of balance barometer, each essentially a siphon barometer with a very long horizontal portion to the tube. One design had an unequal-arm balance, the other an equal-arm balance, but in both the disequilibrium produced by a change of barometric pressure, which transferred

[65] F. Cecchi, *Nuovo Cimento*, Vol. 16 (1862), pp. 233–63.
[66] *Ibid.*, pp. 264–72.

[67] *J. Soc. Physico-Chim. Russe*, Vol. 14 (1882), pp. 213–25. Abstracted in *J. de Phys.*, 2d ser., Vol. 2 (1883), pp. 578–79.
[68] British Patent 17,620 (1908).
[69] *Nov. Comm. Petrop.*, Vol. 2 (1749), pp. 200–3.

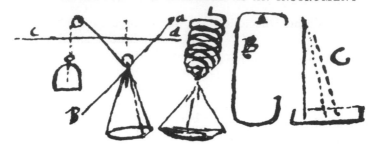

Fig. 6.12. Sketches from Hooke's *Diary*, Aug. 21, 1378.
(By permission of the Corporation of the City of London)

mercury from one chamber to the other, could be canceled by placing weights in a scale pan, or in the equal-arm type, transferring them from one scale pan to the other. Judging by the barometer of the unequal-arm type that is preserved at Florence in the *Museo di Storia della Scienza*, it is doubtful whether much accuracy was attained or attainable with such a large unhandy device.

An entirely different sort of balance barometer, in which the tube, supported in various ways, acts as a moving weight, has been proposed at least ten times. The principle is that the tube is designed so that the center of gravity of the mercury in the system moves horizontally as well as vertically as the atmospheric pressure changes. Some of the designs were intended to expand the scale; others, which will be dealt with in Chapter 9, were meant to give accurate readings. Nor has the principle been forgotten when recording barometers were being considered (Chapter 11).

The first of the balance barometers of this type was possibly adumbrated by Sir Christopher Wren. On Wednesday, August 21, 1678, Robert Hooke was talking to Sir Christopher, and in his diary[70] for that day he made a cryptic entry which is unusual in being illustrated (Figure 6.12). "Sir Chr. Wren . . . told a pretty thought of his about a poysd weather glasse as, at *B*. I told him an other as *C*." That is all, but it seems possible that the one at *B* was a siphon barometer with the top of the tube bent over above the level of the mercury, so that it could be hung on a knife edge or pivoted on an axis. The other is harder to understand, especially as we cannot be sure whether the adjective "poysed" is supposed to apply to both "weather glasses" or only to Wren's. It *could* be a barometer tube pivoted at the top and dipping in a trough of mercury, in which event we might assume a string fastened to the tube near the bottom and passing over a pulley to a weight; but Hooke did not show this. It is, of course, also possible that these devices were not barometers at all, but air thermometers, which were often called "weather glasses."

Barometers on the principle of the "poysd weather glasse" were patented

[70] H. W. Robinson and W. Adams, *The Diary of Robert Hooke, M.A., M.D., F.R.S., 1672–1680* (London, 1935).

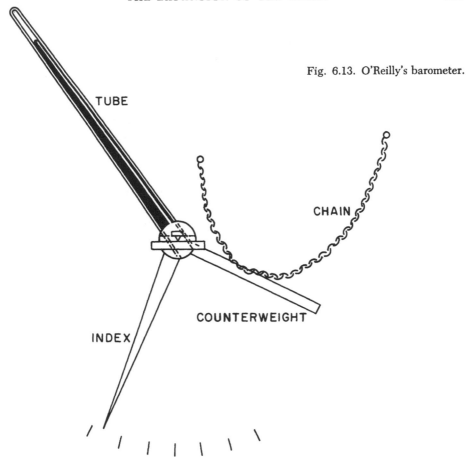

Fig. 6.13. O'Reilly's barometer.

TUBE

CHAIN

COUNTERWEIGHT

INDEX

in 1856 by Mapple[71] and by Wollheim[72] in 1868. Mapple's is a siphon barometer on knife edges, held at a small angle to the vertical by a weighted pointer. Temperature compensation is provided for by mounting a large thermometer sloping the other way—a very ingenious idea.

In 1870 J. P. O'Reilly proposed a large balance barometer, intended for public exhibition.[73] This was to be

[71] British Patent 1029 (1856).
[72] British Patent 529 (1868).
[73] *Proc. Roy. Irish Acad.*, Ser. 2, Vol. 1 (1870), pp. 31–36.

3.73 meters high, and would have been an impressive instrument. As shown in Figure 6.13, the tube dips into a cylindrical cistern, and is counterpoised by a fixed bar, an index, and a variable auxiliary counterweight in the form of a chain, to straighten out the scale. To make this easier, the tube was to be slightly conical. Two steel points on the axis of the cistern, resting in agate cups, support the moving parts. It does not seem to have been made, but it really might have worked. Its simplicity recommends it, and for a display barometer the rectification of the scale would be rather a luxury, so that the chain might even have been left out.

5. THE DIAGONAL BAROMETER

The diagonal barometer is the simplest possible modification of Torricelli's original tube. If we bend the tube sharply through a little less than a right angle at a point about 27 inches or 69 cm. from the level that the mercury will stand in the cistern, then the last few inches of vertical height will correspond to a much greater extent of tube, and in this way the movement of the surface of the mercury may be magnified to whatever extent (within limits) we desire.

Like the balance barometer, the diagonal barometer has long been ascribed, with no documentation, to Sir Samuel Morland; and even in 1951 Dr. C. A. Crommelin of Leiden was obliged to confess that he could not find "when and where Morland published his invention."[74] The matter is made more complicated by the fact that the diagonal barometer was also invented quite independently in Italy.

Let us get this part of the problem out of the way at once. In 1695 Bernardus Ramazzini (1638–1714), Professor of Medicine at Modena, published a work on meteorology[75] in which he made some acute remarks on

the deficiencies of the available barometers. Then

I suppose the whole perpendicular altitude which the mercury in my barometer occupies this year to be divided into 30 parts, or inches; and because my barometer is bent to the side at a height of 28 inches, and the arm extends so as to rise slightly from the horizontal plane, the mercury runs in this so that its motions may be distinguished much more clearly. The remaining two inches of perpendicular altitude, which have been spread out into the greater extent of the tube bent to the side, might well be divided into 90 lines.[76]

This is quite clearly the diagonal barometer. There is no good reason to suspect Ramazzini of plagiarism, as he was in the habit of giving references. For instance, he refers to the well-known little book by D'Alencé;[77] but this does not mention the diagonal barometer. The cryptic reference in the *Philosophical Transactions*, in which William Derham shows a figure of such an instrument, calling it "a tube communicated to me by a friend," did not appear until three years later.[78]

But if Ramazzini invented the diagonal barometer, so did Morland; and Morland has the priority by seven years at least. In 1688 a London clockmaker, John Smith, wrote a sixpenny pamphlet about the barometer in which the following passage occurs:[79]

[74] C. A. Crommelin, *Descriptive Catalogue of the Physical Instruments of the 18th Century*, Rijksmuseum voor de Geschiedenis der Natuurwetenschappen, Communication no. 81 (1951), p. 59.

[75] *Ephemerides Barometricae Mutinenses Anni M.DC.XCIV. und Cum Disquisitione Causae ascensus ac descensus Mercurii in Torricelliana fistulê, juxta diversum aëris statum* (Modena, 1695).

[76] *Ibid.*, p. viii–ix.

[77] Joachim D'Alencé, *Traittez des barométres, thermométres, et notiométres, ou hygrométres. Par M^r. D.*** (Amsterdam, 1688).

[78] *Phil. Trans.*, Vol. 20 (1698), p. 4. Derham's friend has not been identified.

[79] John Smith, *A Compleat Discourse of the Nature, Use and Right Managing of that Wonderful Instrument the Baroscope or Quick-Silver Weather Glass* (London, 1688), pp. 1–2.

Baroscopes have heretofore been made up after divers manners, but chiefly three ways: As first, that now commonly used, with a streight tube, ascribed chiefly to the noble Boyle. Secondly, that with a tube, whose top inclines, devised by Sir Samuel Morland. Thirdly, the wheel-baroscope, invented by the ingenious Mr. Robert Hook (and described in his Micography [sic]) but of these three sorts, the two last are but seldom used, by reason of some inconveniencies either in the shape or charge; Sir Samuel's being such as will not admit of any regular figure, and Mr. Hooks being very dear and costly.

Although this is the earliest reference to the diagonal barometer that I have been able to find, it reads as if it were a familiar instrument in 1688, at least in London. It is, of course, highly unlikely that John Smith's little brochure reached Modena; but the possibility that Ramazzini knew about it cannot be excluded, as it did reach Leipzig and was reviewed in the *Acta Eruditorum*[80] four years later. The following words might have given Ramazzini the idea: ". . . barometrum *Sam. Morlandi*, inclinatum habens verticem, . . . minutiores aeris variationes prodere."[81]

This review does seem to have made the instrument known on the Continent. In the central library of the *Bundeswetteramt* at Offenbach-am-Main there is a dissertation[82] read at Iena in 1701 in which the author quotes Ramazzini's *Ephemerides* and attrib-

utes the diagonal barometer to him, and then says: "Videtur autem hoc barometrum idem esse cum illo cujus inventio in *Actor.* [sic] *Erud. Lips Tom.I.supplem.* Sect. 7. p. 352. tribuitur *Samueli Morlando*, Equiti Anglo, cui alias tubi stentorei insigne inter doctos peperere nomen."[83]

We may safely conclude that Sir Samuel invented the diagonal barometer, probably before 1680, and that Ramazzini reinvented it in 1694 or 1695. But we must not neglect an opinion put forth by Ludwig Achim von Arnim in the *Annalen der Physik*, a century later.[84] Von Arnim quoted from Ramazzini's *Ephemerides* the exact passage that I have quoted above, and then said

It seems obvious to me that Ramazzini is here describing a completely ordinary barometer 30 inches long, and that it is quite naïve of Leupold[85] and others with him to turn things upside down, and instead of translating *ad altitudinem* as "28 inches high," to put "at a height of 28 inches." Ramazzini says that the tube measures 28 inches from top to bottom; then two inches are bent, as in every ordinary tube. But for accuracy he had divided the upper scale into 90 lines. The *ad latum flectitur* holds as good for a bend

---

[80] *Supplementum,* Vol. 1 (1692), pp. 352–56.

[81] *Ibid.,* p. 352. "the barometer of Samuel Morland, having its top inclined . . . to show minute variations of the air."

[82] Georg Albrecht Hamberger (1662–1716), *De barometris.*

[83] *Ibid.,* fol. B4 *recto.* "But this barometer seems to be the same as that of which the invention is attributed (in the *Acta Eruditorum,* etc.) to Samuel Morland, an English Knight, whose speaking trumpets have already made his name famous among the learned."

[84] *Annalen der Physik,* Vol. 2 (1799), pp. 334–35.

[85] Jacob Leupold, *Theatri statici universalis sive theatrum aërostaticum, Oder: Schau-Platz der Machinen zu Abwiegung und Beobachtung aller vornehmsten Eigenschaften der Lufft, etc.* (Leipzig, 1726). On page 257 the diagonal barometer is ascribed to Ramazzini, with no mention of Morland.

at the bottom as at the top. So everything falls on the expressions *ad altitudinem, ad latus*, and consequently *ad* is the whole stumbling-block. Still I note that besides this characteristic, Morland's barometer has a bent upper tube 50 inches long, of which Ramazzini says never a word.

This seems to me to be nonsense of the first quality. Von Arnim seems to have entirely missed the two phrases "& brachium exporrigit à plano horizontali mediocriter assurgens, in quo Mercurius vagatur" and "per latus inflexum fistulae longiori tractu sunt distributae." I can imagine no reason whatever for this seeming aberration on the part of von Arnim.

The diagonal barometer, although never suitable for precise measurements because of the effect of friction and the peculiar shape of the end of the mercury column in the inclined tube, had a long career as a weatherglass. John Smith's objection to the shape of the instrument was quickly overcome, and the shape indeed converted into an asset, by mounting the barometer at the top and left side of a rectangular frame, a large thermometer on the right, "And a looking glass commodiously plac'd on the same frame, between the barometer and thermometer, whereby gentlemen and ladies at the same time they dress, may accommodate their habit to the weather."[86] Instead of the mirror, a perpetual calendar was sometimes provided, as in the elegant instrument in the Science

Museum, London, made by F. Watkins and A. Smith in 1753.[87]

However, the peculiar shape of the diagonal barometer itself did not prevent its wide distribution in its simple form. Some of these barometers are very large, the length of the nearly horizontal part of the tube being of course variable over wide limits merely by varying its slope. In the *Deutsches Museum* at Munich there is a barometer dated 1774 (Figure 6.14) with the sloping part more than a meter long.[88] But the most extravagant instrument of this type must surely be one in the History of Science Museum at Leyden, in which the diagonal tube has a slope of only 1 in 50, and is nearly 3 meters long.[89] It is almost certainly more than a century old, and it still works.

At Munich, again, there is a barometer[90] inscribed "Magnum Barometrum Morlandinum," in which at about 24 inches the tube bends at about 45 degrees to the right for about 4 inches, and then to the left again at a slope of about 1 in 3. Like many other eighteenth-century barometers, this has the tube turned up at the bottom, and terminating in a small pear-shaped bulb.[91] Although it states that the instrument was *Approbata in Accademia Hallae Magaborgicae*, the scale seems completely arbitrary, going up and down from a zero in the middle, where it is marked *Veränderlich, Variable*, and *Variabile*.

This double bend is described by

---

[86] John Patrick, *A New Improvement of the Quicksilver Barometer, Made by John Patrick, in the Old-Baily, London* (Broadsheet, n.d.). This is probably about 1700. See also Harris, *Lexicon Technicum*.

[87] Science Museum inventory no. 1876–814.
[88] Inventory no. 22585.
[89] Inventory no. Th 20.
[90] Inventory no. 6281.
[91] See p. 133 below.

Leupold[92] as a device to give the diagonal barometer a more convenient shape. Such instruments are not common; there is another in the *Landesmuseum* at Zurich.[93] Leupold also remarks[94] that a great deal of mercury flows in and out of the tube in a diagonal barometer, so that the cistern should be relatively large. He recommends a wide upper part and a narrow cylinder below into which the tube dips, thus saving mercury and affording a means of adjustment.

The diagonal barometer was very popular in England, and they are not hard to find at sales and in antique shops. Even great instrument makers like Daniel Quare (1649–1724) made them occasionally; there is a magnificent example in the State apartments at Hampton Court Palace. At the History of Science Museum, Oxford, there is one by the eminent clockmaker John Whitehurst (1713–1788) of Derby, who became a Fellow of the Royal Society. But in general, their scientific importance is negligible, and was never very great.

Another means of reducing the horizontal dimensions of the diagonal barometer is to have two or three sep-

[92] *Theatri statici,* etc., p. 257.
[93] Inventory no. L.M. 7663.
[94] *Theatri statici,* etc., p. 257.

Fig. 6.14. Large diagonal barometer. (*Deutsches Museum, Munich*)

arate tubes, one bent over at 28 inches, for example, another at 29½. The scale on the lower one would extend from about 29 to say 29.6 inches, and on the upper one perhaps from 29.4 inches to 31. I have not found any certain indication of the inventor of this scheme, but Magellan[95] refers it to "feu Mr. Horme, gentilhomme Anglois de Ashby, en Lancastre." Johann Friederich Luz, a cleric at Gunzenhausen, in an interesting book[96] on barometers, repeats this, making "Horme" *ein Edelmann*. Most probably Magellan was thinking of Charles Orme of Ashby-de-la-Zouch in Leicestershire, the friend referred to by Henry Beighton in a paper[97] published forty years earlier, which does describe the filling of diagonal barometers.

Barometers of this sort seem to have been fairly common in England. In the Whipple Museum at Cambridge there is a diagonal barometer with two tubes[98] inscribed "Samuel Lainton Maker Halifax." Interestingly enough a similar barometer in the History of Science Museum at Oxford was also made at Halifax in Yorkshire, this one by Charles Howarth; and I have also seen a left-handed one by Howarth in a London antique shop.

Luz was one of the last writers to take the diagonal barometer seriously, devoting ten pages to its construction and errors.[99] It is clear that he is writing from personal experience, as is claimed in the title of his book. But he admits its deficiencies: that it is hard to boil out the bent tube, and that the mercury tends to stick.

At about the same period, Father Ximenes of Florence thought it worthwhile to calculate the change of slope necessary to compensate for the variation of level in the open limb, as a function of the relative diameters of the open and closed limbs.[100] He even worked out the theory of a diagonal barometer with a pear-shaped bulb and a curved inclined tube to compensate for the fall in the bulb as the atmospheric pressure rises![101]

6. THE "SQUARE" OR L-SHAPED BAROMETER

The "square" barometer, shown diagrammatically in Figure 6.15, usually bears the name of Johann Bernouilli (1667–1748), one of the celebrated

---

[95] *Description et usages*, etc., p. 152.

[96] *Vollständige und auf Erfahrung gegründete Beschreibung von allen sowohl bisher bekannten als auch einigen neuen Barometern, wie sie zu verfertigen, zu berichtigen und übereinstimmend zu machen, dann auch zu meteorologischen Beobachtungen und Höhenmessungen anzuwenden. Nebst einem Anhang seine Thermometer betreffend* (Nürnberg und Leipzig, 1784). We shall have further occasion to refer to this work. The reference here is to p. 35.

[97] "The Imperfections of the common Barometers, and the Improvement made in them by Mr. Cha. Orme," etc. *Phil. Trans.*, Vol. 40 (1738), pp. 248–58.

[98] Inventory no. 1130.

[99] *Vollständige Beschreibung*, pp. 26–36.

[100] *Teoria de difetti*, fol. 83ᵛ–84ʳ.

[101] *Ibid.*, fol. 84ʳ. It seems that the diagonal barometer is the sort of device that inspires the more fantastic sort of inventor. For example, Wilhelm Huch patented a zig-zag tube in 1887 (D. R. Patent 40676), and even in 1954 (British Patent 733,749) a patent was granted on a helical tube, with the additional feature that the whole tube is in bearings at top and bottom, so that it can be rotated in order to bring to the front the part of the scale in use. The same patent covers a tube with diagonal branches alternately on each side of the main stem, a sort of dendritic barometer!

family of mathematicians who for more than a century was associated with the University of Basle.[102] In fact Bernouilli states[103] that when he first described this barometer to the Academy of Sciences in Paris,[104] Jean Dominique Cassini (1625–1712), the director of the Observatory, had said that he had thought of constructing such a barometer several years before (*jam ante complures annos*) but had done nothing about it. This is not quite fair to Cassini, who had in fact installed one at the Observatory of Paris as long ago as January 3, 1673. His journal, in Latin, is preserved, and the following extract is accompanied by a very clear diagram of the instrument (Figure 6.16):

Today I placed the new rectangular barometer in the north window of my east room, so that one arm should be vertical, the other horizontal. The mercury suspended in the vertical arm is extended into the horizontal one, the outermost part of which is empty. Therefore I made a mark beside the end *a* of this so that I could discern any variation from it.[105]

[102] *Johannis Bernouilli opera omnia*, ed. Cramer (4 vols., Lausanne & Geneva, 1742), II, 204–7.

[103] *Ibid.*, p. 207.

[104] He wrote from Basle on March 17, 1710 (*Registres*, fol. 140–43 and 4 figures). There is no record of any comment by Cassini in the *Registres* or elsewhere at the *Académie des Sciences*.

[105] Observatoire de Paris, ms. AD1, Tom. 4, fol. 3ᵛ.

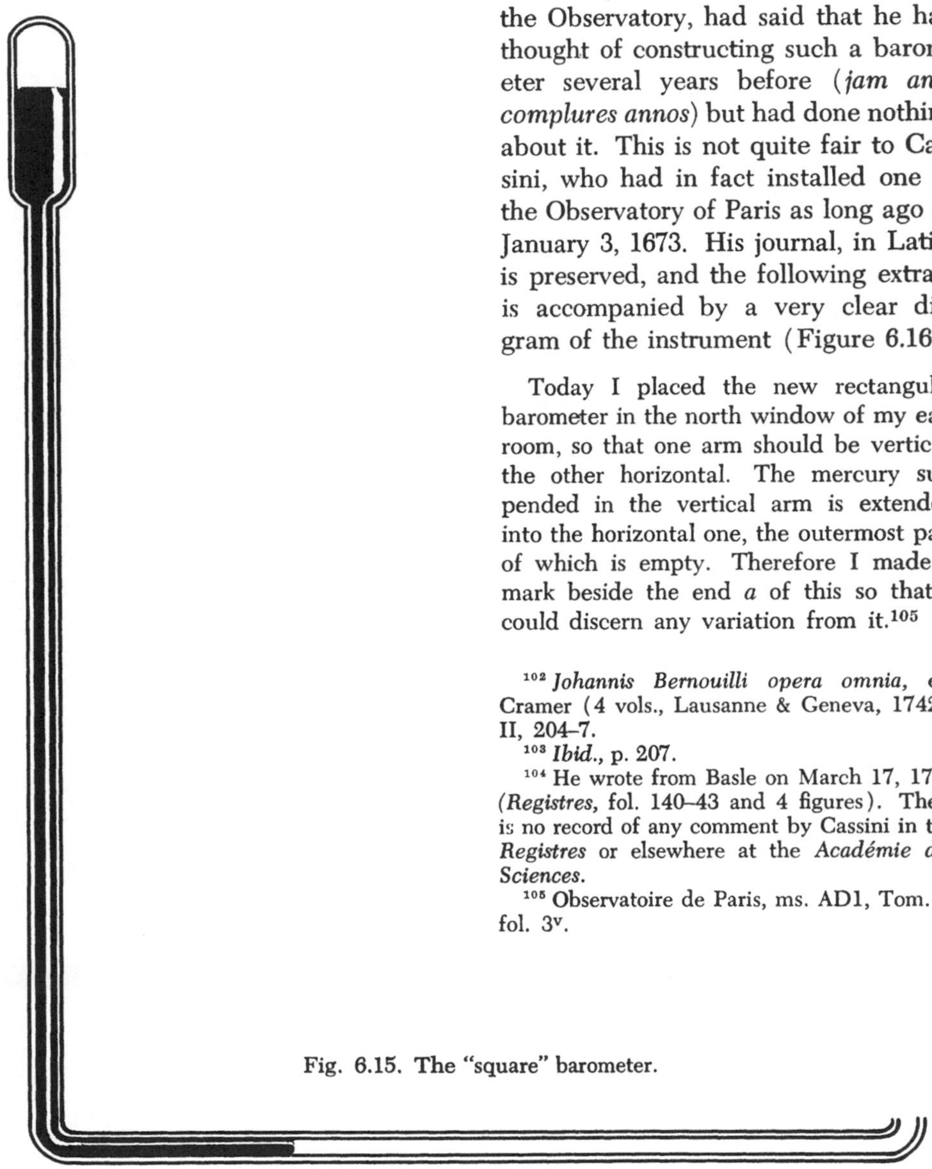

Fig. 6.15. The "square" barometer.

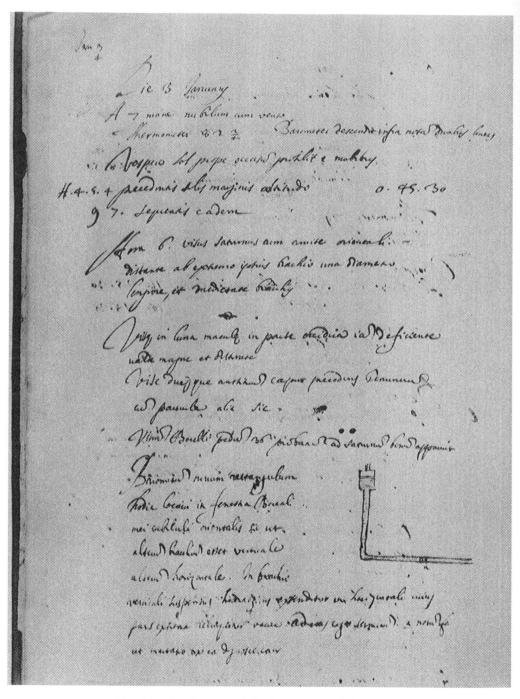

Fig. 6.16. Cassini's journal for Jan. 3, 1673. *(Observatoire de Paris)*

He immediately began observations with the new instrument, finding on January 4 that, while the ordinary barometer had descended *insensibiliter* since the previous day, the index in the rectangular barometer had moved 3⅔ inches.[106] It is quite clear that the instrument should be ascribed to Cassini, and dated back to 1673.

The idea is simplicity itself. If the vertical tube (or the top part of it, which is what matters) is n times as wide as the horizontal tube, the magnification is n^2. But of course the instrument has numerous disadvantages, of which the most serious is the sticking that results from the presence of a mercury-air surface in a tube only about one line (2 mm.) in diameter. The linearity of the scale also presupposes the uniformity of both tubes. This instrument was dealt with at some length by De Luc[107] and also by Luz.[108] De Luc particularly objects to the complicated temperature correction such a barometer must have; this will depend not only on the diameters of the mercury surfaces but also on the total quantity of mercury in the instrument. Luz, on the other hand, emphasizes what he believes to be a great defect in a barometer, the necessity of calibrating it by comparison with another.

There is a large square barometer in the University Museum of the History of Science at Utrecht,[109] with the horizontal tube about 110 cm. long. It is

signed "L. Primaues, Amsteldam"[110] and graduated in both Rhenish and Amsterdam inches, which seem to differ by about 3 per cent,[111] on the vertical tube, and in expanded Amsterdam inches on the horizontal tube.

Bernouilli realized that it might be convenient to bend the narrow tube into folds in a horizontal plane, or to wind it in a horizontal spiral. This latter scheme was again suggested in 1786 by the Brigadier de la Chiche to the Paris instrument maker Assier-Perica.[112] Much later Guthrie[113] made the interesting suggestion that a small bubble of air or (much better) a drop of sulphuric acid should be left in the tube as an index, a small reservoir of mercury being provided at the outer end of the spiral tube.

The brilliant English experimenter

---

[106] *Ibid.*, fol. 4ʳ.

[107] *Recherches sur les modifications de l'atmosphère* (Geneva, 1772), I, 23–25.

[108] *Vollständig und auf Erfahrung gegründete Beschreibung* etc. (Nürnberg und Leipzig, 1784), p. 39.

[109] Catalogue no. G40.

[110] A family of Italian instrument makers, variously calling themselves Primavesi, Primavesy, Primaues, flourished in Amsterdam about the middle of the eighteenth century. We know of Frans, Jean, and "L."

[111] See p. 173 below.

[112] *Bibliothèque physique*, Vol. 5 (1786), pp. 412–13. Antoine Assier-Perica (or Perricat) was one of the most prominent, though not one of the most able, of French instrument makers, and flourished in the last 30 years of the eighteenth century. He managed to get himself made "démonstrateur au Lycée des Arts pour la construction des baromètres" (Daumas, *Les instruments*, [1953], p. 382). He was a consistent self-advertiser; cf. *J. de Phys.*, Vol. 14 (1779), pp. 327–29. In a reference to Assier-Perica, Luz, *Vollständig Beschreibung*, p. 219, writes "But we are already accustomed to the ways of the French, who are pleased to ignore discoveries on purpose, especially those of foreigners, and then, making a small alteration, to pass them off as their own." National prejudice, no doubt; but Daumas (*Les instruments*, pp. 123 ff.) is unenthusiastic about French instrument making in the eighteenth century.

[113] *Phil. Mag.*, Vol. 3 (1877), pp. 139–41.

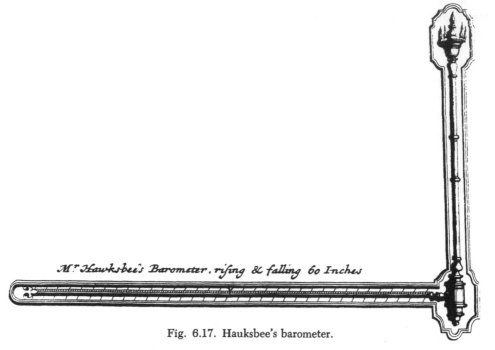

Mr. Hauksbee's Barometer, rising & falling 60 Inches

Fig. 6.17. Hauksbee's barometer.

Francis Hauksbee (d. 1713?)[114] invented a rather similar barometer some time about the year 1700. His instrument, which has some of the characteristics of both the square and the diagonal barometers, is described by John Harris (1667–1719) in his celebrated Dictionary,[115] from which Figure 6.17 is taken.[116] "The ingenious Mr. Hauksbee," wrote Harris, ". . . shewed me a *Baroscope* where the mercury rose and fell 60 inches with very great ease, and without breaking or dividing; and it may very easily be made for 100 or 200 inches, if a strait small thin glass tube can be blown and drawn of that length, and that it were as easily manageable." This barometer is a cistern barometer, in which a small glass tube is joined to the side of the cistern just below the level of the mercury in it, and extends upwards at a small angle to the horizontal, acting as a gauge for the height of the mercury in the cistern. Thus it was more nearly an inverted diagonal barometer than a square barometer, and added to the disadvantages of the diagonal barometer one of those of the square barometer, namely that the small surface of mercury in the tube was in contact with air, not in a vacuum. In spite of this a similar construction was patented in 1898 by T. Horton.[117]

---

[114] He was succeeded as an instrument maker by another Francis Hauksbee (1687–1763), usually referred to as "the younger." It does not seem to be certain that the two were father and son (DNB).

[115] *Lexicon Technicum, or an Universal English Dictionary of Arts and Sciences,* etc. (2 vols., 4°; London, 1704 and 1710). Vol. I (1704), s.v. "Barometer."

[116] The figure is in one of the plates illustrating the word "Air-pump."

[117] British Patent 9132 (1898).

7. THE CONICAL BAROMETER AND ITS DERIVATIVES

The next type of barometer to be invented was one which had no cistern at all. In a sense it does not belong in this chapter, as it was originally intended for use at sea; but as it also had an expanded scale, and as the wish to expand the scale inspired many of its successors, it will be dealt with here.

In 1695 Guillaume Amontons gave a description of "un nouveau baromètre très simple et portatif à l'usage de la mer."[118] This is simply a slightly conical tube, closed at the smaller end, and containing a column of mercury of suitable length, initially filling the smaller end of the tube. When the tube is inverted so that the closed end is upward, the column of mercury will fall in the tube until its length (decreasing as its diameter becomes greater) equals the barometric height. A scale against which the position of one end of the column is read could give any desired magnification, depending on the rate of change of the diameter of the tube. At the C.N.A.M. in Paris there is such an instrument inscribed *Barometre d'Amontons* and probably dating from about this period.[119] It is about 1.8 meters long and the magnification is about five. Only the top of the column of mercury (now of course quite broken up) is provided with a scale.

Amontons noted the precautions that have to be taken. It is to be hung up only when an observation is being made; then it is to be laid flat with great care, or inverted. A drop of mercury frequently separates at the larger end of the column; this is to be reunited with the column by the use of an iron wire. He noted the considerable effect of friction, and realized that this was a defect; but

. . . this barometer [could] serve very well at sea, where the others run too great a risk, for in fact this lack of precision is not considerable enough to prevent the amount by which rains and winds alter the weight of the air being appreciable by this barometer, and we need take no trouble to obtain a greater precision, as it is useless.[120]

This sort of remark concerning marine barometry recurs again and again in the literature for another century and a half. A modern meteorologist, whose training leads him to think in terms of weather maps, may well forget that in the days before the advent of the wireless telegraph the absolute accuracy of a ship's barometer was of little or no importance to anyone, least of all to the mariner; any dangerous weather would be foreshadowed by a marked fall in the "glass," accompanied by typical cloud formations. Amontons was right as far as this was concerned, but he was probably not enough of a sailor to realize that the most important defect of his instrument was the necessity of hanging it up every time an observation was desired. What was needed was an instrument that would tell its story at a glance when the mariner came into his chart room.[121]

---

[118] In his *Remarques et expériences phisiques,* etc. (Paris, 1695), pp. 123–35.
[119] Inventory no. 1580.

[120] Amontons, *Remarques,* etc., pp. 132–33.
[121] Marine barometers are dealt with again in Chapters 7, 9, and 15.

Amontons was puzzled by the friction in his barometer, because ". . . If we make a simple barometer with a narrow tube, the mercury stays up at precisely the same height as in larger tubes, and the movement of the mercury has the same regularity and precision in each, although the same obstacle which seems to be the cause of the inequality in the one, should also introduce irregularity into the movement of the other" (i.e., the simple barometer).[122]

He was obliged to suppose that *la nature cottoneuse de l'air* is a greater obstacle to the entry of the air into narrow tubes than is the viscosity of the mercury to its motion. But it is clear that he was not really satisfied with this explanation. The real reason for the sticking is undoubtedly the dirtying of the mercury and of the glass where the end of the column is exposed to the atmosphere.

This type of barometer, among others, was being made by John Patrick in London shortly after 1695.[123] I do not think, however, that it is certain that Patrick reinvented it. He could easily have heard about it from some English physicist who had seen it in Paris. But William Derham, F.R.S., perhaps somewhat isolated in his rectory at Upminster, wrote[124] about the conical barometer and ascribed its invention to John Patrick. More to the point, so did Edmond Halley (1656–1724), citing as the advantage of the instrument the great enlargement of the scale and seriously proposing it for the measurement of heights.[125]

Of course we cannot exclude the possibility that John Patrick was really the inventor. He was very active in 1695,[126] and since Amontons also imagined for a time that he had invented the three-liquid barometer,[127] might he not also have been in error about the conical type?

This instrument was reinvented in 1880 by Giulio Grablovitz[128] with the lower end turned up in a U, and with scales corresponding to both the top and the bottom of the mercury column for each pressure. Meanwhile both Whiting[129] and Hicks [130] had suggested, as a substitute for the conical tube, two tubes of different diameters joined end to end, one extremity of the mercury column being in each.[131] There were two scales in Hicks's instrument up and down from the center. By this time such apparatus could have no practical value, and it is only to complete the record that we note that a German patent was actually obtained for such an instrument in 1910 by Paul Leiberg of Moscow![132]

---

[122] *Ibid.*, p. 133. We may note in passing that he was unaware of the capillary depression (see Chapter 8).

[123] It is advertised in the curious broadsheet referred to in note 86 above.

[124] Royal Society, *Classified Papers* IV (1), item 58, July 17, 1712.

[125] "A proposal for measuring the height of places, by help of the Barometer of Mr. Patrick, in which the Scale is greatly enlarged." *Phil. Trans.*, Vol. 31 (1721), pp. 116–19. In this we read "Mr. Patrick, who stiles himself the *Torricellian Operator*" (p. 118).

[126] See S. E. Atkins and W. H. Overall, *Some Account of the Worshipful Company of the Clockmakers of the City of London* (London, 1881), p. 244. See also p. 151 below.

[127] See p. 93 above.

[128] Italian Patent 535 (1880).

[129] *Annalen der Physik*, Vol. 117 (1862), pp. 656–58.

[130] *Proc. British Meteorol. Soc.*, Vol. 2 (1864), pp. 206–8.

[131] According to Gehler's *Physikalisches Wörterbuch*, I (1825), 790, this modification was first proposed by [John?] Leslie.

[132] D.R. Patent 236,729, (1910).

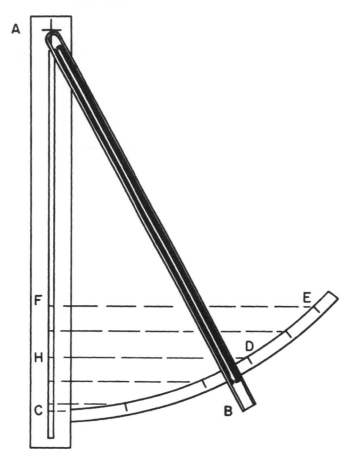

Fig. 6.18. Lana Terzi's barometer.

Earlier than any of these was a device described in 1686 by Francesco Lana Terzi (Franciscus Tertius de Lanis, S.J.).[133] A narrow tube closed at A (Figure 6.18) and open at B can swing about A in a vertical plane. Fill it with mercury to make a column of length AC, equal to the maximum barometric height that is to be measured. Incline it so that the mercury all goes to the closed end, as at AE. On a wide board inscribe an arc CDE of radius AC. Provide a uniform scale (*regula*) FC, and project lines FE, etc. to the arc. To make an observation, let the tube approach the vertical until the mercury just starts to move downward; suppose this occurs when the tube is in the position AD. The barometric height will then be AH, the vertical projection of AD; the arc will, of course, provide an expanded, though non-uniform scale. Perhaps the best that can be said for this highly ingenious idea is that in this way a barometer could be made with the very

[133] *Magisterium naturae et artis* (3 vol. folio; Brescia, 1684–1692), II (1686), 287–88.

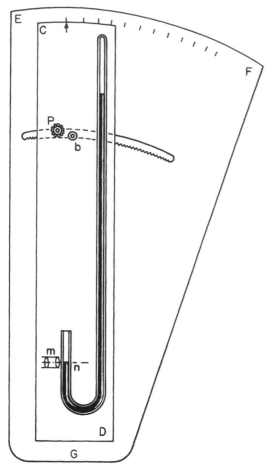

Fig. 6.19. Magellan's sectorial barometer.

to improve the accuracy of reading. The earliest of these, that of Derham,[134] certainly fulfilled the first of these requirements and belongs in this chapter; its success in improving accuracy is doubtful. William Derham's invention was a rack carrying an index, which is set level with the mercury by turning a pinion, as in some barometers before and many since; but this pinion had a pointer on its shaft, moving over a circular scale divided into 100 parts. The pitch circumference of the pinion was made one inch, so that the scale gave the barometric height directly in hundredths.

To make this a real improvement a great change in the index would have been necessary, and also a refinement in workmanship that could scarcely have been available to Derham. The idea was suggested again in 1875 by R. E. Power.[135]

An entirely different suggestion was put forward by Magellan,[136] who invented a new way of reading a siphon barometer (Figure 6.19). The barometer tube is mounted on a board CD which rotates about an axis n against a sector-shaped board EFG, being moved by a pinion p and clamped by a clamp b. The amount of mercury is arranged so that at some maximum reading h_1, the tube and the board CD being vertical, the level of the mercury

minimum of mercury, a serious consideration in the seventeenth century in some places. It would have had all the defects of the conical barometer.

8. MISCELLANEOUS WAYS OF EXPANDING THE SCALE

We must now examine several constructions not easily classified, but intended for the expansion of the scale of the barometer, and sometimes also

[134] *Phil. Trans.*, Vol. 20 (1698), pp. 45–48.
[135] *Quart. J. Meteorol. Soc.*, Vol. 2 (1875), pp. 437-38. See also p. 201 below.
[136] *Description et usages des nouveaux baromètres*, etc. (London, 1779), pp. 154–55. Luz (*Vollständig Beschreibung*, pp. 44–46) says that he finds Magellan's description incomprehensible and describes the instrument as he thinks it must be. I concur in Luz's judgements, including the first. Figure 6.19 is probably a reasonable interpretation.

in the open limb coincides with the axis of a microscope m, which is also on the axis n. If the pressure decreases to h, the mercury will fall in the longer limb and rise in the shorter; but if the tube is tilted, it will rise again in the longer limb; and because there is only a finite volume of mercury in the tube, it will fall in the shorter limb. The angle to which the barometer must be tilted to bring the mercury surface back to n is a function of h and also of the separation a of the two limbs; in fact

$$h = h_1 \cos \theta - a \cos \theta \sin \theta.$$

It is not clear that Magellan noted this, or indeed really understood the action of the instrument.

In the catalogue of the *Museo di Storia delle Scienze* at Florence there is mention of a barometer (no. 1359), somewhat analogous to this but not precisely the same; this instrument did not seem to be available in 1961.

One of the neatest schemes for enlarging the scale of the barometer was invented by Richard Howson,[137] an engineer of Middlesborough-on-Tees (Figure 6.20). It might be called a floating-cistern barometer. A barometer tube A of large diameter is hung from its closed end. The cistern B is a deep glass cup with a hollow glass stalk sealed into the bottom of it. Because of the buoyancy of the stalk the cistern remains suspended, with no other support. As Negretti & Zambra rightly ob-

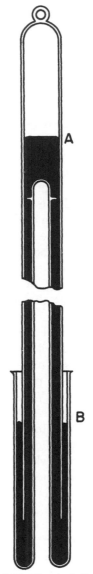

Fig. 6.20. Howson's floating-cistern barometer.

serve, ". . . it seems a perfect marvel. It appears as though the cistern with the mercury in it must fall to the ground."[138] In Howson's original design, the tube was filled by means of

---

[137] First announced by P. J. Livsey, *British Assn. Report* (Manchester, 1861), Transactions of the Sections, p. 64. More completely described in *Proc. British Meteorol. Soc.*, I (1862), 81–83. See also *A Treatise on Meteorological Instruments: Explanatory of Their Scientific Principles, Method of Construction, and Practical Utility* (London, Negretti & Zambra, 1864), pp. 47-48.

[138] *Ibid.*, p. 47.

an air pump. Negretti & Zambra, who offered the instrument for sale, put a cork pad at the bottom of the cistern, which was held against the end of the filled tube while it was being inverted. They also used a glass rod instead of the hollow stalk.

It can be shown that the magnification of this instrument (that is to say the motion of the cistern divided by the change in atmospheric pressure in mercury units) is equal to the internal cross section of the tube at A divided by the cross section of the glass wall of the tube. Of course the top of the stalk must never rise above the top of the mercury column.

It is interesting that Lucien Vidie, better known as the inventor of the aneroid barometer (see Chapter 16), seems to have tried and abandoned the instrument we have just described. In his journal *Les Mondes*, the Abbé F. Moigno refers to a pamphlet published in Paris in which Vidie described this and other experiments.[139] Moigno's report stimulated Tito Armellini to suggest a more complicated barometer on similar principles which does not seem to have had any success.[140] Negretti & Zambra were still offering Howson's barometer for sale in their catalogues of 1873 and 1886.

A barometer in which a tube is floated in a stationary cistern, and therefore the inverse of Howson's invention, was patented by Thomas Telford MacNeill in 1861.[141] It was more complicated mechanically and not really a good solution to the problem, at least as an indicating instrument. Tito Armellini described a rather similar floating barometer in the following year.[142] This was designed in full knowledge of the conditions for vertical stability, having a sleeve of wood or glass round the tube at cistern level, with a cross section greater than that of the vacuum chamber. This made the tube float in the mercury in a cylindrical reservoir; and it will be seen, he said, that this transforms the instrument into a true hydrometer, in which the variations of the line of flotation mark the changes in atmospheric pressure. The lateral stability of the tube would have had to be assured by guides of some sort, but no details were given.

In 1868 Armellini reappeared with an "areometric" barometer, in which a float in glycerine supported a mercury barometer tube. A graduated rod of small diameter, on which the pressure was read, projected above the surface of the glycerine.[143] But this was really only a reinvention of the barometer described 135 years earlier[144] by John Rowning (?1701–1771), rector of Anderby, and author of a once-popular textbook on Natural Philosophy.

In Rowning's instrument, which is shown diagrammatically in Figure 6.21, water is used as a liquid in which to float the barometer tube and cistern, supported by a hollow float MN and extended by a light rod ending in an

[139] *Les Mondes*, Vol. 3 (1863), pp. 24–27. According to Vidie's biographer Auguste Laurant, who reprinted this pamphlet (*Biographie de Lucien Vidie* [Paris, 1867], pp. 261–327), it was published before July 9, 1861. This leaves Howson's priority in doubt; but I have not been able to find the original pamphlet.

[140] *Ibid.*, Vol. 3 (1863), pp. 99–100.

[141] British Patent no. 1733 (1861).

[142] *Correspondenza Scientifica in Roma*, Vol. 12 (1862), pp. 371–72 (May 15).

[143] *Les Mondes*, Vol. 16 (1868), pp. 201–5.

[144] *Phil. Trans.*, Vol. 38 (1733), pp. 39–42.

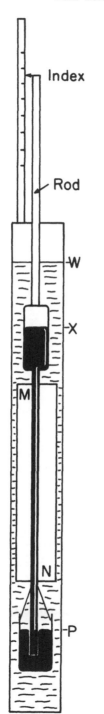

Fig. 6.21. Rowning's floating barometer.

index. The smaller this rod in relation to the barometer tube and cistern, the larger will be the magnification. Rowning would have done better to graduate the rod and observe how high the water came at W, because otherwise any evaporation of water will change the zero; and the theory, more complicated than Rowning suspected, by the way, will depend on the size of the tank. If the internal cross section of the cistern (less that of the tail of the tube) at P is A, that of the tube at X is B, and the cross section of the rod is C, it can be shown that the magnification is

$$\frac{AB\rho}{C[(A + B)\rho - B]}$$

where ρ is the specific gravity of mercury, provided that we read at W and graduate the rod.

Finally in 1922 H. P. Waran, an Indian student in London, described to the Royal Meteorological Society, a special form of siphon barometer (Figure 6.22) in which the scale is read by reflection from the top mercury surface, at an angle of incidence of about 50°.[145] The tube, telescope, and scale are all secured to an iron frame. For portability there could be a piston in the bottom limb. In the discussion of the paper, Dr. C. G. Simpson "would not have thought that there were any more possible ways of making a barometer, but this was certainly a new principle with a lot to recommend it."[146] No one seems to have enquired about vibration, but apart from this, the instrument ought to be easy to read, and we have included it in this chapter

[145] *Quart. J. r. Meteorol. Soc.,* Vol. 48 (1922), pp. 287–91.
[146] *Ibid.,* p. 291.

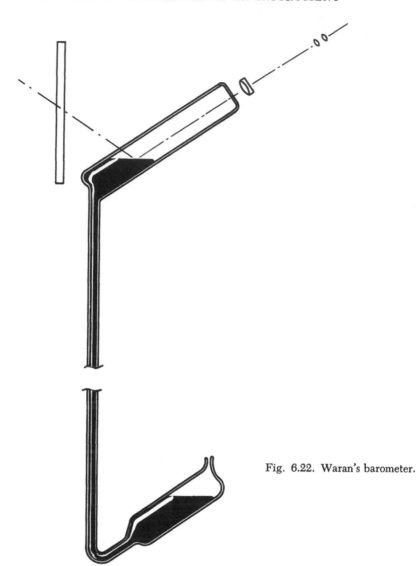

Fig. 6.22. Waran's barometer.

because whatever advantages it may have lie in this direction.

We may now leave these expanded-scale barometers. In the seventeenth and eighteenth centuries they arose in response to a public demand, which indeed persisted to some extent until about 1850. With the invention of the aneroid they were no longer needed,

but nevertheless their development forms part of the history of the barometer, and even a small part of the story of human ingenuity. We shall end this chapter with an account of the legends which have from time to time been engraved on barometer scales, often with much less justification than their authors imagined.

9. THE WORDS ON BAROMETER SCALES

I do not know who first put words on a barometer scale to indicate the weather that was to be expected. The oldest existing barometers seem to have them. In the first enthusiasm for barometric observations, it was naturally surmised that there might be a one-to-one correspondence between the barometer and the weather. On October 6, 1664, Hooke wrote enthusiastically to Boyle:

I have also . . . constantly observed the baroscopical index[147] (the contrivance, I suppose, you may remember, which shews the small variations of the air) and have found it most certainly to predict rainy and cloudy weather, when it falls very low; and dry and clear weather, when it riseth very high, which if it continue to do, as I have hitherto observed it, I hope it will help us one step towards the raising a theorical pillar, or pyramid, from the top of which, when raised and ascended, we may be able to see the mutations of the weather at some distance before they approach us, and thereby being able to predict, and forewarn, many dangers may be prevented, and the good of mankind very much promoted.[148]

But the English winter came on, and dashed Hooke's confidence, for on December 13, 1664, he wrote to Boyle again:

I have lately observed many circumstances in the height of the mercurial cylinder, which do very much cross my former observations; for at this very time the quicksilver is as high as I have a long time observed it, and I don't remember that it has been higher: it has risen a little for these four or five days, and has continued so, notwithstanding the variety of winds, and the multitude of rain, that had lately fallen; and, I think, it rises a little yet, but it is very little. I have taken notice also of two or three other very odd particulars lately in it, which have crossed several other observations.[149]

What would he have thought if he had known that the "odd particulars" would still be receiving the enthusiastic (or at least devoted) attention of meteorologists three hundred years later?

In spite of such little setbacks, the familiar words soon began to appear on the scales of barometers. In 1688 George Sinclair[150] graduated his "weather glass" in "Six half inches, and every half inch into five degrees." Perhaps this was from 28 to 31 inches; the "Degrees" will be tenths of an inch, but there is no feeling that the numbers are important. Then he marks the seven major graduations "Long Fair; Fair; Changeable; Rain; Much Rain; Stormy; Tempests."

This seems to be in the general tradition. It will save repetition if we give a table showing the weather signs on several English barometers of various dates (Table 6.1). Note that there seems to be general agreement about what happens at 29.5 inches, and also that there was little modification between 1753 and 1864. Daniel Quare, whose beautiful barometers are among

---

[147] From the word "index" we may surmise that he is referring to his newly-invented wheel barometer. See p. 94 above.

[148] In *The Works of the Honourable Robert Boyle. To Which Is Prefixed the Life of the Author. A New Edition.* Ed. Thomas Birch (London, 1772), VI, 492.

[149] *Ibid.*, p. 500.

[150] *The Principles of Astronomy and Navigation,* etc. (Edinburgh, 1688), p. 42.

Table 6.1.   *Weather Signs on Barometer Scales*

Inches	Quare, ca. 1700[151]	Graham, 1730[152]	Watkins & Smith, 1753[153]	Negretti & Zambra, 1864[154]
31.0	☿ rising, fair or frost, dry, serene	very dry, great frost	very dry	very dry
30.5		fair, cold	settled fair	settled fair
30.0		clear air	fair	fair
29.5	variable	changeable	changeable	changeable
29.0		rain, snow	rain	rain
28.7		much rain		
28.5	☿ faling [sic] rain, snow, or		much rain, much snow.	much rain
28.3	wind, rainy,	storm		
28.0	stormy	tempest	stormy	stormy

the earliest that have survived, seems to have come to the conclusion that the direction of movement of the mercury surface is of importance. In his barometers the upper and lower sets of remarks do not seem to refer to any particular pressure.

Continental makers seem to have come to similar conclusions. In the *Landes-Museum* at Kassel there is a wheel barometer in a magnificent ormolu case inscribed "Gaudron, Paris," marked on the dial "Beau Temps," "Changeant," and "Pluvieux." The dial has no graduations or figures. According to Daumas,[155] Gaudron flourished about 1675, and this may well be one of the oldest existing barometers. It has an enormous spirit

thermometer with a spherical bulb about 3 cm. in diameter and a scale in no recognizable units.

A typical inscription on nineteenth-century Italian barometers would be "Secco; T. stabile; T. bello; P.[iog]etta variabile; Gran pioggia; Tempesta."

About the middle of the nineteenth century there came a general realization that all these inscriptions were ridiculous and downright misleading. We may well agree with Negretti & Zambra[156] that "if tempests happened as seldom in our latitude as the barometer gets down to 28 inches, the maritime portion of the community at least would be happy indeed." Knowledgeable people had, of course, long been aware of this;[157] but Admiral Robert

---

[151] Science Museum, London, Inventory no. 1948–227.
[152] *Ibid.*, no. 1928–705.
[153] *Ibid.*, no. 1876–814.
[154] *A Treatise on Meteorological Instruments* (London, 1864), p. 10.
[155] Daumas, *Les instruments*, p. 84.

[156] *Treatise*, p. 11.
[157] Even in 1695 Amontons (*Remarques et expériences phisiques*, [Paris, 1695], p. 140–41) was rather blasé about it: ". . . l'on peut juger du peu de fondement de ceux qui marquent sur la graduation d'un barometre de certains endroits affectez au beau ou mauvais

Fitzroy, R.N. (1805–1865), decided to do something about it. Near the end of his life he was able to have barometers issued by the Board of Trade for coastal use engraved with the following:

RISE	FALL
FOR	FOR
N.Ely	S.Wly
NW.–N.–E.	SE.–S.–W.
DRY	WET
OR	OR
LESS	MORE
WIND	WIND
———	———
EXCEPT	EXCEPT
WET FROM	WET FROM
N.Ed. [North-eastward?]	N.Ed.

and barometers for ships as follows:

RISE	FALL
FOR	FOR
COLD	WARM
DRY	WET
OR	OR
LESS	MORE
WIND	WIND
———	———
EXCEPT	EXCEPT
WET FROM	WET FROM
COOLER SIDE	COOLER SIDE

Present-day ships' barometers, of course, have no verbal legends.

Many of the older barometers are mounted on wide boards that give plenty of space for short treatises on meteorology. An early example is an instrument by Giovanni Domenico Tamburini at Florence,[158] on which we are told in very un-literary Italian:

têms [sic], au têms serein ou orageux, &c. comme on a de coûtume de faire aux Barometres construits à la maniére d'Angleterre . . ."
It was necessary to observe the changes.

[158] Museo di storia della scienza, inventory no. 1142. Probably dates from about 1700.

From this instrument we may learn the state of the air. The more this mercury rises the more serene it will be, and the farther it falls, the heavier. Thus we may naturally find out whether it will rain or be fine, and learn where the air may be heavier or lighter.

This is brief compared to the "Historie des Barometers" printed on a very large diagonal barometer at Munich[159] which bears the date 1774.

It was soon noticed that, in western Europe at least, the fluctuations of the barometer are generally wider in winter than in summer. In an anonymous book[160] published at Nürnberg in 1768, this is taken into account very ingeniously by the graphical scale shown in Figure 6.23. Note also the different indications for "rising" (steigend) and "falling" (fallend).

There is another quite different and rather curious kind of legend found on numerous barometers, especially mountain barometers. An excellent example is at Florence,[161] rather difficult to date but probably not later than 1820. At various levels on the scale of the upper limb of this siphon barometer are the following notations:

Opposite 26″- Malous
a little lower Aldorf
At   25.4 Zuric
                Berne
                Covre

[159] Deutsches Museum, inventory no. 22585. There is a bottle barometer by A. Tessa with an even longer "Bericht über das Barometrum" [sic!] in the Landes-Museum at Zurich (LM7678).
[160] Kurze Beschreibung der Barometer und Thermometer auch andern zur Meteorologie gehörigen Instrumenten, nebst einer Anweisung wie dieselben zum Vergnügen der Liebhaber, und zum Vortheil des Publici gebraucht werden sollen (Nürnberg, 1768).
[161] Museo di storia della scienza, inventory no. 1146.

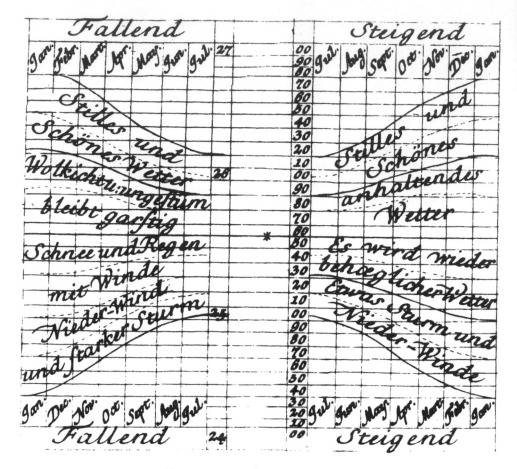

Fig. 6.23. A graphical barometer scale.

24.5   *Montagn du Piemont*
24.1   *Pirenees*
23.75 *Montjur*
22.8   *St. M.^{rie} aux Neige*
22.2   *Sommet de la rygie*
21.6   *Aux Alpes du haut Steg*
       Mont Synie
20.5   *Damons*

and, below the scale altogether,
  Canigou
  Zonne au Mexique
  M^t. Darcis
  Mont Choussa
  Calicou
  Ephis en afrique
  Sommet du Pichintra
  Mont Taurus aux grandes Indes
  Les Chambores auperou

A study of the history of the measurement of heights with the barometer gives some significance to many of the entries in this list. For example *Sommet du Pichintra* is Pichincha, one of the mountains in Peru that Pierre Bouguer climbed and used as an illustration of his rule for barometric hypsometry.[162] *Les Chambores* must be Chimborazo, thought at one time to be the highest peak in the world, and also climbed by Bouguer. *Canigou* refers to a mountain in the Pyrenees measured by Cassini in 1740. But many of the elevations

[162] Pierre Bouguer, *Optical Treatise on the Gradation of Light* (1760), trans. W. E. K. Middleton (Toronto, 1961), pp. 210-11.

indicated by their places on the scale are fanciful, even though this was a barometer intended for mountain use. I have seen several other such lists, nearly all of them including *Pichincha*.

The earliest printed reference I have been able to find to this peculiar sort of inscription is in a curious little book dated 1762,[163] which is, it appears, a translation of a French original published at Mannheim in 1758. It has a folding plate showing a long barometer scale with the inscription:

Barometrum Universale cujus ope non solum tempestatis ratio, set etiam elevatio Atmospherae super horizontem Maris sive altitudo cujuscunque loci diagnosci potest.

The scale has both London and Paris inches, and at about 16 of the latter we find *M. Pitchincha in Peru*, at 14 inches *M. Chimboroso in Peru*. *Quito* is at 20 inches, and there are a number of others.

This, however, is antedated by a reference in an anonymous set of manuscript notes on technical subjects in the Bibliothèque Nationale at Paris[164] to a "Barometre dans le manuscrit de M. Fourcroÿ datté 1749" in which a list of this sort is given, beginning with *Le sommet de chimboraco la plus haute montagne connuë et mesurée* (13 in., 2.9 lines; 3217 fathoms) and continuing with *la pipincha au perou, le sommet du Tenneriffe dans l'isle la plus-considerable des canaries*, and lesser mountains. I have not found Fourcroy's manuscript.

A little later, the famous instrument maker Georg Friedrich Brander (1713–

1783) described such a scale as part of a portable barometer of his invention.[165] A scale similar to that described by Brander may be found on a portable siphon barometer by Hiacint Vaccano of Munich, not otherwise remarkable.[166]

This kind of catalogue apparently became a convention of sorts, and was applied to the scales of very sedentary barometers. In a small hotel at Bruges I saw a very large barometer on a heavy board 25 cm. wide and with a glass cistern 10 cm. in diameter; the sort of thing that is scarcely portable at all. Nevertheless it is graduated in cm. from 5 to 28, from 28 to 65 in cm. and half-cm., and (very roughly) in mm. from 65 to 87 cm. Apart from the usual weather signs in French there are no less than 45 geographical indications from *Petrograd* at 78.5 cm. to *La plus haute cime de l'Himalaya* at 28.8. Evidently the mountains of Peru had lost their pre-eminence by the time this instrument was made, perhaps 1830. It is of some interest that *Hauteur Moyenne du Niveau des Mers* is at 75.8 cm., which is fairly close, though it leaves *Petrograd* and also *Londres* (77.7) and *Copenhague* (76.7) somewhat submerged. *Mont Ben Nevis* keeps company with *Les Pyrénées* at 65 cm.(!); while the most romantic entry is the one at 41.3: *Pic de la Frontière de la Chine et de la Russie*. All in all, quite a conversation piece; something for the guests to study on wet afternoons.

---

[163] Jakob Bianchy, *Das Merkwürdigste vom Barometre* [sic] *und Thermometre*, etc. (Wien, 1762). There is a copy in the library of the *Deutsches Museum*, Munich.

[164] BN, fonds fr. nouv. acq. 1223, fol. 165.

[165] G. F. Brander, *Kurze Beschreibung zweyer besonderer und neuer Barometer, welch sich nicht nur verschliessen und sicher von einem Orte zu dem andern bringen lassen*, etc. (Augsburg, 1772), pp. 6–8.

[166] Munich, *Deutsches Museum*, inventory no. 12.

Chapter 7

The Search for Portability

1. INTRODUCTION

About twenty-five years ago, in the course of my duties, I used to carry a portable barometer with me over a good deal of Canada for comparison with the barometers at the various weather stations that I had to inspect. The barometer I carried was of probably the best design that has ever been developed for such a purpose, and survived over 50,000 miles of travel by every means of transport, from small boat to large aircraft, without breaking and with almost no change in its zero. It is the purpose of this chapter to trace the development of portable barometers from the day in 1648 when Perier and his friends took a glass tube, a scale, and some mercury up the Puy-de-Dôme,[1] down to recent times. One cannot help sympathizing with Perier, or indeed with his countless successors who have carried mercury barometers up mountains in the interest of science.

We must recognize that soon after 1648 it was clearly seen that the barometer might have two uses, entirely independent: to follow the changes of atmospheric pressure with changing weather, and to provide a means of measuring the heights of mountains. For the former purpose it seemed desirable to expand the scale of the instrument in one of the many ways we have described in Chapter 6. For the latter, other devices were gradually introduced.

It must also be understood that a long time had to elapse before it came to seem worth-while to push the accuracy of barometrical observations to its attainable limits. For general meteorological purposes before the days of telegraphy and weather maps, absolute accuracy seemed to be of minor importance.[2] For the measurement of heights, it took more than a century to formulate the mathematical relations between height and pressure precisely enough to demand more than a mod-

[1] See p. 51 above.

[2] For example, the barometers used in the extensive network of weather stations that reported to the *Societas Meteorologica Palatina* in the 1780's were primitive, even by the standards of the time. See *Ephem. Soc. Meteorol. Palat.* (1781), p. 61.

erate accuracy in mountain barometry. Nevertheless, mountain barometers did become more accurate as they became more portable, and the dichotomy implied in the titles of this chapter and Chapter 9 has been adopted only in the interests of clarity and convenience.

The many special constructions devised in the interests of portability are hard to classify. Very roughly, and in approximately chronological order, they may be grouped as follows:

1. The siphon barometer. This is a rather special case which will keep coming up in the next few chapters; from our present standpoint, it was developed into a truly portable instrument by about 1770.

2. An attempt to fold the barometer up, in order to reduce its length. This had little success.

3. Special construction of the tube to minimize the danger of breakage.

4. Special shapes given to the cistern to minimize the risk of air getting into the tube during transport.

5. As a further development of (4), means of varying the volume of the cistern so that it may be full, or almost full, of mercury while being carried about.

6. Means of closing the bottom of the tube during transport; often combined with (5), leading sometimes to very complicated mechanical devices.

7. The barometer tube made of iron; this obviously introduces difficulties of various sorts.

8. The provision of a tripod stand which acts as a carrying case for the barometer when closed for transport.

9. Marine barometers.

10. Arrangements to carry the filled tube separately from the cistern.

11. A revival of the very earliest procedure, i.e., that of filling the tube every time an observation was to be made.

It will be understood, of course, that combinations of some of these means were of frequent occurrence. Meanwhile, a short digression.

2. THE "BOTTLE BAROMETER"

The idea of turning up the lower end of the barometer tube, both to make the instrument easier to carry and to conserve mercury, occurred, as we have seen, to Pascal and to Boyle. In an age when the most obvious defect of the barometer was the smallness of the motions of the mercury, it is not surprising that this construction, which divided the motions by two, should have found little favor. More than a century had to elapse before any significant improvements were made in it. Meanwhile someone, probably an ingenious *émailleur*, discovered that he could save nearly as much mercury, without seriously diminishing the movements in the vacuum space, by blowing a little glass bottle, generally pear-shaped, at the end of the tube before turning it up. By 1690 this practice had become general, as far as barometers sold to the general public were concerned, and persisted well into the nineteenth century.

These "bottle barometers" (*baromètres à bouteille*) had no other advantages than those of being cheap and easy to make. They were not easily or safely portable and the difficulty of defining the zero deprived them of all pretense to accuracy. Nevertheless, there were probably far more of them made than of all other mercury barometers put together, and a history of the barometer cannot pass them by in silence. This seems as good a place as any other to refer to them and let them go.

Outside of England, where the wheel barometer came a good second, the vast majority of the instruments found in private houses were "bottle barometers." The mercury barometer which formed part of the *Contrarolleur*[3] so popular in Holland was nearly always of this type. The tube generally has an inside diameter of 3 mm. or less and the bulb or bottle seldom more than 20 mm. The index is often very crude, the scale most frequently graduated on paper. The best one can say of them is that they were often beautifully decorated, and I suppose they must have been of some assistance in single-observer forecasting, the more so because people soon came to learn that it was more helpful to know whether the barometer was rising or falling than to pay attention to the verbal indications added to the scale.

What is more surprising than their wide distribution is that they were sometimes used in scientific enterprises of importance, the notable example being the great meteorological network organized about 1780 at Mannheim by the Elector Charles Theodore. This network, which extended from Rome to Labrador, was brought into being by Johann Jakob Hemmer (1733–1790), a most energetic cleric who was also the Elector's private confessor. One would have thought that a concern for absolute accuracy would have taken first place in such an enterprise; yet in spite of the availability of much better designs, bottle-barometers were sent out. These had, it is true, larger bulbs than most, 1½ inches in diameter, the tubes being ⅛ inch.[4] Hemmer knew that for every inch the level of the mercury in the tube changes that in the cistern will change by $\frac{1}{81}$ inch, but felt that the error was negligible. The scale, in brass, was in Paris inches and lines, and there was a vernier reading to 0.1 line. That some care was taken in the construction is revealed by the fact that the planks used as bases for these barometers had been seasoned for twenty years. He did not like the ordinary siphon barometer with two scales, because one of the surfaces which have to be read is exposed to the air; but he did admit that other barometers may be compared to such instruments, which can be considered as standard.[5] He was right about the siphon barometer, but it seems strange that he did not send out some of the relatively good cistern barometers that were available.

3. THE SIPHON BAROMETER

It is interesting that Boyle, in describing what we now call the siphon barometer, refers to it as the portable

[3] See p. 90 above.

[4] J. J. Hemmer, *Descriptio Instrumentorum Societatis Meteorologicae Palatinae*, etc. (Mannheim, 1782), pp. 3–8.

[5] *Ibid.*, p. 8. "Ad tamen barometra exegi alia, & idcirco *normalia* dici possunt." See also Chap. 9, p. 234.

barometer[6] and then makes it clear that in its simple form its portability was only relative, with the entire non-portability of the Torricellian tube in mind. It was not really easy to carry about, and, except in a derivative form, the wheel barometer, it was neglected for almost a hundred years. In 1749 Jean André de Luc began to think very hard about barometry,[7] and barometric hypsometry in particular, and decided, on what seemed adequate grounds, to try to perfect the siphon barometer as a portable instrument. Some of the steps he took, as well as his final design, are shown in Figure 7.1, which is a reproduction, unfortunately somewhat reduced, of the beautiful engraving which forms Plate II of the second volume of his book. We shall deal with this development at some length, inasmuch as it had a considerable influence which endured for several decades.

De Luc begins by suggesting that the existing variety of formulas for the measurement of height with the barometer, none of them of general application, owe their errors largely to a misplaced confidence in defective barometers. He insists that the greatest care is necessary in the design and construction of barometers for mountain observations. There is no question at all that De Luc experimented more carefully and at greater length in this direction than anyone else had done, and we may pardon him for feeling, when he had arrived at a design which

satisfied him, that almost the last word had been said on the subject.

He did not at once adopt the siphon barometer. Starting probably from the common *baromètre à bouteille*, he began by two attempts to replace the glass bulb with an elaborate cistern. The first of these, shown in part section in Figure 1 of the plate, was made of ivory and had a square glazed window bearing a short scale against which the level of the mercury could be read. Inside the ivory box was a valve of steel, faced with leather, and this was normally closed by a spring of serpentine form, but could be opened by turning a pin which wound up a chain when the barometer was to be put into use. Through the ivory top of the box, which was screwed on, was a filling hole closed by a steel plug, and there was a similar plug at the bottom for draining out for transport whatever mercury could not be accommodated in the tube. There was no thought of leaving any mercury in the cistern when the barometer was being carried about.

This worked well to begin with, but the spring soon broke. A replacement also broke, and another and another, until finally after using up six springs he found by experiment that the mercury weakened the steel at the points of strain.[8] Then he tried another construction, shown in Figure 2, in which the valve was closed by means of a spiral steel cam and opened by a spring which would be in an unstrained

_____

[6] *Works*, ed. cit. (1772), III, 219–23. See also p. 76 above.

[7] *Recherches sur les modifications de l'atmosphère* (2 vols., 4°; Geneva, 1772), I, 64. This will be referred to as *Modifications*. There was also a second edition (4 vols., 8°; Paris, 1784).

[8] I am informed by Dr. A. R. Bailey of the Imperial College, London, that this phenomenon is known as stress-cracking, and can also be produced by other liquid metals, such as molten solder.

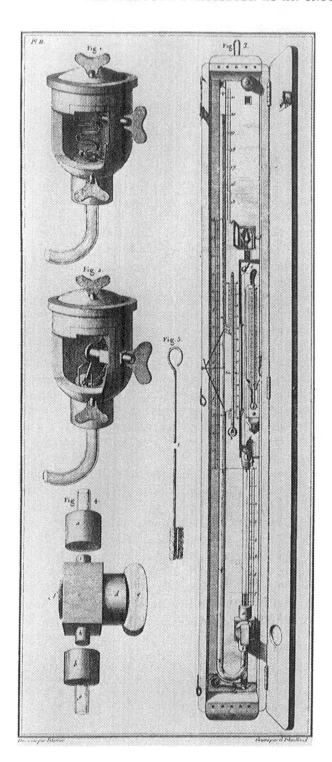

Fig. 7.1. De Luc's
portable barometer.

state when covered with mercury. This worked as far as the valve was concerned, but the steel parts rusted, and the mercury became so dirty that its level in the cistern could not be seen properly. He decided that it was necessary to abandon the use of metal in contact with the mercury.

Then follows an interesting passage: "I was proposing to put a coat of varnish on the pieces of my second barometer, to try to keep them from rusting, *when I found by my experiments that cisterns form an obstacle to the uniformity of the height of the mercury.* I then abandoned this construction in order to find one which might go with a simple turned-up tube. I succeeded; and this third barometer, with which I have made my principal experiments,[9] has lasted for twelve years without my having found any fault in it.[10] (Italics mine.)

I have italicized a phrase in this quotation in order to make the point that De Luc's strong attachment to the siphon barometer was the result, not of reasoning alone, but of careful *ad hoc* experimentation some time shortly after 1750.[11] These experiments will be dealt with more appropriately in Chapter 9, but it is abundantly clear that as far as the portable barometer is concerned, De Luc had thought the matter through with great care. His final barometer was about as easily portable as a mercury barometer can well be; its accuracy was limited in the event by the

grade of instrument making that seems to have been available to him, and in principle only by the circumstance, discovered much later, that the capillary depression on the surface of the mercury is anything but a simple function of the diameter of the tube and nothing else.

He proceeded (as before) to design a valve to isolate the mercury in the tube after it had been inclined so as to fill it, both to prevent the entry of air and to make it impossible for the column of mercury to break the tube by striking against its end. But this time the valve was a stopcock (Figure 4) made of ivory and cork in a very ingenious manner. This had served for twelve years when he wrote.

His barometer is shown assembled in Figure 3 of the plate, and while its general construction is obvious, it is full of fascinating details that show how lovingly he had thought it all out. For example, the bottom of the box holding the tube was grooved so that the tube could be sunk half-way into it and was covered with "un beau papier," on which the scales were drawn. The tube was held in place very simply by a number of twisted loops of copper wire wound with silk. Originally he had felt it desirable to have a short extension tube at the bottom, closed by a thimble of goldbeaters' skin, with a spring pressing against it, as shown in the figure. This was to take care of any expansion of the mercury due to a rise in temperature when the instrument was being carried; but later on he found it superfluous, as a little mercury would leak out between the cork and the ivory of the tap before the pressure became excessive. This

---

[9] I.e., in the mountains.
[10] *Modifications*, II, 5. This was not the end of the cistern on the short arm. Combined with a tap, it was revised by Guerin (*J. de Phys.*, Vol. 53 (1801), pp. 444–48) and by Origo (Pamphlet, 1809). See also Chap. 12.
[11] *Ibid.*, I, 205–8.

was at about the time that he found that the barometer could with advantage be carried upside down (as is nearly always done with modern portable barometers). Even the loop of cord at the top for hanging the instrument up had his personal attention; a piece of wound violoncello string was found to combine the required strength, flexibility, and resistance to wear.

There were two thermometers, the left-hand one permanently attached to the instrument, the other detachable. Apart from the fact that the detachable thermometer was used to measure the temperature of the ambient air, its removal left a space for maneuvering the tool shown in Figure 5, a sponge in a loop of wire, to clean the short limb of the barometer, as was often necessary. The detachable thermometer was graduated in Réaumur and in Fahrenheit degrees; but the other was graduated in a special scale of de Luc's, in which the interval between the freezing and the boiling points of water was 96°, with the zero at 12° above freezing, so that the freezing point was −12° and the boiling point +84°. This scale demands a little explanation.

To explain it, we must first mention the scales of the barometer itself, which were divided up and down from a zero point near the top of the shorter tube, in Paris inches and lines[12] and into quarters of a line (= $\frac{1}{48}$ inch). De Luc said he could estimate $\frac{1}{16}$ of a line without difficulty, and even "par l'habitude que j'ai dans ces observations, je puis saisir même jusqu'au trente-deuxiemes."[13] (He was certainly deceiving himself; this is less than 0.003 inch.) The scale was graduated up and down from a zero because it is easier to add than subtract. Now he had found (anticipating Chapter 8 a little) that with the barometer at 27 inches, the difference in temperature between melting ice and boiling water changed the height of the column by six lines. Thus if this difference in temperature were represented by 96°, each degree would correspond (at 27 inches) to a difference of $\frac{1}{16}$ line. This would make it possible to obtain the temperature correction without calculation. The zero was taken at about the average of the temperatures he encountered in his mountain observations.[14]

De Luc also recognized that the correction for temperature would vary in a linear way with the height of the barometric column; therefore, with extreme ingenuity, he made a trapezoidal scale on vellum, with the barometric height as horizontal co-ordinate, and mounted it like a roller blind on a spring roller, so that it came out of a narrow slot just beside the thermometer. This gave him his corrections at any pressure in sixteenths of a line.[15]

Another excellent example of his acuteness: "the diameter of the bulb of this thermometer should not exceed that of the barometer-tube by much, in order that the two instruments may be equally prompt in responding to changes of temperature. This bulb should be half sunk into the wood, so that like the barometer, it may share in the warmth of the back of the box."[16] As far as I am aware, these important

[12] Concerning units, see Chap. 8.
[13] *Modifications*, II, 21.

[14] *Ibid.*, I, 198–202.
[15] *Ibid.*, II, 31.
[16] *Ibid.*, I, 29.

conditions were not again observed for over half a century, even in the most precise barometers.

Finally, near the top of the instrument, and protected from the wind, there is a rather elaborate plumb bob.

There is one of these barometers in the *Museo di storia della scienza* in Florence, complete except for the mercury, and it may well be the original, as the catalogue speculates.[17]

Looking at this beautiful and ingenious instrument with a cold fishy twentieth-century eye, the observer feels that something is missing, and soon realizes what it is: no verniers, not even indexes. True, the finely-executed paper scales come right up to the walls of the tubes, and De Luc, no doubt reading the position of the edges, rather than the tops of the menisci, felt that that was all he needed.

The fact is that this remarkably successful portable barometer was the effort of a talented amateur who was apparently unwilling to get help from the regular instrument makers. He was gently and respectfully chided for this by one of them, F. W. Voigt, "Court Mechanic" at Jena, in an excellent book[18] which will be referred to again in Chapter 9. The paper scales, no doubt, failed to please him. Similar barometers were made with brass scales and verniers. In the *Instituto Geografico Militare* at Florence, where there is a well-kept collection of surveying instruments, there are two

rather similar instruments of this kind, the older-looking of which (no catalogue number) bears the legend "Barbanti in Torino Nº 2." The other (no. 6243) has no maker's name. They have metric scales, and probably date from about 1830. There is a very similar instrument with brass scales in inches, but no verniers, in the Science Museum, London (no. 1924–51). It was made by Nairne and Blunt, who seem to have been in partnership in London from 1774 to about 1800.

There were also more drastic modifications to the siphon barometer, not specifically to ensure portability, which will be referred to in later chapters.

There were some other more doubtful attempts to improve this instrument. An instrument maker called Loos at Darmstadt tried a three-way tap of iron, which for transport could connect the long tube with a small flexible capsule. But little air bubbles got in the narrow channel of this tap, and eventually found their way into the vacuum space of the barometer.[19]

De Luc's work with meteorological instruments had a good deal of influence, and indeed may have been partly responsible for the remarkable interest in the design of barometers that developed in the last quarter of the eighteenth century. In the introduction to his book Voigt[20] remarks "how

---

[17] Inventory no. 1150.

[18] *Versuch kritischer Nachträge und Supplemente zur Luzischen Beschreibung älterer und neuer Barometer, und anderer meteorologischen Werkzeuge* (Leipzig, 1802), pp. 9–10.

[19] *Gehlers physikalisches Wörterbuch* (2d ed.; Leipzig, 1825–45), I, 777.

[20] *Versuch*, etc., pp. 4–5. And Megnié, Lavoisier's favorite instrument maker, says that no one before de Luc had treated the barometer "avec autant de tact et de clarté; mais on voit . . . que la pratique des bons artistes n'a point accompagné ses recherches." (Quoted by Truchot, *Ann. Chim. et Phys.*, Vol. 18 [1879], pp. 305–6.)

greatly the number of barometers . . . has increased! Really it is astonishing." Most of these designs are "unnecessary and superfluous," resulting from the egotism of instrument makers or made by mechanical amateurs (*von mechanischer Liebhaberen*). But nearly all of them were portable cistern barometers, so that we shall concern ourselves later with those which are worthy of attention.

The siphon barometer, in fact, never seems to have been popular with travelers, except for 20 or 30 years after 1816, when the famous chemist Joseph Louis Gay-Lussac (1778–1850) set out to improve it, and indeed with some success.[21]

Gay-Lussac, like Voigt, starts out by remarking that a "prodigious" number of portable barometers have been proposed, especially in England. Of the few that have been generally adopted, that of Fortin[22] is one of the best. But like all cistern barometers it is a little heavy, and it suffers from the formation of a black powder on the mercury in the cistern, formed (he says) by the friction of the mercury on the leather bag, and making it difficult to set the zero. ". . . la poudre noir dont je viens de parler n'est point de l'oxide, mais bien du mercure *éteint* par un corps gras."[23] For these reasons, he says, he likes the simple siphon barometer, and also because it is exempt from a capillary correction. We shall see later that this last expectation was not justified. The glass part of his barometer is

shown in Figure 7.2. The two equal tubes *NF* and *CD* are joined by a heavy but narrow-bored tube *FBC*. The larger tubes are nearly, but not quite, coaxial; the center of gravity should be on the axis of *NF*. A very small hole is made at *E*, and the end *D* is sealed off after filling the instrument with mercury. The idea is to get just the right amount of mercury in, so that when the barometer is inverted for transport the portion of the mercury that remains in *CD* will not reach as far as *E*. In filling, the mercury is boiled in the tube before the tube is bent at *B*. There need be no mercury at all in the short arm when the barometer is inverted, as long as the narrow tube is only 1 or 2 mm. in inside diameter.

Such barometers were made in considerable numbers and were usually mounted in slender brass tubes, often less than 2 cm. in diameter, sometimes having two slots for reading the levels of the mercury, with an outer tube, also slotted, which can be turned to expose the glass tube to view. To lessen the cost (and also to reduce the size and weight) the scales can be engraved on the tubes themselves. Without a vernier one can then read to $\frac{1}{8}$ or $\frac{1}{10}$ mm. with the naked eye, says Gay-Lussac, "pourvu que l'on observe l'origine de la courbe du mercure" (i.e., the outer edge of the mercury surface).[24]

Once such a barometer has been filled and the end *D* closed, it is, of course, quite impossible to clean it. This was admitted by Gay-Lussac, who evidently thought it unimportant to be

---

[21] *Ann. Chim. et Phys.*, Vol. 1 (1816), pp. 113–19.

[22] See p. 210 below.

[23] *Ann. Chim. et Phys.*, Vol. 1, p. 114. (Italics by Gay-Lussac.) This idea seems rather surprising from him.

[24] *Ibid.*, p. 119. It seems strange nowadays to hear about eighths of a millimeter.

N

F

D

E

C

B

Fig. 7.2. The tube of Gay-Lussac's siphon barometer.

able to do so. He himself had made a voyage of more than 500 leagues with it; his friend Descostils more than 1,200, mostly in a postal coach. The mercury remained "aussi net que le premier jour."[25] There is no doubt that it was portable; but it would be time, and not motion, that would soil the mercury.

Barometers of the Gay-Lussac type were made by numerous excellent instrument makers throughout the nineteenth century. (A word of warning: in the catalogues of some museums the term "Gay-Lussac type" is used for almost any portable siphon barometer. It should really be applied only to such as have the tube of the form shown in Figure 7.2, or something similar, and with two fixed scales.) The most beautifully made example that I have seen is at Kew Observatory.[26] It is signed "Thomas Jones, 62 Charing Cross,"[27] and has scales in inches, tenths, and fiftieths, the verniers reading directly to $\frac{1}{500}$ inch. Other English makers of such instruments were Troughton & Sims[28] and later, Negretti & Zambra.[29] The latter instrument has a tube as an Figure 7.2, but the shorter limb is stopped by a glass rod wound with what looks like cotton thread, presumably a device for cleaning the open limb. The inch scales are en-

[25] *Ibid.*, p. 117.

[26] Inventory no. 154.

[27] Thomas Jones (1775–1852) became an apprentice of Jesse Ramsden in 1789, and later carried on his own business, first at 21 Oxenden St., London, and then at other London addresses. He became well known for his large astronomical instruments, and was elected a Fellow of the Royal Society in 1835. (DNB)

[28] London, Science Museum. Inventory no. 1907–65.

[29] Science Museum, Inventory no. 1924–50.

graved directly on the tubes, as on a burette. The date is uncertain, but probably after 1860.

Excellent barometers of the Gay-Lussac type were made by the Parisian instrument maker Bunten, who applied his air trap to them.[30] At the C.N.A.M.[31] there is an elegant instrument inscribed "Bunten quai pelletier 30 Paris 1841," which has the scales engraved on silver and verniers reading to 0.1 mm., the reading slides being moved by means of pinions working in racks cut directly in one side of each of the slots through which the mercury surfaces can be seen. There are slots at the back so that the tops of the menisci can be seen against the light. The greatest diameter of the tubular brass case is about 2 cm., the upper part being somewhat narrower. There is a similar instrument, dated 1840, in the History of Science Museum, Oxford. Another French instrument maker who seems to have made numerous barometers of this type is J. Salleron.[32] In Germany, Geissler of Bonn also made such instruments.[33]

While the best portable siphon barometers were on the lines of those developed by De Luc and by Gay-Lussac, other types appeared from time to time. The principle of graduating the two tubes in opposite directions was actually patented in 1765 by Henry Pyefinch and John Hyacinth de Magal-

haens.[34] Instead of the stopcock, which was hard to make, Gottfried Erich Rosenthal provided his ordinary siphon barometer with a slight constriction at the bottom of the shorter limb, into which a cork on the end of a thin piece of whalebone could be put for transport.[35] Half a century later this idea was taken up again and considerably improved by Johann Georg Greiner (1788–1860) a Berlin instrument maker of high reputation.[36] In order that the mercury might have somewhere to go if it should expand, Greiner replaced the piece of whalebone by a strong tube of capillary bore in which a number of widenings had been blown. This was later figured by E. E. Schmid[37] of Jena, who states that these barometers were used in the Prussian meteorological stations at the time. Meanwhile Moritz Meyerstein of Göttingen reported[38] the design of a portable siphon barometer in which a spring-loaded piston can be fitted in the open limb for transport.

Some at least of Admiral Fitzroy's barometers referred to above (p. 129) had a sort of iron check valve, probably to assist in transport, at the bottom

---

[30] See p. 144 below.

[31] Inventory no. 2628.

[32] In the *Instituto di fisica* of the University of Bologna there is a very neat instrument inscribed "T. Salleron 24 rue Pavée (du Marais) Paris." Near the bottom it is engraved with the number "180," which may or may not indicate a thriving trade in barometers.

[33] Florence, *Instituto Galileo Galilei*, Inventory no. A 110.

[34] British Patent 825, (1765). Magalhaens later used the name Magellan, and has been referred to above (p. 122).

[35] Rosenthal, *Beyträge zu der Verfertigung, der wissenschaftlichen Kenntnis, und dem Gebrauche meteorologischer Werkzeuge.* (2 vols.; Stuttgart, 1782 & 1784), I, 30–33.

[36] Greiner, "Ueber einen neuen Verschluss von Heberbarametern," *Verh. Ver. Gewerbefleiss Preussen* (pamphlet, 1835). Greiner's business was taken over by the firm of Fuess, which became and remains very well known in Europe.

[37] *Lehrbuch der Meteorologie* (Leipzig, 1860), p. 820. (*Allgemeine Encyklopädie der Physik*, XXI Band.)

[38] *Annalen der Physik*, Vol. 46 (1839), pp. 620–21.

of the short limb, which was very wide.[39]

4. "FOLDED" AND FOLDING BAROMETERS

An early attempt to make the barometer portable which need not detain us long was made in 1688 by Amontons.[40] Obviously a barometer would be easier to carry if it were shorter; so Amontons proposed splitting up the height of the mercury column into two or more lengths, with either air or a lighter liquid between (Figure 7.3). At the same time the scale could be expanded. A concise and final judgment has been made on this barometer (as well as on the conical barometer described in the preceding chapter) by Daumas:[41] "The barometers of Amontons serve as an example of these short-lived attempts which were so faithfully described in every eighteenth century physics text. The makers produced them regularly to meet the demands of collectors."[42] It seems almost incredible that H. Goold-Adams should have been granted a patent[43] in 1896 for such a barometer, with a partial vacuum between the mercury columns.

A rather more serious if not practical attempt to shorten the barometer for traveling is represented by an instru-

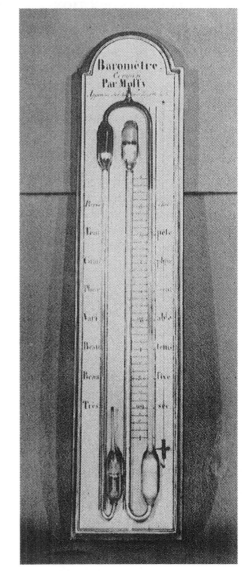

Fig. 7.3. Amontons' "folded" barometer. (C.N.A.M., Paris)

ment in the Leiden museum. This barometer (of which I have found no published description) really folds, being provided with a ground joint in the middle of the tube! It looks fragile.

---

[39] There is one at the National Maritime Museum, Greenwich, inventory no. B–11.

[40] *Hist. Acad. Sci.* (Paris, 1733), II, 39.

[41] *Les instruments scientifiques aux XVII*e *et XVIII*e *siècles*, p. 272, note.

[42] The barometer in the C.N.A.M. (Inventory no. 1579), from which our illustration is taken, was made by Mossy as late as 1768. There is a similar one inscribed "Torre fecit" at the History of Science Museum, Oxford.

[43] British Patent 25,517 (1896).

5. SPECIAL CONSTRUCTION OF THE BAROMETER TUBE

One of the earliest accidents that must have happened to the barometer is the breaking of the tube at its upper extremity by the impact of the mercury column when the instrument is being carried. The better the vacuum, the more likely this is to happen. Even when barometers were made so that they could be carried upside down, the tube could easily be broken during inversion; and it may be taken for granted that many seventeenth-century tubes were not too carefully annealed after they had been closed at one end.

The earliest attempt, and a very effective one, to get over this difficulty seems to have been made by Daniel Quare (d. 1724), whose beautiful portable barometers[44] had a constriction just below the top of the tube, forming a small roughly spherical chamber into which the mercury could not flow very quickly. It may be worth recording that a German patent on this device was granted as recently as 1942.[45]

In 1765 Edward Spry[46] added some glass beads to the little chamber. And in 1817 Kennedy[47] introduced a small bell-shaped piece of glass attached to a spiral spring which was anchored to the end of the tube. This was certainly not as good a solution.

A more subtle danger than the outright breaking of the tube was the introduction of a little air, which might pass unnoticed and falsify the readings.

As we shall soon see, a great deal of mechanical ingenuity was devoted to the prevention of this; but one of the simplest and most effective of these ideas is a small modification to the barometer tube itself.

This is the air trap invented by the Paris instrument maker Bunten, certainly before April, 1828, when a very high-powered committee, composed of Savart and Arago, reported enthusiastically on it to the Academy.[48] In its original form it was intended to improve the Gay-Lussac siphon barometer already described,[49] of which Bunten was a well-known maker. This barometer was praised by the committee; "Le peu de poids et de volume de l'ingenieux baromètre qu'il [Gay-Lussac] a imaginé, sa commodité, l'exactitude dont il est susceptible, ont été justement appréciés." But air can get into its column, and nearly always does if it is carried in a horizontal position. Bunten has remedied this defect without subtracting from the excellent properties of the barometer. "Il lui a suffi, pour cela, de former dans le grand tube une cloison vitreuse, *du centre* de laquelle descend perpendiculairement un tube capillaire d'une certain longueur," as in Figure 7.4, which we have drawn from Schmid.[50] Air would be most unlikely to go up the middle. It is pleasant to read that the Academy adopted the report, and this was equivalent to congratulating Bunten.

This air trap was one of those happy inventions with almost no drawbacks

---

[44] For further details see p. 150 below.

[45] D.R. Patent 745,892.

[46] *Phil. Trans.*, Vol. 55 (1765), pp. 83–85.

[47] *Quart. J. Sci.*, Vol. 1 (1817), p. 295.

[48] *Acad. des Sci., Paris, Procès Verbal*, (14 April, 1828), p. 52.

[49] P. 140 above.

[50] *Lehrbuch der Meteorologie*, p. 820.

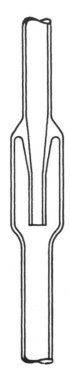

Fig. 7.4. Bunten's air trap.

save for the trifling cost of making it, and since 1828 it has been incorporated into thousands of barometers of various types. A recent Japanese design for a portable siphon barometer has no less than three such air traps,[51] an extravagance that would certainly have been embarrassing before the days of high-vacuum pumps, because of the difficulty of filling such a tube.

6. SPECIAL CONSTRUCTION OF THE CISTERN (WITH FIXED VOLUME)

If a cistern barometer is to be carried about at all, some form of closed cistern

is obviously essential. This necessity is in apparent conflict with the requirement that the air must have free access to the surface of the mercury, so that the apparatus may act as a barometer. Two solutions suggest themselves: to provide a closed cistern with a plug or valve which can be removed when the barometer is set up for observation, or to make part of the cistern of some material that is pervious to air but impervious to mercury.

It was discovered very early in the history of the barometer that certain hardwoods, especially along the grain, possess this property to perfection. Who first found this out I do not know; but by 1688 it was textbook knowledge, or at least it was the sort of information that would go into a charming little book on meteorological instruments written for fashionable people and decorated with the sort of copper engravings which it is now customary to call "baroque."[52] On page 35 of this delightful volume, after describing a barometer with a boxwood cistern, D'Alencé writes: "On a ensuite éprouvé par expérience, qu'il n'est pas besoin que ce baromètre ait aucuns trous ni aucunes vis, les seuls pores du bois étant suffisans, pour lui donner la communication avec l'air, qui doit agir sur la superficie du vif-argent contenu dans la boite."

Others found this useful property of wood incredible. William Derham, for instance, in a long letter[53] to the Royal Society dated July 17, 1712, wrote that he did not believe it wise to trust in the porosity of the wood. He preferred

[51] H. Watanabe, *Tokyo, J. Meteorol. Soc. Japan,* Vol. 32 (1954), pp. 11–15.

[52] D'Alencé (see note 77 in Chap. 6).
[53] Royal Society, *Classified papers* IV (1), item 58, 8 p. (autograph); pp. 1–2.

holes through the top of the cistern, closed at their inside ends by leather. He had no doubt that leather was pervious to air, a property which he had indeed made use of in 1698[54] in a portable barometer with a cistern of glass or "close-grained wood" having a "notch" (groove) around the outside. A leather cover was tied to the tube (which presumably passed through a hole in the leather), and then, after the tube had been filled and inverted into the cistern, the leather was lashed to the cistern. The tube and cistern were held in their proper relation by a case, which in turn was hung from a tripod.

To return to our boxwood; it may not now be entirely superfluous to mention that this close-grained yellow wood, from the trees *Buxus sempervirens* and *B. balearica*, was used in instruments, up to quite recent times, for rather the same purposes for which we now use some of the artificial substances so inaptly named "plastics" (though not in the same way). It is possible to cut durable threads in it, even with a pitch of one millimeter or less. It was used as a major material in the cisterns of many excellent barometers, either with or without recourse to its permeability to air. As an example of the latter we may cite a portable barometer presented to the *Académie royale des Sciences* in 1758 by J. B. L. de Hillerin de Boistissandeau (1704–1799).[55] I have tried to reconstruct from the brief description a probable cross section of the cistern (Figure 7.5), as it seems to me to be a characteristic use of mate-

rials. The cistern is made of two pieces of boxwood, screwed together with a leather gasket between them. There is a conical hole in the upper part, into which the barometer tube is cemented with mastic; and also a small air hole, which was really superfluous. The filled tube having been cemented in, with everything upside down, the air hole could be temporarily closed, enough mercury poured in, and the remaining pieces of the cistern screwed on over its gasket. As there is no means of adjustment, one can only assume that the scale was calibrated by comparison with some other barometer.

Assuming that there was a plug to stop the air hole, such a barometer would be reasonably portable if inverted with care and carried upside down. In the C.N.A.M. at Paris there is a mountain barometer[56] of this

[56] Inventory no. 1583.

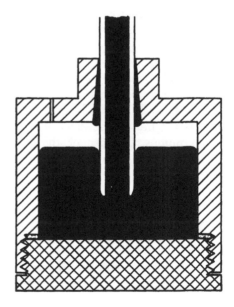

Fig. 7.5. Boistissandeau's barometer cistern.

---

[54] William Derham, *Phil. Trans.*, Vol. 20 (1698), pp. 2–4.

[55] *Hist. Acad. r. des Sci.* (1758), pp. 105–6.

general sort (except that there is no air hole) in which the vacuum still seems to be very good. It is attributed to Mossy; but the scale is inscribed "Bodeur à Paris à [??] Quai de l'horloge (Année 1819)." It is in a turned wooden case, the upper part having a wooden semi-cylinder about 3 mm. thick which can be rotated between ivory mounts to cover up the scale and thermometer. Modern fixed-cistern barometers are often essentially similar except for the materials and for the addition of the Bunten air trap.[57]

As I said, such barometers can (and should) be carried upside down. If they are not, then the mercury may not get out of the cistern, but air is likely to get into the tube. Numerous more-or-less successful attempts have been made to lessen the probability of this if, for example, the instrument is shipped in a long box, lying horizontally. As early as 1688 D'Alencé pointed out that if the cistern is spherical and more than half-full of mercury, and if the tube ends at its center, then the instrument can be placed in any position with no danger of the end of the tube being exposed.[58] The critical word is *placed*. What happens if the mercury splashes about is another matter.

By 1726, when Leupold wrote the remarkable book to which I referred[59] in Chapter 6, it is evident that a number of ingenious cisterns had been devised for portable barometers. Leupold describes and figures two closed, non-adjustable cisterns with spherical cavities, made from two pieces of box-

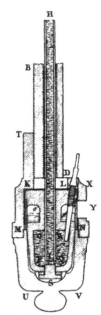

Fig. 7.6. Maigné's portable barometer.

wood which were glued, rather than screwed, together after filling.[60] A further feature of these cisterns will be noted later, but we should remark here that Leupold trusted in the porosity of the wood. He also describes a heavy glass cistern, spherical in form, with the tube fastened to a cork stopper having two narrow iron tubes through it.[61] There must have been some very ingenious trick for filling this, but it is not described.

Another approach is to make the barometer safe to carry right way up, or nearly so. In 1725 Johann Georg Leutmann suggested a boxwood cistern with two chambers, the lower the smaller by far, with the tube dipping well into it.[62] This would have had to

[57] See p. 144 above.
[58] D'Alencé, p. 35.
[59] *Theatri statici*, etc.

[60] *Ibid.*, p. 253, and Plate III, figs. XI and XII.
[61] *Ibid.*, p. 254, and Plate IV, fig. II.
[62] *Instrumenta meteorognosiae inservientia* (Wittenberg, 1725), pp. 19–24.

be carried somewhat inclined, or the mercury would have bounced up and down in the tube. A solution of *that* problem in terms of a double chamber had to wait until P. Maigné, an instrument maker in the rue Aumaire, Paris, invented a cistern made of ivory and divided into two parts by a *"double fond."*[63] (Figure 7.6) The two chambers communicate with each other and with the external air by a tapered hole through the double bottom and a threaded hole through the top of the cistern; both holes can be closed at the same time by a plug, which is removed when the barometer is hung up for observation. To make the instrument portable, it is inclined toward the holes, and the tube fills; the plug is screwed home, and that is all there is to it, as long as the joints are mercury-tight.

The same general principle was adopted by John Newman, who described his portable barometer in 1833.[64] But design (see Figure 7.7) and materials were far superior.

The cistern consists of two separate compartments; the top of the lower and the bottom of the upper, being perfectly flat, are pivoted closely together at the centres,

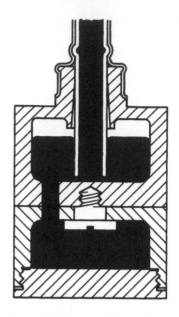

Fig. 7.7. Newman's iron cistern.

so that the lower can move through a small arc, when turned by the hand. This movement is limited by two stops. The top of the lower compartment and the bottom of the upper have each a circular hole, through which the mercury communicates. When the instrument is required for observation, the cistern is turned close up to the stop marked "open" or "not portable." When it is necessary to pack it for travelling, the mercurial column must be allowed to fill the tube by sloping the barometer gently; then invert it, and move the cistern to the stop marked "shut" or "portable." In this condition, the upper compartment is completely filled with mercury, and consequently that in the tube cannot move about, so as to admit air or endanger the tube. Nor can the mercury pass back to the lower compartment, as the holes are not now coincident, and the contact is made too perfect to allow the mercury to creep between the surfaces. The tube does not enter the lower compartment, which is completely full of mercury when

---

[63] Described by J. N. P. Hachette in *Ann. Chim. et Phys.*, Vol. 47 (1803), pp. 213–16. Cotte (*Traité de météorologie* [Paris, 1774], I, 516) ascribes this idea to Lavoisier, as is indeed quite likely.

[64] *British Association, Report for 1833*, pp. 417–18. John Frederick Newman set himself up in business at 7 & 8 Lisle St., Leicester Square, London, in 1817. He was the third Newman to be in the business in London, following James (*fl.* 1798–1827 in the Strand), who was apprenticed to T. Newman in 1793. John Newman later moved to 109 Regent St., then in 1826 to 122 Regent St. Some time after 1850 the firm became Newman and Son, which was absorbed by Negretti & Zambra in 1862.

the instrument is arranged for observation. The spare capacity of the upper cistern is sufficient to receive the mercury which descends from the tube to the limit of the engraved scale, which in these barometers generally extends only to about 20 inches. A lower limit could of course be given by increasing the size of the cisterns, which it is not advisable to do unless for a special purpose. This barometer may be had mounted in wood, or in brass frame. If in wood, it has a brass shield, which slides round the scale part of the frame, so as to be easily brought in front of the tube and scale as a protection in travelling; the vernier screw, in this case, being placed at the top of the instrument.[65]

Newman himself said that "the object of this construction is to make barometers portable without the use of a leather bag, which has always been a defective part of the instrument."[66] As these leather bags have been known to last over a century, this is probably only a statement of prejudice.

There are three of these Newman portable barometers at the Meteorological Office, Bracknell,[67] and one at Kew Observatory.[68] These barometers seem to have been remarkably successful in spite of the fact that there does not seem to be anywhere for the mercury to go if it expands. It is possible that the position of the eccentric hole is such that a small air bubble usually remains in the upper chamber. For mountain work they had the advantage that plenty of room could be provided for the mercury which comes out of the tube at low pressures; but they must have been rather heavy.

There is a Newman portable barometer at the Utrecht University Museum[69] which, according to the card catalogue prepared by the late Dr. P. H. van Cittert, has a small cup in the upper chamber, surrounding the end of the tube, as an additional safeguard against the entry of air. It also has the bulb of the thermometer in the cistern, an unfortunate idea of which Newman seemed very fond, his larger barometers having this feature.[70]

A barometer, the cistern of which falls into this class, was devised by J. J. F. W. von Parrot, a professor of physics at Dorpat who in 1829 made a journey with it across the Caucasus and finally up Mount Ararat, a mountain which plainly had a special place in his emotions.[71] As this instrument is of interest for other reasons, it will be dealt with in Chapter 9.

The fixed-cistern barometer is excellent for use at a fixed station, and as it has to be taken to the station, some effort must be expended in making it portable. Barometers of this class will be considered at greater length in Chapter 9, but we may note two twentieth-century patents which deal with this matter of portability. The first[72] calls for an inner glass vessel into

[65] Negretti & Zambra, *A Treatise on Meteorlogical Instruments*, (London, 1864), p. 38. By this time the firm of Negretti & Zambra had taken over the manufacture of these barometers.

[66] *British Association, Report for 1833*, p. 417.

[67] Inventory nos. 10 and 4110B have mahogany frames, the third, inventory no. 4108B, a brass frame. This will be referred to again in Chap. 9.

[68] Inventory no. 152. This has the brass frame.

[69] Inventory no. G 31.

[70] See p. 218 below.

[71] Parrot, *Reise zum Ararat* (Berlin, 1834), II, 1–8.

[72] German Patent 366,566 (1921), to Allgemeine Physicochemische A.G.

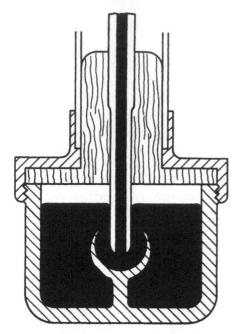

Fig. 7.8. Bilham's barometer cistern.

which the barometer tube is fastened, having a small hole at the bottom and near the top a plug of some stuff (leather, ceramic material, etc.) which is impervious to mercury but not to air. The level of the mercury comes nearly up to this plug. Such a device ought to be very efficient, and is entirely possible to manufacture. A simpler solution is due to E. G. Bilham.[73] This is a small capsule or thimble on a stalk rising from the bottom of the cistern (Figure 7.8). The tube ends in this capsule and leaves an annular aperture which is below the mercury level in all positions of the barometer. The provision of an internal flange halfway up the cistern, with small holes near its edge, is the most usual device to damp the violent motion of the mercury. I have not been

able to discover who first employed it. In spite of all such devices, barometers of this type are transported in an inverted position and with great care.

7. SPECIAL CONSTRUCTION OF THE CISTERN (VARIABLE VOLUME)

The first patent ever issued on the construction of barometers was dated August 2, 1695, and given to Daniel Quare (d. 1724). The Patent Office later assigned it the number 342. A manuscript copy of the original is in the Bodleian Library, Oxford,[74] beginning "William the third by the Grace of God . . ." and finally giving protection to "A portable weather glass or barometer, which may be removed and carried to any place though turned vpside downe without spilling one drop of the quicksilver or letting any air into the tube, and that nevertelesse the air shall have the same liberty to operate vpon it as on those comon ones now in use with respect to the atmosphere."

The news that the patent was being applied for disturbed the Clockmakers' Company, one of the City Guilds. On June 3, 1695:

Upon a motion and discourse concerning Daniell Quare and his endeavoring to obtayne a Patent for the sole makeing of portable Weather-Glasses, it was resolved and ordered that the Company doe forthwith endeavour to put a stop to the passing of that Patent.[75]

---

[73] British Patent 379,834 (1931).

[74] Rawlinson MS. A 241, fol. 90r–91r.

[75] S. E. Atkins and W. H. Overall, *Some Account of the Worshipful Company of the Clockmakers of the City of London* (London, 1881), p. 244.

But this "lobby" failed in its object, and on September 30:

"The Master and Wardens made a report of their proceedings against Mr. Quare's Patent for portable Weather-glasses, and that the patent was passed, and there may be suits of law or trouble to some Members that make or sell those Weatherglasses. It was unanimously voted and ordered that the Company will defend any Members of the Company or their servants, and also Mr. John Patrick (who assisted the Company) in any actions or suits that may be brought against them on that account.[76]

I do not know whether there were actually any lawsuits, but the mention of John Patrick is of interest. This instrument maker, who was not a clockmaker, called himself "The Torricellian Operator," and seems to have had a great reputation for barometers, perhaps more than he deserved. In the British Museum copy (537.a.34) of John Smith's *Compleat Discourse*[77] there is an entry in a shaky hand on the page facing the title: "the best Baroscopes are made by John Patrick of the Bull head inn [?] in Jermin [?] Street London." Smith was a clockmaker, and so, interestingly enough, was Quare. It would obviously pay John Patrick to oppose the granting of the patent.

The patent itself contains no specification or other indication of how Quare proposed to make barometers which would do what he promised. Fortunately many of these beautiful barometers survive.[78] While differing in mate-rial and slightly in detail they all have an unmistakable appearance, and are recognizable at a glance with their spiral-turned cases of wood or ivory, their scales engraved on silver or brass and enclosed in a box with a glass front, and their three little metal urnshaped knobs at the top, the right- and left-hand ones turning screws which operate simple indexes in grooves in the scale plate. Many of these instruments have decorative folding feet so that they could be placed on a table instead of being hung on a wall.

But it is not my intention to discuss the aesthetic qualities of these barometers. They are referred to in this section because of the way in which they were made portable, apparently with success. Besides the constriction near the top of the tube already referred to, Quare seems to have invented a cistern of variable volume, which could be made smaller so that, for transport, the tube and cistern would be completely full of quicksilver.

The cistern is a hollow cylinder of boxwood,[79] with the tube cemented into the top. The bottom of the cistern is a leather bag. This rests on the padded end of a screw which works in a nut set into the bottom of the outer wooden or ivory case of the cistern. When the barometer was to be made portable, the screw was advanced,

---

[76] *Ibid.*

[77] See note 79 of Chap. 6.

[78] A partial list: An ivory-cased one at Hampton Court Palace, very magnificent, with all the weather indications in French. A similar but slightly less gorgeous one at the Victoria and Albert Museum, London. One at the Meteorological Office, Bracknell, Bucks (inventory no. 1). One at the Science Museum, London (inventory no. 1948–227). An ivory-mounted one at the *Museo di Storia della Scienza,* Florence (inventory no. 1135), with the inscription "Faits Portatifs par Dan Quare A. Londres."

[79] Or sometimes at least partly of ivory. See *Horolog. J.,* Vol. 43 (1900), pp. 23–24.

raising the leather bag and diminishing the volume of the cistern until both it and the tube were entirely full of mercury. The instrument would then be inverted carefully, and carried upside down. When the instrument had been reinstalled in its new location, the screw was simply lowered as far as it would go. It was tacitly assumed that it would always leave the cistern with the same volume, but, in fact, the volume enclosed by the leather bag must have been somewhat dependent on the humidity. These were not mountain barometers, as is evident from the shortness of their scales. Quare's idea was to make first-rate "weather glasses" for domestic use which could be made in London and sent almost anywhere. The soundness of the idea and the excellence of his workmanship is abundantly proved by the condition of the surviving examples.

In spite of his patent, Quare's cistern was widely imitated and improved upon. In the Florence museum there is a most unusual barometer[80] signed "Dollond, London," and on stylistic grounds probably dating from about 1780. The tube of this is in a rectangular mahogany frame, hung by a Cardan suspension on a neat four-legged stand. A heavy lead weight encloses the cistern, which is of Quare's form. It is an interesting commentary on the durability of this construction that although the bottom of the cistern housing and the adjusting screw have been lost, leaving the whole weight of the mercury in the cistern pressing on the leather bag, which is exposed to view, the instrument seemed to have its

proper quantity of mercury when I examined it in March, 1961.

Mountain barometers were also made on this principle. There is an early nineteenth-century example in the Science Museum, London.[81] This has a scale in inches and a vernier reading to 0.002 in., giving an unjustifiable impression of accuracy; but the instrument is well conceived from the standpoint of portability, with its brass sheath that can be rotated about the axis of the cylindrical wooden case in order to protect the tube, scale, and thermometer.

It will be noted that barometers of this type were effectively fixed-cistern barometers when in use. Some time in the 1930's the late Dr. John Patterson, then Director of the Meteorological Service of Canada, impressed from personal experience by the difficulty of transporting the existing fixed-cistern barometers in an immense country, designed an instrument with a leather bag. Apart from the shape of the upper part, the essential difference between this cistern and Quare's was that the bag and its surroundings were carefully made so that when the screw was removed the weight of the mercury would press the leather into contact with the metal everywhere, ensuring a definite volume to the cistern.

The leather bag forms part of a large number of designs which, although many of them were also portable, it will be more appropriate to consider in Chapter 9.

The Tonnelot barometer used in the French Meteorological Service[82] makes

[80] Inventory no. 1160.

[81] Inventory no. 1893–134.

[82] Inquiry at the offices of the *Météorologie Nationale* failed to adduce a reference to an original published description.

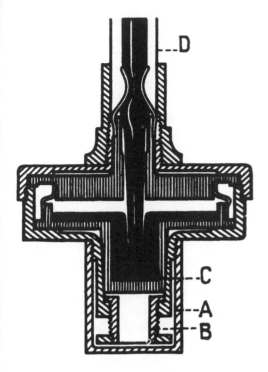

Fig. 7.9. Cistern of Tonnelot.

an iron tube having a screw thread at its upper end, and the barometer tube is cemented into this iron tube in such a position that when the cover is screwed on to the cistern over a leather washer, the mouth of the barometer tube is nearly at the bottom. This is the position for portability, for the dimensions are arranged so that the tube and cistern are then entirely full of mercury, any excess escaping through a very small hole in the iron cover. For use, the cover is unscrewed and the tube raised, being held in its elevated position by screwing the upper end of the iron tube into a threaded bracket provided for that purpose.

A development of the same idea was described eight years later at Zurich.[84] To use this barometer, the cistern is unfastened from its threaded boxwood support C by unscrewing the member F, and is then raised or lowered by the screw Q until the level of the mercury in the cistern coincides with an index b (Figure 7.10).

Another idea was to make the cistern cylindrical and smooth inside, like a pump cylinder, and to provide it with a piston.[85] This is probably much more difficult to make mercury-tight than might be imagined; Horner's solution was a cork piston covered with leather on top and side, working in a glass cylinder.

Another way of avoiding the use of the leather bag (to which end a great deal too much ingenuity seems to have

use of an ingenious principle of construction in order to have the cistern full of mercury during transport but still allow it to act as a fixed-cistern barometer when finally installed. In this instrument the lower portion of the cistern can be moved up by a screw as in the Patterson barometer, but it is of boxwood or similar material, and joined to the upper part by a flexible leather ring (Figure 7.9).

But there are other ways of making a cistern of variable volume. In 1758 J. G. Sulzer described[83] a portable barometer with a very deep cistern, like a test tube, with an iron ring at the top on to which an iron cover can be screwed. This iron cover is pierced by

[83] *Acta Helvetica,* Vol. 3 (1758), pp. 259–65.

[84] Christoph Jetzler, *Abh. der Naturf. Gesell. in Zurich,* Vol. 3 (1766), pp. 383–98.
[85] *Gehlers physikalisches Worterbuch,* 1 (1825), 784–87. The idea is ascribed to Horner, but the journal reference is incorrect. See also p. 214 below.

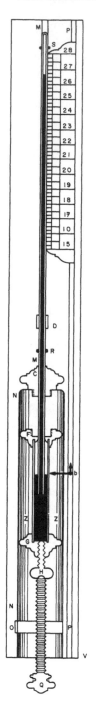

Fig. 7.10. Jetzler's portable barometer, 1766.

been devoted) was to provide the whole cistern with a screw thread near the top, which could run up and down on a corresponding thread cut on the edge of a plate in which the tube was mounted. Something like this seems to have been suggested by J. H. Hurter in 1786,[86] and again by Bunten in 1839.[87] The same idea forms part of a patent[88] issued in 1861 to Enrico Angelo Ludovico Negretti and Joseph Warren Zambra;[89] but the instrument described does not appear in the book *A Treatise on Meteorological Instruments*, published by Negretti & Zambra in 1864. In Hurter's barometer, however, the mercury was poured out of the cistern into a bottle for transport, after the end of the tube had been closed by a spring-loaded, leather-covered plunger. The screw thread was used for adjusting the level of the mercury, made visible by a float.[90]

A very ingenious means of making a "fixed" cistern barometer portable,

[86] *J. de Phys.*, Vol. 29 (1786), pp. 345–49. Hurter is described as a court painter in England and also the "Propriétaire d'une Manufacture d'instrumens de mathématiques, physiques, optiques & astronomiques à Londres." (*Ibid.*, p. 346.)

[87] *Compt. Rend.*, Vol. 9 (1839), pp. 501–2.

[88] British Patent 238 (1861).

[89] According to the records of the present firm, "Henry (or Enrico) Negretti started trading, probably about 1840, in conjunction with John Ronketti. Joseph Warren Zambra probably started trading about the same time; for some time he had as partner [Cesare?] Tagliabue." Negretti and Zambra became partners in 1850, going into business in Hatton Garden, London, and moving to Holborn Viaduct in 1870. E. Negretti died in 1879, J. W. Zambra, in 1897, the business continuing in the families. In 1862 the firm of Newman & Son was absorbed, and their premises at 122 Regent Street (now the head office) taken over. The Holborn Viaduct building was destroyed in 1940 by enemy action.

[90] See Chap. 9, p. 207.

which I have not seen described, is represented by a barometer bearing the number 4116F in the collection of the Meteorological Office, Bracknell. This is a "Baromètre Marin" bearing the crown and anchor device of the *Marine Impériale* (presumably that of Louis Napoleon) and signed "Ernst a Paris N°. 202."[91] It is a sturdy instrument in a mahogany housing, with a scale reading from 72 to 80 cm. and a thermometer in Réamur and centigrade degrees. Fig. 7.11 is a cross section of the parts of the cistern, which is made of hard wood, the interior pieces of box and the outer shell of mahogany. The volume of the cistern depends on the position of the piece A, as will be seen from the following, copied, as far as its condition permitted, from a paper label which surrounds the outer case of the cistern. A few words are conjectural and the spelling has been preserved.

Précautions à prendre pour le transport et la conservation des Baromètres Marins. Evitez tout choc, tout movement brusque qui pourraient cassez le tube en y introduire l'air.—Suivant que [le] Baromètre doit voyagez ou être mis en observation, la cuvette doit être fermée par un couvercle dit de voyage ou par un couvercle dit d'observation.—Ces deux couvercles sont les deux côtés d'une même

rondelle de buis.—Lorsque l'on voudra adaptez l'un ou l'autre, le moyen le plus sûr est de s'asseoir, posez doucement le sommet du baromètre par terre entre les pieds, maintenir l'instrument à peu près vertical avec les genoux, dévissez le cylindre d'acajou qui enveloppe la cuvette, dévissez le [couv]ercle de buis qui termine la cuvette, le renversez pour le vissez sur l'autre côté.—Ne jamais renversez le Baromètre brusquement, et surtout quand il est en observation.—Lorsque le vide est [imparfait au] sommet du tube, il faut, dans ce cas, l'incliner [douc]ement sans arrivez jusqu'à le mettre horizontal, de manière que [le mercure r]este toujours au sommet du tube; attendre, dans cette [pos]ition que le vide soit comblé. On peut alors renversez totalement le Bar[omètre] soit pour y adaptez la cuvette de

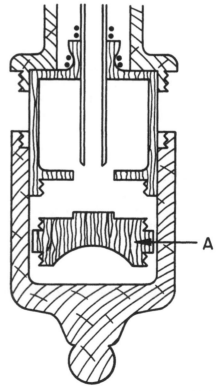

Fig. 7.11. Ernst's cistern, partly disassembled.

[91] The instrument maker Ernst had his shop at 11 rue de Lille, Paris, in the 1830's and 1840's. There is a reference to two barometers by Ernst being referred to a Commission of the *Académie des Sciences* in 1839 (*Compt. Rend.*, Vol. 9, p. 477); one was a modified Fortin, the other a "baromètre à siphon concentrique." In 1841 Bravais and Martin used barometers by Ernst in extensive international comparisons (*Acad. r. de Belgique, nouv. Mém.* 14:31-78). Delcros (*Annuaire mét. de France* [1849], pp. 140–42) praised his workmanship.

voyage, soit pour le [tr]ansporter à la main en un lieu peu éloigné; soit pour le tenir à [l'abri des] saluts ou exercises à feu.

Paris, Imp. Goyer. Port Dauphine 7.

I cannot help feeling that this is a very simple and strong construction which ought not to have been forgotten.

8. CLOSURE OF THE END OF THE TUBE

The patent referred to in note 108 also claims protection for the closure of the tube by a cushion or pad when the instrument is made portable. But the idea of closing the barometer tube goes back at least to 1726, when it was described by Leupold.[92] He used a leather bag, as shown in Figure 7.12, which was pushed up by the screw after the barometer had been inclined so as to fill the tube. The bag is formed into the shape shown and securely glued to the part *CD* of the cistern. It may not be inappropriate to mention here that a good deal of the prejudice which developed against the leather bag was due to attempts to secure it by an adhesive alone.

In 1786 J. H. Hurter described[93] a form of closure that was spring-loaded. The construction is not very clear, but evidently the lower portion of the cistern could be screwed up for transport after inclining the barometer, pressing a spring-loaded member covered with

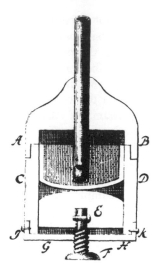

Fig. 7.12. Leupold's closure.

leather against the bottom of the tube. When this had been done, the rest of the mercury was poured out of the cistern into a boxwood container (*une boîte de buis*). This dodge made it unnecessary to worry about the tightness of the rest of the cistern. In 1793 Haas also made use of the spring-loaded closure,[94] but in a better design which did not involve emptying the reservoir. Later Körner[95] used it in a portable barometer of extreme complexity which, although it was his own, derives much from that of a departed fellow-townsman, F. W. Voigt.[96]

There were various other barometers in which the tube could be closed. Some of these demanded the mercury-

---

[92] Ref. 85 in Chap. 6; p. 255 and Plate IV, figs. VI and VII. Jacob Leupold (1674-1727) was a mechanic in Leipzig, and from 1701 to 1725 the Saxon Commissioner of Works and Mining.

[93] *J. de Phys.*, Vol. 29 (1786), pp. 345-49.

[94] *Gren's J. der Phys.*, Vol. 7 (1793), pp. 238-40. The situation is a little confused by the fact that Haas and Hurter were partners for some time after 1782 (Cavallo, *Phil. Trans.*, Vol. 73 [1783], p. 436).

[95] Friedrich Körner, *Anleitung zur Verfertigung übereinstimmender Thermometer und Barometer für Kunstler und Liebhaber dieser Instrumente* (Jena, 1824), pp. 152-62.

[96] See p. 225 below.

tight fitting of a screw in a nut, which is not a simple thing to do. Gödeking's barometer was of this type,[97] as was a portable barometer introduced by J. Newman[98] a few years before he thought of his divided iron cistern.[99] A diagram of this cistern of 1823 is given in Figure 7.13, which is self-explanatory. There are two of these barometers at the Meteorological Office, Bracknell.[100] In quite recent times a similar device has been patented in Germany by Fuess,[101] with the addition of an arrangement to prevent too much force being exerted on the end of the tube in screwing it up, although it is entirely positive in unscrewing it.

9.   IRON BAROMETERS

From the standpoint of the early scientific traveler, the main trouble with a barometer was the fragility of the tube. You could not load it on a pack mule and reasonably expect it to arrive intact. Why, then, not make barometer tubes of iron? True, they would be opaque, so that you could not see the mercury; but a little ingenuity would provide a substitute for vision.

The first account of experiments of this sort seems to be that of Guillaume Amontons,[102] who in 1705 took a musket barrel of medium caliber and had it welded shut at one end. After it was

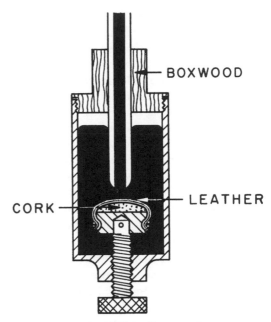

Fig. 7.13. Newman's earlier iron cistern.

set up as a barometer tube, the mercury remaining in it could be poured out and weighed. The results puzzled him because, I suspect, the tube was being slowly outgassed, even if the weld was really sound. He concluded that one must assume the atmosphere to be made of particles of air of various sizes, and that iron had larger pores than glass.

The first practical iron barometer was probably that devised by one Blondeau, a professor of mathematics at Brest. In the *Journal de Marine* which he founded, he published a series of eight *Memoires sur les Barometres Marins* during the years 1778–1780. In the second of these[103] he described his

---

[97] *Scherers allgem. J. der Chem.*, Vol. 2 (1799), pp. 93–96.

[98] *Quart. J. Sci.*, Vol. 16 (1823), pp. 277–79.

[99] See note 65 above.

[100] Inventory nos. 4102B1 and 4102B2.

[101] D.R.P. 718,192 (1940), and 740,761 (1942).

[102] *Mém. Acad. Roy. des Sci.* (1705), pp. 267–71.

[103] *J. de Marine* (1779), 4e Cahier, pp. 112–18. Blondeau does not appear either in Poggendorf's *Biographisch-litterarisches Handwörterbuch* or in the *Dictionnaire de Biographie française*.

siphon barometer, completely of iron except for an ivory float. This was very ingeniously designed so that the long limb may have the mercury boiled in it, and with a variable aperture (which may be closed entirely) at the bottom of the shorter arm. It was probably a fairly advanced piece of machining for the 1770's. The barometer was read by noting against a scale the position of the end of an iron wire projecting from the float and passing loosely through a bushing at the top of the shorter limb. The scale, of course, had to be divided in half inches and half lines.

In his third memoir[104] he described an improvement. To prevent his barometer being strained by changes in temperature, he added an extra tube to the bottom, horizontal part of the siphon, closing this with a piece of pig's bladder pressed upon, for transport, by a spring-loaded button.

He is not sure, he admits later, of the permanence of the vacuum in iron barometers. The interesting reason he gives is that

... iron is ... very odorous; it may therefore happen that the emanations of the iron tube carry into the vacuum of this tube an aëriform fluid which could even excite the emanations of the mercury, and alter the barometer little by little.[105]

The reference to "les emanations du mercure" is connected with a theory of his own that the presence of air is necessary for the evaporation of mercury: "... où il n'y a plus d'air, il n'y a plus d'évaporation."[106] He had no idea how much evaporated, nor, of course, of the

relatively enormous amounts of gas that may be adsorbed on any metal surface, to be released slowly into a vacuum.

There is an iron siphon barometer at the C.N.A.M. in Paris, which came into the museum in 1849, but is not otherwise dated.[107] The float, wire, and scale arrangement is just like the one described by Blondeau, but there is an iron stopcock at the bottom of the shorter limb, and an ivory fitting of some sort at the base of the longer limb. The compressed and inverted scale reads from 24 to 31 (presumably Paris) inches, so that it can scarcely be intended for marine use, and the form of its hinged wooden case suggests overland journeys; but it must have been very heavy for such a purpose.

Nearly a century later two iron siphon barometers were described by Thomas Stevenson.[108] One, ascribed to his friend Edward Lang, had an ingenious arrangement of three stopcocks and a plug, so that the vacuum could easily be renewed at any time. But there was no place for such things by 1875, or even by 1867, when one Faà de Bruno had an iron barometer made by Salleron in Paris and hopefully submitted it to the *Académie des Sciences*.[109]

Meanwhile the eighteenth century afforded one more example before it ended, a complicated barometer invented by Nicolas Jacques Conté, the

---

[104] *Ibid.* (1779), 5e Cahier, pp. 150–53.
[105] *Ibid.* (1780), 2e Cahier, p. 57.
[106] *Ibid.*, p. 56.

[107] Inventory no. 4249.
[108] *J. Scottish Meteorol. Soc.*, Vol. 4 (1875), pp. 265–66.
[109] *Compt. Rend.*, Vol. 65 (1867), p. 613. The construction cannot be deduced from the brief notice of this instrument.

director of the École Aërostatique (i.e., school of ballooning) at Meudon. It was used, according to the card on the original instrument in the C.N.A.M.,[110] on Napoleon's expedition to Egypt. Whatever else it might do, it would not break, for it was beautifully made from solid steel.[111] It had a tube *ABCOD* (Figure 7.14) and a cistern *EFGH*. This could be closed at the bottom by a plug *TRU* and contained a rather remarkable structure in two parts, *IL* and *NM*. *NM* could be turned by means of the projection *R*, which is like a screwdriver, so that the channels *O* and *Q* might communicate or not, as desired. *V* is a plug through which the cistern could communicate with the air.

When the instrument is being filled, *TU* forms a piston with which a vacuum may be produced, to get air out of the mercury. (All the parts must have been fitted to close tolerances.) The barometer being filled, *LMNP* is turned to close *OQ*, and—the brief account is none too clear at this point—the mercury was presumably emptied out of the cistern and the barometer, or the mercury weighed. It was then put together again, carried up the mountain, *OQ* opened and shut, and the process repeated. This seems like the sort of idea that would occur to a military man who was going to have barometric observations and damn the expense! He also needed a good deal of transport for all the steel and mercury. We shall meet Conté briefly again in Chapter 16.

[110] Inventory no. 8767.

[111] Conté, *Bull. des Sciences, par la Société philomathique*, 11 Floreal an., VI (31 April 1798), pp. 106–7.

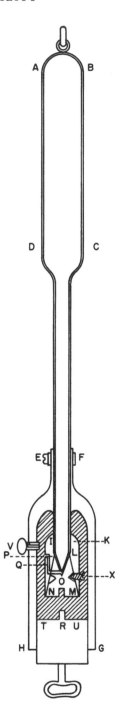

Fig. 7.14. Conté's steel barometer.

10. DESIGN OF THE ENTIRE BAROMETER FOR PORTABILITY

Apart from special tubes and cisterns, a great deal could be and was done by careful design of the entire barometer. Daniel Quare's barometers, for example, were more than adequate for occasional moves, though not intended for voyages. But after about 1750 instruments began to be designed for traveling, with some care in mounting and protecting the tube. In 1755 Brisson[112] presented to the French Academy a portable cistern barometer which he had tested on a voyage of more than 200 leagues, most probably by stagecoach. The tube was encased in wood except for a few inches near the top; this was probably somewhat novel at the time, but became standard for cistern barometers before the end of the eighteenth century. By the time Sir H. C. Englefield described his portable barometer in 1806,[113] a rotating brass (or sometimes wood) cylinder had been added which could be turned to expose the scale and the thermometer for observation or to protect it for carrying. Englefield's barometer was a simple fixed-cistern barometer with a tube only $\frac{1}{10}$ inch inside diameter and a cistern one inch in diameter and completely of boxwood (which, he notes, is porous enough to let the pressure of the air act on the mercury), in two parts, screwed together with a leather washer between. The mercury was boiled in the tube while filling. The case was a mahogany tube "of the size of a common walking-stick." Two opposite slits were cut in the upper part of this, one with its sides beveled, a thermometer on one bevel, the scale and vernier (reading to 0.002 inch) on the other. It was thus possible to see the top of the mercury column against the light. The wooden tube was 38 inches long, giving space for stowing a second detachable thermometer in a chamber at the top.

Englefield was determined to have these barometers properly made, and ". . . employed . . . Mr. Thomas Jones, of No. 120, Mount-street, Berkeley-square, pupil of the late Mr. Ramsden, and who will furnish them at the price of two guineas and a half without the attached thermometer, three guineas with it, and three guineas and a half with the attached and detached thermometer."[114]

This barometer, of which the portability, if not the accuracy, was undoubtedly enhanced by the slenderness of its tube, was a tremendous success. Its maker wrote in 1817 that since its introduction in 1806 he had sold between 300 and 400 of these instruments.[115] Yet I have not come across a single one in any of the museums I have visited, and indeed only two barometers of any kind by this maker. Perhaps we may assume that when their usefulness as barometers was at an end, they made excellent walking sticks.

Meanwhile De Luc had made his famous portable siphon barometer already described. With its tube dexterously bedded down in a paper-lined channel, and the absence of any large

[112] *Hist. Acad. r. des Sci.* (1755), p. 140.
[113] *J. Nat. Philos., Chem., and the Arts*, Vol. 14 (1806), pp. 1–12.

[114] *Ibid.*, p. 12. See also note 27, p. 141.
[115] Thomas Jones, *A Companion to the Mountain Barometer*, etc. (London, 1817), p. 2.

volumes of mercury, this was very successful for mountain work.

One of the difficulties of mountain barometry, and indeed out-of-doors barometry almost anywhere, is that the instrument has to be supported in a vertical position somehow or other. The mountaineer is therefore likely to be burdened with a tripod. Why not make the tripod do duty as a case for the barometer?

This innovation is probably, though not certainly, due to Jesse Ramsden.[116] "Mr. Ramsden's portable barometer" is described by William Roy, an engineer officer who became a Major General, in a long paper[117] dealing with hypsometry. The legs of the tripod are enlarged and hollowed out near the bottom to accommodate the cistern of the barometer. This construction was copied or reinvented with changes in detail by J. B. Haas[118] and Friedrich Körner.[119] Gehler[120] wrongly ascribes the tripod construction to Nicolas Fortin, who indeed made use of it (Figure 7.15). There is a traveling barometer in the *Deutsches Museum*, Munich, signed "Rumpf in Göttingen 1823,"

with such a tripod,[121] and another,[122] much earlier, marked "Haas & Hurter, London N° 38." This instrument has a sensitive plumb bob built into it, though it is difficult to see how verticality was maintained, as the center of gravity of the instrument appears to be above the point of support.

11. MARINE BAROMETERS

The marine mercury barometer offers a peculiar problem, not only of portability but of observability, because it has to be read in a ship, a platform that is in a complicated state of translation and rotation. I do not know who first took a barometer to sea, but he probably got a surprise. Nor do I know when it was; but either it was before the end of 1667, or the ineffable Robert Hooke was indulging in one more of his characteristic feats of abstraction at about that time. On January 2, 1667/8, he read a paper to the Royal Society and deposited a copy[123] with the Secretary.

At the top of this interesting document someone, I think Oldenburg, has written, "Some ways of Discovering yᵉ various Pressure [*sic*] of yᵉ Air at Sea, by Mʳ Hook, read before yᵉ Soc Jan 2. 1667/8." The rest is in Hooke's writing. After referring to the desirability of measuring the pressure at sea, and to the difficulty caused by the motion of the ship, he continues:

---

[116] Jesse Ramsden (1735–1800), one of the most famous of all instrument makers, was apprenticed to Burton in the Strand, London, in 1758, and set up for himself in 1762. He became internationally noted for his astronomical instruments, was elected to the Royal Society in 1786, and received its Copley medal in 1795. He was the first really large-scale entrepreneur in instrument making; at one time he had sixty workmen, but could not keep up with the demand for his instruments, especially from the Continent. His barometers were a side line, but were highly praised.

[117] *Phil. Trans.*, Vol. 67 (1777), pp. 653–788. The barometer is described on page 658, and illustrated on plate XV.

[118] *Gren's J. der Phys.*, Vol. 7.

[119] *Anleitung zur Verfertigung*, etc.

[120] *Physikalisches Handworterbuch*, I (1825), 789.

[121] Inventory no. 3949.

[122] Inventory no. 25. Probably between 1780 and 1800; but I have not found Haas in the London directories. Hurter appears briefly near 1800.

[123] Royal Society, *Classified Papers* XX, item 48. Autograph.

I contrived severall ways for the preventing that inconvenience in good part.
Now though I found very much of that
might be taken [*sc.* care] of by having an
exceeding small hole in the open end of
the tube to admitt the air, whereby the
air on the top of the mercury was as it
were imprisond, or w[ch] is much better by
having a small stop cock placed just in the
bending of the two pipes soe as after they

Fig. 7.15. Fortin barometer and tripod.

have been fild and inverted according to the usuall manner the key of the stopcock may be just soe farr turn'd as to leave the least imaginable passage for the mercury between the two stemms, yet the inconvenience of the vibration was[124] not wholy removed; though I believe by this last expedient a wheel Barometer may be very well used at Sea by a judicous man and therefore for such a person I think it preferrable before any other way what soever.

But realizing that not everyone is "judicous," he then describes his other "marine barometer" consisting of an air thermometer and a spirit thermometer, which has been reinvented so many times and will be dealt with in Chapter 15.

At any rate, Hooke invented the constricted tube for the marine barometer, and no one remembered. As far as I have been able to determine it seems to have been reinvented in an improved form by Edward Nairne[125] more than a century later.

In 1773 Constantine John Phipps, second Baron Mulgrave, published an account of a voyage to the north.[126] This was a scientific voyage, and Captain Phipps described his principal instruments.

The Marine Barometer was made by Mr. Nairne, from whom I received the following description:

"The bore of the upper part of the glass tube of this barometer, is about three-tenths of an inch in diameter, and four inches long. To this is joined a glass tube, with a bore about one-twentieth of an inch in diameter. The two glass tubes being joined together, form the tube of this barometer. . . .

"In a common barometer, the motion of the mercury up and down in the tube is so great that it is not possible to measure its perpendicular height; consequently, cannot shew any alteration in the weight of the atmosphere: but in this marine barometer, that defect is remedied. The instrument is fixed in gimmals, and kept in a perpendicular position by a weight fastened to the bottom of it."[127]

It seems to have been satisfactory. Magellan, a few years later busy as usual picking up chips from the instrument makers' floors, was delighted to learn that Nairne and Blunt had worked at the marine barometer with great success, and that several such, of their new construction, had been perfectly successful when tried at sea.[128] He himself had suggested[129] an ivory fitting in the tube, with a very small hole in it, which gets right back to Hooke, though he did not know this. The long narrow tube is a better, because more controllable, idea.

Blondeau's iron barometer[130] was, of course, intended for marine use. He was aware of the idea of constricting the glass tube; but it is likely to be broken by the firing of the ship's

[124] Was! Had Hooke taken a barometer to sea?

[125] Edward Nairne (1726–1806) early interested himself in scientific studies. He was apprenticed to Matthew Loft in 1741, and later established a shop at 20 Cornhill, London, as an "optical, mathematical, and philosophical instrument maker." From 1774 to about 1800 there was a firm, Nairne and Blunt. In 1776 Nairne was elected a Fellow of the Royal Society.

[126] A Voyage Towards the North Pole Undertaken by His Majesty's Command (London, 1773).

[127] Ibid., p. 123–24.

[128] Magellan, Descriptions et usages, etc., p. 145.

[129] Ibid., p. 143.

[130] See p. 157 above.

Fig. 7.16. Marine barometer of about 1830.

guns.[131] This was, and is, a serious matter in naval vessels, and the problem was attacked with some success about 1860 by Admiral Fitzroy,[132] who designed an improved marine barometer with the tube shock-mounted in rubber. In firing tests, the Fitzroy barometers survived the discharge of heavy guns close to the instruments, while barometers of the usual type were shattered.

But this is getting too far ahead. The marine barometer developed rapidly between about 1845 and 1860, especially in England. The later part of this development was encouraged by the Kew Committee of the British Association, at that time under the chairman-

ship of J. P. Gassiot,[133] which administered the affairs of the famous observatory from 1842 to 1871. About 1830 the marine barometer ordinarily used had its frame largely of wood, a boxwood cistern with a leather bottom, and the scale and vernier in a glass-fronted case. It seems[134] to have been developed into practically the present-day form of marine barometer by 1855, largely by the co-operation of John Welsh of Kew Observatory with the instrument maker Patrick Adie.[135] Our Figures 7.16 and 7.17 show the extent of the change,[136] and it is remarkable how little the instrument has altered in general appearance in the century since 1860. One change, adopted about 1850, which does not show in the figures, was the provision of the Bunten air trap[137] below the narrow part of

---

[131] *J. de Marine* (1778, 1er Cahier) pp. 13–14.

[132] See p. 128 above.

[133] See British Association, *Reports* (1853), p. xxix–xxxi; (1854), xxviii–xxxi. The committee is often referred to as the "Gassiot Committee."

[134] F. J. W. Whipple, *Quart. J. r. Meteorol. Soc.*, Vol. 63 (1937), pp. 127–35.

[135] This business appears to have started about 1850 at 395 Strand, London, moving later to 15 Pall Mall and in 1884 to an address in Westminster. It must be distinguished from that of Alexander Adie, Edinburgh. Patrick Adie was still making instruments in 1876, when he had seven barometers in a special loan collection of scientific apparatus at the South Kensington Museum. The firm continued in business until 1942. There is a marine barometer dating from 1855 by Adie in the Peabody Museum, Salem, Mass. (Inventory no. M.10485).

[136] They are both taken from Negretti & Zambra, *A Treatise on Meteorological Instruments* (London, 1864), p. 15 and p. 17. A barometer similar to the earlier one is in the Science Museum, London (inventory no. 1908–83). It is by Thomas Jones.

Marine barometers with elaborate wooden frames continued to be made until at least 1850, as may be seen in the Peabody Museum, Salem, Mass. (Inventory no. M.751, a barometer by James Bassnett, Liverpool).

[137] See p. 144 above.

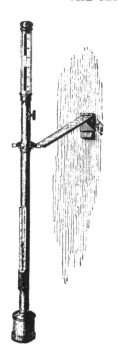

Fig. 7.17. The "Kew Pattern" marine barometer of 1860.

tively."[139] None of them seemed to be constricted quite enough, and after a further trip in May with five barometers, a "contraction" of 18 to 25 minutes was recommended. It was also noticed that when observed ashore, the marine barometer always lags behind the ordinary barometer.[140] Indeed it must. These "Kew pattern" marine barometers were such convenient instruments that many of them began to be used at land stations. According to Whipple[141] the disadvantage of the constricted tube for this purpose was "pointed out by Alexander Buchan,[142] acting as an inspector for the Meteorological Office." They were then made without the constriction for land use, as they still are.

It is rather remarkable that John Welsh did not seem to realize that the definition which he adopted for the "contraction" was a very unsatisfactory one. The speed at which the mercury surface falls under such conditions is proportional to the distance it still has to fall, so that (assuming a steady pressure) it finishes by falling infinitely slowly, and in fact never reaches its destination. Thus the end point of such a measurement involves a completely arbitrary judgment on the part of the observer.

A better idea, which was adopted in 1868,[143] was to measure the time of fall from, say, 1.5 inch to 0.5 inch above the actual reading. This should be 4 or 5 minutes. The rather simple theory

the tube. This was, of course wrongly, referred to as the Gay-Lussac air trap, or pipette.[138]

It was seen to be of importance that the constriction in the tube should be standardized, so that it would damp out the oscillations engendered by the motion of the ship, but would not cause the readings of the barometer to lag unduly behind the atmospheric pressure. As a measure of the constriction, the mercury was raised in the tube by tilting it, and the time required to recover the original reading was determined. In March, 1854, Welsh and Adie took three of Adie's barometers to sea, with "the tubes contracted in different degrees, viz. in the proportions of five, ten, and fifteen minutes respec-

[138] Negretti & Zambra, A Treatise, p. 17.

[139] British Association, Report (1854), p. xxviii.

[140] Ibid., p. xxx.

[141] Quart. J. r. Meteorol. Soc., Vol. 63 (1937), p. 132.

[142] The famous Scottish meteorologist.

[143] E. Gold, Great Britain, Meteorol. Office, Prof. Note. No. 48 (London, 1928).

seems first to have been published by the physicist Sir George Stokes in 1881.[144] He also noted other effects; when a barometer is swinging with the motion of the ship it is, on the average, inclined to the vertical, and so will read too high. This is opposed by the centrifugal force developed by the mercury column in rotating about its point of support.[145] The relative magnitude of these errors depends on the construction of the barometer, its manner of support, its location in the ship, and the ship's period of roll; this was shown by M. A. Giblett,[146] a brilliant young meteorologist who perished in the disaster to the dirigible R-101 in 1931. Giblett wrote out and solved the differential equations of motion for all these effects. Finally in 1925 S. N. Sen showed that the falling time could be calculated to within 2 or 3 per cent by applying the theory of viscous flow.[147]

But the days of the marine mercury barometer are numbered, because of the development of ever more reliable aneroid barometers.[148]

12. THE FILLED TUBE CARRIED SEPARATELY

There was one school of thought which reasoned that in mountain work you were going to break barometer tubes anyhow, so why not have plenty of them? Pierre Charles Le Monnier (1715–1799), member of the *Académie royale des Sciences*, gave an interesting description of mountain barometry in 1740.[149] He filled seven or eight tubes, using the new process of boiling the mercury in the tube,[150] and carried them upside down. They were of various lengths and diameters; at the time, it seemed unsafe to assume that the reading would be independent of the length of the tube, let alone its bore. On each tube a fine line had been scratched "with the corner of a gun-flint" about an inch from the open end, and another exactly 24 inches above the first, "& nous avions toujours une provision de petites cartes longues d'un, deux, ou trois pouces, exactement divisées en lignes & quarts de lignes."[151] At the various stations they set up a tube vertically in a cup of mercury, and

. . . we put into the same position one of our little cards which, floating on the surface of the mercury in the cup, showed us how high the first mark on the tube was, above that surface. This quantity was always added to 24 inches. In the same way we determined how much the top of the mercury column was above or below the upper mark, and this quantity was added to or subtracted from the sum of the first two, according to whether the mark was higher or lower than the extremity of the column. This method is very easy to use, and I do not believe that we can get a more accurate one.[152]

In other words the geodesists of 1740 had no confidence in the portable ba-

[144] *Great Britain, Meteorol. Office, Rep. Meteorol. Council for 1879–80* (London, 1881), p. 28–32.

[145] See W. G. Duffield and T. H. Littlewood, *Phil. Mag.*, 6th ser., Vol. 42 (1921), pp. 166–73.

[146] *Phil. Mag.*, Vol. 46 (1923), pp. 707–16.

[147] *Great Britain, Meteorol. Office, Geophys. Mem.*, no. 27 (London, 1925).

[148] See Chap. 16.

[149] *Suite des Mém. Acad. r. des Sci.*, année 1740, pp. clxxi–clxxvii.

[150] See p. 243 below.

[151] *Suite des Mém.*, 1740. p. clxxii.

[152] *Ibid.*, p. clxxiii.

rometers then available and felt that they were making an advance on the older technique of filling the tube *in situ.*

The next reference I have found to this technique of carrying the filled tube, the cistern, and the mercury separately deals with a traveling barometer invented by no less a user of such things than Baron F. H. A. von Humboldt (1769–1859),[153] described in the *Index biographique* of the *Académie* as "Naturaliste, geologue, polygraphe, voyageur scientifique."

The cistern of this barometer, which is stated to have served Humboldt well "even in the deserts of Thibet," was a deep tube of square section. The tube was closed by a two-part iron fitting consisting of a threaded bushing cemented to the tube and an iron bolt with a square head that was a loose fit in the cistern. The cistern having been partly filled with mercury, and the scale assembled to it, the tube was inverted into it and turned counterclockwise to loosen the bolt. The zero was established by letting mercury overflow from a small tap in the side of the cistern. When the observations had been made, the tube was inclined until full, turned clockwise to tighten the bolt, and removed from the cistern. There is little doubt that the parts of such a barometer could be carried with little risk in a properly constructed case.

In connection with this barometer there occurred an interesting example of the sort of scientific feudalism that was current on the continent of Europe

at the time. The instrument maker F. W. Voigt of Jena, in his excellent book on meteorological instruments,[154] claims rather bitterly that he wrote the article on Humboldt's barometer in Scherer's Journal and got no credit for it from the grand Professor. This was four years after Jesse Ramsden, F.R.S., received the Copley Medal![155]

Two American barometers of this sort will be mentioned in Chapter 12.

13. FILLING THE TUBE ON THE SITE

It may seem to the reader inappropriate to introduce a section with the above heading at this point in the discussion of portable barometers. The justification for doing so lies in the fact that in the last half of the nineteenth century there were some suggestions for filling barometer tubes easily and quickly with the possibility of attaining a fairly satisfactory vacuum.

With the primitive methods of filling, and especially with the narrow tubes usually employed, a good vacuum was almost never obtained. The usual procedure was to expel the larger bubbles by inverting and reinverting the tube

---

[153] *J. de Phys.*, Vol. 4 (1798), pp. 468–70. *Scherers allgem. J. der Chemie*, Vol. 2 (1799), pp. 96–99.

[154] *Versuch kritischer Nachträge und Supplemente zur Luzischen Beschreibung älterer und neuer Barometer, und anderer meteorologischen Werkzeuge* (Leipzig, 1802), p. 267.

[155] See note 116, p. 161. Daumas (*Les instruments scientifiques aux XVIIe et XVIIIe siècles*, [Paris, 1953], p. 138) lists the superior social status of many English instrument makers as one of the important causes of their superiority in the eighteenth century. "En France, le constructeur est resté pour les hommes de science, un simple ouvrier manuel ou un boutiquier. . . . En Angleterre, au contraire, les constructeurs les plus réputés sont admis très tôt à la Royal Society."

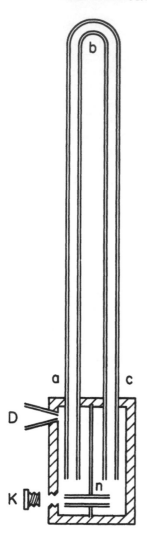

Fig. 7.18. Leupold's easy-to-fill barometer.

used for several decades, for barometric hypsometry.

As a matter of fact two schemes for filling a barometer tube easily were available to Bouguer as early as 1726, though they are unlikely to have come to his attention and are mentioned here merely as scientific curiosities. In that year Leupold,[157] reasoning that the barometer tube is hard to fill because it is closed at one end, devised the scheme shown in Figure 7.18. A hair-pin-shaped tube *abc* is sealed into a double cistern, the two chambers of which have a connecting tube *n*, while one has a screw plug *K*, and a removable funnel *D*. To fill it, the instrument is turned on its side with the funnel upward and the plug *K* removed. Mercury is poured in. Cistern *a* fills, then the tube, then cistern *c*. The plug is screwed home and the instrument stood up, the excess mercury draining out to a fixed level through the funnel *D*. It would probably work, but might not be very easily portable, even without the mercury.

If the vacuum does not need to subsist for long, perhaps the upper end of the tube might be closed, after filling, by a suitable valve. So thought the ingenious Leupold,[158] who made the "walking-stick barometer" shown in Figure 7.19. The upper end of the tube is drawn down, thickened, and ground off flat, leaving a tiny hole at *e*, which can be closed by the screw *b*, the end of which is covered "with a little bit of leather dipped in wax."[159] To fill the instrument it is laid horizontally and

and then to try to gather the smaller ones by moving a thin iron wire up and down in the tube. This technique was used by Bouguer in the famous expedition to the Andes, 1735–1744,[156] in which he established a rule, widely

[156] See the note by the Abbé de la Caille in Bouguer's posthumous *Traite d'Optique* (Paris, 1760), p. 323.

[157] *Theatri statici.*, etc., p. 266
[158] *Ibid.*, pp. 254–55.
[159] *Ibid.*, p. 254 ". . . mit einem Stücklein in Wachs geduncktes Leder."

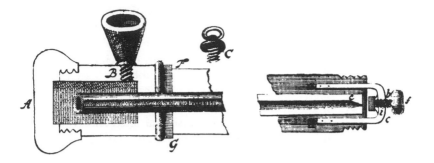

Fig. 7.19. Leupold's walking-stick barometer.

the funnel D screwed into the hole B. The hole at e is opened and mercury poured into the funnel until the tube is full; e is then closed, the cistern filled, and the instrument turned right way up. The funnel being removed, the surplus mercury runs out to establish a proper level, and the hole is closed with its screw plug. The barometer is of course carried upside down after a cap has been screwed on over the end at b. Leupold felt that the great advantage of this construction is that a narrower tube can be used than is otherwise possible. He did not realize that the accuracy of such an instrument would be severely limited by variations in the capillary depression.

As far as I know, this idea, which could of course be applied to larger tubes, has never been revived in this simple form.

A more practical idea of this sort had to wait until 1872, when Gustave Uzielli described a traveling barometer of entirely original design.[160] The tube, which may be quite large, say 8 mm.

inside diameter, is longer than usual, and has a constriction and a glass plug of special shape, which may include a thermometer (Figure 7.20). The tube is graduated, after the manner of a burette. Several of these tubes are packed in a traveling case with an iron cistern provided with leveling screws, a long rod which screws into the base of the cistern and has clips to support the tube, a clip-on index and vernier, and a clip-on steel point to which the level of the mercury may be adjusted by a screw plug in the cistern. And, of course, a flask of mercury and a funnel.

To bring the instrument into operation, the tube is filled as well as possible and inverted into the cistern, already partly filled with mercury. Then it is inclined until it is full (except for any air) to the very top. Next it is *slowly* brought upright, then inclined again until the mercury is at some distance above the constriction. Finally it is *quickly* brought upright again, when a little mercury will remain in the upper chamber and act as a seal, trapping any air.

There is a rather similar barometer in the *Museo Copernicano* on the

[160] *Nuovo Cimento*, ser. 2, Vol. 7 (1872), pp. 98–104. Also as a pamphlet, *Baromètre hypsométrique* (Florence, 1872).

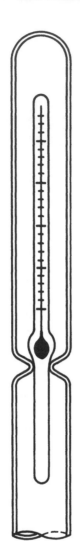

Fig. 7.20. The top portion of Uzielli's barometer tube.

Monte Mario at Rome. It differs from the description in that the scale is not graduated on the tube, but on a brass tube which loosely surrounds the barometer tube itself. This brass tube terminates below in a steel point, and the whole thing can be moved up and down by means of a screw.

In 1875 Staff Commander C. George, R.N., described to the Meteorological Society a very simple way of setting up a barometer, especially for travelers.[161] There is a tube, graduated on the glass, and a cistern made up of a short, wide glass tube and two rubber stoppers. An essential element is a length of twisted catgut with a "small strip of calico" at one end, and "the upper part of a crow's feather" at the other. When the barometer is empty, this is carried in the tube. To set the instrument up, the stopper at the bottom of the cistern is removed, and the tube first wiped out with the calico; then the twisted catgut is replaced, feather end inward, and as mercury is poured in (the cistern acting as a funnel, the feather brush is moved up and down to trap any enclosed air, which is guided to the top by the helically-twisted catgut as it is gradually pulled out.

Before the reader smiles too broadly he must be told the sequel, which is in two installments. The first took place in 1878, when Frederick Bogen described to the Meteorological Society an instrument which he called a "Standard cistern-siphon barometer."[162] This had an iron cistern, into the top of which a short open glass tube was cemented and a long closed tube fitted by means of a ground joint. There was a filling plug at the bottom of the cistern. All the readings were to be taken in the short limb. This barometer was simply filled, in Torricelli's manner. Bogen, who must have been an extraordinary character, recom-

[161] *Quart. J. Meteorol. Soc.*, Vol. 2 (1875), pp. 29–36.
[162] *Ibid.*, Vol. 5 (1878), pp. 137–42.

mended his instrument in the following terms:

Thirty-five years of practical experience on the West Coast of South America, in the capacity of mining engineer and director of scientific undertakings, enable me to form a tolerably correct idea of what is required in this line; and I hope the Fellows of this Society will admit the superiority of this new barometer over other forms . . .[163]

They did not. Whipple, Brooke, Symons, and Strachan, in the discussion, refused to accept a barometer in which the mercury had not been boiled. Ellis objected to the system of reading by the shorter limb only. Obviously in a bad temper by this time, Bogen denied the advisability of boiling the mercury in the tube. It is evident that he did not realize that the absence of

a bubble when the tube is inclined does not exclude the possibility that condensable vapors may be present.

The rest of the story transpired in 1881. G. M. Whipple (1832–1893) of Kew Observatory, possibly annoyed by the pretensions of Bogen, thought of Commander George's simple barometer, which by this time had been tidied up a little by replacing the rubber bungs with threaded caps, etc. He tried comparing both George's and Bogen's barometers with the Kew Standard[164] and found George's much better than Bogen's. When Bogen's barometer was filled by George's method, the results were better still!

George's barometer is preserved at Kew.

---

[163] *Ibid.*, p. 140.

[164] *Quart. J. Meteorol. Soc.*, Vol. 7 (1881), pp. 185–89.

Chapter 8

The Corrections to the Mercury Barometer

1. INTRODUCTION

A MERCURY barometer may be described as a device by which the vertical distance between two surfaces of mercury, one in a vacuum and the other in the air, can be measured. In order that the pressure of the atmosphere in some recognized units may be derived from such a measurement, certain corrections must be applied to it. Some of these, the correction for scale errors, zero error, and capillarity, appertain to the particular barometer. Then there is a correction for temperature, which is a function of the temperature of the various parts of the barometer and also of its construction. Finally there is the so-called gravity correction, a function of the position of the barometer in latitude and elevation above or below sea level, and not depending at all on the particular barometer under consideration.

We must first discuss the units in which barometric height, and later atmospheric pressure, has been measured.

2. UNITS OF MEASUREMENT

It may be remembered that in Torricelli's description of his experiment he stated that the mercury in the tube remained "at a height of an ell and a quarter and a finger more."[1] The *braccio*, the seventeenth-century Florentine unit of length, was about 54 cm. As we have seen, the further development of the barometer took place mainly north of the Alps, and barometers came to be graduated in other units. As far as I can remember, the only barometer I have seen with a scale in *braccia* is at Florence[2] and is almost certainly nineteenth-century. It has, however, three other scales, English inches, French inches, and centimeters, and may have been made for illustrative purposes.

Over most of Europe from 1650 to 1800 and later, the unit of length was the "foot," divided into 12 "inches,"

[1] P. 23 above: "all 'altezza d'un braccio e l. q[uarto] e vn dito di più."
[2] *Museo di storia della scienza*, inventory no. 1132.

172

each inch divided into twelve "lines." This duodecimal system was almost universal, but in England inches were from the very first often divided into tenths, lines being rare by 1800. French inches and tenths are uncommon, but there is a barometer divided in this way at Munich, signed "Breitinger à Zuric," and probably dating from about 1800. English inches are occasionally divided into sixteenths. When verniers are used with scales in lines, they are also usually duodecimal, less often in 10ths of a line.

Unfortunately the foot was different in every country. The following Table 8.1, derived from a long article[3] which dates from 1831, will give an idea of this diversity. One suspects that the number of significant figures is excessive, except perhaps for the French and English units. These various kinds of feet seem to have persisted for a good deal of the nineteenth century; or at least it seemed worth-while to insert a table of "Mesures étrangers" in the 1849 volume of the *Annuaire météorologique de la France*, which includes all the above except the last two, and about eight more, not to mention four kinds of feet said to exist in China.

The adoption of the metric system in France was one of the better results of the French Revolution. The new units became compulsory in France in 1801; but such is the force of custom that another law had to be passed in 1837 providing severe penalties for the use of any other weights and measures after January 1, 1840. There seems to have been a tendency to refer to centi-

meters as "French inches." In his private card catalogue of the Utrecht University Museum, the late Dr. P. H. Van Cittert wrote that in a previous catalogue, prepared in 1839, the scale of a certain barometer was said to be in French inches (Franse duimen); but "They are true centimeters, which makes it evident that these were called "French inches" in the early days of the metric system."[4]

Table 8.1

Values of one meter in terms of various old measures

1 meter	= 3.07844 French feet
"	= 3.28083 English feet
"	= 3.16353 Viennese feet
"	= 3.18680 Prussian feet
"	= 3.36813 Swedish feet
"	= 3.18715 Danish feet
"	= 3.49052 Würtemburg feet
"	= 3.4263 Bavarian feet
"	= 4.000 Hessian feet
"	= 3.333 feet of Baden

It is interesting that as early as 1802 the metric system had begun to catch the imagination of people in other parts of Europe. In an excellent book on the barometer, F. W. Voigt of Jena presented a very well-written and obviously heartfelt appeal for its adoption, concluding:

I have been speaking about a matter which ought to be close to the heart of every real scientific man. It follows from all I have said that the new French system of measurement is to be strongly

---

[3] Gehler, *Physikalisches Wörterbuch* (2e Aufl.; Leipzig 1825–45), s.v. "Mass," VI (1831), 1218–1391.

[4] I am greatly obliged to Dr. J. G. Van Cittert-Eymers, who is enthusiastically carrying on her husband's work, for making available to me photocopies of the cards dealing with barometers.

recommended for the barometer. It is desirable that it should soon be adopted.[5]

But in other parts of Europe barometers were to be graduated in Paris inches and lines for a long time after 1800, either with or without a corresponding graduation in centimeters and millimeters. Even in 1857 Bauernfeind had barometers made to his own design by Peter Rath in Munich which were graduated only in Paris lines.[6] In the *Deutsches Museum* there are two barometers,[7] signed by J. G. Greiner, Jr., Berlin, which are graduated in inches and lines, and which probably date from after 1850. It is interesting that the Paris, rather than the English, inch seems to have been the most general unit almost everywhere outside of England and the United States for the graduation of barometers up to the time when the metric system became common. But one finds barometers with scales in both kinds of inches, especially by Italian makers. In the *Istituto di Fisica* at the University of Bologna, for example, there is a very handsome cistern barometer in a wooden frame, signed "Guiseppe Brusca, 1806." This has two separate scales and verniers, one in English inches with the word "London" at the top, the other in French inches and bearing the superscription "Paris." Even English makers sometimes did this, as for example in an instrument in the *Museo Copernicano* at Rome, signed

"Troughton London" and probably dating from about 1800. This has scales in English inches and tenths, French inches and lines.

As the nineteenth century approached its end, only two units survived for the graduation of barometers: the English (then called Imperial) inch,[8] and the millimeter. These held sway in international metrology until the 1914–1918 war.

In 1889 James Asher, who had already suggested a millesimal scale of temperature, made the proposal that the unit of pressure should be obtained by dividing 760 mm. of mercury into 1000 parts.[9] Nothing came of this, but shortly before 1914 E. Gold of the Meteorological Office, London, became convinced that a fundamental change was desirable.[10] We do not really want to measure the height of a column of mercury, he reasoned; we want to measure the pressure of the atmosphere; then why not express the results of barometric measurements in the c.g.s. units of pressure? It must be admitted that the graduation of an aneroid barometer in inches or millimeters of mercury does seem a little absurd.

The c.g.s. unit of pressure is the dyne per square centimeter, roughly one-millionth of the pressure of the atmosphere at sea level. In 1888 a committee of the British Association (Gold pointed

---

[5] *Versuch Kritischer Nachträge und supplemente zur Luzischen Beschreibung älterer und neuer Barometer,* etc. (Leipzig, 1802), p. 98.

[6] Carl Maximilian Bauernfeind, *Elemente der Vermessungskunde,* (2d. ed.; München, 1862), pp. 351-52. A barometer (no. 2702) in the *Deutsches Museum* may be one of these.

[7] Inventory nos. 8935 and 54905.

[8] The United States inch was very slightly different from the Imperial inch. The exigencies of engineering have now brought them into agreement, and into a simple relation with the millimeter: 25.4 mm. = 1 inch.

[9] *Scientific American,* 2d ser., Vol. 60 (1889), p. 244.

[10] E. Gold, *Quart. J. r. Meteorol. Soc.,* Vol. 40 (1914), pp. 185–201.

out) proposed the name *barad* for the dyne cm.$^{-2}$. In 1900 a congress of physicists at Paris called it the *barye*. About 1906 the famous Norwegian theoretical meteorologist V. Bjerknes adopted the name *bar* for 10^6 dynes cm.$^{-2}$. Gold thought that one-thousandth of this would be a convenient unit, and called it the *millibar*. The word has stuck, and the millibar is now one of the three units of pressure officially recognized,[11] the others being the "millimeter of mercury" and the "inch of mercury," the qualifying phase "under standard conditions" being assumed. It will become clear what this means as we proceed.[12]

When the graduation of barometers to read directly in millibars was under discussion, Sir Napier Shaw suggested the term *baromil* for the actual length of a column of mercury under standard conditions which would produce a pressure of one millibar on its base. But there does not seem to have been any need for a word to express this idea, and the term did not survive.

3. THE DENSITY OF MERCURY

At this point a short note on the density of mercury is in order. Torricelli gave it as 14, and this was confirmed by the *Accademia del Cimento* in an experiment in which they added water above the mercury in the vase of the Torricellian experiment.[13] At about the same date (1660) Boyle[14] obtained 13.75 by a rather similar method, and 13.68 by weighing. The method used by the Florentine Academy ought to cancel out errors due to capillarity, and their three-per-cent excess over the modern value probably means that they were giving only a round number. Mariotte's experiments[15] with a water barometer yielded the value 13½, twenty years later.

All through the eighteenth century and much of the nineteenth there was a lack of confidence in the constancy of the density of mercury. In 1783 N. von Beguelin[16] discussed its supposed great variability and suggested it as a good reason for the variation in the readings of different barometers. In 1800 one Citizen Pugh quoted Lord Charles Cavendish to the effect that "la gravité spécifique du mercure pur, après une ou plusieurs distillations, varie de 13.500 à 13.600."[17] But the mercury used by the "barometer-merchants" is not so pure, and its specific gravity varies from 13.00 to 13.60! If this seems excessive, let us refer to Abraham Rees' *Cyclopaedia* of 1819–20.[18] The writer of the article on barometers in this large work is prepared to believe that the specific gravity of mercury can vary between 14.11

[11] World Meteorological Organization, *Guide to Meteorological Instrument and Observing Practices* (2d ed. [WMO No. 8. TP 3] Geneva, 1961), p. III, 1.

[12] Meanwhile for the curious non-technical reader: 1 mm. mercury (under standard conditions) = 1.333224 mb. (millibar). 1 inch mercury = 33.8639 mb.

[13] *Saggi* (1841 ed.), p. 39.

[14] *Works* (2d. ed.; 6 vols.; London, 1772), I, 88.

[15] *Hist. Acad. r. des Sci.* (1683), I, (Paris, 1733), 361.

[16] *Nouv. Mém. Acad. Berlin* (1783), pp. 148–58.

[17] *Observations sur la pésanteur de l'Atmosphère*, etc. (Rouen), An VIII., p. 10.

[18] *The Cyclopaedia* (39 vols.; London, 1819; with 6 vols. of plates, 1820), *s.v.* "barometer."

[!] and 13.00. "The heaviest is commonly reputed to be the purest mercury." (It would probably be a gold amalgam.) As a sufficient comment on this sort of pyrrhonism we may note that in 1847 Regnault[19] obtained the value 13.596 at 0°C., which, as far as it went, was left unchanged in the normal value 13.595930 adopted in 1881 by the *Bureau International des Poids et Mesures*.[20] It has since been recognized that small variations in the density of mercury can result from variations in its isotopic composition, but this presents a problem only for the most precise barometry, and even then is only one of several uncertainties of similar magnitude. The present official value for the density of mercury at 0°C. is 13.5951 grams per cubic centimeter; "and, for the purposes of calculating absolute pressures by means of the hydrostatic equation, the mercury in the column of a mercury barometer is regarded conventionally as an incompressible fluid."[21] This last consideration actually affects only barometry of unusual precision.

4. THE "INDEX CORRECTION"

It is only an exceptional barometer that will yield absolute values of atmospheric pressure, or barometric height, independent of any other instrument. Such barometers will be dealt with in Chapter 10. All barometers which do not fall in this privileged class must have their inherent errors investigated by comparison with such "standard"

instruments, either directly, or through the intermediation of another instrument that has been so compared. The so-called "index correction," therefore, really has no history separate from that of the barometer itself and from that of the correction for capillarity. We shall only note that this correction is, or should be, made first, before the other corrections now to be discussed.

5. THE CORRECTION FOR TEMPERATURE

Torricelli's original tube was found to be greatly affected by temperature, as we saw in Chapter 2. This was undoubtedly due to the imperfect vacuum, in all probability to the presence of a good deal of water vapor. Subsequent experimenters had some of the same trouble, though not to such a disastrous extent. It was not at first realized that the thermal dilatation of the mercury itself would cause an appreciable error, and when this was noticed it appears to have been felt that it was really air contained in the mercury that was expanding. Denis Papin (1647–1714) seems to have thought this, for in the *Journal de Sçavans* for April 9, 1685, there is an "extrait du Journal d'Angleterre" which says that "M. Papin Doct. en Med. & l'un des membres de la même Société [the Royal Society] prétend faire un Barométre qui ne sera sujet ni au froid ni au chaud, & ce en épurant d' air le vif argent qu'il y vent employer."[22]

---

[19] *Mém. Acad. r. des Sci.*, Vol. 21 (1847), p. 162.
[20] O. Broch, *Travaux et mémoires du B.I.P.M.*, I (1881), p. A.44.
[21] WMO, *Guide*, p. III, 3.

[22] P. 111. I have not discovered the original of this. But in 1738 Henry Beighton (d. 1743) still thought that it was air in the mercury that expanded (*Phil. Trans.*, Vol. 40 [1738], p. 249).

Guillaume Amontons (1663–1705) was aware that there really should be such a correction, and tried to produce a table in 1704.[23] The year before,[24] he had devised a sort of constant-volume air thermometer and a scale of temperature based on it. In 1704 he measured the expansion of mercury in a large bulb provided with a calibrated tube, measuring the temperature with his air thermometer. Now the correction to the readings of a mercury barometer cannot be found by measuring the expansion of mercury relative to glass in this way; but Amontons did not see this; or he thought that glass would not expand. He died while still considering this subject.

The next episode in this story is one of those unfortunate accidents which from time to time have retarded the progress of science. In 1709, Philippe de la Hire (1640–1718) wrote a paper on the weight of the air. In this he recounted how he had placed in a room an ordinary barometer, a two-liquid barometer, and a thermometer which had been made by Amontons. Every day for three years he had observed these instruments with the idea of determining the corrections to the barometers, using all possible care. He had read them in the great heat of summer, and at all temperatures down to freezing point. "J'ay observé que dans le baromètre simple le mercure ne change pas sensiblement de hauteur, soit qu'il soit exposé au grand soleil même en été, ou à l'ombre dans un lieu mediocrement frais."[25]

To make sure that this would not go unnoticed, the Secretary of the Academy (Fontenelle) referred to it in the *Histoire*, or annual report, and added, "M. de la Hire . . . a supposé, *comme il est vrai*, que le mercure du Barometre simple ne se dilatoit ni ne se condensoit sensiblement par le chaud ou par le froid."[26]

De la Hire obviously had the misfortune to have a barometer with just about enough air in the vacuum space to annul, by its expansion, the dilatation of the mercury that would otherwise have been manifest. It was a long time before his apparently authoritative result was challenged. Even Charles François de Cisternay du Fay (1638–1739), an excellent experimenter[27] whose barometers must have been much better than de la Hire's,[28] thought that the barometer, when properly made, was not sensitive to changes of temperature,[29] but in a paper by the mathematician Joseph Saurin (1655–1737) dealing chiefly with the two-liquid barometer, it was assumed as self-evident that the temperature correction of the ordinary barometer is linear in temperature and in the height of the column of mercury.[30] But the correction of such an instrument is

[23] *Mém. Acad. r. des Sci.* (1704), pp. 164–172.

[24] *Ibid.* (1703), pp. 50–56.

[25] *Ibid.* (1709), pp. 181–82.

[26] *Hist. Acad. r. des Sci.* (1709), p. 2 (Italics added). It was the custom to summarize in the *Histoire* those papers in the *Mémoires* which the Secretary felt to be of importance.

[27] For an appreciation of his scientific work see I. Bernard Cohen, *Franklin and Newton* (Philadelphia, 1946) (*Mem. Amer. Philos. Soc.*, Vol. 43, whole vol.), pp. 371–76.

[28] See Chapter 13.

[29] *Mém. Acad. r. des Sci.* (1723), pp. 295–306.

[30] *Mém. Acad. r. des Sci.* (1727), pp. 282–96.

rendered futile, he thought, "par l'in-exactitude même, qui est inévitable dans la construction de ces sortes d'in-strumens."[31] And again, "On auroit lieu sans doute d'être très satisfait du ba-rometre, si l'on pouvoit s'assûrer de la justesse de cet instrument à un tiers de ligne près. Je suis fort éloigné de croire que dans l'usage on doive compter sur une si grande exactitude." It is instructive to contrast this pes-simism with the optimistic inventive-ness shown by Leupold in his *Theatri statici* (1726),[32] though it is improb-able that Leupold actually obtained a much greater accuracy. The difference seems to have been that at Leipzig someone was actually giving thought to the barometer as a scientific instru-ment, while at Paris its construction was in the hands of *émailleurs*.[33]

Also in Germany, C. F. Ludolff wrote of the correction of the barometer for temperature in 1749,[34] the year in which De Luc began the two decades of research into barometric hypsometry which furnished the material for the greater part of his most important book.[35]

De Luc was quite well aware of the mistake that Philippe de la Hire had made, and understood how it had oc-curred. He also saw clearly that it would not do to measure the dilatation of mercury as it appears in a thermom-eter, for the glass also dilates; and in

the barometer the length of the scale and its mounting is also affected by temperature. Thus experiments must be performed on actual barometers, having as good a vacuum as it is pos-sible to attain.[36] De Luc was a great believer in the *ad hoc* experiment, the field experiment; his methods were nothing if not direct.

So he installed several barometers in a room in which he could vary the temperature, and one more, as a con-trol, in another room where he kept the temperature as steady as he could. He had three Réaumur thermometers, one near the top, one near the middle, and one near the bottom of the barom-eters.[37] He thought it well to note that he had found by experiment that pine boards, such as those on which the scales of his barometers were mounted, were affected very little by temperature or humidity in the direction of the grain.

By a very considerable extrapolation outside of the rather limited range of temperature available to him, he de-duced a convenient rule, as follows: when the barometer stood at 27 Paris inches, the difference in temperature between melting ice and boiling water would alter its reading by exactly six lines (half an inch).[38]

---

[31] *Ibid.*, p. 284.

[32] See note 85, p. 111 above.

[33] See p. 93 above.

[34] *Mém. Acad. Sci. Berlin* (1749), pp. 33–37.

[35] J. A. De Luc, *Recherches sur les modifi-cations de l'atmosphère* (2 vols. 4°; Geneva, 1772; 2d ed. 4 vols. 8°; Paris, 1784).

[36] *Ibid.*, I, 191. As late as 1765 A. M. Lorgna (*Della graduazione de' termometri a mercurio e della rettificazione di Barometri semplici*, [Verona, 1765]) determined the ap-parent expansion of mercury in glass and applied it directly to the correction of the barometer.

[37] *Ibid.*, p. 197.

[38] We have already seen in Chapter 7 (p. 138) how he made use of this information in the construction of his barometer.

Rather similar experiments were made by Sir George Shuckburgh,[39] William Roy,[40] Gottfried Erich Rosenthal,[41] and Johann Friederich Luz.[42] Of these, the researches of Roy were by far the soundest. Roy built an apparatus for establishing a working barometer in a water bath that could be heated by six lamps, the barometer cistern being heated by a seventh. By this means he could vary the temperature of the mercury column over a wide range. He found that the rate of increase in the height of the column was a decreasing function of temperature. "Mr. Ramsden" (Jesse Ramsden, the famous instrument maker) suggested that the vapor pressure of mercury might be the cause of this. Later he used a longer tube, with 14½" of vacuum space, and found about four per cent difference in the total expansion between 32°F. and 212°F., depending on whether the vacuum space was "frozen" or "boiled." Finally he published an extended table of corrections in inches for various barometric heights from 15 to 31 (English) inches and for temperatures between 0°F. and 102°F.[43] This was the first such table derived, without serious extrapolation, from experiment.

In 1787 Guarinus Schlögl, an ecclesiastic of Rothenbuch, published a table in Paris lines and degrees Réaumur,[44] based on a study of the work of De Luc, Shuckburgh, Roy, Rosenthal, and Luz. It is interesting to see how their various results compare. Between the freezing and boiling points of water, a mercury column of 27 inches expands by 6 lines (De Luc); 5.91 (Shuckburgh); 5.46 (Roy); 5.56 (Rosenthal); and 5.52 (Luz). Schlögl adopts 5.5 lines, as being the mean of the results of Roy and Luz; he praises Roy for the wide range of temperature he used, and it is hard to see why he paid much attention to Luz, whose experiments seem scarcely as extensive as those of De Luc. He agrees with the latter that the correction should be a linear function of both the height of the column and the difference in temperature; and in the choice of 10°R. as a standard temperature. But in his Table I, which gives the correction for every line from 240 to 348 (20 to 29 inches), and every tenth of a degree [!] from 0.1 to 35.0°R., the standard temperature is 0°R. The table occupies 110 pages!

Besides these rather *ad hoc* researches, there were a number of measurements of the absolute dilatation of mercury in the succeeding forty years, mostly by measuring its dilatation in glass, and then allowing for the glass. A list of these is given in a set of tables published in 1820 at Halle.[45] A tre-

[39] *Phil. Trans.*, Vol. 67 (1777), pp. 513–97.

[40] *Phil. Trans.*, Vol. 67 (1777), pp. 653–788.

[41] *Beyträge zu der Verfertigung, der wissenschaftlichen Kenntnis, und dem Gebrauche meteorologischer Werkzeuge* (2 vols.; Stuttgart, 1782 and 1784).

[42] *Vollständige und auf Erfahrung gegründete Beschreibung von . . . Barometern*, etc. (Nürnberg & Leipzig, 1784).

[43] Roy, *Phil. Trans.*, Vol. 67 (1777), pp. 683–87.

[44] *Tabulae pro reductione quorumvis statuum barometri ad normalem quemdam caloris gradum publico usui* (Ingolstadt, 1787).

[45] Carl Ludwig Gottlob Winckler, *Tafeln, um Barometerstände, die bei verschiedenen Wärme-Graden beobachtet worden sind, auf jede beliebige normal-temperatur zu reduciren* (Halle, 1820). There is a copy in the library of the *Bundeswetteramt* at Offenbach-am-Main.

mendous experimental advance was made in 1817 by P. L. Dulong and A. T. Petit,[46] and it is rather interesting that as the most urgent reason for determining the absolute dilatation of mercury they adduce the measurement of heights with the barometer. Their method depended only on the hydrostatic equilibrium between the masses of mercury in two vertical tubes, joined at the bottom by a narrow horizontal tube, one being heated. Temperatures were measured with an air thermometer.

The great advantage of this method is shown by the fact that they considered their results accurate to four significant figures, while previous determinations differed by several units in the second place. The mean dilatation per degree C. between 0° and 100°C. was found to be 1/5550, all the values in "un grand nombre de mesures" lying within the range 1/5547 and 1/5552. It is interesting that of the earlier determinations those of De Luc and Shuckburgh lay closest to this figure. Regnault[47] later refined these measurements, which were again corrected by Broch at the *Bureau International des Poids et Mesures*;[48] but the differences were relatively small.

At the beginning of the nineteenth century few people seem to have realized that the temperature correction to a barometer is to some extent a function of the design and materials of the barometer itself. The recognition that the dilatation of the scale of the barom-

eter (or of the frame, if the scale is fastened to it) might be important was probably hindered to some extent by De Luc's book. It will be remembered that his famous barometer had paper scales glued to a pine board, the change in length of which he decided to ignore. Among those who were clear about this was Laplace, who with Lavoisier had measured with great care the dilatation of brass and other solids.[49] But even in 1823 the extensive tables of Hahn[50] make no attempt to take the expansion of the scale into account. The earliest table that I have found which takes account of the relative expansion of mercury and brass is that published by Bouvard in an article chiefly devoted to the capillary correction.[51] Francis Baily published a correct formula and tables in 1837,[52] when he described "a new barometer, recently fixed up in the apartments of the Royal Society." All the important tables published from that time forward of course included the effect of the expansion of the scale. One of the most influential sets was compiled by Guyot[53] and was widely used for many years. It had separate tables for barometers in English inches, with brass scales, the same "with Glass or Wooden Scales," for metric barometers with brass scales,

---

[46] *Ann. de Chim. et de Phys.*, Vol. 7 (1817), pp. 113–54.

[47] *Mém. Acad. r. des Sci.*, Vol. 21 (1847), pp. 271–328.

[48] B.I.P.M., *Trav. et Mém.*, Vol. 2 (1883), part 2, pp. 1–21.

[49] Lavoisier, *Oeuvres* (Paris, 1862), II, 739–59.

[50] Hahn, *Barometrische Tafeln*, etc. (Leipzig, 1823).

[51] Paris, Bureau des longitudes, *Connaissance des Temps pour l'an 1829*, pp. 303–9.

[52] *Phil. Trans.*, Vol. 127 (1837), pp. 431–41. The tables are ascribed to Professor Schumacher.

[53] Arnold Guyot, A *Collection of Meteorological Tables, with Other Tables Useful in Practical Meteorology* (Washington, Smithsonian Institution, 1852). These are collected from various sources, not always properly identified.

and for barometers with brass scales graduated in Paris lines. The appropriate temperature scale (Fahrenheit, Centigrade, or Réaumur) is used for each table.

When the thermometer attached to the barometer is read for the purpose of correcting the reading of the instrument for temperature, it is implied, or at least hoped, that the temperature indicated by the thermometer is actually the average temperature of the mercury column. If the temperature of the room is changing at the time, this may not be so. In 1902, R. T. Ormond confirmed by direct experiment[54] what many people had suspected. He took an ordinary barometer of a type used at many meteorological stations, and replaced the tube by a glass tube of similar size, but closed at the bottom and full of mercury. A thermometer dipped into this mercury, and its readings could be compared with the thermometer ordinarily attached to the frame of the instrument. After a fire had been lit in a cold room, or a window opened in a warm one, the difference might rise to 3°F., which could lead to errors of nearly 0.01 inch in the corrected height of the mercury column. Ormond was led to suggest that the morning readings at stations in cold climates might be systematically in error, for in Scotland at least, central heating was yet to come. An elaborate theoretical and experimental study of this phenomenon was also made in 1947 by J. C. Evans, with equally pessimistic results.[55]

Through a historical accident, the standard yard (and hence English ba-

rometer scales) had its normal length at 62°F.; the mercury, on the other hand, was standard at 32°F., like the meter; as a result, the tables for the metric scales cannot be converted for use with English barometers merely by substituting units. In the English tables for brass scales the correction becomes zero not at 32°F., but at 28.5°F. Also, in making a barometer having scales in both inches and millimeters, the relation between the two has to differ slightly from what would result from the official conversion factor.[56] This peculiarity was not confined to the United Kingdom; the Russian barometers were graduated in half-lines, each apparently equal to $\frac{1}{24}$ English inch and standard at $13\frac{1}{3}$°R.[57] And it is another indication of the wide use of the Paris inch, forty years after the introduction of the metric system, that A. Bravais and Ch. Martins, in a long paper on the international comparison of barometers,[58] found it desirable to point out that in barometers divided in French lines, the scale ought to be exact at $16\frac{1}{4}$°C. Barometers divided in this way seem to have been in use at some observatories as late as 1879,[59] according to Hellman, who also gave a general formula for the corrected height B of a barometer, which reads:

$$B = b \cdot \frac{1 + m\,(\tau - \theta)}{1 + q\,(t - T)}$$

[56] L. Casella, *British Meteorol. Soc. Proc.*, Vol. 2 (1865), pp. 417–18. For example, 30 in. should be opposite 761.75 mm. instead of 762.00 mm.
[57] A. T. Kupffer, *Tables psychrométriques et barométriques à l'usage des observatoires météorologiques de l'Empire de Russie* (St. Petersbourg, 1841).
[58] *Acad. r. de Belgique, Bruxelles, Nouv. Mém.*, Vol. 14 (1841), pp. 31–78.
[59] G. Hellmann, *Repertorium f. Meteorol.*, Vol. 6 (1879), no. 8.

[54] *Trans. Roy. Soc. Edin.*, Vol. 42 (1902), pp. 541–42.
[55] *Phys. Soc., Proc.*, Vol. 59 (1947), pp. 242–56.

where q is the (volume) expansion-coefficient of the mercury, m the linear coefficient of dilatation of the scale (both for one degree of whatever thermometer is used), T and θ the standard temperature of the mercury and of the scale respectively, t and τ their actual temperatures at the time the barometer is read. In general the scale and the mercury are, justifiably or not, generally considered to be at the same temperature, so that $\tau = t$. In the metric barometer $\theta = T = 0°C.$, so that the formula reduces to

$$B = b\,(1 + mt)/(1 + qt).$$

In the same year, an International Congress of meteorologists in Rome decided to ask the International Bureau of Weights and Measures for help in preparing a set of international meteorological tables by means of which a tendency toward standardization might be established. Eleven years later these were published[60] in the sumptuous style of the *Travaux et Mémoires* of the bureau. No more odd units; it must have been under English and American pressure that feet and Fahrenheit degrees were left in as an alternative to the metric system. The latest figures for the constants were used, some specially measured, like the coefficient of dilatation for brass scales.[61]

Not much change has occurred since that time in the temperature corrections of barometers graduated in millimeters or inches, which are now standardized, at least for meteorological pur-poses, by the World Meteorological Organization, an agency of the United Nations Organization,[62] if we except the corrections to barometers with non-adjustable cisterns, which we must now consider. We shall be obliged to anticipate to some extent a portion of the next chapter (p. 238).

In a barometer with a fixed cistern the mercury rises in the cistern when it falls in the tube and vice versa. The magnitude of the error to which this may give rise depends on the ratio of the cross section of the tube to the area of the mercury surface in the cistern. There are two ways in which this error can be eliminated: we may use a standard scale and calculate a correction, or we may fit the barometer with a scale appropriately contracted. In the first case we must know the point on the scale at which the instrument has been adjusted to read correctly; in both we must know the ratio of the cross sections.

Fixed-cistern barometers have, of course, been in use since the time of Torricelli, and the idea of contracting the scale dates from 1792.[63] Yet it would seem that the first sound attempt at a theory of their temperature correction was made by Carl Jelinek in 1867, and this for the so called "neutral-point" barometer, without the contracted scale; in fact, Torricelli's barometer.[64] In the translator's introduction to the English version, it is

---

[60] Comité météorologique international, *Tables météorologiques internationales publiées conformément à une décision du congrès tenu à Rome en 1879* (Paris, 1890).

[61] J. R. Benoit, *Trav. et Mém. B.I.P.M.*, Vol. 2 (1883), pp. C.161–C.166. Also Reference 60, p. B.26–27.

[62] The current rules for metric barometers are in its *Guide to Meteorological Instrument and Observing Practices* (2d ed.; Geneva, 1961), pp. III, 7.

[63] See p. 240 below.

[64] Carl Jelinek, *Wien., Akad., Sitzungsber.*, Abt. 2, Vol. 56 (1867), pp. 655–62. *Trans.* W. T. Lynn in *Meteorol. Soc. Proc.*, Vol. 4 (1868), pp. 93–100.

pointed out that such barometers were no longer used in England for scientific purposes. Yet they were on the Continent.

The correction to a standard temperature of barometers with fixed cistern and contracted scale does not appear to have been worked out until about 1914, apparently at the National Physical Laboratory,[65] though there seems to have been no official publication from NPL.[66] It turns out that the additive term depends on the ratio of the total volume of the mercury in the barometer and the effective cross section of the cistern.

This theory seems to have been developed independently by K. Irgens in 1928.[67] The net result is that it is possible to add a constant quantity to the barometric height and then use the standard tables, with sufficient accuracy for most meteorological purposes. For special purposes this will not be good enough, and one answer is to avoid fixed-cistern barometers under such circumstances.[68]

It is remarkable that the need for an additional correction seems to have been largely disregarded in Germany. E. Kleinschmidt, although he commented on Irgens' paper in 1934,[69] entirely ignored the matter in his monumental *Handbuch der Meteorologischen Instrumente*,[70] a year later. A summary of the history of this correction has been given by R. Sneyers.[71]

This is an appropriate place to refer to the almost simultaneous discovery in 1782 by Gottfried Erich Rosenthal[72] and by Paul de Lamanon[73] that a siphon barometer with a uniform tube could be used to measure its own temperature. Suppose it has two scales, graduated upward from the lowest point, and suppose that a and b are the readings of these. Then their sum $a + b$ will vary linearly with temperature, and can be calibrated by varying the temperature. The difference $a - b$ is, of course, the barometric height, which can be corrected to a standard temperature by use of the data obtained in the calibration.

This property of a siphon barometer was rediscovered in 1821, to the applause of the Academy of Dijon, by one Goubert,[74] in 1850 by Victor Pierre,[75]

[65] R. Glazebrook (Ed.), *Dictionary of Applied Physics* (London, 1923), III, 154. The theory of Paul Schreiber *(Meteorol. Zeits.*, Vol. 36 [1919], pp. 349–50) should be mentioned, but it was not adequate.

[66] I am indebted to Mr. P. H. Bigg for this information. The *Computer's Handbook* (M.O. 223, 1916, p. 13) states that the temperature correction for the adjustable barometer must be increased by 1/20 of its own magnitude to make it appropriate to the "Kew pattern" fixed-cistern barometer.

[67] *Meteorol. Zeits.*, Vol. 45 (1928), pp. 441–44. Summarized in W. E. K. Middleton and A. F. Spilhaus, *Meteorological Instruments* (3d ed.; Toronto, 1953), pp. 33–35.

[68] See W. G. Brombacher, D. P. Johnson, and J. L. Cross, *Mercury Barometers and Manometers*. National Bureau of Standards monograph 8, Washington, 1960.

[69] *Meteorol. Zeits.*, Vol. 51 (1934), pp. 194–95.

[70] Berlin, 1935. Karl Gödecke (*Ann. Meteorol.*, Vol. 3 [1950], pp. 103–5) said that even in 1950 fixed-cistern barometers in Germany were being corrected by the ordinary tables.

[71] *Inst. r. météorol. de Belgique, Miscellanées*, no. xxxvii (Brussels, 1951).

[72] *Beyträge zu der Verfertigung, der wissenschaftlichen Kenntnis, und dem Gebrauche meteorologischer Werkzeuge* (2 vols., Stuttgart, 1782 & 1784), I, 10–16.

[73] *Obs. sur la phys.*, Vol. 19 (1782), pp. 7–8.

[74] *Acad. des Sci., Arts & Belles-Lettres de Dijon*, séance publique du 24 aôut 1821 (Dijon, 1822), pp. 110–16.

[75] *K. Akad. d. Wiss., Wien, Sitzungsber.*, II Abt., Vol. 5 (1850), pp. 281–89.

and once again by Alois Handl in 1867.[76] Handl demonstrated that the sensitivity is adequate for some purposes.

6. THE CORRECTION TO STANDARD GRAVITY

To us, the atmospheric pressure corresponding to a mercury column of given height obviously depends on the acceleration due to gravity, although this cannot have been clear to Torricelli, nor to Boyle. Therefore it is necessary to reduce the reading of the barometer to some agreed standard value of this acceleration.

The first suspicion that this quantity might differ in different parts of the world arose when Jean Richer (1630–1696) who was sent by the French Academy on an expedition to Cayenne in 1672–1673, found that the pendulum of his clock had to be shorter at Cayenne than at Paris, if it was to keep time.[77] At first disbelieved, this observation was later interpreted on the basis of Newton's *Principia* as an indication that the earth bulges somewhat at the equator, and led to many lengthy and arduous field experiments to determine the "figure of the earth"; experiments which are still being refined. It also came easily out of Newton's theories that the acceleration due to gravity must vary with height. Apart from local anomalies, this variation has been known with enough precision for purposes of barometry since the time of Laplace.[78] It was taken into account in

the practice of barometric hypsometry, however, long before any attention was paid to it in the interpretation of the actual barometer readings themselves. There is nothing about it in the tables of Baily or of Guyot.[79] On the other hand, it was treated from a rather sophisticated standpoint by Jacques Babinet (1794–1872) in 1848;[80] this contrasts with the elementary discussion given fourteen years later by a German professor of surveying, Carl Maximilian Bauernfeind. In his impressive textbook, which includes a dozen pages on the barometer, the "gravity correction" is dismissed without even a formula.[81]

It is probably fair to say that the introduction of the gravity correction into meteorological practice owes a great deal to such perfectionists as Heinrich Wild (1833–1902), the Director of the Russian meteorological service, who commenced a long paper "Ueber die Bestimmung des Luftdrucks" with a quantitative discussion of the various errors of barometry.[82] Wild was also very active at international conferences, and may have been instrumental in the movement which resulted in the publication of the *International Meteorological Tables* in 1890.[83] The form of the function given in these tables, relating the acceleration of gravity at latitude ϕ and height H above sea level ($g_{\phi,H}$) with the standard value ($g_{45.0}$), is

---

[76] *Ibid.*, Vol. 57 (1868), pp. 109–14.
[77] *Observations astronomiques*, etc. (Paris, 1679) (*Hist. Acad. Sci.* [Paris, 1733] VII, Part I, 320.)
[78] *Mécanique céleste*, I (1799).

[79] Notes 52 and 53 above.
[80] *Compt. Rend.*, Vol. 26 (1848), pp. 265–66.
[81] *Elemente der Vermessungskunde* (2 Aufl.; München, 1862), p. 345.
[82] *Repertorium für Meteorol.*, Vol. 3 (1874), no. I, pp. 1–145.
[83] *Tablis Météorologiques internationales.*

$$g_{\phi,H} = g_{45.0}(1-0.00259 \cos 2\phi)\,(1-$$
$$0.000000196\ H)$$

if H is in meters.

This formula, with the value $g_{45.0} = 980.62$ cm. sec.$^{-2}$, lasted in meteorological practice until January 1, 1955, when the World Meteorological Organization fell into step with physicists and engineers, who for half a century had used a standard value of 980.665 cm. sec.$^{-2}$, not tied to any latitude.[84] The WMO also adopted a much more elaborate method of calculating the local acceleration of gravity,[85] with a $\cos^2\phi$ term, and devices for taking into account the height of the land and depth of the sea in the region around the station.

7. THE CORRECTION FOR CAPILLARITY

The reader who expects to find in these pages a discussion of the history of the theory of capillary attraction will be disappointed. I have no intention of trying to compete with James Clerk Maxwell, whose article in the 9th edition of the *Encyclopaedia Britannica* (largely reprinted, with additions by the 3rd Baron Rayleigh, in the 11th edition) is a masterpiece. Here we shall merely refer to some researches which arose out of, or were applied to, the practice of barometry.

In 1660 Robert Boyle noted that a mercury surface, in contradistinction to a surface of water, is convex, and that "if you dip the end of a slender pipe in it, the surface of the liquor (as it is called) will be lower within the pipe, than without."[86] Hooke, in a pamphlet written in 1661, attempted to explain these experiments.[87] Noting that "there is a much greater inconformity or incongruity . . . of air to glass, and some other bodies, than there is of water to the same,"[88] he at length "explains" the curvature of a water surface in a glass tube by a combination of gravity, which would make it flat, and "the greater congruity of one of the two contiguous fluids, than of the other, to the containing solid."[89] The second force would, he says, make the surface spherical. But if the glass tube be greased, the water surface will be convex upward, and will not rise as high in small pipes as in large ones. "Quicksilver also which to glass is more incongruous than air (and thereby being put into a glass-pipe, will not adhere to it, but by the more congruous air will be forced to have a very protuberant surface, and to rise higher in a greater than a lesser pipe). . . ."[90] Clearly Hooke had a qualitative idea of the balance of forces which more than a century later was made into an exact theory; but having got this far, he unaccountably abandoned this idea, forgot all about the behavior of mer-

[84] World Meteorological Organization, Fourth session of the Executive Committee, Geneva, 1953. *Abridged Report with Resolutions.* (WMO No. 20, RC5.) (Geneva, 1953), p. 88.

[85] World Meteorological Organization, *Guide to Meteorological Instrument and Observing Practice* (2d ed. [WMO No. 8, TP3.] Geneva, 1961), pp. III.14 to III.16.

[86] *Works* (London, 1772), I, 81.

[87] Robert Hooke, *An Attempt for the Explication of the Phaenomena Observable in an Experiment Published by the Honourable Robert Boyle, Esq.*, etc. (London, 1661).

[88] *Ibid.*, p. 7.

[89] *Ibid.*, p. 18.

[90] *Ibid.*, p. 19.

cury, and concluded that since it takes more pressure to force air through a small pipe than through a large one, as he thought, the sufficient cause of the rising of water in narrow pipes is that "those degrees [of pressure] that are requisite to press it in, are thereby taken off from the air within, and the air within left with so many degrees of pressure less than the air without; it will follow, that the air in the less tube or pipe, will have less pressure against the superficies of the water therein, than the air in the bigger. . . ."[91] A moment's thought about the opposite behavior of quicksilver would have saved Hooke from this disastrous hypothesis. Fabre, too, who made many such experiments with water, oil, spirit, and mercury, and found the rise (or fall) greater the smaller the tube, thought the changed pressure of the air to be responsible; or, because the experiments worked in a vacuum, it could be the pressure of the "subtle matter."[92] Sturm[93] agreed with Fabre.

There was a return to an attraction hypothesis in the work of Louis Carré,[94] who did not, however, deal with mercury, and in that of Francis Hauksbee, famous for his early experiments on frictional electricity. In 1709, after describing many experiments on capillarity, he wrote, "It appears evident to me, that *the principle we ought to have recourse to in this case, is no other than that of attraction.*"[95] The law must, however, be different from that of Newtonian gravitational attraction, and he did not see how it was to be determined. His experiments with two tubes of equal bore but very different thickness showed him that the surface is mainly, if not solely, involved in the phenomenon.

Dr. James Jurin found that the capillary rise depends only on the cross section of the tube at the surface of the liquid and deduced the (inexact) law that the rise is inversely proportional to the radius of the tube,[96] from the hypothesis of the attraction between the edge of the liquid and the wall of the tube. By this time it was widely recognized that barometers with tubes of different bores gave different readings, and Gabriel Philippe de la Hire (1677–1719) had in 1711 advised tapping a barometer before reading it, "afin de faire couler le mercure à sa vraye hauteur."[97]

It was at about this period that the barometer began to be used seriously in survey operations as a means of leveling, and one of the people most interested was the Director of the Paris Observatory, Jacques Cassini (1677–1756).[98] In 1733 Cassini, in a paper

[91] *Ibid.*, p. 25.

[92] Honoré Fabre, *Dialogi physici in quibus de motu terrae disputatur*, etc. (Leiden, 1665), pp. 157–206.

[93] Johann Cristophorus Sturm, *Collegium experimentale, sive curiosum*, etc., part I (Nürnberg, 1676), pp. 44–45.

[94] *Mém. Acad. r. des Sci.* (1705), pp. 241–54.

[95] *Physico-Mechanical Experiments on Various Subjects*, etc. (London, 1709; 2d [posthumous] ed., 1719), p. 156. The italics are in the original.

[96] *Phil. Trans.*, Vol. 30 (1718), pp. 739–47; (1719), pp. 1083–96.

[97] *Mém. Acad. r. des Sci.* (1711), p. 4.

[98] The second of the famous family of astronomers who dominated the affairs of the *Observatoire* for more than a hundred years: Jean Dominique (1625–1712), who was born in Sardinia and brought to Paris by Louis XIV; Jacques; César François (1714–1784); and Jean-Dominique (1748–1845).

entitled "Réflexions sur la hauteur du baromètre observée sur diverses montagnes,"[99] reported that De Plantade had found that up to an elevation of 1000 fathoms, barometers with narrower tubes stood lower than those with wider ones; but that above 1000 fathoms they all stood at the same height, whether they were wide or narrow. Cassini seems to have accepted this statement quite uncritically. The mistake arose, no doubt, from an interpretation of a few faulty observations; De Plantade's larger instrument may have got a little air in it on its way up the mountain. It caused a little trouble until in 1745 the Bologna Academy disposed of the mistake by enclosing three barometers, having tubes 1, 1½, and 2 lines in bore, in a chamber that could be evacuated.[100] In the same report the idea was put forward that the capillary depression in a barometer is influenced by the kind of glass as well as by the size of the tube. In another paper[101] it was noted that the capillary depression also depends on temperature; but it may be observed that the direction of the variation would accord with the explanation that a little air or water vapor had been left in the vacuum space.

The capillary depression in the barometer was one of the first matters considered by the Academy of Turin.[102] J. F. Cigna, reporting as secretary, said that the Academy doubted the conclusions of Balbi. They supposed that it might be more difficult to get rid of the air in a small tube. They made several inconclusive experiments, and then "while these were being tried, the cavalier Salutius indicated a new kind of experiment"[103] that would settle the whole matter. This was to make two barometers of different sizes, drawn from the same glass, with but one vacuum space, i.e., joined at the top. They boiled the mercury in these tubes, and found a smaller difference in height than had been found at Bologna; but they agreed with Balbi that a *vis repulsiva tuborum* must cause the phenomenon. Applying ice or heat, however, the change in level was the same in either tube.

This paper was brought to mind when I saw a barometer in the storeroom of the *Deutsches Museum* in Munich which had such a hairpin-shaped tube, the two limbs differing somewhat in size. But although it is a rough apparatus it was probably not used for this experiment, for it has the markings traditional on mountain barometers (see p. 000). It must remain a mystery.

Meanwhile theory was not making much progress, although there was qualitative agreement on a good many points. In 1736 Josias Weitbrecht, one of the learned Germans whom Peter the Great had attracted to his new capital, set out the main assumptions on which he thought a theory of capillarity would have to be based.[104] They are as follows:

[99] *Mém. Acad. r. des Sci.* (1733), pp. 40–48.

[100] P. B. Balbi, *Comm. Acad. Bononiensis*, Vol. 2 (1745), part I, pp. 308–9.

[101] *Ibid.*, pp. 353–60.

[102] *Miscellanea Taurinensis*, Vol. 1 (1759), pp. 7–18.

[103] *Ibid.*, p. 11. Salutius was Guiseppe Angelo Saluzzo, Conte di Menusiglio (1734–1810).

[104] *Comm. Acad. Sci. Imp. Petrop.*, Vol. 8 (1736), pp. 261–309.

1. All bodies attract when brought close together.

2. Water and glass attract each other.

3. Mercury is more strongly attracted to itself than to glass.

4. The radius of attraction is extremely short.

Weitbrecht saw clearly that the rise of water in a tube must be the result of an equilibrium between gravity and the attractive forces. The cylinder of water is uniquely suspended by the margin. He does not seem to have said that the cylinder of mercury is held down by the edge in a similar way.

It may not be superfluous to point out here that the capillary depression once had an importance to the scientific user of barometers which it has lost today. Nowadays the meteorologist, for example, never thinks of it. It is not that his barometer is free from it; but because the instrument has been compared directly or indirectly with an absolute barometer, the correction for capillarity has been incorporated into the "index correction," and has lost its identity. The physicist, needing more accurate results, builds a barometer with negligible or at least calculable capillary depressions. It was not so in the eighteenth century; the capillary error and the imperfections of the vacuum were probably the factors which most severely limited the accuracy of barometric observations, and the production of an adequate vacuum was achieved many years before the capillary depression was adequately handled (if, indeed, it can ever be so in a small barometer).

In view of this it is rather surprising to find very little about the capillary error in De Luc's *Recherches sur les modifications de l'atmosphère* (1772), among the hundreds of pages which he devoted to the barometer and its use. I think the reason is because he was convinced that with his siphon barometer he need pay no attention to capillarity, for (he believed) the depression of the mercury was the same in the two arms.[105]

De Luc's preference for the siphon barometer was not shared by the Hon. Henry Cavendish (1731–1810); nor was his habit of observing the edge of the mercury surface.[106] Cavendish preferred to use a cistern barometer and to observe the summit of the convex surface. "This manner of observing appears to me more accurate than the other; because if the quicksilver should adhere less to the tube, or be less convex at one time than another, the edge will, in all probability, be more affected by this inequality than the surface."[107] But it is well to correct the readings of the cistern barometer for capillarity; to make this possible, he gave a table of corrections obtained experimentally by his father, Lord Charles Cavendish.[108] This was obtained in air, not in a vacuum; it was not known at the time whether this was important. Lord Charles Cavendish's table, which was quoted for half a century, is reproduced in Table 8.2.

---

[105] *Recherches*, I, 207.

[106] *Phil. Trans.*, Vol. 66 (1776), pp. 375–401.

[107] *Ibid.*, p. 381.

[108] *Ibid.*, p. 382.

Table 8.2. Capillary Depression According
to Lord Charles Cavendish

Diameter of bore	Depression
0.6 in.	0.005 in.
0.5	0.007
0.4	0.015
0.35	0.025
0.30	0.036
0.25	0.050
0.20	0.067
0.15	0.092
0.10	0.140

It is likely that Henry Cavendish's preference for reading the top of the meniscus may have been occasioned by the improvement in the reading index introduced a few years earlier by Jesse Ramsden.[109]

At about this time there was a flurry of interest in the possibility of making barometers with the mercury surface flat. This seems to have been first achieved by Dom Casbois of Metz.[110] A barometer tube with a bulb at the top was heated very hot and filled with boiling mercury; then the mercury was cooled and reboiled in the tube several times, and finally the bulb was removed with a blow lamp and the mercury boiled once again. The method was investigated by Lavoisier,[111] who took a two-tube barometer of Megnié's construction[112] and filled one tube in the ordinary manner, the other in this special way, in order to compare the capillary depressions. In this way he noted that the one with the flat surface tends to stick. It is really not a good idea to have the surface flat, as was noted by Laplace[113] in 1812.

The reason for the flat surface was discussed from time to time. P. L. Dulong[114] came to the conclusion that the mercury became oxidized by all the boiling, and that the oxide dissolved to some extent in the mercury and changed its surface tension. He suggested boiling the mercury in hydrogen, a somewhat hazardous idea in 1831. This article was translated and reprinted by Poggendorf in the *Annalen* the following year,[115] but in a long footnote, Poggendorf said he doubted the explanation given by Dulong. On the other hand, Heinrich Buff (1805–1878), a professor of physics at Giessen, made a series of experiments which led him to conclude that the oxide is responsible for the adhesion of the mercury to the glass.[116] It is possible that the formation of oxide may be a sufficient, but not a necessary, cause. I remember very well an occasion in the 1930's when a number of barometer tubes which had been filled by distillation *in vacuo* at the Meteorological Office in Toronto were found to have flat mercury surfaces. These had been out-gassed under high vacuum,

[109] See Chap. 9, p. 196 below.
[110] *Encyclopaedie méthodique; Physique*, t. I, p. 107* (1793). (The asterisk is significant, as there is another page 107.)
[111] *Oeuvres* (Paris, 1865), III, 753–58.
[112] See Chap. 9, p. 250. In 1879 Truchot (*Ann. Chim. et Phys.*, Vol. 18 [1879], p. 309) noted that a century later this barometer, then at the Chateau de la Canière (Puy de Dôme) still had "une surface parfaitement plane, très nette et très mobile, et la différence de 2/3 de ligne entre les niveaux du mercure dans les deux tubes existe comme il y a cent ans."

[113] Paris, Bureau des longitudes, *Connaissance des temps pour l'an 1812*, pp. 315–20.
[114] In S. D. Poisson (1781–1840), *Nouvelle théorie de l'action capillaire* (Paris, 1831), pp. 291–93.
[115] *Annalen der Physik.*, Vol. 26 (1832), pp. 455–58.
[116] *Ann. der Chim. u. Pharm.*, Vol. 36 (1840), pp. 113–17.

probably at a rather higher temperature or for a longer time than usual. There would be no possibility of the formation of oxide. The late Dr. John Patterson, then Director, was greatly interested in this phenomenon, which had never before come to his attention; and I recall that he seemed a little disappointed when I exhumed the reference to Lavoisier.

Let us now return to the year 1802, when John Leslie (1766–1832) showed that the rise of water in a tube could be explained by an attraction between the liquid and the tube which was everywhere normal to the surface of the latter.[117] He pointed out that the neglect of the capillary correction in height measurement with the barometer will automatically lead to a positive error. In 1805 Thomas Young (1773–1829), famous for his contributions to optics, published an essay on the "Cohesion of Fluids," in which he showed that the phenomena could be explained on the principle of surface tension, combined with the assumption of the constancy of the angle of contact between the solid and the liquid.[118] Young showed that his theory gave good agreement with the observations of Lord Charles Cavendish above referred to.

Meanwhile Pierre Simon Laplace (1749–1827) had been busy analyzing the universe mathematically. In the supplement to the tenth book of his *Mécanique céleste*,[119] published in 1806, he obtained results very much like those of Young. While from a purely theoretical standpoint his conclusions were later modified by Gauss and by Poisson, the numerical results given by his theory were scarcely affected,[120] and have formed the basis of many tables of the capillary correction of the barometer, one of the first being by Laplace himself.[121] Laplace assumed that the surface tension and angle of contact are the same in the vacuum as in air. He knew that the angle of contact can vary with conditions, but happily the effect is small if the tube is large. This last observation demonstrates a fine disdain for practical affairs, for Laplace must have known that the tubes of the barometers used in field work were very small.

It was not long before the promise offered by analysis was found to be an illusion. In 1826 Johann Gottlieb Friedrich von Bohnenberger (1765–1831) of Tübingen set up a very large barometer with a tube 1.2 Paris inches in bore,[122] and a very large cistern into which several smaller tubes could be simultaneously inverted. He also made several siphon barometers. In his large tube the surface of the mercury seemed to be flat except for a distance of two lines from the wall, and he concluded that the depression was zero, as it certainly was to within the precision of the "mikroskopischer Apparat" which

---

[117] *Phil. Mag.*, Vol. 14 (1802), pp. 193–205.

[118] *Phil. Trans.*, (1805), pp. 65–87. (This volume was not numbered.)

[119] *Mécanique céleste* (Paris, 1799–1825), T. IV, pp. 389–552.

[120] See J. C. Maxwell in *Encyclopaedia Britannica*, (11th ed.; 1910), V, 256–58.

[121] *Connaissance des temps*, etc. These are for tubes 2 to 20 mm. in bore.

[122] *Annalen der Physik.*, Vol. 7 (1826), pp. 378–85. Bohnenberger seems to have been an excellent experimenter, one of the small number of physicists of the period with a real flair for instruments. Indeed the filling and boiling of this large tube must have offered problems.

he used to measure his elevations.[123] He soon found that the capillary depression *in vacuo* was less than that in air. He also investigated various kinds of glass without finding much difference.[124]

John Frederic Daniell (1790–1845) experimented with barometer tubes which had been filled both with and without boiling, and as a result concluded that the capillary depression in "boiled tubes" is about half that in tubes which had not been boiled.[125] In 1840 the Royal Society published a set of meteorological tables[126] in which there is a table referring to "unboiled" and "boiled" tubes, after Daniell. The same table is reproduced by Guyot.[127]

We saw that the capillary depression is a function both of the surface tension and of the angle of contact. Probably nothing could be done in practice about possible variations of the former, but it became evident that the latter varies, and that its variations could be measured. A. Bravais showed how to do this (Figure 8.1).[128] The eye is placed at an angle h (positive downward) below or above the horizontal plane through the edge of the meniscus,

and an opaque screen gradually raised behind the tube, lighted from the back. When the whole of the meniscus is just darkened, the angle H is measured, and the angle of contact V is then given by the formula $V = (H + h)/2$. Bravais also showed how to measure the bore of a barometer tube *in situ*, and gave a double-entry table of the capillary depression in terms of the bore of the tube and the angle of contact. He realized that an equivalent procedure would be to measure the height of the meniscus, but seemed to prefer the angle of contact.

J. Delcros, on the other hand, felt that it would be easier for the meteorological observer to measure the height of the meniscus, and in a long article on barometry at weather stations,[129] he published a double entry table in which the arguments were the internal *radius* of the tube (from 1.0 to 7.0 mm.) and the height of the meniscus. This table was also reproduced by Guyot[130] and was widely used.

The subsequent history of the capillary correction is one of gradual frustration, as it became evident that the unmeasurable surface tension could not be supposed invariable even in the same barometer. J. Pernet in 1886 had suggested enclosing a small open-ended tube in the barometer, with its upper end in the vacuum and its lower end in the mercury, and measuring the difference in level and the radii of curvature of the two menisci;[131] this sounds like

[123] According to J. Stulla-Götz, *Physik. Zeits.*, Vol. 35 (1934), pp. 404–7, the depression in a tube 31 mm. in diameter is one micron.

[124] The astronomer F. W. Bessel (*Astr. Nachrichten*, Vol. 9 [1831], col. 111–14) thought that the capillary constants should be investigated for the kind of glass used.

[125] *Meteorological Essays and Observations* (London, 1823), pp. 360–62.

[126] Royal Society, *Report of the Committee of Physics, including Meteorology, on the Objects of Scientific Inquiry in Those Sciences* (London, 1840), p. 57 and Table I, p. 81.

[127] Guyot, *A Collection of Meteorological Tables*, Table X, p. 90.

[128] *Ann. de Chim. et de Phys.*, Vol. 5 (1842), pp. 492–508.

[129] *Annuaire météorol. de la France* (1849), pp. 139–74.

[130] Guyot, *A Collection of Meteorological Tables*, p. 91.

[131] *Zeits. für Instrumentenkunde*, Vol. 6 (1886), pp. 377–83. Pernet was at the *Normal-Aichungs Kommission* (Standards Laboratory), Berlin.

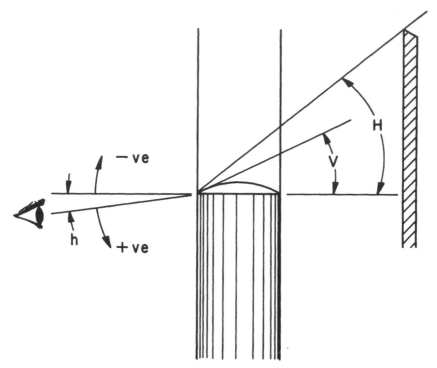

Fig. 8.1. Bravais' method of measuring the angle of contact.

a counsel of despair. In 1894 C. Maltézos, after still another theoretical discussion, gives up: "We must therefore put aside all tables and formulas about capillarity . . . and be content with the practical method, which consists in taking the mean value of the depression found by comparing the barometer in question with a normal barometer for a fortnight. It would be best if this comparison were to be made during days of varying temperature and atmospheric pressure, in order to be able to form an empirical table, for each barometer, of the mean depression corresponding to each value of the height of the meniscus (for each tenth of a millimeter)."[132] The first part of

this program is what is done today, when the "normal" barometer is one which has such large tubes that the capillary error (let alone its variations) may be neglected entirely.

The history of the subject since 1900 concerns mainly the efforts of physicists and physical chemists to improve the precision of manometers used in the laboratory, and is thus somewhat out of our domain.[133] We must, however, refer to a suggestion by H. Christoff,[134] who thought it likely that the vibration

[132] Compt. Rend., Vol. 118 (1894), p. 585.

[133] The interested reader is referred to W. G. Brombacher, D. P. Johnson, and J. L. Cross, Mercury Barometers and Manometers, National Bureau of Standards Monograph 8 (Washington, 1960), for an excellent account.

[134] Ann. der Hydrogr., Vol. 66 (1938), pp. 473–74.

of a ship would effectively alter the surface tension of the mercury in a marine barometer, and thus cause the capillary error to differ from the value it had when the index correction was determined on shore.

8. MECHANICAL DEVICES FOR THE CORRECTIONS

For a large fraction of the human race, even of literate people, almost anything is preferable to the necessity of making a computation; and inventors have constantly endeavored to produce devices to save them this trouble. As early as 1749 C. F. Ludolff[135] applied to a barometer a brass plate with diagonal rulings, having a scale of temperature along the bottom. This plate slid sideways in guides, just behind the barometer tube. Each diagonal line was marked with the appropriate barometric height at its left end, and the equivalent pressure in pounds per square inch, calculated from the data then available, at its right end. This was probably the first barometer that could be read in pressure units.[136]

We have already described the "roller blind" scale attached to the thermometer of De Luc's famous portable siphon barometer.[137] This did not entirely avoid calculations, but it did save looking anything up in a table. In 1910 De Luc's device was "improved" by Dr. Wilhelm Schocke of Kassel, who patented various ways of moving the scale

transversely behind the thermometer by a system of levers as the reading index of the barometer was moved up and down.[138] But long before this a device was described by J. H. Müller of Darmstadt which was, in effect, the inverse of Schocke's scheme.[139] In Müller's barometer the scale itself could be moved up and down by a greatly reducing lever system when a pointer was set opposite the end of the thermometer column. This would give the exact correction at a mean pressure, but the accuracy could be improved by having a nomogram behind the thermometer, with verticals representing pressure and nearly horizontal lines for temperature. The pointer was then replaced by a horizontal bar, which would be adjusted to pass through the appropriate intersection on the nomogram. It is worth noting that this elaborate apparatus was applied to the ordinary "bottle barometer" with the turned-up tube and pear-shaped bulb.

For the correction of marine barometers graduated in millibars, Gold[140] showed in 1914 that it was feasible to make a single correction for index error, gravity, temperature, and height above sea level (to the heights possible in a ship) by a suitable addition to, or subtraction from, the temperature of the attached thermometer before the table of temperature corrections is entered. It was therefore found possible to replace the ordinary attached thermometer of the "Kew" pattern marine barometer with a thermometer

---

[135] *Mém. Acad. Berlin* (1749), pp. 33–37.

[136] A barometer of this period at Utrecht (inventory no. G–40) also has a pressure scale.

[137] See Chap. 7, p. 138.

[138] German Patent 234,556 (1910).

[139] *Annalen der Physik.*, Vol. 5 (1800), pp. 17–32.

[140] *Quart. J. r. Meteorol. Soc.*, Vol. 40 (1914), pp. 185–201.

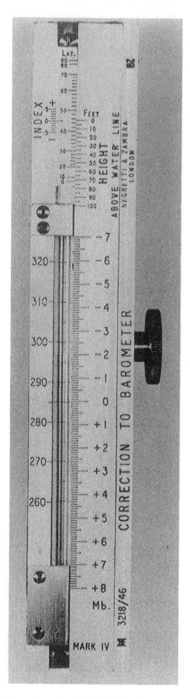

Fig. 8.2. The "Gold slide," recent form. *(By permission of the Controller, H. M. Stationery Office)*

built into a special slide rule (Figure 8.2). In this "Gold slide" there is a small movable scale of latitude which is clamped once and for all with its fiducial mark opposite the calibrated index correction of the particular barometer. A larger slide carries a scale of "height above water line" which can be moved by a rack and pinion to make any height coincide with any latitude; it also carries a scale of barometer corrections in millibars which is beside the tube of the thermometer. The observer has only to set the latitude opposite the height of the barometer above water line and read the correction off opposite the end of the mercury in the thermometer. In spite of several approximations, the accuracy is sufficient.[141]

In 1921 S. P. Fergusson described[142] an attachment to the barometer which is a combination of De Luc's diagonal scale (now a transparent plate) and the "Gold slide." Index and gravity corrections are taken care of by displacing the transparent plate in a vertical direction. This device makes one simplifying assumption fewer than Gold's, and can be used over a wider range of pressures. It does not attempt to reduce the reading to sea level.

A very recent reading mechanism for a barometer, which will be described in Chapter 9, includes means of compensating for the effect of temperature, exactly at one pressure, approximately at neighboring pressures.[143]

[141] Great Britain, Meteorological Office, *Handbook of Meteorological Instruments; Part I, Instruments for surface observations* (London, 1956), p. 33.

[142] Washington, *Monthly Wea. Rev.*, Vol. 49 (1921), pp. 289–93.

[143] P. Du Pasquier and S. Letestu, Swiss Pat. 307036 (1955).

Chapter 9

The Improvement
of Accuracy

1. INTRODUCTION

IN ANY scientific instrument there are really only two requirements for accuracy. Firstly, the instrument should have and maintain a calibration under given conditions; and secondly, the errors under other conditions should be known and constant in time. Provided the first of these requirements is satisfied, it is really only a matter of *convenience* to have the readings of the scale correspond exactly with some scale widely recognized for the specification of the quantity being measured.

In the mercury barometer, the improvement of accuracy has consisted in the gradual evolution of designs which permit the measurement of the vertical distance between two mercury surfaces more and more exactly, and which ensure adequate constancy of calibration, including, as an important requirement for this, constancy of the error due to capillarity.

Seen from this point of view, the evolution of the barometer seems even more complex than it appeared from the standpoints of Chapters 6 and 7. We cannot ignore all the false starts

and dead ends, although some of the more fantastic aberrations need not be recorded. Compared to the astronomical telescope, for example, the barometer has had a checkered career, even though notable telescope makers have from time to time busied themselves with its improvement. The reason is, I think, that while everyone knew that to make a good telescope all the resources of technique and ingenuity would be needed, the barometer appeared to be a simple instrument that could be managed by any country craftsman. Moreover, from the time of Galileo, there has been no doubt of the desirability of making bigger and more perfect telescopes; but it was not until almost the middle of the nineteenth century that the requirements for accuracy in barometers became at all clear. When this happened, it also became clear that there was a need for a few barometers of a much higher order of accuracy than the majority. In this chapter we shall stop short of these special instruments, which will be dealt with in Chapter 10.

Some sort of arrangement is needed, and it cannot be in any way chronological. We shall (quite arbitrarily) begin with mechanical improvements, first the means of reading or adjusting the top and bottom of the mercury column, then the question of the scale itself. Thirdly, we shall discuss the matter of stabilizing the capillary error, including in this the history of the long rivalry between the siphon and the cistern barometers. This will lead naturally to the history of the fixed-cistern barometer. Next we shall deal with the means used to secure a good vacuum over the mercury. We shall end the chapter with an account of several special constructions which, although not always successful, have been devised with accuracy in mind.

2.   THE LEVEL IN THE TUBE

When barometers were first observed, the most obvious source of inaccuracy, though by no means really the worst, was the difficulty of reading the level of the mercury in the tube. This was complicated by the two possible definitions of the level: the extreme summit of the mercury meniscus, or its intersection with the tube. Sometimes it is quite easy to be sure that the barometer maker intended the latter alternative, as in the famous siphon barometer of De Luc,[1] where the scale lines ended right up against the tube, and there was no index. Where there is an index, the decision is more difficult. In early barometers the index was usually a small triangular pointer, sometimes with its top horizontal, sometimes with its bottom edge level, sometimes isosceles, with neither edge level.

In the last case, it is fairly certain that the edge of the meniscus was read, and where the top of the index is level there is a strong presumption in favor of this conclusion, especially when the index curves round, close to the tube. If the bottom edge is horizontal, the twentieth-century observer may assume that the intention was to read the summit of the meniscus; but this may well be an erroneous idea, for some people preferred to read the edge even as late at 1860.[2] The practice of reading the summit of the meniscus seems first to have become common in England, and the reader may remember that in 1776 Henry Cavendish declared himself strongly in favor of it.[3] It is more than likely that this is due to the great improvement of the reading index introduced, apparently by Jesse Ramsden, some time about 1775. This improvement was twofold: first, the index was in the form of a ring which enclosed the tube and had its lower edge turned accurately flat and horizontal; second, the mounting of the barometer had a slot cut in it behind the tube, so that the surface of the mercury could appear as a curve against an illuminated background. The flat end of the ring could then be set with its plane tangent to the curve. It is probable that this simple device reduced the setting error by almost an order of magnitude over the plain index, merely by making it easy to eliminate the effect of parallax. My reason for attributing this invention to Ramsden is that Lavoisier did so:

. . . nous nous sommes servis du moyen dont Ramsden paraît s'être servi le premier: il consiste à faire descendre, par

---

[1] P. 136 above. De Luc, of course, recorded this intention.

[2] C. M. Bauernfeind, *Elemente der Vermessungskunde* (2d ed.; München, 1862), p. 355.
[3] See p. 188 above.

le moyen d'une vis adaptée à une crémaillère, un corps plan jusqu'à ce qu'il soit en contact avec la surface du mercure, et d'observer au transparent.[4]

It is noteworthy that from that day to this the great majority of high quality barometers have incorporated this simple invention. It has been found advisable to have the ring considerably larger than the barometer tube.

It may be noted that Louis Cotte (1740–1815) attributes the invention of the ring-shaped index to the Paris instrument maker Megnié.[5] This claim will not hold water at all, for Megnié was the other half of the *nous* in the above quotation from Lavoisier. Moreover, we have Megnié's own words, quoted by Truchot a century later in a description of Lavoisier's surviving instruments: "Ramsden, artiste d'Angleterre, fit . . . un baromètre très ingénieux; c'est le premier que j'aie vu qui mesure le mercure au transparent."[6] He goes on to say that this is a valuable discovery.

The index having been set to the level of the top of the mercury, its posi-

tion must be read against the scale. One of the simple aids to doing this is the vernier, invented about 1630 by Pierre Vernier (c. 1580–1637) of Ornans in Burgundy.[7] I have not been able to determine when, or by whom, the vernier was first applied to the barometer. It comes as a surprise not to find any mention of the vernier in the great book by De Luc,[8] in spite of the long historical note on the barometer with which it opens. Again it may have been Ramsden at about the date of the barometer mentioned in note 4; there is also a barometer of which the details are illustrated in Figure 9.1,[9] dating from about 1770. This has scales in English inches and tenths (near the tube) and French inches and lines, with a vernier to 0.01 in. for the former, and to 0.1 line for the latter. The index, which is apparently flat on the bottom edge, is stated to be duplicated behind the tube to avoid parallax, so that the top of the meniscus must have been read. It is also noteworthy that this instrument had a thermometer graduated in Fahrenheit degrees, provided with a movable index and a second scale of temperature-corrections in hundredths of an English inch calculated for a pressure of 30 inches of mercury. The zero of this scale is opposite 55°F. The reviewer (probably the Abbé Rozier) mentions that there

[4] Lavoisier, *Works* (Paris, 1865), III, 755–56. There is a barometer by Ramsden (Inventory no. 1893–143) in the storeroom of the Science Museum, London. This must have been a little earlier; instead of a ring there are index plates behind and in front of the tube, the front one being hinged to a bracket on the vernier slide. They were designed with flat tops, the front one with a small semicircular notch opposite the center line. There is no slot for light, and no rack and pinion.

[5] Cotte, *Mémoires sur la météorologie. Pour servir de Suite et de Supplément au Traité de Météorologie, publié en 1774* (2 vols.; Paris, 1788), I, 500. According to Daumas, *Les instruments*, pp. 360–61, this is probably Pierre Bernard Megnié, but the story of this important *atelier* is in some confusion. At any rate, we shall have occasion to refer again to the barometers Megnié made for Lavoisier.

[6] *Ann. Chim. et Phys.*, Vol. 18 (1879), p. 305.

[7] It is quite incorrect to call this device a Nonius, as is done in some countries. The arrangement described by Pedro Nuñez or Nonius (1492–1577) in his *De Crepusculis* (Lisbon, 1542), fol. e1^v–e3^v, is quite different and much more complex.

[8] *Recherches sur les modifications de l'atmosphère* (Geneva, 1772).

[9] "Description du baromètre de Ramsden." *Obs. sur la phys.*, Vol. 1 (1772), pp. 509–12.

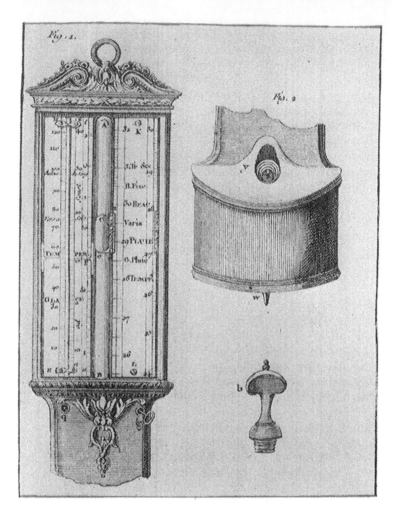

Fig. 9.1. A barometer by Ramsden about 1770. *(Courtesy of the Trustees of the British Museum)*

will be an error at other pressures. We shall hear more about this barometer.

In the Science Museum there is another early barometer with a vernier, marked "J. Sisson, London,"[10] which is also interesting for other reasons.[11] It is

unlikely on stylistic grounds to be the work of the elder Sisson, Johnathan, but if it were, he would have to be given the priority. There is also a fascinating passage in Roger Pickering's description of his meteorological station.[12] The following, from page 5, is his entire description of his barometer:

I have found those with open cisterns more sensible than the portable ones. That with which I make my observations,

[10] Inventory no. 1927–1910. There is also one in the Whipple Museum, Cambridge (Inventory no. 281). Johnathan Sisson (1690–1760) is stated to have handed over his shop to his son Jeremiah in 1747, who was apparently still in business in 1788 (see Daumas, *Les instruments,* p. 306). In *Hist. Acad. r. des Sci.* (1751), p. 173 there is a reference to a "Barometre portatif Anglois de la construction de Sisson," but no details.

[11] See p. 222 below.

[12] *Phil. Trans.,* Vol. 43, no. 473 (1744), pp. 1–17. (No. 473 is erroneously numbered from page 1.)

is with an open cistern, furnish'd with a micrometer, that divides an inch into 400 parts; by which I am capable of perceiving the most minute alteration of the gravity of the air: it was made by Mr. Bird of the Strand; whose accuracy in graduation deserves, I think, notice and encouragement.

Now what was a "micrometer"? There is just a possibility that it may have been a vernier.

There is a barometer in the C.N.A.M. at Paris[13] by Mossy, dated 1768, with a vernier reading to twelfths of a line, or $\frac{1}{144}$ inch, and numerous other eighteenth-century barometers with verniers still exist. One of the most elegant, by Nairne & Blunt, London, is in the office of Dottoressa Bonelli, the Director of the Museo di storia della scienza in Florence; there is another by the same makers in the Whipple Museum, Cambridge.[14] In the same collection there is an instrument by "B. Martin in Fleet Street London"[15] which shows that even in purely decorative barometers a vernier was soon demanded.

Gottfried Erich Rosenthal applied diagonal scales to a siphon barometer instead of verniers,[16] but he does not seem to have persuaded anyone of their superiority.

Once the idea of the slotted frame was adopted, it was a short step to make the frame out of a brass tube, carrying the scales, with the ring index in the form of a short piece of tubing that would slide easily inside or outside it. The pioneer in this seems to have been one Horner, whose excellent portable barometer is described in Gehler's Physikalisches Wörterbuch[17] as dating from 1799. In this instrument there is a slotted ring of spring brass to grip the frame tube at any desired position. Another piece of tube carries the vernier and its flat lower end forms the index. The two tubes can be moved relatively for fine adjustment by means of a threaded, knurled ring.

A similar but simpler construction was used in the portable cistern barometer of Nicolas Fortin (1750–1831), described in 1809 by J. N. Hachette.[18] Fortin's barometer was even more interesting for other reasons, and we shall meet it again.

Since Fortin's time the vernier on many types of barometers has been mounted in this way, with differences in the manner of moving it. In Fortin's original construction an outer tube with a knurled ring and spring friction fingers was provided, the index and vernier being merely pushed up and down. Later on it became more usual to provide a rack and pinion, the rack being attached to an internal sliding tube, the lower edge of which formed the index. In this construction the vernier is engraved on a rectangular piece of tube which slides nicely in the front slot, thus performing the double function of keeping the index tube from

---

[13] Inventory no. 1582.

[14] Inventory no. 747.

[15] Inventory no. 811.

[16] Beytrage zu der Verfertigung, der wissenschaftlichen Kenntnis, und dem Gebrauche meteorologischer Werkzeuge (2 vols., Stuttgart, 1782 & 1784), I, 30–33.

[17] I (1825), 784–85. The reference by Gehler in a footnote is wrong.

[18] Programmes d'un cours de physique, etc. (Paris, 1809), pp. 221–25. Nicolas Fortin was one of the most important French instrument makers, who contributed mightily to the renaissance of his craft in France after the Revolution. For many details see Daumas, Les instruments, passim.

rotating and bringing the surface of the vernier flush with the scale.

This mechanism probably reached its final form, and almost its present-day appearance, in the hands of Patrick Adie about 1850. Very soon after that the larger London makers were making large numbers of barometers, especially marine barometers, all of very similar appearance. Some time after 1850 the practice of enclosing the entire scale in a glass tube arose,[19] and has continued to be standard British practice, though less popular elsewhere.

The slotted tubular frame was also applied to the siphon barometer of Gay-Lussac,[20] and in this instrument the whole mechanism has to be duplicated. One of the most elegant of these instruments is in the *Conservatoire National des Arts et Metiers* at Paris,[21] bearing the inscription "Bunten quai pelletier 30 Paris 1841." This has scales at top and bottom engraved on silver in millimeters, and verniers reading to 0.1 mm. To economize weight and reduce the diameter, Bunten cut racks through part of the thickness of the tube along one side of each front slot, mounting a pinion with a small knurled knob on each of the index slides. The greatest diameter of the brass tube is only about 20 mm., and the top part is narrower than this. A design like this requires the superb workmanship it received.

I have seen only one type of barometer in which the traditional action of scale and vernier is reversed, i.e., the scale moves and the vernier is fixed. This was made by John Newman some time after 1830. There is one at Kew[22]

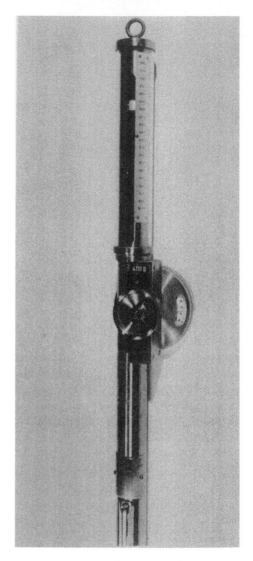

Fig. 9.2. H. Darwin's scale and dial, 1911.

and one at the Meteorological Office, Bracknell.[23] It is very difficult to see what possible advantage Newman hoped to gain from this construction. Against any such one must be set the serious disadvantage that the scale is wrong way up, the higher readings being at the bottom. I should imagine

[19] As in our Figure 7.17.
[20]   See p. 140 above.
[21] Inventory no. 2628.
[22] Inventory no. 152.

[23] Inventory no. 4108 B.

that the chance of an observer making a mistake would be greatly increased.

For a stationary barometer the index seldom needs to be moved far from one observation to the next. The firm of Fuess made a barometer, at some time about 1880, in which the entire motion was provided by a long screw, operated by a knurled collar. This idea was later patented by F. L. Halliwell[24] with the important addition that a vernier is dispensed with, the knurled collar being graduated. This was put on the market by Negretti & Zambra, but does not seem to have won general acceptance. Another way of avoiding the use of a vernier is by making the pitch circumference of the pinion (in a rack-and-pinion drive) some convenient fraction of an inch or centimeter, and attaching a graduated, circular scale to the pinion shaft. This was originally suggested in 1698 by William Derham.[25] In 1911 it was patented[26] and greatly improved by H. Darwin of the Cambridge Instrument Co. An instrument of this design (Figure 9.2) may be seen at the Meteorological Office, Bracknell.[27] It was presented to the Meteorological Office in 1921; but apparently the vernier is hard to improve on for this purpose.

We must again go back to the year 1698. In that year Stephen Gray, also concerned about the difficulty of reading the top of the mercury column, wrote to the Secretary of the Royal Society "about a way of measuring the

heighth [sic] of the mercury in the barometer more exactly,"[28] sending a sketch (Figure 9.3), which to us scarcely needs explanation, though we should note that there was a cross hair at the focal point of the objective of the compound microscope, and that the pitch of the long screw was 0.1 inch,

[28] Phil. Trans., Vol. 20 (1698), pp. 176–78.

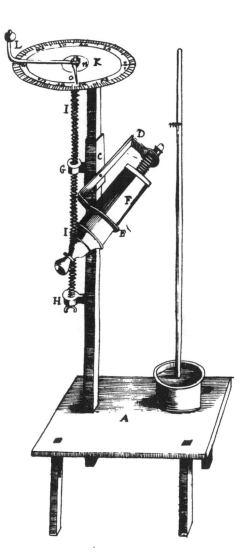

[24] British Patent 139,029 (1919). The instrument may be seen at the Science Museum, London (Inventory no. 1921–321).

[25] Phil. Trans., Vol. 20 (1698), pp. 45–48.

[26] British Patent 12,326 (1911).

[27] Inventory no. 4111B.

Fig. 9.3. Gray's micrometer, 1698.

so that with his 100-division dial he could read directly to 0.001 inch. There was a spring at the top of the screw to prevent backlash.

If the reading of the position of the top of the column had really been the only, or even the chief, difficulty, Stephen Gray's idea would have sufficed for a long time. This use of a microscope is very attractive, especially to presbyopic old gentlemen. In view of the other errors, why not attach the microscope to the vernier slide? This is what Sisson did in the instrument preserved in the Science Museum (p. 198). The bracket by which the microscope is supported does not really look stiff enough for its purpose.

In a catalogue of their instruments published[29] in 1829, the German firm of Pistor and Schiek advertise both siphon and cistern barometers with compound microscopes for reading. I have not seen one of these, but in the *Deutsches Museum* at Munich there are two beautifully made siphon barometers with brass scales, and microscopes at both ends, by J. G. Greiner jun. of Berlin.[30] There is a rather similar instrument by the same maker in the University Museum at Utrecht,[31] dated 1834 by the late Dr. P. H. Van Cittert, and at any rate included in an inventory of 1838.

Microscopes are much used in primary barometers, but these will be dealt with in Chapter 10.

Once you start with optical devices the sky is the limit. In 1907 Adalbert Deckert of Rees-am-Rhein was granted a patent for a periscope to enable both ends of a barometer to be brought into the same field of view,[32] and in 1952 P. P. Suárez-Cobián of the Spanish meteorological service, on the debatable assumption that the siphon barometer would be preferable to the fixed-cistern barometer if both sides could be read at once, suggested another rather similar optical method of doing this.[33]

For the sake of completeness, I should record a couple of tricks for avoiding parallax. They are both "done with mirrors," as tricks, in popular legend, frequently are. The first, which I have not seen described, is represented by a barometer of unknown origin in the Munich *Sternwarte*. This has a siphon tube of 10 mm. bore throughout, and a Bunten air trap. The scale is ruled on the front surface of a piece of mirror glass, about 8 cm. wide and a meter long, in millimeter intervals from 0 to 900, each line passing right across the entire mirror behind both the tubes. It is numbered in centimeters. As the graduations are reflected in the mirror surface at the back of the glass, the observer has only to place his eye at such a height that the line nearest the top of the mercury column appears to coincide with its reflection.

The second trick was described by the celebrated E. E. N. Mascart (1837–1908).[34] In this a scale is seen by reflection in a beam-splitting mirror (half-gilt, Mascart suggests), scale and mirror being placed so that the virtual image of the former coincides with the

[29] In *Astr. Nachr.*, Vol. 7 (1829), col. 93–102.

[30] Inventory numbers 8935 and 54905. The second is a much larger instrument.

[31] Inventory no. G.35.

[32] D.R.P. 205,178 (1907).

[33] Madrid, Servicio Meteorológico Nacional, *Publicaciones*, Serie A, núm. 22 (1952).

[34] *J. de Phys.*, sér. 2, Vol. 2 (1883), p. 343.

axis of the barometer tube. This is an attractive idea, but its technical realization would be somewhat costly, and the requirements for lighting rather severe.

If cost is not a consideration, a barometer can be made to set itself; this was done in 1948 by a firm in Wyoming.[35] A beam of light only 0.002 inch high is projected across the top of the mercury surface and falls on a photoelectric tube, controlling a servo system which causes the light beam and the tube to follow the mercury surface, if necessary to within 0.0001 inch. A counter, geared to this motion, gives an instantaneous reading of the height of the mercury column in easily legible figures, and as the whole instrument is enclosed in a thermostat, the gearing can be arranged to show the actual atmospheric pressure.

3. THE LEVEL IN THE CISTERN

In the use of the cistern barometer it is important for accuracy, either that the level of the mercury in the cistern should be at the zero of the scale, or that adequate allowance should be made for any departure from this condition. The second alternative is adopted with the fixed-cistern barometer, which was dealt with at some length in Chapter 7 and will appear again farther on; but at the moment it is our duty to explore what at first sight looks like an immense jungle of devices designed to ensure that the level of the mercury in the cistern remains as near as possible to the zero of the scale.

We shall try to reduce the subject to a kind of order by recognizing that almost all these mechanisms are in fact a combination of one of the devices to the left of the vertical rule in the following Table 9.1 with one of those on the right. The numbers in parentheses are the approximate dates at which, as far as I can determine, each device came into use. Some of the improvements which appeared from time to time consisted in a better use of materials rather than a novel mechanism.

Table 9.1   Classification of Cisterns

Means of indication	Means of alteration
Overflow (Before 1726)	Pouring (Before 1700)
Fixed index (1726)	Leather bag (1695)* (1772)
Pool of mercury (1736)	Plunger in cistern (1744)
Floating index (1775)	Movable scale (1779)
Scale on cistern (1799)	Cistern screws up (1810)

* In 1695, but not until 1772 in connection with an index.

At the time Jacob Leupold wrote his *Theatrum machinarum* (1726),[36] the only means of adjusting the level of the mercury seems to have been by overflow. Figure 9.4 shows the sort of simple barometer cistern that was in vogue at that time. It was made of boxwood. After the barometer tube had been filled and glued into the piece

---

[35] Ideal Laboratory Tool & Supply Co., *Instruments*, Pittsburgh, Vol. 21 (1948), p. 596. See also Chap. 12, p. 353 below.

[36] *Pars III, Theatri statici universalis, sive theatrum aërostaticum*, etc. (Leipzig, 1726). See especially plates III and IV and p. 253.

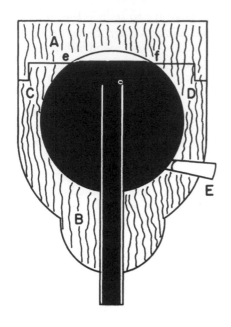

Fig. 9.4. Simple barometer cistern, after Leupold, circa 1720, as in process of filling.

BCD, it was filled to the level *ef* with mercury, and the piece *A* glued on. When the glue had dried, the instrument was carefully up-ended, and the plug *E* removed in order to let the mercury run out to its final level, and of course the amount of mercury in the instrument would depend to some extent on the atmospheric pressure at the time. Barometers on this principle were probably made in larger numbers than any other type except the "bottle barometer" for another hundred years, with the important improvement that the plug became an ivory screw. In 1825 Gehler[37] stated that barometers of this sort were sold in Paris under the name "Geneva barometers." They were intended to be portable; when such a

barometer was to be moved, it was carefully laid horizontal with the ivory screw upward, the screw removed, the cistern filled with mercury, and the screw replaced. When the barometer was again erected and the screw taken out, the surplus ran out until a certain level had been attained. Réaumur's nephew, Mathurin Brisson (1723–1806), presented such a barometer to the Académie in 1755, stating that he had tested it on a journey of more than 200 leagues, no light matter in those days.[38]

In the *Deutsches Museum* there is a portable barometer[39] of this type, obviously intended for mountain use, bearing the inscription "Haas & Hurter, London N⁰ 38."[40] This has scales in French inches and lines, and in English inches and twentieths. The vernier on the French scale reads to 20ths of a line, and on the English scale apparently to $\frac{1}{1200}$ inch! This recalls the animadversions of F. W. Voigt,[41] who wrote in 1802 that he had been to workshops where cistern barometers reading to 0.001 inch were being made, without finding a mechanic who knew that such a barometer ought to be "corrected" by comparison with a siphon barometer. Obviously the adjustment by overflow would not justify any such precision of reading, but the makers

[37] *Physikalisches Wörterbuch*, Vol. 1, p. 781, note 1.

[38] *Hist. Acad. r. des Sci.* (1755), p. 140.
[39] Inventory no. 25.
[40] Haas & Hurter do not appear as partners in the London directories, as far as I can determine. In fact I have not seen Haas mentioned there. I am informed by Mr. Adams of the Science Museum, London, that H. B. de Saussure took a barometer by Hurter when he went up Mont Blanc in 1787, and that there is a portable barometer by Haas in the *Landes-Museum* at Darmstadt, and a barometer of De Luc's pattern at Haarlem.
[41] *Versuch Kritischer Nachträge*, etc. p. 66.

who had a balanced view of the requirements were few.

In the last decades of the eighteenth century this scheme of adjustment by overflow was often combined with the leather bag which enabled the volume of the cistern to be adjusted. This type of barometer, not always intended to be portable, was made very carefully by the best makers. There is one in the C.N.A.M., Paris,[42] inscribed "Dollond London" on its silvered brass scale, and probably dating from about 1780, and another signed "R. Bianchy Rue St. honoré No. 252."[43] Of the same type is a very handsome barometer in the *Istituto di fisica* of the University of Bologna, with the inscription "Giuseppe Brusca, 1806." This is graduated in both English and French inches. As we shall see, much better ideas for adjust-

[42] Inventory no. 4245.

[43] Inventory no. 8517. Bianchy flourished c. 1785. See Daumas, *Les instruments*, p. 195.

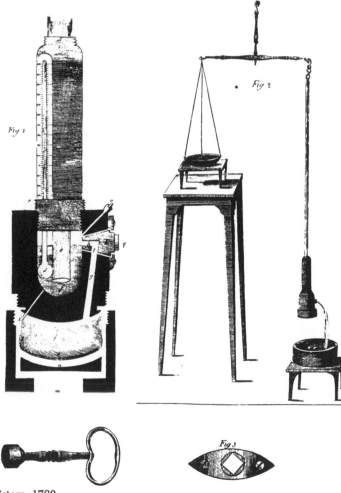

Fig. 9.5. Austin's overflow cistern, 1790.

ing the level of the mercury were available by 1806.

There are two overflow barometers by William Cary (c. 1759–1825) in the History of Science Museum at Oxford. In these a short ivory tube projects from the top of the adjustable cistern; there is a horizontal saw cut in this tube, and presumably the mercury was raised until it just overflowed through, or perhaps until its surface could just be seen in, this saw cut. The tube had an internal thread into which a plug could be screwed for portability. The workshop founded by Cary seems to have been in business in London for a century after 1790.

An interesting variation on the overflow cistern was described to the Royal Irish Academy by the Reverend Gilbert Austin in that year.[44] The plate which illustrated his paper is reproduced in Figure 9.5. A supply of mercury is contained in a leather bag, and when the stopcock is turned as shown, the cistern can be overfilled by pressing the bag, and the mercury then allowed to run out into the bag again. For portability the cistern could be filled and the stopcock turned with the key.. The narrow channel shown from the bottom of the cistern down to the left was used for the initial filling and then closed. Using the apparatus shown to the right, Austin established by numerous experiments that "four grains of mercury cannot be added to the quantity which reaches the standard height in the basin without overflowing."[45] He believed, possibly with reason, that the floating gauges then in use could

give greater errors. At any rate, his design avoided a separate supply of mercury.

Incidentally, he put the bulb of the thermometer in the cistern, an unfortunate practice that seems to have been copied by (or to have occurred independently to) other makers, especially John Newman.[46] It is really the temperature of the mercury in the tube that we need to know.

One of the most sensitive ways of adjusting the level of a liquid is by means of a point, either under the surface, or above it. Leutmann used one in a simple and interesting way[47]

[46] See p. 219 below.
[47] Described by Leupold, *Theatri statici* etc., p. 254 and Table IV.

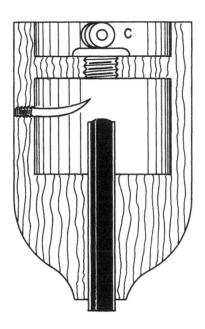

Fig. 9.6. Leutmann's iron point, 1726; cistern as inverted for filling.

[44] *Trans. r. Irish Acad.*, Vol. 4 (1790), pp. 99–105.
[45] *Ibid.*, p. 104.

(Figure 9.6). When the barometer tube had been filled, mercury was poured in until its surface coincided with the point, and the screw C fastened up. The level in the cistern when the barometer was turned right way up would, of course, depend on the atmospheric pressure at the time, and the instrument would act as a fixed-cistern barometer.

A much better way of using a point was developed nearly half a century later by Ramsden.[48] Let us refer again to figure 9.1. When the ivory screw b was removed from y, it was possible to see the little ivory point z, and by raising or lowering the leather bottom of the cistern, to alter the level of the mercury so that it would just touch the point. If the mercury is clean, this can be done with great enough precision to make other errors much more important; but we may wonder about the difficulty of seeing the contact in Ramsden's barometer in anything but a very strong light.

This difficulty was removed by Fortin thirty years later; but I prefer to postpone consideration of Fortin's remarkable instrument until after we have looked at another means of indicating the level of the mercury in the cistern. I refer to the ivory float.

There are two descriptions[49,50] of a barometer devised by Ramsden, apparently a short time after the one with the point. The two accounts agree, but the second is the clearer. There was a hole in the side of the cistern, closed by a strong ivory screw when the instrument had to be moved. Open, it gave a view of the interior, in which were "deux morceaux d'ivoire fixes sur lesquels est tracée une ligne horizontale, & entre ces deux morceaux d'ivoire qui sont séparés, on a placé un petit cylindre d'ivoire qui a toute la liberté de monter et de descendre sur la surface du mercure contenu dans le reservoir."[51] On this ivory cylinder had been drawn a black line which is made to coincide with the other lines by adjusting a screw [which moves a leather bag?] at the bottom of the cistern. One of these barometers was brought to Montmorency in 1775 by Sir George Shuckburgh. "Il m'a paru fort exact, & fait avec tout le soin, la précision & la propreté que les Artistes Anglois savent mettre à leurs ouvrages."[52] A handsome tribute!

On the basis of these publications, it would seem reasonable to ascribe the invention of the ivory float to Ramsden. The matter is rendered more complicated by the existence of a barometer with an elegant silvered scale signed by the famous John Bird (1709–1776).[53] Out of the top of the cistern of this barometer projects a short tube terminating in an inverted glass vial. In this is an ivory rod with a mark around it, and a vertical ivory plate bearing a horizontal scratch. The cistern is adjusted by means of a screw at the bottom, and presumably a leather bag. If John Bird, who retired in 1773, really made this barometer, it would bring him into competition with Ramsden

[48] *Obs. sur la phys.*, Vol. 1 (1772).

[49] William Roy, *Phil. Trans.*, Vol. 67 (1777), p. 658.

[50] Louis Cotte, *Mémoires sur la météorologie* (Paris, 1788), I, 509–10.

[51] *Ibid.*, p. 509.

[52] *Ibid.*, p. 510.

[53] In the History of Science Museum, Oxford, Inventory no. 33,25.

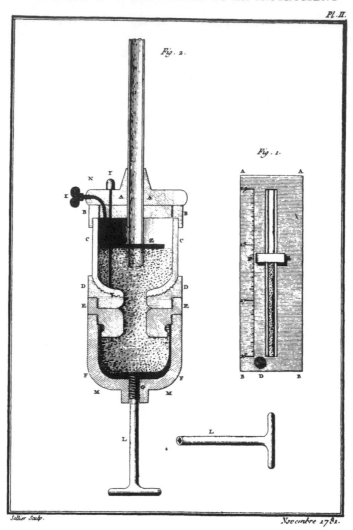

Fig. 9.7. Assier-Perica's barometer, 1781. (*Courtesy of the Trustees of the British Museum*)

for the priority; but the instrument has been extensively repaired and it is not unlikely that only the scale is in fact by John Bird. In any event, this design of float is much easier to use.

There is a nineteenth century barometer at Oxford signed "Carpenter & Westley 24 Regent St. London," very like the one by Bird in design, but with a tube of about ¾ inch bore and a huge cistern.[54] It would seem that the idea was prevalent for a long time.

Soon after the Ramsden design was known on the continent, Antoine Assier-Perica devised what was probably the best cistern for a portable

[54] Carpenter & Westley were in business at 24 Regent St. from about 1840 to 1855.

barometer invented in the eighteenth century (Figure 9.7).[55] The figure is fairly clear; *AA, BB, DD, EE,* and *FF* are of boxwood, *CC* of glass, *X, YY,* and *ZZ* of ivory. But (1) apparently *BB, CC* and *DD* are held together only by some adhesive, which seems risky; (2) the engraver has not shown the leather bag as being lashed to the part *EE*; (3) the screw *G* at the bottom must have been a good deal longer. The fitting *X* was of course unnecessary, but Assier-Perica, who had peculiar ideas about some of the properties of matter, would not trust boxwood to be porous. The float *ZZ* was free to move up and down, and there was a mark on the ivory rod *YY,* to which the top of *ZZ* could be adjusted.

Hurter[56] and later Haas[57] developed barometers, already referred to in Chapter 7, in which the stem of the float protruded through the top of the cistern. The only other actual descriptions of barometers with ivory floats that I have seen are by Georg Winkler,[58] an Austrian professor of mathematics, and by Friedrich Körner[59] of Jena. In Winkler's barometer the stem of the float has a mark round it, which is brought level with the top of a small ivory boss on the top of the boxwood

cistern. Adjustment is by means of a leather bag and screw. There is a screw cap for the place where the stem of the float comes out, to prevent loss of mercury when the instrument is carried upside down. In Körner's barometer the cistern is extremely complicated, and seems to be derived from the designs that his brilliant fellow-townsman Voigt had developed a quarter of a century earlier, just before his early death. By 1824 there were much better and simpler designs.

Barometers with floats seem to have been especially popular in Italy, even well into the nineteenth century. In the *Linceo Giulio Beccaria* at Milan there is one made locally by Marelli and dated 1818. Many museums seem to have them.[60]

Floating an index on the cistern is one thing; floating the entire scale is another. In the Museum of the University, Utrecht, there is a barometer, as far as I know unique, in which this is done.[61] From the weather signs on the scales it is probably of Dutch make; Dr. P. H. Van Cittert dated it about 1800. The scales, in French, English, and Rhenish inches, are engraved on thin silvered brass, fastened to a light wooden lath. This lath terminates in a wooden ring which floats on the mercury in the large cistern. There are, of course, guides for the lath, and it seems

[55] *Obs. sur la phys.,* Vol. 18 (1781), pp. 391–94. Also described by V. Ueber in *Magazin für das Neuste aus der Physik und Naturgeschichte,* Gotha, Vol. 1, part 3, (1782), pp. 98–100.
[56] *J. de Phys.,* Vol. 29 (1786), pp. 345–49.
[57] *Gren's J. der Phys.,* Vol. 7 (1793), pp. 238–40.
[58] *Beschreibung eines verbesserten, bequemen und einfachen Reise-Barometers, nebst praktischer Anleitung zum Gebrauche desselben,* etc. (Wien, 1821).
[59] *Anleitung zur Verfertigung übereinstimmender Thermometer und Barometer für Kunstler und Liebhaber dieser Instrumente* (Jena, 1824).

[60] Florence, *Museo di storia della scienza,* one by Nairne & Blunt (referred to on p. 451); *Istituto Geografico Militare,* no. 154, by Barbanti, Torino, graduated in millimeters (and therefore nineteenth-century); no. 6242, probably made in Florence; Faenza, *Società Torricelliana,* one by Lenvie, Paris, with both inch and metric scales; Rome, *Museo Copernicano,* no. 1659, a mountain barometer, origin unknown.
[61] Inventory no. G.39.

unlikely that the results would have been very encouraging.[62]

Let us now return to the portable barometer of Nicolas Fortin (1750–1831), whose surname is probably known to more people in connection with the barometer than any other except, perhaps, Torricelli. Indeed, there seems to be a tendency even in what should be well-informed quarters to attribute to Fortin the invention of the leather bag, the ivory point, and the use of a glass tube in the construction of the barometer cistern. He invented none of these, as we have seen. But he did more; he combined them; and the result was a design of adjustable barometer cistern that has endured with scarcely any change for a hundred and fifty years.

Our Figure 9.8, from Hachette's book,[63] shows it well enough. The glass cylinder is held between its end plates, padded with thin leather gaskets, by four long screws. From the top of the cistern, about half way between its wall and the barometer tube, depends an ivory point. By compressing the leather bag at the bottom of the cistern, the mercury can be brought up until the point just seems to touch its image in the shining surface. With good lighting and good eyesight, this adjustment can be repeated time and again to within 2 or 3 hundredths of a millimeter.

The ivory point was put at the summit of the annular meniscus. In 1829 Laplace suggested that if the point were somewhat down the slope of the

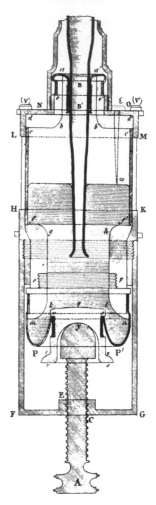

Fig. 9.8. Fortin's barometer cistern, from Hachette, 1809. *(Courtesy of the Trustees of the British Museum)*

meniscus, the capillary depression in the tube could be compensated,[64] and Bouvard[65] calculated a table for various sizes of barometer tube. It would seem that the scheme would be awkward except for very large tubes, since even with a tube of 12 mm. bore the point

[62] Much later, the idea of floating the scale was patented (DRP 223,230 [1908]; U.S. Patent 984,972 [1911] by Dr. Wilhelm Schocke of Kassel.

[63] *Programmes d'un cours de physique*, etc. (Paris, 1809).

[64] Paris, Bureau des Longitudes, *Connaissance des Temps pour l'an 1829*, pp. 301–2.
[65] *Ibid.*, p. 308.

should only be 3.13 mm. from the wall. It appears that some were made in this way, for in 1841 A. Bravais and C. Martins[66] stated that in practice the desired compensation was seldom reached and recommended that the point should be at the top of the meniscus, where it has been ever since.

From about 1840 almost every barometer maker of note has made this instrument, and usually under the name of "Fortin barometer." In France, beautiful barometers of various sizes have been made by Deleuil, Ernst, Secretan, and others. At the *Museo Copernicano* in Rome there is a particularly large and well-made example by Secretan, which appears to have been used by Father Secchi (see Chapter 11) and probably dates from about 1850. It has a very beautifully engraved scale in mm. and half-mm. on silver, the vernier reading to 0.02 mm. The tube must be at least 10 mm. in bore, the cistern about 40 mm., with a beautiful ivory point, and in 1961 the mercury was quite clean and the whole instrument in very good condition. Indeed, the length of time these barometers last makes some of the eighteenth-century objections to the leather bag look very ill-advised.

In England, innumerable "Fortin" barometers were made, especially by Adie, Casella, and Negretti & Zambra. They were sent all over the world. In 1879, for example, G. Hellmann[67] examined Adie no. 1019 at Nicolaew in Russia. I remember cleaning one of Adie's, dating probably from the 1880's, at South-west Point, Anticosti Island, in 1936. It had a very low index error.

These barometers were never as popular in Central Europe, which was enamoured of the siphon barometer through most of the nineteenth century; but they were made by Fuess and by Pistor and Martins of Berlin. H. Wild, then Director of the Russian Meteorological Service, had one, made by the latter firm, at St. Petersburg in 1874.[68]

There were only minor variations in the design and appearance of these cisterns. The exact design of the boxwood parts has some effect on the ease and safety of cleaning the cistern, which must be done occasionally (it is really very easy if you know how); and in my opinion a modification made in the United States is outstanding in this respect (see Chapter 12). With the idea of minimizing the amount of cleaning required, a large barometer was constructed in Vienna near the end of the nineteenth century in which, after the reading had been made, the mercury was lowered into a chamber beneath the cistern and communicating with it only through a small hole. This hole could be closed by a screw if desired. In 1908 J. Liznar of the *Zentralanstalt für Meteorologie und Geodynamik* referred to this instrument, which does not seem to have been described in print, at the end of his description of a smaller barometer on the same principle, which he calls a modified Fortin.[69]

---

[66] *Acad. r. de Bruxelles, Nouv. Mém.*, Vol. 14 (1841), p. 32.

[67] *Repertorium für Meteorol.*, Vol. 6, No. 8 (1879), p. 42. Hellmann (1854–1939) later became director of the *Reichswetteramt* and a noted historian of meteorology.

[68] *Repertorium für Meteorol.*, Vol. 3, No. 1 (1874), p. 63.

[69] *Meteorol. Zeits.*, Vol. 25 (1908), pp. 76–78.

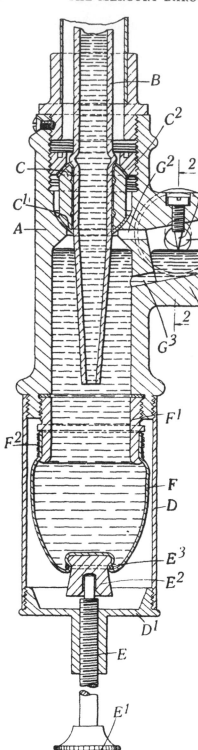

Fig. 9.9. Abraham's patent barometer, 1937.

Attempts have been made from time to time to preserve the principle of adjusting to a point while making the bottom of the cistern entirely of metal. In a patent[70] issued in 1861 to Enrico Angelo Ludovico Negretti and Joseph Warren Zambra, there is described a rigid cistern attached to the frame of the barometer by a screw at its upper part, so that it can be raised and lowered. At the bottom of the cistern there is a cushion or pad on a swivel, which closes the end of the tube when the instrument is made portable. An ivory point is provided, which is used in the ordinary way. A somewhat similar barometer was ascribed to Alvergnat nine years later.[71]

A radically new design of barometer using a point (stainless steel this time) was patented in 1937 by R. M. Abra-

[70] British Patent 238 (1861).
[71] C. Mêne, *Rev. hebd. Chim. sci. et ind.* (1869–1870), pp. 257–60.

ham[72] and put on the market by the firm of Casella, of which he was managing director. It is still called a "Patent Fortin barometer!" Figure 9.9 shows a diagram of the cistern; the glass tube has been replaced by an eyepiece and a window in a side tube, and there is an air filter under the cap at the right. It is very easy to set, but must be carefully installed so as to be vertical, for it will not assume that position when hung up. In the Canadian Service a successful attempt has been made to remove this last disadvantage, and at the same time to reduce the weight of the instrument, while keeping the ease of setting (Figure 9.10). The air filter is not retained.

According to Benzenberg,[73] an instrument maker called Loos at Büdingen used two ivory chisel edges to indicate the zero, instead of a point.

We must now go back to about 1772, when a glass tube seems to have been described as part of a barometer cistern for the first time by Georg Friedrich Brander of Augsburg (1713–1783), one of the most celebrated instrument makers of that, or indeed any other, time.[74] It is interesting to find that even the great makers of astronomical instruments did not neglect the barometer. Brander even found time to write a small book about it,[75] in which he described two new forms of the instrument, the second of which (Figure 9.11) had a glass tube as part of the cistern, a highly original design. The

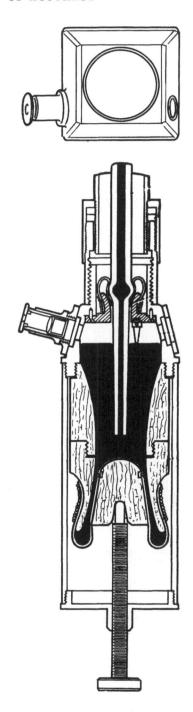

Fig. 9.10. Canadian inspection barometer. (*Dept. of Transport, Meteorological Branch*)

[72] British Patent 497,025 (1937).
[73] *Annalen der Physik*, Vol. 6 (1810), p. 353.
[74] See Ernst Zinner, *Deutsche und niederländische astronomische Instrumente des 11.–18. Jahrhunderts* (München, 1956), pp. 256–63 and *passim*.
[75] *Kurze Beschreibung zweyer besonderer und neuer Barometer*, etc. (Augsburg, 1772).

glass tube was clamped between two sturdy boxwood ends by a sort of clevis and a screw, and in the bottom block there was a filling plug, shown rather diagrammatically in our figure. There was a short scale in lines engraved on the glass tube, for the proper adjustment of the amount of mercury. For transport the barometer would be carefully inverted and the cistern filled almost full; then, when it was reinstalled, mercury would be allowed to run out until the level in the cistern reached the appropriate mark.

The next notable cistern incorporating a glass tube was that of Horner, described in Gehler's dictionary but apparently dating from just before 1800.[76] In this the glass tube not only permitted the level of the mercury to be seen, but also acted as a cylinder in which a piston could be moved to alter the level (Figure 9.12). The glass tube g is held between the steel top a and a projection h' on the outer brass tube h by the threaded cap c, with suitable gaskets. The piston is a rather complicated structure of cork covered with leather and compressed between the steel fittings i and l. It is moved by means of a screw q working in a nut in the bottom plate p. Two rectangular windows are cut opposite one another in the outer tube as shown by the dotted line, extending a little above the bottom edge (ef) of the top cap. This level is made the zero of the scale. The windows were probably made long so that the condition of the plunger could be observed. Note that there is a simple little valve v to let in the air, operated by a nut u.

This is a very complicated structure and in all probability it was dictated by a mistrust of the leather bag. Yet it is the latter, and not the piston, which has survived.

Another way of altering the level of the mercury in a cistern is to plunge a small cylinder into the mercury. The first use of this which has come to my notice was reported to the *Académie des Sciences* in 1744.[77] Father Le Clerc

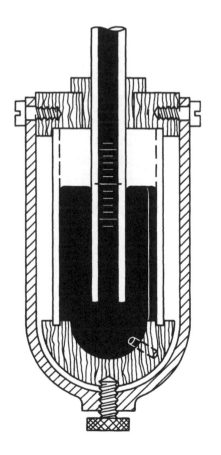

Fig. 9.11. Brander's barometer cistern, 1772.

[76] *Physikalisches Wörterbuch*, I (1825), 784–85 and Figs. 154 and 155.
[77] *Registres* (ms.), 1744, 8 Feb. (pp. 53–54).

had submitted a barometer to the Academy, and it had been referred to a Committee, as usual. This Committee consisted of Réaumur and Buffon, and on February 8 they made their report. "Ce petit Tuyau se meut par le moyen d'une petite Machine assez Simple, et de laquelle il donne une description fort claire." Unfortunately the Committee gave no further description, clear or otherwise, and the Academy withheld its approval, feeling that an ordinary barometer with a relatively large cistern would be better.[78]

According to Blondeau,[79] Lavoisier had a barometer in which the level of the mercury in the cistern was varied by a cylinder of ivory about an inch in diameter which could be raised and lowered by a screw. Later John Gough (1757–1825) suggested[80] a plunger to get over the change in level in the bulb of a "common weather glass," or "bottle barometer." A piston could be used to place the mercury surface at the level of a line engraved round the bulb at the zero of the scale.

Another idea is to have a pinion moving two racks in opposite directions, one carrying the reading index of the barometer, the other "a plug or plunger the exact size of the internal diameter of the tube . . . so that whatever the displacement that had taken place in the cistern, owing to the rise

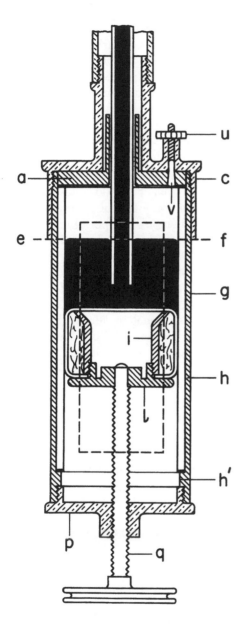

Fig. 9.12. Horner's barometer cistern, circa 1799.

[78] The indicating part of this barometer seems also to have involved some mechanism ". . . l'auteur rend les divisions du Barometre fort sensibles par la transposition qu'il fait de ces divisions sur un cadran mobile." (*Ibid.*, p. 54).

[79] *J. de Marine*, 6e cahier (1780), p. 188.

[80] *J. Nat. Phil., Chem. & the Arts*, Vol. 18 (1807), pp. 81–85. John Gough was a Quaker of Kendal, Westmorland, a blind man of great talent, who taught mathematics to John Dalton and William Whewell. He was the blind man in Wordsworth's *The Excursion*.

or fall of the mercury, it is exactly compensated by the plug being more or less immersed in the mercury."[81] This invention seems to have been presented to Negretti & Zambra by one Wentworth Erk. There is some mystery here; the illustration accompanying the description is almost certainly of a large barometer with a movable scale, a type of instrument which we must now consider.

The sole reason for being able to adjust the level of the mercury in a cistern barometer is to bring it to the zero of the scale. On the Mahomet-and-mountain principle, it may be easier to move the scale to bring its zero to the level of the mercury in the cistern. This can be done just as well in the siphon barometer, substituting a setting and a reading for two readings. As far as the cistern barometer is concerned, the invention of setting to a point was an invitation to make the scale movable, because the point may go directly on the end of the actual piece of material on which the scale is engraved.

It is true that the traveling barometer of Christopher Jetzler (Figure 7.10) had a movable scale; but this was stated to be for adjustment at very high mountain stations where the vertical motion possible to the narrow cistern was insufficient to accommodate the rise in level of the mercury. The first person to make use of the principle as the sole means of adjustment was, as far as I can determine, J. F. Luz.[82] This is rendered more probable by the categorical statement to that effect by

the usually well-informed F. W. Voigt.[83] This would place the invention somewhere about 1780. It was applied to the siphon barometer, and it is interesting that Voigt prefers to have the tube, rather than the scale, movable, because this disturbs the mercury more safely than tapping it before a reading. The two procedures are otherwise equivalent, of course, and I suppose it is a much easier mechanical problem to move the scale. At any rate, I have never seen one with the tube movable.

Nor have I seen a movable-scale barometer that I would date before 1820 without any fear of contradiction, although there are two portable barometers at Munich[84] which might be a little earlier, and one at Milan[85] which might well belong to the eighteenth century, though it is impossible to be sure. This last is the sort of barometer that an instrument maker might have constructed after reading De Luc's *Modifications*[86] and thinking for a while; it is the same shape as De Luc's siphon barometer, but has the movable scale.

The next description of such a barometer that I have found is that of W. Hisinger,[87] who produced a slender siphon barometer enclosed in a two-part mahogany "walking stick," and used it in travels through Scandinavia. Besides the movable scale and a vernier, it was interesting in that the tube was constricted over much of its length.

---

[81] Negretti & Zambra, *A Treatise on Meteorological Instruments* (London, 1864), p. 8.
[82] *Vollständige . . . Beschreibung*, etc. (1784).

[83] *Versuch Kritischer Nachträge*, etc. p. 147.
[84] *Deutsches Museum*, inventory nos. 17 and 61214.
[85] *Linceo Giulio Beccaria*, inventory no. 444.
[86] Geneva (1772).
[87] *Annalen der Physik.*, Vol. 7 (1826), pp. 33–40.

In 1829 J. J. F. W. von Parrot, Professor of physics at Dorpat, made a journey over the Caucasus mountains to Mount Ararat, the ascent of which seemed to afford him some peculiar emotional satisfaction. Perhaps as a form of self-justification for this indulgence, he took levels and made other scientific observations, designing a barometer specially for the purpose and having it made locally in Dorpat. In 1834 he described his journey and his instruments.[88] The barometer was a movable-scale cistern barometer in a flat wooden case. The cistern A (Figure 9.13), rectangular in section, was made of layers of paper glued together around a hardwood base B. About halfway up the cistern was a dividing plate C of cork, just pressed in, with a hole to fit the tube and a smaller hole for a rod D, which could be lowered to form a stopper. There was just enough mercury in the barometer to fill the tube, which was 80 cm. long, and the lower part of the cistern, below the cork. When the barometer was set up, the stopper was removed and the mercury from the tube formed a pool on top of the cork. The movable scale, graduated from 350 to 720 units,[89] was extended by an iron rod E, terminating in an ivory socket F which held the stem of an ivory float G that floated, without any vertical constraint, on the pool of mercury. The proper position of the scale was indicated by the coincidence of index marks H on the socket and float. The vernier slide had two

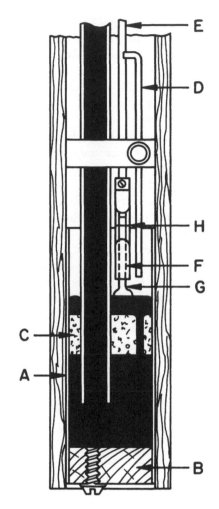

Fig. 9.13. Parrot's barometer cistern, 1829.

silvered plates, one behind the tube, one in front, each with a horizontal line. It was designed to read the top of the meniscus, and the necessity of illumination from the rear was avoided by having the front plate extend only to the center line. There was a plumb-bob in a closed glass tube, to protect it from the mountain winds, and a thermometer with a bulb of the same diameter as the barometer tube.

[88] J. J. F. W. von Parrot, *Reise zum Ararat* (2 vols.; Berlin, 1834). The barometer is described in Vol. 2, pp. 1–8.

[89] Half-lines, i.e, units of 1/24 (Paris) inch. This was fairly common in Russian barometers at the time.

Von Parrot got this barometer safely over the Caucasus mountains and part way up Mount Ararat, when the tube was broken in an accident. He returned to base, installed a new tube, and took it up the mountain again. He still had it when he wrote.

I have described this barometer at length because this way of using a float has not, as far as I am aware, been repeated, and because the entire instrument seems so very ingenious and suitable for its purpose, although, apart from the graduation of the scale, it could be constructed by a mechanic of moderate ability. Also, it survived two journeys over the Caucasus in the days before railways. It is very interesting to learn that, with a somewhat more professional cistern of steel, it became and remained very popular in Russia, especially for journeys in that immense country.[90]

The barometer designed by Francis Baily[91] for the Royal Society had a movable scale, but will be treated later under another heading.

We now come to one of the most successful and durable station-barometers the world has ever known. I refer to the famous movable-scale barometer by John Frederick Newman.

A barometer for permanent installation at a fixed observatory need not be portable, but unless it is to be made there, it must at least be transportable. When Captain Edward Sabine, F.R.S. (1788–1883) persuaded the British government to establish a number of magnetic and meteorological stations in widely separated parts of the British Empire,[92] he wished to equip each one with as accurate a barometer as could possibly be sent out. The design was entrusted to Newman, then on Regent Street, who in 1833 had devised the portable barometer described on page 148 above. Something much more accurate was called for, and it is a measure of Newman's ability and self-confidence that he was able to develop such an outstanding instrument out of his earlier design.

A cross section of the lower part of this barometer is shown in Figure 9.14.[93] If the reader will refer to Figure 7.7 and the discussion on page 148, it will not be necessary to describe the way in which the barometer is made transportable. We need only note the way in which the movable scale was adapted to the previous design, an iron rod tipped with ivory passing through a mercury-tight packing gland in the boxwood top of the cistern into one of the two square brass tubes which form the sides of the barometer frame. This rod and the scale at the other end of it could be raised and lowered by the rack and pinion, the contact of the

---

[90] H. Wild, *Repert. f. Meteorol.*, Vol. 3 (1874), No. 1, p. 70. "This instrument, almost entirely unknown elsewhere but very widely distributed in Russia, and used almost exclusively for travelling. . . ."

[91] *Proc. Roy. Soc.*, Vol. 4 (1837), pp. 1–3; *Phil. Trans.*, Vol. 127 (1837), pp. 431–41.

[92] A committee, of which Sabine was a member, worked at this from 1836 to 1839, and several observatories were established in 1840 and 1841. Sabine, who became a major general in 1856 and was later a K.C.B., was probably the most active British geophysicist of all time, besides holding important offices in the Royal Society (of which he was president from 1861 to 1871) and the British Association for the Advancement of Science.

[93] This figure is based on a drawing made by Messrs. Negretti & Zambra, obtained for me through the kindness of Mr. A. L. Maidens of the Meteorological Office, Bracknell. The drawing is only approximately to scale.

ivory point with the mercury being observed through the thick glass tube which forms the upper part of the cis-

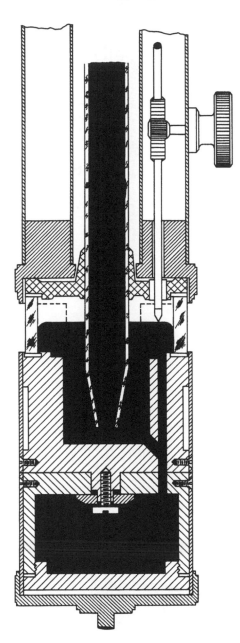

Fig. 9.14. Lower part of Newman barometer, circa 1839.

tern. A scale of this form can be measured directly before being installed in the barometer, which is an advantage in that it permits an absolute calibration to the accuracy required for meteorological purposes, though international comparisons are rightly preferred. The part of the scale carrying the graduations was of platinum in these barometers. The tubes had an inside diameter of a little over half an inch, which ensured a reasonable constancy of capillary correction.

Probably the least satisfactory feature of this design is that the thermometer (not shown in the figure) has its bulb dipping into the mercury of the cistern. In many rooms there is a gradient of temperature from ceiling to floor, and if one thermometer is to indicate the average temperature of the mercury column, it had better have its bulb about halfway up the barometer tube, and as near it as possible.

Barometers of this type were sent to the main meteorological and magnetic[94] observatories all over the British Empire. J. M. Sil[95] has given the history of No. 58, installed in 1841 at the Colaba Observatory, Bombay. It had been used every day from then till 1940 and had "never been dismantled for cleaning or repairs." In 1866 it became hard to see the reflection of the ivory point (after 25 years in the tropics!) so the scale was adjusted once and for all and the barometer used, with suitable "corrections for capacity" as a fixed-cistern barometer. In 1935 it was compared with the Kew Observa-

[94] Sabine was especially interested in terrestrial magnetism.

[95] India. Meteorol. Dept., *Sci. Notes,* Vol. 8, No. 88 (1940).

tory standard through the intermediacy of six Fortin-type barometers sent out from England, and found to have an error of only 0.001 inch.

No. 33 was sent to Toronto in 1841 and used almost every day until 1946 for all the official observations of atmospheric pressure at standard hours. The cistern was cleaned at long intervals and it was always used as a movable-scale barometer. Occasional international comparisons suggest little alteration in its vacuum.

No. 34 is at Kew Observatory. According to F. J. W. Whipple,[96] it was put into use for the standard observations in January 1851. In 1915 a millibar scale and a thermometer in absolute degrees were added. In 1916, as the cistern needed cleaning, "a new tube with a quill air trap was fitted by Messrs. Negretti & Zambra." In its revised form, it was still in use at Kew in 1961.

The most remarkable thing about these barometers, especially to an "old hand" like the writer, is that they arrived at their overseas destinations not only undamaged but with their corrections almost unaltered. Furthermore, they were taken out, set up, and tended by soldiers—in Toronto, the Royal Engineers. It is evident that Newman was a superlative workman with a flair for designing barometers.

Shortly after the Swiss meteorological network was set up in the 1860's, movable-scale barometers were obtained from the firm of Hermann & Studer of Berne (later Hermann & Pfister and still later J. H. Pfister). There are four such instruments bearing these various labels

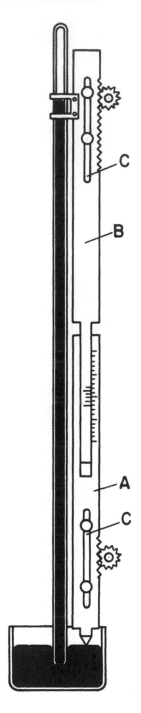

Fig. 9.15. Diagram of Dulong and Arago's barometer.

[96] Letter dated Oct. 13, 1927, to the Director of the Meteorological Office, London. Dr. Whipple was then the Superintendent at Kew.

in the storeroom of the *Zentralanstalt für Meteorologie* at Zurich. They are practically identical, with tubes of about 7 or 8 mm. bore and such extremely large glass cisterns (12 cm. diameter) that the movable scales seem almost superfluous. But they are there, all of brass except for steel points to touch the mercury, graduated from 570 to 765 mm. (some of the stations were high) and with verniers reading to 0.1 mm. With a thermometer near the middle, each was mounted on a heavy plank. Assuming that the tubes were filled, sealed, and transported separately, they must have been entirely practical, and scarcely merit their retirement.

At the C.N.A.M. in Paris there is an interesting movable-scale barometer that is stated to have been used by Dulong and Arago in their experiments on the gas laws.[97] It is inscribed "Lecomte à Paris," and looks as if it may have been made specially. The tube is probably 10 or 11 mm. in bore, and the cistern about 60 mm. and rather shallow. The scale is peculiar, and must be described by reference to Figure 9.15. *A* and *B* are two strips of brass about 30 mm. wide, cut as shown, and each movable by rack and pinion, being guided by pins in the slots *C, C*. The scale is on *A*, which has an ivory chisel edge, the vernier on *B*. I am unable to see the advantage of this complicated construction.

Most nineteenth-century movable-scale barometers are siphon barometers. Three[98] have already been referred to. Such barometers nearly always have

similar reading devices at top and bottom, the bottom one being fixed at the zero of the scale and set to the lower mercury surface, the upper one attached to a vernier slide. They are usually the kind of instruments that one might find in a laboratory. There is one which certainly falls into that category in the *Museo Copernicano* at Rome, made by the firm "Tecnomasio Milano." Its upper and lower chambers have an inside diameter of 28 mm., which should avoid capillary troubles, but the reading indexes scarcely justify the size of the tubes.

However, there are occasional really portable barometers, such as the one at Florence made "according to the ideas of G. B. Amici,"[99] which have movable scales. This instrument bears the inscription "Galgano Gori fece in Firenze l'Anno 1846." It has an elegant brass case with slots front and back to permit reading it against the light.

After all this it will appear as comic relief to read that in 1863 one W. Symons (*not* the famous English meteorologist G. J. Symons) obtained a patent for the application of a movable scale to a siphon barometer![100]

In Munich there is a twentieth-century cistern barometer with a movable scale, made about 1920 by Wilhelm Lambrecht of Göttingen.[101] This differs from others in that instead of terminating the scale by a point, Lambrecht has provided a rectangular strip of metal which dips into the mercury, and

---

[97] Inventory no. 12008. This reference would date it about 1830.

[98] *Deutsches Museum*, 8935 and 54905; Utrecht University, G.35. See p. 202 above.

[99] *Museo di storia della scienza,* inventory no. 1131. Dr. Bonelli is sure that there is documentary evidence for connecting this barometer with the great astronomer and microscopist Giovanni Battista Amici (1786–1863).

[100] British Patent 813 (1863).

[101] *Deutsches Museum*, inventory no. 22338. This is described in Lambrecht's list 6 (1925).

carries on its front surface a black triangle. The superiority of this design is at least debatable.

The barometer by Sisson referred to on page 198 has an arrangement which is distantly related to the movable scale, and as far as I am aware it is unique. A short ivory scale is fixed to the frame of the instrument just above the reservoir; against this scale slides a vernier, also of ivory, the lower end of which has a chisel edge that can be made to touch the mercury in the cistern. When this has been accomplished, the correction for the departure of the mercury from its standard level can be read off on the scale and vernier. It may be significant that in order to do these things the wooden cover of the cistern must be lifted off, so that it was probably not intended that they should be done at each observation.

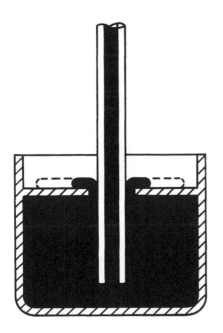

Fig. 9.16. The "pool-of-mercury" barometer.

We now return to the middle of the eighteenth century and consider the invention of a form of constant-level barometer cistern quite unlike all others. This depends on the peculiar behavior of mercury when poured carefully on to a horizontal plane. A little will form a drop, and if more is added, the drop will not increase in height, but will spread out over a greater portion of the plane, the level remaining the same until the plane (assumed to be limited in area) is full. We shall call the barometers which were designed to take advantage of this phenomenon "pool-of-mercury" barometers. The general principle of their design is shown in Figure 9.16. Enough mercury is used to make a small "collar" around the barometer tube when the atmospheric pressure is highest, and as the pressure falls the excess mercury spreads out on the plane, as shown by the dashed line.

This sort of barometer was almost certainly invented in the Netherlands. There are two possible inventors, Pieter Eizenbroek of Haarlem and Hendrik Prins of Leiden. Eizenbroek described it in 1761,[102] claiming to have discovered the principle: "I have availed myself of a property of quicksilver which I came across by accident; for some time ago, having poured out some quicksilver on a horizontal plane for a special purpose, I found that however far it spread out, it kept the same thickness or depth."[103] A diagram shows the

---

[102] "Bericht over eene merkelijke verbetering in het maaken van barometers." *Verh. Holl. Mij. d. Wet.*, Vol. 6 (1761), part 1, pp. 353–57. I have spelled the author's name as in this article; it can also be found in the forms Isenbroek, Ijsenbroek, and Eysenbroek, or even Eisenbroog. He flourished between about 1735 and 1760.

[103] *Ibid.*, p. 354.

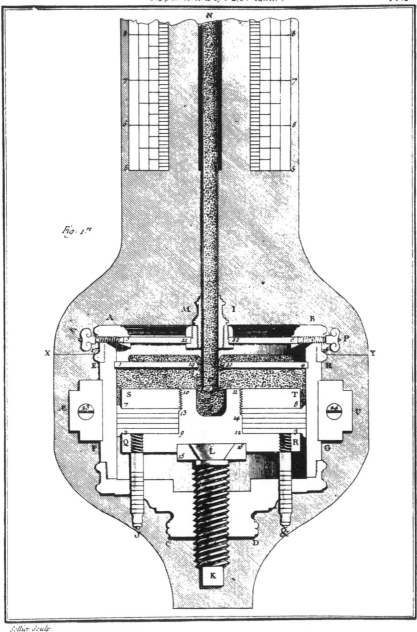

Fig. 1.ʳᵉ

Fig. 9.17. Anonymous barometer of 1782. *(Courtesy of the Trustees of the British Museum)*

application of this property to the barometer.

Louis Cotte ascribes this sort of barometer to "Pierre Eisenbroog."[104] Other authors, on the contrary, speak of it as due to Prins, beginning with De Luc in his famous book,[105] which must have been read by nearly everyone who had

[104] *Mémoires sur la météorologie,* etc. (Paris, 1788), Vol. 1, p. 514.

[105] *Recherches sur les modifications de l'atmosphère* (Paris, 1772), Vol. 1, p. 35.

anything to do with barometers at the time.

Not much appears to be known about the life of Hendrik Prins;[106] in 1736 he was making barometers that were praised by the great Petrus van Musschenbroek (1692–1761), who wrote that one "can obtain rather perfect (*vry volmaakte*) barometers from my brother,[107] or from the honest artists G. Fahrenheit and H. Prins, who have vied with each other in their endeavors to bring this instrument to the summit of perfection."[108] We may discount the civic pride, which was and is one of the charms of Leiden, and note only that Prins is mentioned on equal terms with Fahrenheit and Jan van Musschenbroek.

Eizenbroek published his paper in the proceedings of the national scientific society, and one would think it would have caused argument if his claim had been unfounded. At any rate, this principle was thought highly of by a number of people. De Luc thought that the reason that it was not generally adopted was simply that such barometers were difficult to construct and maintain.[109] The author of a *Dissertation sur le baromètre* in the Abbé

Rozier's *Journal de physique* (probably the Abbé himself) quotes De Luc with approval.[110] He is convinced of the excellence of this type of barometer, especially after making accurate measurements with a little vernier gauge, which showed the constant height of pools of mercury of various sizes. Endeavoring to perfect this instrument, he has constructed the instrument shown in Figure 9.17. The plane on which the mercury spreads and the top of the cistern are both of glass, the remainder of boxwood or ivory, except for the leather washers in the piston. It is a most impressive design, but I must confess that I do not see how the lower glass plate is inserted. It should be noted that the line XY does not indicate a joint of any kind; it is the zero of the scale, the level at which the mercury stands.

One of those who were obviously most impressed with the "pool of mercury" barometer, which he ascribed to Prinz [*sic*], was F. W. Voigt. Between 1799 and 1802 he described no less than

[110] *J. de phys.,* Vol. 21 (1782), pp. 436–50.

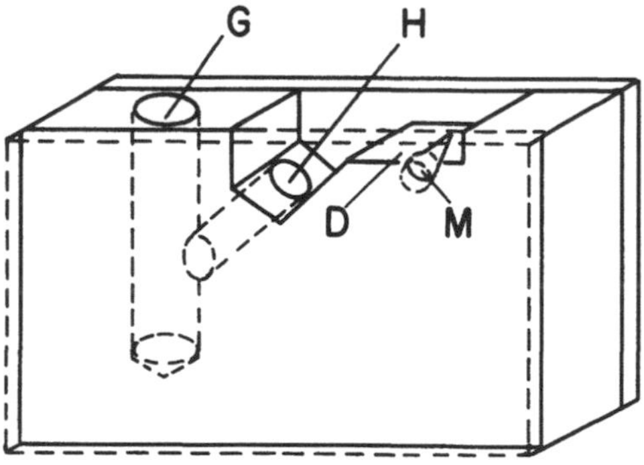

[106] I am informed that a German historian is working on a biography.

[107] Jan (1687–1748). At the back of the *Beginselen* there is a catalogue of Jan's instruments, covering 16 columns of this large quarto; it includes three kinds of barometer: An Amontons, or marine barometer; a barometer on a walnut plank; and "a barometer and thermometer together on a plank, decorated with neat carving."

[108] *Beginselen der Natuurkunde,* etc. (Leiden, 1736), p. 599.

[109] De Luc, *Recherches,* etc. Vol. 1, p. 35.

Fig. 9.18. Voigt's barometer of 1799. The front plate is shown by dashed lines.

three radically different ways of constructing such barometers.[111] The first and simplest had a cistern (Figure 9.18) made of a block of iron and two iron plates. The barometer tube goes in at G, and (for transport) a ground or screwed plug at H. Enough mercury is poured in to fill the V-shaped trough and spread out over most of the face D (the *Prinzsche Ebene*), the surplus flowing away through the drain M. A little cup of triangular section is carried, which can be hung below M. There were other modifications of this basic idea.

In 1800 Voigt passed from the simple to the extremely complex and described the barometer cistern of which a cross section is shown in Figure 9.19.[112] The plane on which the mercury spreads out is the top surface of the block A. This cistern is designed for portability and is arranged so that the lower part may be filled with mercury and the tube sealed for transport. In the drawing it is shown in a position that makes it almost ready to be moved, the cistern sealed by the leather-covered plunger B, but the knurled wheel E not turned quite far enough to allow the spring-loaded plunger m to push the leather bag o against the end of the barometer tube R, which is cemented into B.

To put the barometer into service it was hung up, the hand-wheel E turned

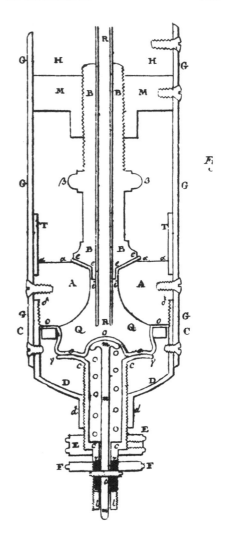

Fig. 9.19. Voigt's barometer of 1800.

[111] F. W. Voigt, *Beiträge zur Verfertigung und Verbesserung des Barometers* (2 Parts; Frankfurt 1795 and Leipzig 1799).

[112] *Magazin für den neuesten Zustand der Naturkunde*, Vol. 2 (1800), pp. 185–89. Also described in F. W. Voigt, *Versuch kritischer Nachträge und Supplemente zur Luzischen Beschreibung älterer und neuer Barometer, und anderer meteorologischen Werkzeuge* (Leipzig, 1802), pp. 288–300.

to let the leather bag descend, and the piece B screwed up, carrying the tube R with it, until the shoulder β was stopped by the nut M. The wheel E was then carefully turned until a pool of mercury filled about half the plane top of A.

It should perhaps be remarked that the members G were merely wide strips of brass joining the upper and lower parts of the structure, so that the pool of mercury was quite in the open.

There is a third design by Voigt, but we shall not go into details. I know of no further development of this principle, which was clearly unable to compete with Fortin's relatively simple device.

Barometers of this sort are now rare. There are four *prinsendozen* at Utrecht,[113] three of them simple affairs with a plane large enough to accommodate the whole range of pressure if the barometer tube is small, the fourth a very ingenious iron one in which the plane can be flooded from an adjustable second reservoir and part of the mercury then drained back through a hole, reminding one of the overflow cistern of Austin (p. 206 above).

4. THE APPLICATION OF ELECTRICITY TO THE BAROMETER

The success of the electric telegraph after 1845 set off a veritable chain reaction of invention, so that by 1856 Theodose Du Moncel's *Exposé des applications de l'électricité* had started into a second edition of over 2,000 pages.[114] The barometer is not forgotten. Du Moncel himself had replaced the ivory point of the Fortin barometer by a platinum one, and used

[113] University Museum, inventory nos. G36 (2) and G37 (2). On his catalog card, Dr. Van Cittert explains that "a *prinsendoos* [doos = box] is a mercury reservoir according to Prins, Fahrenheit's successor and apprentice (ca. 1750)." This does not square with the passage quoted above from Musschenbroek. I begin to favor Eizenbroek.

[114] Paris 1856–62, 5 vols. Count Th. Du Moncel (1821–1884) was elected to the *Academie des Sciences* in 1874.

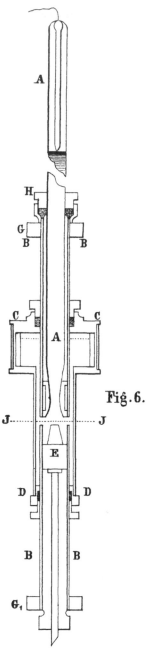

Fig. 6.

Fig. 9.20. Müller's barometer, 1878.

an electric circuit to indicate the exact point of contact between it and the mercury. This was much more certain, he thought, than setting the point by eye.[115] He next reported on an invention by Masson, who had built a barometer in which both surfaces were set electrically. A platinum wire came down into the vacuum chamber, sealed into the glass, and there was another at the lower end of a micrometer screw in the cistern, which was deeper than usual. The electric circuit included a battery, an indicator, the two platinum contacts, and the mercury, all in series. To read this barometer the mercury in the cistern was raised until contact was just made with the point in the vacuum chamber; then the micrometer screw was backed off until the circuit was just broken at the lower contact.[116] The remainder of the applications reported by Du Moncel are to barographs, and will be noticed in Chapter 11.

A much more sophisticated design was produced twenty years later by F. C. G. Müller.[117] This barometer, invented, he says, in 1874, is arranged so that the upper surface of the mercury is always at a constant level, the cistern being moved up and down. As a cylindrical steel tube (which is part of the barometer tube) passes entirely through the cistern, there is no change in the mercury level in the cistern relative to the cistern itself when the level of the upper mercury surface is restored to its standard level. In Figure 9.20, AA is the glass barometer tube, fastened airtight to the steel tube BB, which is of uniform cross section, and communicates with the cistern CCDD by means of two holes J,J. E is a

stopper which can be screwed up if the barometer is inverted for transport. H and D are stuffing boxes.

The proper height of the upper mercury surface is established electrically through a platinum wire sealed through the glass, and the reading made on a scale having a vernier attached to the cistern.

This barometer was also made into a barograph, and will be referred to again in Chapter 11.

The barometer of J. J. Boguski and L. Natanson of Warsaw[118] is almost adequately described by Figure 9.21. It is, in principle, exactly similar to that of Masson, but this is a siphon barometer with an extra vessel into which a plunger can be screwed in order to

[118] *Ibid.*, Vol. 36 (1889), pp. 761–63.

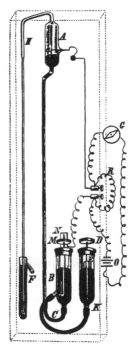

Fig. 9.21. The barometer of Boguski and Natanson, 1889.

[115] *Ibid.*, II, 405–6.
[116] *Ibid.*, pp. 407–8.
[117] *Annalen der Physik.*, Vol. 4 (1878), pp. 286–94.

raise the level of the mercury. The two points, as in Masson's instrument, are in series with a battery O, a galvanometer G, and a high resistance R.

Much later, R. Goldschmidt made an "electrical barometer" which could be compensated for temperature.[119] The vacuum chamber of this contains a hairpin-shaped carbon filament, dipping into the mercury and connected to two leads sealed through the glass at the end of the tube. The filament is thus more or less shorted out according to the height of the column of mercury. The compensation is done by another similar tube nearly full of mercury and sealed at both ends, containing a second filament. The two filaments are in a Wheatstone bridge. If the temperature changes at constant pressure, the balance of the bridge—supposing the values of the remaining resistors to have been chosen correctly—will not alter. If the pressure changes, the balance is restored by altering a resistance in series with the filament in the barometer tube, and the value of this resistance is a function of the pressure. Goldschmidt claimed to have found that capillary phenomena between the carbon filaments and the mercury could be neglected and stated that a sensitivity of 10^{-4} mm. of mercury could be obtained.

Other electrical indicating arrangements have been described for barometers of high precision, and will be dealt with in the next chapter. Platinum-to-mercury contacts in air were soon found to be troublesome, and in any event the complication was not found worth-while for indicating ba-

rometers. Recording barometers are another matter.

5. CISTERN-SIPHON BAROMETERS

The barometer illustrated in Fig. 9.21 is an example of those barometers which are in a sense a hybrid of the siphon- and the cistern-barometer. The general term for such barometers is *cistern-siphon barometer* (G., *Gefäss-Heberbarometer*), and they are, in effect, siphon barometers with the addition of some sort of reservoir of mercury by means of which the level in both limbs of the siphon can be simultaneously varied.

The first of these seems to have been due to Lavoisier or someone who worked with him. When Truchot was examining Lavoisier's barometers, then at the Chateau de la Canière, he found an empty barometer tube of unusual form; the lower part, curved into a siphon, carried a supplementary reservoir. This contained a mass of glass that could be submerged more or less in order to bring the mercury levels to a given point.[120] I do not know whether this tube was ever part of a barometer.

The next one was from Russia, and due to Johann Heinrich Lorenz Panzner or Pansner, (1777–1851); it was a traveling barometer.[121] With slight changes, it was described again in

---

[119] *Bull. Soc. des sci. med. et nat. de Bruxelles,* Vol. 66 (1908), pp. 125–29.

[120] *Ann. de Chim. et de Phys.,* Vol. 18 (1879), p. 310. I imagine that some purists may object to the inclusion of barometers with a separate chamber and a plunger among cistern-siphon barometers; but the principle is the same, and there seems no point in dealing with them separately.

[121] *Das Reisebarometer des Dr. L. Panzner* (St. Petersburg, 1808). (Pamphlet, 26 p., 1 plate.)

1811.[122] It had a long closed tube and a short open one, both ending in a cistern which could be adjusted by means of a leather bag and a screw. There was an iron tap at the bottom of the shorter limb, which made it possible to invert the instrument for transport. There seems to be some discrepancy between the descriptions and the plates, and it is not entirely clear how the surface in the short tube was observed; but as there seems to have been only one vernier, it is most probable that it was brought to a fixed mark, the zero of the scale. It is noteworthy that Panzner (careful man!) had the brass scale compared at Moscow with a standard scale from Paris, the error being only 0.03 line in 28 inches.

This instrument really was portable; its inventor made a long journey in Asia, with the barometer fastened upside down in the corner of a wagon; a severe enough test, one would imagine.

There is a barometer by Dollond in the Science Museum, London[123] of which the above might well be a description, except that in this case there is no doubt that the mercury in the lower limb was brought to a fixed mark. It is unfortunately not dated, but I notice that before I had seen Panzner's papers I had tentatively assigned the date 1800 to it. It is not impossible, though unlikely, that someone at Dollonds had seen Panzner's pamphlet or the Memoirs of the Moscow Society of Naturalists; but at any rate here is the physical embodiment of Panzner's idea.

In 1829 John Adie of Edinburgh published an "Account of a new cistern for barometers."[124] In this construction the mercury is brought to the proper level in the short limb by moving a glass plunger in the cast-iron cistern. The plunger is made mercury-tight by means of a gland packed with leather and is moved by a screw. The zero of the scale is at the upper edge of a rectangular opening through the brass tube which encloses the short glass tube, and the plunger is to be depressed "until the surface of the mercury cuts off the light." At about this time A. T. Kupffer[125] at St. Petersburg designed a rather similar barometer, which was developed by T. Girgensohn,[126] instrument maker to the Academy, and widely distributed in Russia. It could not, however, be transported filled, and Kupffer laid great stress on the possibility of determining the state of the vacuum by observing the height of the mercury column with more than one volume for the vacuum space. Outside of Russia these barometers do not seem to have been used; the only cistern-siphon barometer that I have seen that I would date between 1820 and 1870 is in Florence,[127] beautifully made with a mounting completely of brass and scales in four systems of units. It has a low tripod with leveling screws on all three legs.

A cistern-siphon barometer that would seem to have distinct possibilities was suggested in 1833 by the

[122] Mém. Soc. Natur. Moscou, I (1811), 58–64. (Here Pansner)

[123] Inventory no. 1893–133.

[124] Edinburgh J. Sci., new ser., Vol. 1 (1829), pp. 338–40.

[125] Mém. Acad. imp. des. Sci. St. Petersbourg, VI Sér., Sci. math., phys. et nat., Vol. 1 (1831), Bulletin, 25 Aug. 1830, pp. xxvi–xxviii.

[126] Ibid., Vol. 3 (1835), Part I, Bulletin, 31 Oct. 1834, pp. xiv–xvi.

[127] Museo di storia della scienza, inventory no. 1132.

famous Cambridge mineralogist W. H. Miller (1801–1880),[128] in a brief description with no illustrations. Apparently there was to be a fixed point near the top of the closed tube, and a movable one, rigidly attached to a vernier, which could be set to the lower surface of the mercury. It would have had all the defects of siphon barometers unless made very large, but is of interest because it seems to be the first suggestion of setting both mercury surfaces to points.

In the 1870's the cistern-siphon barometer was revived with great energy by Dr. Heinrich Wild (1833–1902), the dynamic director of the Russian meteorological service. According to Wild, it had two advantages: the mercury could always be *brought up* to the mark on the shorter limb, which should give both menisci a more regular and stable form, and by having two or more index marks, the vacuum could be investigated by a technique which will be discussed later (p. 249). The first advantage, of course, is shared with all adjustable cistern barometers; but I think Wild was considering only the siphon type.

Not satisfied with Kupffer's and Girgensohn's barometers, Wild designed one of his own, and had a number of them manufactured by Turretini in Geneva.[129] This design is shown in Figure 9.22, and in view of what has gone before no special description is needed, but the large thermometer—its bulb was 8 mm. in bore, just like the barometer tubes—should be noted, neatly stowed in the surplus part of the second tube, so that one might reason-

[128] *B.A.A.S. Report for 1833*, p. 414.
[129] H. Wild, *Repert. Experim. Phys.*, Vol. 11 (1875), pp. 389–97.

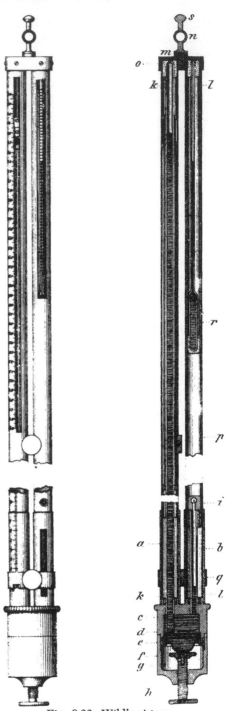

Fig. 9.22. Wild's cistern-siphon barometer, 1875.

ably expect the thermometer and the tube to change their temperature at the same rate. The short limb could be closed for transport by a small tap *i*, operated from outside by a key.

There is one of these barometers at the Meteorological Office, Bracknell,[130] apparently in excellent condition. It is unsigned.

After two or three years' experience with this instrument, Wild found that it was a little subject to damage in traveling, and required very exact workmanship. He therefore designed one with only one brass tube, with the two limbs vertically over one another, the long one being bent and then passing into and through the lower part of the short limb on its way to the cistern (Figure 9.23). This involved a fairly expert piece of glassworking. Wild described this as a "Control-Barometer," it being intended primarily for meteorological inspectors.[131] He had the first one made by Turretini, but also communicated the idea to R. Fuess of Berlin, who lost no time in developing a line of cistern-siphon barometers that were extremely popular in Central Europe for the next half-century or more, altering relatively little except for a few mechanical refinements. Fuess had one ready for the Berlin Industrial Exhibition the same year.[132]

The larger form, usually though perhaps unjustifiably called a "Normal-

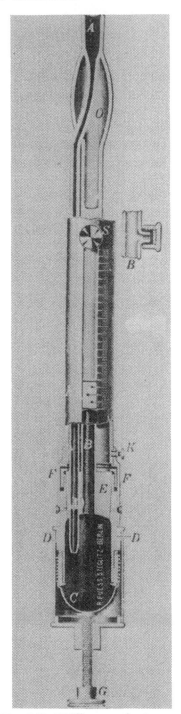

Fig. 9.23. Wild's second cistern-siphon barometer, 1879. (*R. Fuess*)

---

[130] Inventory no. 4113F.

[131] H. Wild, *Repert. für Phys.-Technik*, Vol. 15 (1879), pp. 399–408.

[132] L. Löwenherz in *Bericht über die wissenschaftlichen Instrumente auf der Berliner Gewerbeausstellung im Jahre 1879* (Berlin, 1880), pp. 218–28.

Barometer," was designed to fulfill Wild's conditions, stated in 1877:[133]

1. The tube must have an inside diameter of at least 12 mm.

2. The mercury must be movable up and down the tubes.

3. Both reading devices must be of the same construction.

4. The scale and reading devices must be very durable.

5. The thermometer must be installed in such a way as to indicate the mean temperature of the mercury column.

6. The barometer ought to be transportable when filled.

7. If a tube breaks, the installation of a new tube must not change the absolute correction.

Wild had an accuracy of 0.01 mm. of mercury in mind, and seems to have thought, at least in 1877, that it might be possible to make a transportable barometer which would have this accuracy. In actual fact, the capillary depression in the open limb can vary by several times this amount in a 12 mm. tube.

There was also a smaller form for ordinary meteorological stations, and in the 1879 exhibition a less costly design in which the scales were engraved directly on the glass tube, the verniers on the *inside* of short pieces of glass tube that slid on the main tube. This was an ordinary siphon barometer, but a second version had an additional short limb into which a thermometer (killing two birds with one stone!) could be dipped to a greater or lesser extent by operating a rack and pinion,

so as to use various portions of the scales. Löwenherz[134] thinks this construction fragile and not really portable, but it is interesting that it had another thermometer near the top of the instrument.

Apart from special precision barometers (Chapter 10) this almost ends the history of the cistern-siphon barometer. Fuess' later versions[135] have been mechanically extremely elegant, and have an interesting device at each end, due to E. Kleinschmidt,[136] for measuring the height of the meniscus, on which the capillary correction depends. This consists of a plane-parallel piece of glass which can be brought in front of half the meniscus, and tilted until the edge of the meniscus as seen through the plate coincides with the summit as seen directly. The height of the meniscus is then read on a drum.

6. THE BAROMETER SCALE ITSELF

The accuracy of the scale itself has worried careful barometrists at least since the time of De Luc, even though his scales were drawn on paper. This, incidentally, was not quite as bad as it sounds, because they were glued to a pine board. Even before De Luc, Le Monnier (see p. 166) had been worried about scales, and had had two scratches 24 inches apart made on his tubes, using only short paper scales from these fixed points. Lavoisier, whose ideas about desirable accuracy were some-

---

[133] *Bull. Acad. imp. des Sci. St. Petersbourg,* Vol. 23 (1877), p. 87.

[134] *Bericht... wissenschaftliche Instrumente,* etc., p. 228.

[135] See for example their list 111.0 of 1953.

[136] *Meteorol. Zeits.,* Vol. 46 (1929), pp. 344–46.

what advanced for his time, objected not only to scales mounted on wood, but even to long brass scales, preferring to have reference marks scratched on the barometer tube with a diamond, to which the zeros of movable scales could be set.[137] In 1839 Moritz Meyerstein of Göttingen described a portable siphon barometer which had scales on this principle.[138]

But De Luc had another trouble with his scales, because at that time (about 1760) it was rare to get scales from different makers, even in Paris, which agreed in length. This he found intolerable.

That is why, not knowing which to prefer, I begged someone who had relations with M. de Mairan to get me, through him, an official foot (*pied de Roi*), exactly equal to that which he used in his very careful détermination of the length of the seconds pendulum.[139] When I obtained this authentic scale, I made one 27 inches long, marking this distance by two points on pieces of brass let into a ruler of pinewood. . . .[140]

After this, things gradually improved with the progress of standardization, until, by the middle of the nineteenth century, barometers by reputable makers probably had scales as accurate as there was any need for. By this time ordinary barometers were, in any event,

being compared with more elaborate instruments, the scales of which had been specially calibrated by comparison with national standards.

7. THE CISTERN BAROMETER VERSUS THE SIPHON BAROMETER

Although the siphon barometer dates, as we saw in Chapters 3 and 4, from the time of Pascal, Boyle, and Borelli, almost all barometers (except wheel barometers) with any pretense to accuracy made before 1770, and certainly almost all portable barometers, were of the cistern type. De Luc, who developed the first good portable siphon barometer (p. 136), started experimenting with cistern barometers, or more correctly, modified "bottle barometers"; but his own observations and experiments led him to prefer the siphon barometer. He observed that the shape and height of the meniscus in the cistern varied according to the cleanness of the mercury and of its containing vessel. To observe the level of the top of the meniscus was, with the means then available, very difficult; plainly the observation of the edge could lead to errors.[141] This is a serious matter for the measurement of heights, he points out, since $1/16$ line (about 0.13 mm.) corresponds to a height difference of about 5 feet. Any twentieth-century reader familiar with barometers or with chemical apparatus can appreciate the difficulty of reading the level in a wide vessel to such a precision without special aids.

[137] *Oeuvres* (Paris, 1865), III, 762.
[138] *Annalen der Physik*, Vol. 46 (1839), pp. 620–21.
[139] Mairan, *Mém. Acad. r. des Sci.* (1735), p. 157.
[140] J. A. de Luc, *Recherches sur les modifications de l'atmosphère* (Geneva, 1772), I, 212. J. F. Luz, *Vollständig und auf Erfahrung gegrundete Beschreibung von . . . Barometern,* etc. (Nürnberg & Leipzig, 1784), pp. 52 ff., also refers to the chaos in scales.

[141] De Luc, *Recherches*, etc., I, 203–4.

Not satisfied with conjecture, or with previous observations on the capillary depression in tubes of various sizes, he then embarked on a series of experiments *ad hoc*; almost naïve experiments, which consisted in setting up a number of barometers with upper and lower surfaces of different absolute and relative sizes (even with the upper surfaces the larger) and comparing their readings. He found it impossible to arrive at a general rule, or rather he doubted whether the game would be worth the candle, but

. . . I contented myself with the positive knowledge that barometers made of a single tube bent back at one end and of uniform diameter from one end to the other are the only ones in which the height above the level represents immediately that of the column of mercury sustained by the weight of the atmosphere; and consequently they all stand at the same height.[142]

He is aware, he goes on to say, that the open branch, being in communication with the air, gradually becomes dirty, and that this can cause a small error—a quarter of a line, he says—but the tube can be cleaned. Nevertheless he is willing to admit that the siphon barometer is not very convenient for daily observations at a fixed station, and while this is not really his subject, he will not omit saying a word about it.

This "word" extends to three quarto pages and seems to have had a great influence. As far as the construction of barometers is concerned, he had four recommendations:

1. They must be filled by boiling.
2. There must be a thermometer with his special scale[143] near the tube.
3. Barometer makers should have a standard siphon barometer with which to compare all the barometers they make.
4. The cisterns of barometers should be large enough to permit the change of level in them to be neglected.

The third point was the novel one. De Luc took the matter extremely seriously, as is evident from a footnote which is worth quoting in full:

This is my purpose [*i.e.*, to have all barometer-makers adopt his scheme] in a notice which I propose to distribute to all barometer-makers who come to Geneva; it contains, in four quarto pages, everything essential in the construction of the *standard*, the manner of applying the scale to ordinary barometers, and the use of the thermometer for correcting for the effects of heat on the height of the mercury. As far as I can, I shall also see that these instructions reach the chief towns in which physics is cultivated, so that amateurs may be able to insist that workmen construct their barometers according to this method.[144]

This has the proper evangelical ring, perhaps appropriate to a citizen of Geneva; one can almost imagine him handing out these tracts. But of course this was a most fruitful suggestion, and beyond reproach except for one unfortunate fact: he did not know that the capillary depression in air was not the same as in vacuum, and that therefore his *étalon* would not give the answer he thought it gave; not even when it was clean. When it is remembered that the tube of his siphon barometer was

---

[142] *Ibid.*, p. 207.

[143] See Chapter 7, p. 138 above.
[144] De Luc, *Recherches*, etc., I, 211, note.

only five or six millimeters in bore, it will be realized that his standard was far from adequate, and yet his insistence on the necessity of such a standard bore fruit which ripened when theory and technique had progressed far enough.

It also fired the opening gun in the great battle of cistern versus siphon. Henry Cavendish replied in no uncertain terms, in his "Account of the meteorological instruments used at the Royal Society's house:"[145]

I prefer the cistern to the siphon barometer, because both the trouble of observing and error of observation are less; as in the latter we are liable to an error in observing both legs. Moreover, the quicksilver can hardly fail of settling truer in the former than in the latter; for the error in the settling of the quicksilver can proceed only from the adhesion of its edge to the sides of the tube; now the latter is affected by the adhesion in two legs, and the former by that in only one: and, besides, as the air has necessarily access to the lower leg of the siphon barometer, the adhesion of the quicksilver in it to the tube will most likely be different, according to the degree of dryness or cleanness of the glass. It is true, as Mr. De Luc observes, that the cistern barometer does not give the true pressure of the atmosphere; the quicksilver in it being a little depressed on the same principle as in capillary tubes. But this does not appear to me a sufficient reason for rejecting the use of them. It is better, I think, where so much nicety is required, to determine, by experiment, how much the quicksilver is depressed in tubes of a given bore, and to allow accordingly.[146]

Luz[147] and Voigt,[148] both of whom had carefully studied De Luc, echoed his ideas. It should be noted that none of these three thought the siphon barometer the best for the ordinary meteorological station, and only De Luc preferred it for a traveling barometer. All three insisted only that cistern barometers should be compared with a siphon barometer and calibrated in his way.

But in 1816 Gay-Lussac expressed a general preference for the siphon barometer because, he said, it is exempt from a capillary correction; and this opinion seems to have carried great weight. It was also pointed out rather vaguely by Arago in 1826[149] and clearly by Kupffer in 1832[150] that a siphon barometer in which *both* levels could be changed at will (i.e., a cistern-siphon barometer) offered the possibility of checking the excellence or otherwise of the vacuum.[151] Either because of the influence of Arago and Gay-Lussac or for some other reason, the rather uncritical acceptance of the siphon (especially the cistern-siphon) barometer for meteorological stations lasted for a long time in Central Europe. It never spread to Great Britain. It survived in Europe in spite of mounting evidence of the unreliability of small siphon barometers. As early as 1839 a committee of the Paris Academy composed of Cordier, Savary, and

[145] *Phil. Trans.*, Vol. 66 (1776), pp. 375–401.

[146] *Ibid.*, pp. 381–82.

[147] *Vollständig . . . Beschreibung*, etc. p. 166.

[148] *Versuch Kritischer Nachträge*, etc., p. 62. "Ein richtiges Barometer muss . . . die Heberform haben."

[149] *Ann. Chim. et Phys.*, Vol. 33 (1826), p. 395.

[150] *Annalen der Physik*, Vol. 26 (1832), pp. 450–51.

[151] This will be discussed later (p. 249).

Arago, reporting on a new cistern barometer by Bunten, remarks that small siphon barometers had been disappointing because of the uncertainty in the capillary correction to the open limb.[152] Much earlier, in 1826, J. G. F. von Bohnenberger (1765–1831) of Tübingen, objected to the siphon barometer on the same grounds:

. . . the siphon barometer is not very suitable for obtaining accurate results, owing to the variability of the capillary depression in the shorter limb, as the inner surface of the tube is covered with more or less moisture in various states of the atmosphere.[153]

He seems to have been ahead of his time; even in 1862 C. M. Bauernfeind was still recommending a siphon barometer with a 5 mm. tube for surveying.[154] In 1858 the Paris instrument maker Jules Salleron (1829–1897) listed a simple siphon barometer with a tube 16 to 18 mm. in bore and claimed an accuracy of 0.01 mm. for it.[155] It had two microscopes that could be moved along a prismatic bar, and turned to a separate scale. In 1855 a barometer was made for the Observatory of Milan, very similar except that the scale was

on the bar carrying the microscopes, an inferior idea, but felt to be quite adequate.[156]

In the 1870's and 1880's stations all over the Germanic countries, Russia, and even Italy, were being equipped with cistern-siphon barometers, largely because of the tremendous influence of Heinrich Wild of St. Petersburg and his co-operation with the firm of Fuess, who produced a durable and mechanically excellent instrument. Meanwhile it had become clear that the capillary depression was different in air and in vacuum.[157] This probably disposed of any idea that a siphon barometer would or could give accurate results without correction. But the variability of the correction was a worse trouble. Guyot was aware of this in 1850; on July 9 of that year he wrote in French from Cambridge, Mass., to Professor John F. Fraser, "Now in this case [referring to the Gay-Lussac siphon barometer as made by Bunten] the worst trouble is the variation of the menisci. This variation is so great that in certain cases the length of the mercury column, measured, as is usual, between the summits of the two menisci, can show differences of 3 to 4 tenths of a millimeter, even though the mercury is at the same temperature. . . . If the diameter of the tube is very large, this source of error may become very small."[158]

In 1885 H. F. Wiebe of the *Normal-Aichungs-Kommission* (Standardization

[152] *Compt. Rend.*, Vol. 9 (1839), p. 501.

[153] *Annalen der Physik*, Vol. 7 (1826), p. 380. This opinion was echoed in 1835 by Georg Breithaupt (*Ann. der Phys.*, Vol. 34 [1835], p. 41).

[154] Elemente der Vermessungskunde (2d ed.; Munich, 1826), pp. 351–52.

[155] *Notice sur les instruments de précision* —*Météorologie* (Paris, 1858), p. 1. Salleron at first worked at 1 rue du Pont-de-Lodi, but in 1860 moved to 24 rue Pavée du Marais. In 1888 he sold his business to his partner Jules Dujardin; the firm of Dujardin & Salleron is still active at 3 rue Payenne, but makes no meteorological instruments. Salleron's barometers date chiefly from about 1858 to 1862.

[156] Francesco Carlini, in *Milano, Effemeridi astronomiche per l'anno 1856* (Milan, 1855).

[157] E.g., to J. Delcros, "Instructions sur le choix, la construction, l'observation, la vérification et les erreurs du baromètre propre aux observations météorologiques sédentaires." *Annuaire Météorol. de la France* 1849, p. 150.

[158] Amer. Philos. Soc., Philadelphia, ms. B/F865. Quoted by permission.

Commission) in Berlin, who was not unduly influenced by Heinrich Wild, gave chapter and verse for this,[159] unfortunately in a journal of small circulation. From 1870 to 1882, he reported,—no hasty conclusion, this—a Greiner siphon barometer with 9 mm. tubes had been compared with other barometers in the laboratory of the Commission. Its readings were found to vary irregularly from time to time, to the extent of no less than 0.5 mm., "chiefly from large variations in the effect of capillarity."[160] It should be noted that 9 mm. was a large bore for station barometers at that date.

Second, a comparison of two siphon barometers by Fuess, an old one with a dirty short limb, the other new, showed a change of 0.38 mm. after the dirty one was carefully cleaned by wiping the short limb with nitric acid on filter paper. The height of the lower meniscus changed from 1.52 to 0.88 mm. All but 0.12 mm. of the change in difference could be explained, according to Delcros' table,[161] by this alteration in the height of the meniscus. This second example was given rather better publicity a year later by J. Pernet, also of the Commission,[162] who added that within a space of three months the difference in the readings of two siphon barometers varied over a range of 0.8 mm., even though one had tubes with a bore of 9 mm., and the other of 11 mm. The levels of one of these could be moved up and down at will (cistern-

siphon barometer), the other not, so that this observation might make one wonder—Pernet did not bring this out—whether the trouble lay in the ability or in the inability to adjust the level. Pernet did remark that as the capillary depression in such a barometer may be 10 times the error of reading the instrument, its variation can be very noticeable. He suggests that it must be measured every time, even in the vacuum chamber. This could be done by enclosing a small open-ended tube in the barometer, and measuring the difference in level inside and outside it, and the two radii of curvature. He even gives the necessary theory. Except for the effect which working at a standardizing laboratory probably had on a nineteenth-century German scientist, one would suspect him of trying to laugh the siphon barometer out of court—that is, if one could forget that it was in this same decade that Cleveland Abbe in the United States devoted thirty-seven pages of a treatise on *meteorological* instruments to a discussion of the errors of thermometers.[163]

Ten years later the *Deutsche Seewarte* (naval observatory) at Hamburg made a frontal attack on the problem and reported their results at the century's end.[164] Besides two "normal" barometers, one with 25 mm. tubes made by Fuess in 1886[165] and one with 14 mm. tubes, made by Adie, they had eight other instruments: two movable-cistern barometers, three cistern-siphon barometers (*Heberbarometer mit be-*

[159] *Metronomische Beiträge*, Vol. 4 (1885), pp. 13–43.

[160] *Ibid.*, p. 13.

[161] *Annuaire Météorol. de la France* (1849).

[162] *Zeits. für Instrum.*, Vol. 6 (1886), pp. 377–83.

[163] Cleveland Abbe, *Ann. Rep. Sec'y of War for 1887* (Washington, 1888), Vol. IV, Part 2, pp. 53–89.

[164] *Archiv. d. dtsch. Seewarte*, Vol. 23 (1900), No. 2.

[165] See Chapter 10.

weglichem Boden), and three fixed-cistern barometers with contracted scales. After thousands of observations and the application of some elementary statistical procedures, they came to the following very definite conclusions:[166]

1. Siphon barometers are unsuitable for practical use because of the progressive variation of the height of the meniscus in the open limb, and because when this becomes dirty the reading tends to be lower with rising pressures than with falling.

2. Just after they are cleaned such barometers are very good; it takes six months to a year for them to become dirty.

3. To an accuracy of 0.1 mm., fixed-cistern barometers with contracted scales are very good.

4. In general, the Fortin construction gives the best accuracy for a station barometer.

5. Nevertheless the fixed-cistern barometer, when made by an instrument maker who has had experience with this type, is to be preferred for regular observations at meterological stations.

The third of these conclusions agrees with what had been known in England for thirty or forty years, and the fifth has since been concurred in almost everywhere. It is beyond question that the accuracy and comparability of barometer readings over a large part of Europe had been less than they might have been for decades because of a rather stubborn devotion to the siphon barometer in the face of clear warnings. This is one of those object lessons that

Clio is always offering, for ever in vain. I feel sure that De Luc would have regretted this outcome of his experiments.

This report from the *Seewarte* should have settled the matter for good, but the guerillas of the Prussian Weather Service kept on sniping for years. The last gun of any notable caliber seems to have been fired in 1916 by the great meteorologist R. Süring,[167] who said that on the basis of thirty years' experience with siphon barometers he was sure that

a well-made cistern-siphon barometer is at least equal (*mindestens ebenbürtig*) to a cistern barometer under the assumption that the heights of the menisci are measured and the changes due to these taken into account according to the tables of Schleiermacher and Delcros.[168]

The cistern-siphon barometer would have to be a great deal better than its competitor to make up for five extra operations every time it is observed.

8.   THE FIXED-CISTERN BAROMETER

And how did the fixed-cistern barometer come to be so deservedly praised? Although much was said about it in Chapter 7, it remains to refer to some developments which were directed to improving the accuracy of this relatively simple instrument.[169]

---

[166] *Archiv. d.d. Seewarte*, Vol. 23, No. 2, p. 10.

[167] *Ber. ü d. Tätig. d. Kgl. Preuss. Meteorol. Inst. 1916*, Anhang, p. (24)–(42).

[168] *Ibid.*, p. (42).

[169] To avoid misunderstanding, it should be made clear that by "fixed-cistern barometer" I mean a cistern barometer in which no account whatever has to be taken, by the observer, of the position of the lower mercury surface. Thus, although the cistern of a movable-scale barometer may be physically fixed, such an instrument is not included.

In any such barometer the level in the cistern goes down as the level in the tube goes up. To ensure the accuracy of such an instrument we must take this into account.

When the requirements were moderate, all that was needed was to make the cistern fairly large in relation to the tube. In 1688 John Smith noted that if the tube is more than ¼ inch in bore "it will cause the mercury in the cistern to rise and fall too much, when that in the tube does on the contrary either fall or rise (except due proportion be observed)."[170] Smith suggested a cistern 3 inches in diameter, while admitting that there were few as large. Later he describes a "Cistern-gage" for finding out how much the level in the cistern changes when the amount of mercury in the tube between the 28- and 31-inch marks is added. This gauge "need only be a pin driven into the streight edge of a ruler, and then cut off to such a length, that when the ruler is laid cross the brims of the cistern, the pin may reach down to near the middle of its depth, or so deep as you judge the mercury will rise when the instrument is justed."[171]

Later makers (perhaps when mercury was less expensive) sometimes made relatively enormous cisterns, as in the barometer at Paris inscribed "Mossy breveté du Roi quai Pelletier à Paris 1789."[172] This instrument has a glass cistern about 15 cm. in diameter, and as the tube has a bore of only about

6 mm. the error caused by neglecting the variation of level was only about $\frac{1}{625}$ of the maximum excursion from whatever mean pressure it was set to be right at. This might attain 32 mm., say, giving an error of not over 0.05 mm. from this cause, negligible for many purposes even today, and quite insignificant in 1789. Mercury could be saved by making the cistern with an axial cross section in the form of a T, since it is only the area of the surface that matters.

In the storeroom of the Science Museum, London, there are parts of a barometer which appears to have been made for the Meteorological Society by "Robert Carr Woods, 47 Hatton Garden London," probably about 1830. This has a heavy glass cistern about 20 cm. in diameter, and a tube about 15 mm.[173] The very interesting barometer at Bruges, described on page 131, also has a large glass cistern.

Such barometers were unhandy. Since the error is the product of the relative area of the mercury surfaces in tube and cistern and the departure of the barometric height from some particular value, it is linear in barometric height, and can be reduced to zero by contracting the scale so that one unit is equal to $A/(A + S)$ standard units, where A is the area of the mercury in the cistern (i.e., the cross section of the cistern at the level of the mercury, minus the entire cross section of the tail of the tube), and S the internal cross

---

[170] John Smith, *A Compleat Discourse of the Nature, Use and Right Managing of that Wonderful Instrument the Baroscope or Quicksilver Weather Glass* (London, 1688), pp. 10–11.

[171] *Ibid.*, p. 22.

[172] C.N.A.M., inventory no. 19948.

[173] Robert Carr Woods, an instrument maker, was one of the original members of the Meteorological Society. There seems to be no published account of this barometer, and the original minutes of the Society (which preceded the British Meteorological Society) cannot be found.

section of the tube at the upper mercury level. The suggestion that this might be done, usually attributed to the famous chemist John Dalton (1766–1844),[174] was actually made by Hugh Hamilton (1729–1805) a Fellow of the Royal Society and at the time Dean of Armagh,[175] in a discussion[176] of a very ordinary barometer described at great length by a namesake.[177] He showed how the contracted scale could be calculated. It was again suggested, probably quite independently in view of the political climate of the intervening years,[178] by J. Guerin of Avignon.[179]

In spite of this simple possibility, fixed-cistern barometers continued to be made with scales in standard units through most of the nineteenth century, and these were in official use in some countries at least until 1880. The famous Newman portable barometers[180] were often called "neutral-point barometers," because Newman followed the excellent practice of stating on the frame of the barometer the point on the scale at which it was set (this he called the neutral point) and the "correction for capacities," S/A in the notation above. For example, a Newman instrument intended for mountain use[181] bears the legend "Correction for capacities $\frac{1}{50}$ Neutral point 30.194 Capillary action + .040 Temperature 62°." The last is the temperature at which the scale represented standard inches. The inclusion of the "capillary action" must mean that the observer should add 0.040 inch to the reading before taking its difference from 30.194, increasing this by $\frac{1}{50}$ and adding it (algebraically) to 30.194 inches. Nowadays the capillary depression would have been taken care of in adjusting the barometer, and it seems likely that Newman calibrated his barometers by actual linear measurement, rather than by comparison with any sort of standard instrument.

Fixed-cistern barometers with contracted scales became common in British ships and stations after 1850 under the name of "Kew barometers" which indeed they still bear. Except that they are now graduated to read in millibars, they have changed little in general appearance in almost 100 years. In 1925 it was suggested[182] that the bottom of the cistern of this barometer should be made as a plunger with a leather packing, so that the instrument might be exactly adjusted by the makers or the testing laboratory. The simpler solution of a screw and locknut, positioning the entire cistern in relation

[174] Meteorological Observations and Essays (London, 1793), pp. 7–8.

[175] In Ireland, the seat of a famous observatory.

[176] R. Irish Acad., Trans., Vol. 5 (1792), pp. 117–27. The paper was read on December 1, 1792. But see Chapter 6, p. 97 above.

[177] Rev. J. A. Hamilton, R. Irish Acad., Trans., Vol. 5 (1792), pp. 95–116.

[178] But it is amazing how hard it is to find any trace of war or revolution in reading the scientific journals of the period, except perhaps the French ones. I remember noticing this with something of a shock about 15 years ago when looking carefully through Gren's Journal der Physik and Gilbert's Annalen, which coincided in time with the French Revolution and the Napoleonic wars.

[179] J. de Phys., Vol. 53 (1801), pp. 444–48. It is almost incredible that in 1904 a patent (British patent 26,702) could have been granted to A. Sutherland on the contraction of the scale.

[180] See p. 148 above.

[181] Meteorological Office, Bracknell. Inventory no. 10.

[182] S. N. Sen, M.O., London, Geophys. Mem. No. 27, 1925.

to the outer tube which carries the scale, does not seem to have occurred to Sen.

9. THE IMPROVEMENT AND MAINTENANCE OF THE VACUUM

The very first experimenters with the "Torricellian tube" found, as we have seen in Chapters 3 and 4, that it was both necessary and difficult to get a good vacuum at the top of the tube. As early as 1649 Zucchi noted the difficulty of filling a tube with mercury without introducing bubbles.[183] He also found that in his experiments, the mercury rose if the space at the top of the tube was cooled and fell if it was warmed. This was just what Torricelli and Viviani had found,[184] and was of course a sign of a poor vacuum, probably of the presence of water. Boyle[185] found it hard to get a good vacuum and described dodges such as letting a bubble run to the end and back, clearing out small bubbles with an iron wire, etc. He was seldom able to fill a tube so well at this period (1660) that no bubble was visible at the end of the tube when this was inclined. In 1666 Oldenburg quotes a letter from Dr. John Beal[e] (1603-ca. 1683) regarding the new wheel barometer: "My wheel-barometer I could never fill so exactly with mercury, as to exclude all air; and therefore I trust more to a mercurial cane, and take all my notes from it."[186] In a marginal note Oldenburg adds:

"The exclusion of all air is here necessary, because air being subject to the operation of heat and cold, if any of it remain in the barometer, it will cause it to vary from shewing the true pressure of the air."[187]

There is no evidence in such practical books as those of Smith[188] and Dalencé[189] that any more advanced techniques had been developed by that time. It was generally thought that air was always contained in mercury.[190]

The sort of thing that could result is illustrated by an episode reported in 1705 by Guillaume Amontons to the Academy of Sciences. "Monseigneur le Chancelier" [Pontchartrain?] had a simple barometer, "monté à la manière d'Angleterre, c'est à dire, de ceux qui ont deux petits platines de cuivre sur lesquelles sont marquées les differentes dispositions qui peuvent arriver dans l'air, comme beau tems, changeant, pluïe, &c."[191] But the Chancellor found that it always indicated "stormy," and asked Amontons about it. Amontons inclined it, but found no air above the vacuum, although measurement showed it to read 1½ inches too low. Puzzled, he changed the mercury, with no success. Then he put in a new tube and corrected the Chancellor's barometer. After that he began to make experiments with a number of tubes inverted into the same cistern, and finally concluded that the spirits of wine which

[183] Nicolaus Zucchius, *Nova de machinis philosophia* (Rome, 1649), p. 106.
[184] See p. 24 above.
[185] *Works*, (London, 1772), I, 38–39.
[186] *Phil. Trans.*, Vol. 1 (1665/6), pp. 155–56.

[187] *Ibid.*, p. 156.
[188] John Smith, *A Compleat Discourse*, etc. (London, 1686).
[189] [Joachim Dalencé (or d'Alencé)], *Traittez des barométres, thermométres, et notiométres, ou hygrométres. Par M^r. D°°°* (Amsterdam, 1688).
[190] See Chapter 8, p. 176 above.
[191] *Mém. Acad. r. des Sci.* (1705), p. 230.

had been used to clean the tube was to blame; but (not realizing that the vapor would condense when the barometer was tilted) he supposed that the spirit had unstopped the pores in the glass so that "une plus grande quantité des plus petites parties de l'air"[192] could pass through. This explanation was echoed by Nicolas Hartsoecker in his *Eclaircissements sur les conjectures physiques* five years later.[193]

This question of dryness bedeviled the early experiments on the luminescence of the barometer (see Chapter 13).

Maraldi made elaborate experiments on barometer tubes, washed in various ways, which were reported on by Fontenelle in 1706.[194] When the tubes were washed with alcohol, he concluded, some drops remained which became "extremely rarefied" and lowered the mercury, either because they filled the space themselves in their rarefied state, or because the air they contained was liberated. If a tube was filled immediately after washing with alcohol and wiping several times with different cloths, the depression was greater than if it was left for some time before being filled. In dry weather tubes that had been washed with spirits of wine could be dried perfectly by leaving them exposed to the air for several days. Washing with alcohol ("l'esprit de vin") caused a greater depression than washing with brandy ("l'eau de vie"); this again a greater depression than washing

with water. In the absence of any clear idea of vapor pressure, this offered a puzzle, and the solution arrived at could scarcely have been right:

And as to the theory, we can imagine nothing but that the little drops of liquid, which have wetted the inside of the tube, being rarefied in the vacuum, or the air contained in this liquid being set free, the mercury is lowered. The first idea is the less likely, because if spirits of wine lowered the mercury by itself, it would lower it less than brandy would, for it is lighter. Brandy, lighter than water, would lower it less; but the contrary is true. Therefore, in conformity with the second idea, there should be more air contained in spirits of wine than in brandy, or else it escapes more easily; and it will be the same thing with brandy as compared with water. Now these hypotheses look rather satisfactory.[195]

Things are not always what they seem. To illustrate the complete absence of a real theory we need only consider the following sentence from the same report: "It is no use washing and rubbing the *outside* of a tube with spirits of wine; the mercury does not descend at all."[196] No doubt a refutation of Amontons' hypothesis of the year before. It had become a rather prevalent habit to explain away any pneumatic difficulty that proved intractable by postulating some degree of porosity in glass.[197]

There was need of a way of drying the inside of the tube and a way of getting all the bubbles of air disentangled from the mercury. Both these ends were served to perfection by the practice of boiling the mercury in the tube.

---

[192] *Ibid.*, p. 234.
[193] Amsterdam, 1710, p. 154.
[194] *Hist. Acad. r. des Sci.* (1706), pp. 1–3. Giacomo Filippo Maraldi (1665–1729), Cassini's nephew, and the first Royal Geographer, was one of the Italians imported by Louis XIV.

[195] *Ibid.*, p. 3.
[196] *Ibid.*, p. 2. The italics are mine.
[197] See Chapters 3 and 4, *passim*.

I do not know who first boiled the mercury in the barometer tube. The first description of the process which I have found is that of Du Fay, in a paper primarily devoted to the luminescence of the barometer,[198] and he says that he learned of the method from a German glassworker. This man, says Du Fay, passed an iron wire, with a cotton swab on one end, through the glass tube, removed the wire and sealed one end of the tube with a blowlamp. When it had cooled, he put the iron wire back, without the cotton, and filled the tube with mercury (previously poured through a paper funnel with as small a hole as possible) to a third of its length. Then he lit some charcoal in a small stove and, holding the tube in an inclined position, gradually brought the end near the glowing embers until the mercury boiled, turning the tube and moving the iron wire up and down until no more bubbles emerged; after which he moved the tube gradually across the stove until he had boiled all the mercury in the tube. When it had cooled down, he added enough mercury to fill another third of the tube and boiled it as before; but he did not think it necessary to boil the remaining third. This omission would be inexcusable nowadays.

It seems that in Germany itself the technique was more advanced, to judge by the results obtained by J. G. Leut-

mann, whose book[199] will be referred to again in Chapter 13.

It took more than a decade for the method to become generally used. It was reported as an innovation by Charles Orme, already mentioned as the inventor of the diagonal barometer with multiple tubes,[200] in the *Philosophical Transactions* of 1738.[201] This account is doubly interesting because it also refers to the preliminary purification of the mercury by distillation. The mercury was then poured into the tube, and "all the air got out by the methods used [*sc.* up to that time] in filling tubes."[202] Every part of the tube was then heated for long enough to boil the mercury, ". . . which curious and fatiguing operation is continued for the space of four hours."[203] In case this seems excessive, it turns out that this description applies to a diagonal barometer, with a tube 49 inches long. Familiarity with electric ovens and such things makes it hard for us to imagine the difficulty of handling such a tube over a charcoal fire without breaking it or burning the operator's hands or both.

In 1740, Le Monnier in France mentions boiling the mercury in barometer tubes as if it were a fairly well-known procedure.[204] The technique developed

---

[198] *Mém Acad. r. des Sci.* (1723), pp. 295–306. Charles François de Cisternay Du Fay (1698–1739) was a brilliant soldier, administrator and experimentalist who in his short life made valuable contributions to half a dozen sciences. For an appreciation, see I. Bernard Cohen, *Franklin and Newton*, Philadelphia, The American Philosophical Society, 1956, p. 371.

[199] *Instrumenta meteorognosiae inservientia.* (Wittenberg, 1725).

[200] See page 114 above.

[201] "The imperfections of the common barometers, and the improvement made in them, by Mr. Cha. Orme of Ashby-de-la-Zouch in Leicestershire, where they are perfected and rectified." *Phil. Trans.*, Vol. 40 (1738), pp. 248–65.

[202] *Ibid.*, p. 250.

[203] *Ibid.*, p. 251.

[204] *Suite des Mém. Acad. r. des Sci.*, Année 1740, p. clxii.

steadily, and in 1768 Paul d'Albert, Cardinal de Luynes (1703–1788), with the help of an artisan named Santinello-Cappy, was able to boil out a tube 13¼ lines (about 29 mm.) in diameter, a record which stood for about sixty years. The arrangements for doing this were very well-conceived.[205] A cylindrical furnace was made from two cylinders of heavy iron wire netting, the inner one being open at both ends. Burning charcoal could be put between the cylinders. They boiled 3 or 4 inches of mercury at a time, stirring it with an iron wire, and letting the tube cool before adding more. When it was full, a boxwood collar with a small plug in the middle was cemented on to the open end of the tube, the plug hole quite filled with mercury, the plug (probably conical) pushed in, and the whole inverted into a wooden cistern, with complete success.

The Cardinal also knew that the vapor of mercury is dangerous. He advises anyone who wants to fill a large barometer to do it in a large room with no gold or silver about, and no gilded furniture; "& sur-tout, dans un petit appartement, cette vapeur pénétrante pourroit intéresser la santé, c'est par cette raison que j'ai fait charger mon gros baromètre dans une très-vaste orangerie, dont la porte et toutes les fenêtres étoient ouvertes."[206]

Naturally De Luc wrote a good deal about filling barometers in his famous book.[207] He starts with the warning that the glass must not be too thick, or it will crack. Then he describes the process of boiling, and it seems that he was able to operate with a tube full to within about two inches of the open end, beginning the heating at the closed end and manipulating the tube over a small stove placed at the edge of a table.

De Luc also discovered that it was as important to outgas (as we should now say) the tube as to get air out of the mercury itself, and that once this has been done, air attaches itself to the tube only slowly. But let him tell about it:

The greater part of the air which comes out of the tube during this operation comes off the glass walls; and what is very remarkable about this is that when this layer of air has once been removed from a tube, and when the mercury that replaces it has remained there for some time, we can empty the tube, let the air back in, and then put back new mercury which has not boiled, without the air attaching to the glass. . . . When the new mercury is made to boil, it does not begin, like that which preceded it, by covering itself with this great quantity of little air-bubbles, though the operation is similar in the two cases, for in both, the air which at first filled the tube is replaced by un-boiled mercury. In the first case, then, these little bubbles must be produced by a layer of air which lined the tube, and which, once removed, re-establishes itself only very slowly. . . . From this I conjecture that new tubes, or those which have been long unused, are lined inside with impalpable particles of dust and damp, around which form little atmospheres which dilate with heat, and that the glass itself has little cavities on its surface into which air insinuates itself.[208]

---

[205] *Mém. Acad. r. des Sci.* (1768), pp. 247–69.

[206] *Ibid.*, p. 255.

[207] *Recherches sur les modifications de l'atmosphère* (Geneva, 1772), I, 192–97.

[208] *Ibid.*, I, 194–95.

De Luc has a long and fascinating foot-note describing the process of drawing tubes in an eighteenth-century glass-works, by way of explaining how the inside surface of *new* tubes comes to be damp and dusty.

By 1772 the practice of boiling the mercury in barometer tubes was universal in the manufacture of instruments for serious scientific purposes, but "unboiled" barometers continued to be made for several decades, so that even in 1840 the Royal Society felt it desirable to publish tables of the capillary correction for both "boiled" and "unboiled" tubes.[209]

Not everyone had as much success as Cardinal de Luynes' mechanic in filling large barometers. In 1823 several half-inch tubes were broken before success was attained in the construction of a barometer for the Royal Society.[210] In the years 1853–1854, under the super-intendence of the Kew Committee, several attempts were made "to prepare, by the usual method of boiling, a ba-rometer tube of large dimensions."[211] The problem was passed to Enrico Negretti, who finally succeeded in making two or three tubes of one inch bore; but even though the mercury had been thoroughly cleaned, it quickly became contaminated, and rings of dirt formed on the glass. Tubes filled under a partial vacuum without boiling be-came dirty in the same way. Welsh then "had fresh tubes made under [his] own inspection, and sealed at the glass-works immediately after being

drawn."[212] This was better, but not good enough. Then, wrote Welsh, "About this time I had the advantage of consulting Mr. John Adie of Edin-burgh, who informed me that he had also experienced the same inconveni-ence, and that he had removed it by thoroughly cleaning the tubes by sponging with whiting and spirits of wine."[213] Welsh followed Adie's direc-tions and was successful with a tube of 1.1 inch bore, which was built into a barometer. It is probably fortunate that he was using a vacuum pump.

After the pioneering effort by Gue-ricke[214] in 1654, the first account of an air pump being used in filling barom-eters is probably that of Poleni in 1709.[215] However, this was only a means of drawing mercury into a nar-row tube, one end of which was then sealed up with cement. The next use of it to remove air from the tube and the mercury seems to have been made by Daniell,[216] who employed a vacuum of about half an inch of mercury in filling the instrument, as a preliminary to boiling. Karl Friedrich Mohr of Kob-lenz (1806–1879) thought that the use of a vacuum pump while shaking the tube gently was a satisfactory substi-tute for heating,[217] but experience with his barometers is not reported. New-man seems to have had a much better idea, noted with approval by the Jury at the Great Exhibition: "Mr. Newman has adopted the method of filling tubes in vacuo, and boiling them under

[209] See page 191 above.

[210] J. F. Daniell, *Meteorological Essays and Observations* (London, 1823), p. 350.

[211] John Welsh, *Phil. Trans.*, Vol. 146 (1856), p. 507.

[212] *Ibid.*, p. 508.

[213] *Ibid.*, p. 508.

[214] See page 61 above.

[215] Johannis Poleni, *Miscellanea, hoc est, dissertatio de barometris*, etc. (Venice, 1709).

[216] Daniell, *Meteorological Essays*.

[217] *Dingler's Polytech. J.*, Vol. 79 (1841), pp. 194–96.

diminished pressure, at a temperature which obviates all oxidation of mercury."[218] This is the modern way of doing it, although it is even better to outgas the tube for some time in a high vacuum and at a fairly high temperature before the mercury is (preferably) distilled into the tube from a vacuum still.

Heinrich Wild evolved an elaborate method of filling barometer tubes *in situ* at the Russian weather stations.[219] He set up a simple air pump, a drying tube, a spherical flask with two outlets, and the barometer tube, all in series, with rubber-tube connections. The apparatus was pumped out several times, alternating with letting air in through the drying tube. Finally he brought mercury to the boil in the flask under vacuum, warmed the tube, and tipped the mercury into it. One may wonder whether the barometer tube would be sufficiently outgassed, but Wild, who was a hard man to please, seems to have been satisfied.

The purification of mercury has concerned many barometer makers and other scientists.[220] Orme, as we have seen, distilled his mercury in 1739. In 1781 Tiberius Cavallo, F.R.S. (1749–1809) wrote[221] ". . . quicksilver is very easily distilled, by which means it is separated from lead, tin, and other

metals, with which it is often combined."

Traces of lead and tin quite spoil the bright, clean appearance of mercury,[222] and in 1831 C. von Riese wanted to ascribe most of the troubles in the open limb of the siphon barometer to impure mercury.[223] Indeed it must often have made them worse.

But it became evident that distillation alone (particularly at very low temperatures under a high vacuum) would not necessarily get rid of these metallic impurities. Heinrich Wild[224] used a very complicated chemical process to clean mercury, beginning with ferric chloride solution, which acts on copper and some other metals. Such processes always get the mercury wet, and Wild finally dried his over concentrated sulphuric acid *in vacuo*.

One of the advantages of vacuum distillation is that it results in dry mercury. It was at about this time that the valveless mercury pump due to Sprengel[225] came into use, making the attainment of rather high vacua (~5 x 10⁻⁴ mm. of mercury) possible.[226] In 1873 A. Weinhold described an apparatus

[218] Great Exhibition of 1851, *Reports of Juries* (London, 1852), p. 657.

[219] *Repertorium für Exp.-Phys.*, Vol. 7 (1871), pp. 256–59. (Abstract in *J. de Phys.*, Vol. I [1872], pp. 265–66).

[220] C. L. Gordon and E. Wichers, *Ann. New York Acad. Sci.*, Vol. 65 (1957), pp. 369–87, give 132 references to the purification and properties of mercury, beginning with Pliny the Elder (A.D. 23–79).

[221] *A Treatise on the Nature and Properties of Air and Other Permanently Elastic Fluids* (London, 1781), p. 81.

[222] According to E. Wichers (*Rev. Sci. Instr.*, Vol. 13 [1942], pp. 502–3) mercury which remains clean and bright after filtration through a pinhole in a paper cone contains less than one part of base metals per million parts of mercury.

[223] *Astron. Nachr.*, Vol. 8 (1831), col. 287–94.

[224] *Repert. f. Meteorol.* (1871), pp. 258–59.

[225] *Chem. Soc. J.*, ser. 2, Vol. 3 (1865), pp. 9–21.

[226] For the early history of the mercury pump, see Sylvanus P. Thompson, *Telegraphic J. and Electrical Rev.* (London), Vol. 21 (1887), pp. 556–58; 587–90; 610–13; 632–34; 659–65. A satisfactory valveless pump was described by J. Mile in 1828 (*Archiv. f. die gesammte Naturlehre*, Nürnberg, Vol. 15 [1828], pp. 1–9), but entirely forgotten.

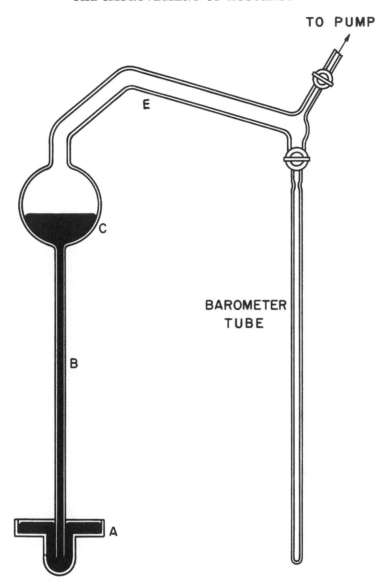

Fig. 9.24. Waldo's barometric still, 1884.

using a modified Sprengel pump and a barometric still for distilling mercury under high vacuum.[227] The action of the barometric still may be illustrated by Figure 9.24, after a description by Frank Waldo of a rather similar apparatus.[228] It consists essentially of a tube with a flask at the top, and dipping

[227] *Repertorium für phys. Technik,* Vol. 9 (1873), pp. 69–74.

[228] *Amer. J. Sci.,* ser. 3, Vol. 27 (1884), pp. 18–19.

into a reservoir at the bottom, the vertical distance between the center of the flask (or somewhat below it) and the mercury in the reservoir being equal to the height of the barometric column at the place where it is installed. Waldo used a Sprengel pump; when it was operated, the mercury in A was drawn up the tube B and C and the whole apparatus, including the barometer tube, evacuated. Then heat was applied to C, and the mercury vapor that resulted condensed in the large tube E and ran down into the barometer tube. With the Sprengel pump it was advisable to have the mercury just about dry to begin with, but with modern pumps this is less critical.

As mentioned above, distillation had been found inadequate to free mercury from all other metals, and in 1911 G. A. Hulett gave a well-documented demonstration[229] that three distillations are necessary. It appears that silver is the most difficult metal to leave behind.

I do not know whether he had read Hulett's paper (though this is quite likely), but about 1935 Dr. John Patterson (1872–1955), then Director of the Canadian Meteorological Service, who was a perfectionist about barometers, had a barometer-filling system installed in the attic at the Meteorological office in Toronto which gave him and everyone else great satisfaction. It is shown schematically in Figure 9.25. By means of this apparatus, any number of tubes up to eight could be filled in one operation with mercury triply distilled in the barometric stills S_1, S_2, and S_3. Before filling, the tubes were outgassed by heating for several days under vacuum, each tube being surrounded by a tubu-

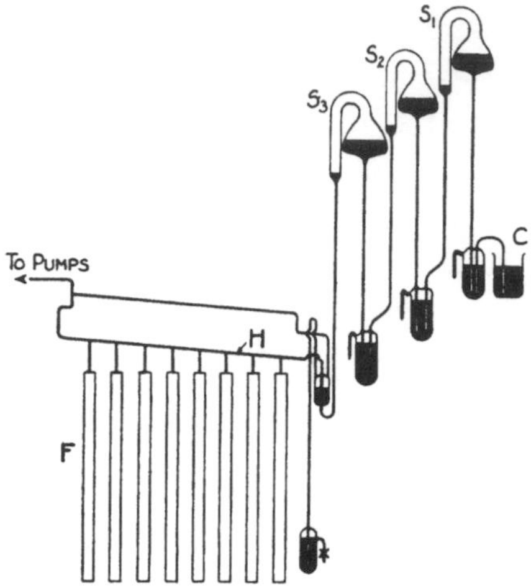

Fig. 9.25. Patterson's barometer-filling apparatus.

lar electric heater F for that purpose, the direct descendant of Cardinal de Luynes' ring-shaped charcoal stove.[230]

From time to time doubts have been expressed about the permanence of the vacuum even in well-made barometers, even after it ceased to be possible to believe that glass is porous. One of these doubts, which remains somewhat of a mystery, arose from an observation by Michael Faraday,[231] who filled, over mercury, some bottles four-fifths full of a mixture of two parts hydrogen and one part oxygen. They were stoppered with well-fitting glass stoppers under the mercury, and left inverted, with mercury covering the stoppers and the necks of the bottles, for fifteen months. It was then found that ordinary air had largely replaced the mixture in all the bottles; in one, completely.

[230] For a bibliography of the most recent methods of purifying mercury, see W. G. Brombacher, D. P. Johnson, and J. L. Cross, *Nat. Bur. Stds. Monograph* 8, (Washington, 1960).

[231] *Quart. J. Sci.*, Vol. 22 (1827), pp. 220–21.

[229] *Phys. Rev.*, Vol. 33 (1911), pp. 307–16.

The natural conclusion of a skeptic would be that Faraday had been the victim of a practical joker or an inquisitive laboratory assistant. But J. F. Daniell[232] claimed to have confirmed this surprising result and believed that barometers gradually deteriorate by the penetration of air around the end of the tube between the mercury and the glass. Observing that platinum is wetted by mercury when boiled in it, Daniell suggested that a short piece of platinum tube should be sealed to the open end of the barometer tube before the latter is filled. Few people appear to have taken much notice of this suggestion.[233] Daniell believed he had confirmation of his theory in the records of the *Societas Meteorologica Palatina*, 1781–1792. At all their stations, the mean pressure for the years 1787–1792 is lower than that for 1781–1786. An examination of the differences shows, however, that the probable error of many of them is so large that they are not statistically significant. And as we saw in Chapter 7, the barometers used were rather elementary, and their vacua might have deteriorated because of imperfect filling.

The conclusion reached by Daniell is scarcely tenable in view of the many well-made barometers which have retained their accuracy almost unchanged for many years, even for a century, as for example the Bombay and Toronto barometers referred to on page 219.

However, if not in this manner, then in other ways, the vacuum can deteriorate. A test of its excellence or otherwise would be highly desirable. Such a test was first suggested by Arago.[234]

If a barometer were constructed, said Arago, so that the size of the vacuum space could be greatly varied, then a correction could be made for the remaining air, because the pressure exerted by this would vary inversely as its volume. Probably inspired by this, Kupffer designed a cistern-siphon barometer which, as modified by Girgensohn,[235] was widely distributed in Russia. In a later paper,[236] Kupffer set out the required simple theory more clearly than Arago had done. Apparently the correction was in danger of being attributed to Kupffer; and Arago, not a man to let a well-deserved priority slip from him, found it necessary in 1844 to call the attention of the Academy of Sciences to the matter, using as an excuse his desire to bring to their notice a barometer, very easily portable, made for him by the instrument maker Gambey.[237]

Arago originally intended his correction to catch relatively large errors in portable barometers. In fact he connected it with a vague project for a barometer which could be filled *in situ* by a traveler. But it was taken up by the makers and users of much more accurate instruments, and in 1886 Pernet found it desirable to give warning that the method is rendered of doubtful value by the fact that the part of the open limb generally used, and the part seldom used, are very likely to have

---

[232] *Elements of Meteorology* (London, 1845), Vol. 2, pp. 290–96.
[233] But see p. 343 below.

[234] *Ann. Chim. et Phys.*, Vol. 33 (1826), p. 395.
[235] *Mém. Acad. imp. des Sci.*, Vols. 1 and 3.
[236] *Annalen der Physik*, Vol. 26 (1832), pp. 446–51.
[237] *Compt. Rend.*, Vol. 19 (1844), p. 703. Unfortunately this instrument was not described.

different values of capillary depression. A less important error will be caused by any variation of temperature in the vertical, unless the cross section of the tubes is everywhere the same.[238]

10.   BAROMETERS WITH TWO TUBES

As an insurance against having a defective vacuum without knowing it, Lavoisier had Megnié[239] make barometers with two independent tubes dipping into the same cistern. In 1778 or 1779 Lavoisier had eight of these instruments made, and presented them to friends in various places. One of the friends was Cotte.[240] Truchot[241] published most of Megnié's description of these a century later; the two tubes and the scale were moved up and down as a system in order to adjust the zero.

There are three of these barometers at the C.N.A.M. in Paris, one[242] exhibited as a part of the fine collection of barometers and the other two[243] in a different gallery among a fascinating collection of Lavoisier's apparatus. This dependence on tubes demands a trustworthy instrument maker, as Megnié undoubtedly was; in other circumstances one might well end with two tubes filled equally badly.

There was another famous barometer with two tubes, set up in 1837 by Francis Baily in the Royal Society's apartments.[244] This had an entirely different purpose, one tube being of flint glass, the other of crown, in order to see whether, in the course of time, one would have a greater chemical effect on the mercury than the other.

[238] *Zeits. für Instrum.*, Vol. 6 (1886), p. 380.
[239] See note 5 in this chapter.
[240] *Mémoires sur la météorologie* (Paris, 1788), Vol. 1, p. 500.
[241] *Ann. Chim. et Phys.*, Vol. 18 (1879), pp. 304–8.

[242] Inventory no. 19949. This is inscribed "Barometre de Megnie pour M^r. de Lavoisier de l'Academie R.^le. des Sciences &c N°. 2 1779."
[243] Inventory nos. 7658 and 8761.
[244] *Proc. Roy. Soc.*, Vol. 4 (1837), pp. 1–3; *Phil. Trans.*, Vol. 127 (1837), pp. 431–41. Francis Baily (1774–1844) was a successful business man who retired in 1825 and became an astronomer of great distinction.

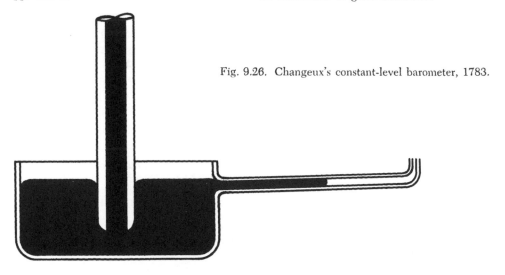

Fig. 9.26. Changeux's constant-level barometer, 1783.

The bore of the flint-glass tube was 0.594 inch, of the crown-glass tube 0.658 inch, and Baily noted that the capillary corrections would be +0.004 and +0.003 inch respectively, according to Laplace's theory. The thermometer, in accordance with the unfortunate fashion of the time, was in the cistern; nevertheless Baily gave an excellent discussion of the temperature corrections of such a barometer, the brass scale of which was movable, with an agate point that could be made to touch the surface of the mercury.

11. AUTOMATIC MAINTENANCE OF THE ZERO

From time to time attempts have been made to devise means of keeping the level of the mercury in the cistern constant either by floating a suitably proportioned cistern in a bath of liquid, usually mercury, or by suspending the cistern on springs, which would elongate when more mercury descended into the cistern.

It is scarcely necessary to pay much attention to these devices, none of which have proved worth the complication that they entail. Of the floating cisterns we may mention those of Cummins[245] and Schocke.[246] The latter's patent included the idea of suspending the cistern by a spring, but in this he was preceded by C. T. Coathupe.[247]

An entirely different idea which attained some celebrity at the end of the

eighteenth century was due to P. N. Changeux (1740–1800)[248] who suggested joining a tube to the side of the cistern, inclined very slightly upward, just at the level of the mercury (Figure 9.26). If more mercury came into the cistern, it would merely increase the length of a mercury column in this side-tube, with little change of level.

12. AUTOMATIC COMPENSATION FOR TEMPERATURE

In Chapter 8 we described a number of devices, which were essentially nomograms, for facilitating the application of the temperature correction to the mercury barometer. The use of these still involved reading a thermometer and adding or subtracting the quantity indicated by the nomogram. By automatic compensation, we here understand that the necessity of doing any arithmetic of this sort is entirely done away with.

Samuel B. Howlett invented a "compensating barometer" in 1839, which was intended to reduce both the "capacity-" and temperature-corrections to zero.[249] The principle is shown in Figure 9.27. There is an additional tube of the same diameter as the barometer tube, but closed at the bottom and open at the top. It is supported by a float, and filled with mercury to a height of 28 inches above the level of the mercury in the cistern. A scale slides up and down so that the difference in level of the mercury in the two

---

[245] Charles Cummins, British Patent 8462 (1840).

[246] Wilhelm Schocke, D.R.P. 223,230 (1908).

[247] *Quart. J. Meteorol. & Phys. Sci.*, Vol. 1 (1842), pp. 106–7.

[248] *Description de nouveaux baromètres à appendices, qui ont un niveau constant,* etc. (Paris, 1783), pp. 3–8.

[249] *Proc. Royal Soc.*, Vol. 4 (1839), p. 133.

tubes can be measured. This would, of course, suffer from the tendency of the open surface to get dirty, but there seems to be no reason why the extra tube should not be evacuated and closed. It also seems that the lower end

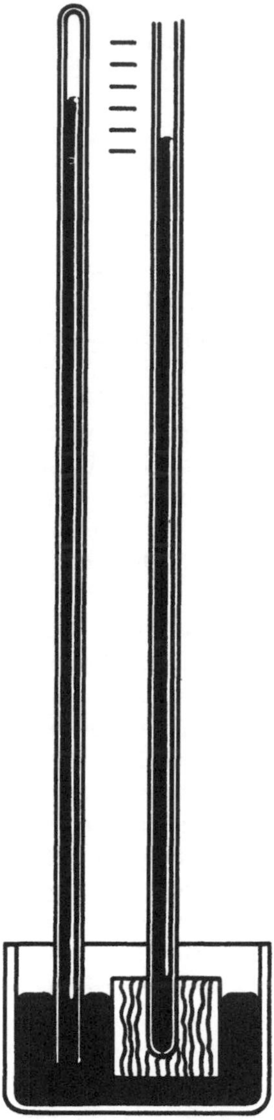

Fig. 9.27. Principle of Howlett's barometer, 1839.

of the extra tube ought to be at the level of the mercury in the cistern rather than below it.

In the last hundred years several people have discovered, quite independently, that a siphon barometer *of which only the lower limb is read* can be compensated exactly for temperature at any one pressure, and approximately at neighboring pressures, by giving appropriate dimensions to the tube.[250-257] Radau,[251] for example, stated without proof that the total volume of mercury in such a barometer to make the reading independent of the temperature at some barometric height b is

$$V = qbc/(q-3e)$$

in which c is the cross section of the vacuum chamber, q the absolute (volume) expansion coefficient of mercury, e the (linear) coefficient of dilatation of glass. Rysselberghe[253] seems to have found this hard to believe, but he had barometers of various forms made, containing the appropriate volume of mercury, and verified the formula even when the two mercury surfaces were of different sizes. Having convinced him-

[250] P. Volpicelli, *Atti r. Acad. Lincei,* Vol. 23 (1869), pp. 168–96.

[251] R. Radau, *Moniteur Scientifique,* Vol. 9 (1867), p. 712.

[252] G. W. Hough, *Annals of the Dudley Obs'y, Albany, N.Y.,* Vol. 1 (1866), pp. 88–90.

[253] F. Rysselberghe, *Oesterr. Zeits. f. Meteorol.,* Vol. 10, (1875), pp. 205–8.

[254] C. M. Goulier, *Compt. Rend.,* Vol. 84 (1877), pp. 1315–17.

[255] P. Czermak, *Zeits. für Instrum.,* Vol. 11 (1891), pp. 184–89.

[256] H. Sentis, *J. de Phys.* (ser. 3), Vol. 1 (1892), pp. 77–79.

[257] J. Hartmann, *J. Sci. Insts.,* Vol. 13 (1936), pp. 323–30.

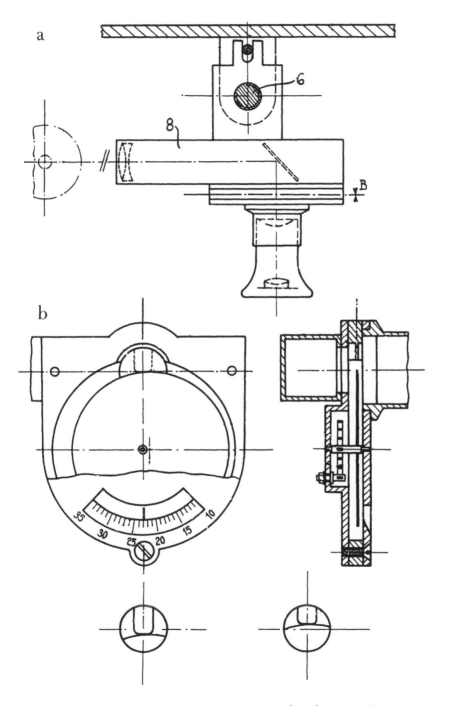

Fig. 9.28. Du Pasquier and Letestu's reading device, 1955.

self, he then gave a general proof of Radau's formula. The other writers gave formulas applicable to special forms of the barometer. This principle was of service to the designers of certain barographs, as we shall see in Chapter 11.

Volpicelli[250] also suggested a means of compensation for a cistern barometer, in which a plunger would be moved up and down in the cistern by the differential expansion of metal rods. He gave an elaborate mathematical design. O. de Candia[258] described a siphon barometer with a fiducial mark on its open limb. The mercury is brought up to this by dipping in an aluminum cylinder, graduated in units of S/s millimeters, where s is the cross-sectional area of the cylinder, S the interior cross section of the tube. This gives an enlarged scale; and by a proper choice of S/s the temperature error can be made zero at some mean barometric pressure and very small at neighboring pressures.

We shall end this section with a reference to a somewhat complex, but entirely practical device which can be applied to any fixed-cistern barometer,[259] and is actually in use in the Swiss Meteorological Service. As will be seen from Figure 9.28a, an angled telescope 8, in which the mercury surface can be observed, is driven up and down by the screw 6, which is geared to two counters, one giving the local barometric pressure, the other the pressure reduced to sea level.[260] It

[258] *Nuovo Cimento*, Vol. 2 (1895), pp. 115–19.
[259] Pierre Du Pasquier and Serge Letestu, Swiss Patent 307,036 (1955).
[260] This last obviously involves some assumptions.

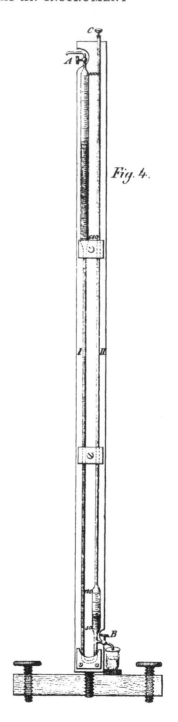

Fig. 4.

Fig. 9.29. Bohn's barometer, 1877.

seems that with quite simple gear ratios this can be done in millibars (at Geneva) to one part in ten thousand.

The temperature correction is taken care of by having an eccentric disk in the focal plane of the telescope objective, the edge of which the inverted image of the mercury meniscus is made to touch (Figure 9.28b). This disk, which is very light, is rotated by a bimetallic spiral. The compensation is, of course, strictly accurate only for some one pressure, but a change of temperature of ±10°C. and a change of pressure of ±30 millibars, acting together, are stated to lead to an error of only 0.04 mb. The only mental reservation I have about this device is that some care would have to be taken to have the thermal lag of the bimetallic spiral in its box equal to that of the barometer tube.

13. "CHEMICAL LABORATORY" BAROMETERS

It is natural that a chemist, finding that he needs a barometer, should think of making one in the way other glass apparatus is constructed. A number of designs for such barometers appeared in the last quarter of the nineteenth century, and some of them have interesting features. In all cases they are designed to avoid the necessity of boiling the mercury in the tube.

The barometer shown in Figure 9.29 was described by J. Konrad Bohn of Aschaffenburg in 1877.[261] The figure is self-explanatory, but it should be emphasized that the top of the tube at C

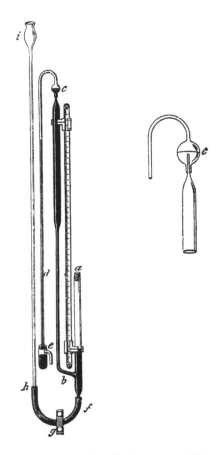

Fig. 9.30. Diakanoff's barometer, 1884.

is higher than the outlet A. By pouring mercury in at C, and alternate use of the taps A and B, the vacuum in the upper chamber could be made as good as desired, and renewed at will. Bohn, who thought so well of his instrument that he sent it to the Exhibition of Scientific Instruments in London (1876), engraved the scales on the tubes, as in a burette, but it could, of course, have been used with a cathetometer.

Diakanoff's barometer (Figure 9.30)[262]

[261] *Annalen der Physik*, Vol. 160 (1877), pp. 113–18.

[262] *J. de Phys.*, sér. 2, Vol. 3 (1884), pp. 27–29.

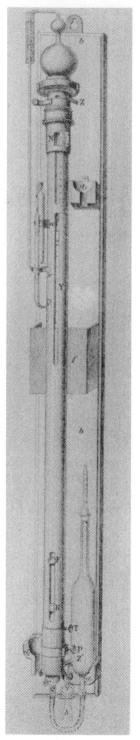

Fig. 9.31. Sundell's barometer, 1887.

is not very different in principle. It is filled by temporarily plugging a, pouring plenty of mercury into i and manipulating the pinchcock g. The tube cde forms a mercury seal. The author recommends that the tube should be chemically cleaned and then dried by passing pure dry hydrogen through it.

A. F. Sundell of the Meteorological Service of Finland described a rather similar barometer[263] which was improved by the addition of drying tubes where air enters the trap and the reservoir. He made comparisons with a large Wild-Fuess cistern-siphon barometer[264] which do indeed suggest that such an instrument can be filled successfully without boiling. He then had one constructed in a form suitable for traveling (Figure 9.31) and made a trip around all the main observatories of Europe, obtaining very good results in his comparisons.[265] This barometer might, I suppose, have been described in Chapter 7, but it seems more appropriate to deal with it here.

In 1877 C. Kraewitsch described in Russian a barometer which could be filled without boiling. His paper was abstracted, but not very clearly, in the *Journal de Physique*;[266] and ten years later, finding that it seemed quite unknown, he decided to describe it again, in German.[267]

The construction will be evident from Figure 9.32, with the remark that the tube C is very narrow ($\frac{1}{3}$ to $\frac{1}{4}$ mm. bore), D 4 to 6 mm. The tubes

[263] *Acta Soc. Scient. Fennicae,* Vol. 15, (1885), pp. 389–98.
[264] See page 231 above.
[265] *Acta Soc. Scient. Fennicae,* Vol. 16 (1887), pp. 431–92.
[266] Vol. 6 (1877), pp. 197–98.
[267] *Repertorium für Phys.,* Vol. 23 (1887), pp. 339–48.

A and B can advantageously be very large, and Kraewitsch suggests 60 mm., which seems to come into the region where the law of diminishing returns holds full authority. To fill the instrument, a rubber tube and a funnel are attached at O, the tap R closed, and A filled with dry mercury. The barometer is inclined (D to the left), R opened, and the whole thing filled right up to

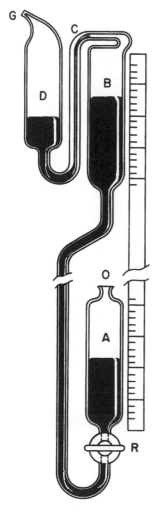

Fig. 9.32. Kraewitsch's
barometer, 1877 (1887).

G. The tap R is then closed, and G sealed off. The rubber tube and funnel being removed, the instrument is then inverted to pour the excess mercury out of A. It is then put right way up and R opened, the mercury falling to about its proper height in B. Any small amount of air in B can be pushed into D by repeatedly inclining the instrument and bringing it vertical again; while for transport it is inclined and R closed. Kraewitsch very fairly emphasized that the success of the instrument depends on careful cleaning of the glass, clean glassblowing, and thorough purification of the mercury.

The most ingenious of these designs, though not, I think, the most advantageous, is that of W. J. Waggener.[268] It is a loop of glass made as shown in Figure 9.33, the lower chamber C serving as the "cistern," when the barometer has been filled. S and S' are stopcocks. I shall leave the method of filling it as a puzzle for the reader, with the clue that at one stage it is turned round and round in the plane of the drawing. The difficulty of observing the position of the lower mercury surface would have greatly reduced the utility of this barometer.[269]

14. MISCELLANEOUS BAROMETERS

To end this chapter we must refer to a few miscellaneous designs which do not seem to fit into any of the ordinary categories.

[268] *Amer. J. Sci.*, ser. 3, Vol. 42 (1891), pp. 387–88.
[269] There are no doubt many other barometers of this general sort in the immense chemical literature, which I cannot claim to have searched.

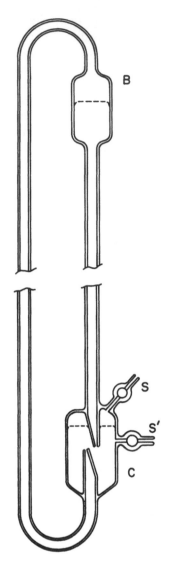

Fig. 9.33. Waggener's barometer, 1891.

Gotha to a demonstration of the superiority of weighing, rather than measuring, the barometric column, and the next seven to a description of his barometer, which is simply a uniform tube hung from an "English hydrostatic balance" on a tall stand with a screw adjustment for its height, so that before weighing the tube a mark on it can be brought exactly to the level of the mercury in the cistern into which the tube dips. As a modification he has a small side tube near the top of the tube, drawn out to a fine capillary. Before filling the tube he hangs it in the cistern with mercury up to the mark both outside and (the capillarity serving to equalize the pressures) inside. He then balances the tube and fittings, and after sealing the capillary he fills the tube and hangs it on the balance again, when the weight he has to add to the counterpoise gives him an absolute measure of a sort. He notes that it is a good idea to counterbalance the tube for a mean pressure and then add weights to either pan when making an observation.

This use of weights was envisaged in an elaborate balance barometer devised by Prony in 1799.[271] This is of the type in which a change in barometric pressure results in a horizontal displacement of the center of gravity of the mercury. The restoration of equilibrium by putting weights in scale pans is determined by observing marks on the beam through a microscope. The motion of the beam is limited by stops.

A similar but less elegant balance barometer "in wood or metal" is re-

The balance barometer offers possibilities of accurate measurement, besides the mere enlargement of the scale dealt with in Chapter 6. In 1741 Petter Olof Berghman devoted the first five pages of his doctoral dissertation[270] at

[270] *Dissertatio academica de Barometro nomini & omini,* etc. (Gotha, June 15, 1741).

[271] *Annalen der Physik,* Vol. 2 (1799), pp. 311–13 (Trans. from *Bull des Sci.,* Brumaire An VII, p. 156.)

ported in 1815 as having been devised by "M. Jecker."[272] This consisted of a siphon tube in the form of three sides of a rectangular frame 2½ times as high as it was wide. A crossbar carried a knife-edge to support the whole thing near its center of gravity. According to Cadet, this barometer was intended to be portable, two taps being placed in the tube with this in view; and it was stated to be so sensitive that it would indicate a difference of elevation of one foot. One can scarcely conceive of anyone carrying such an unwieldy device up a mountain.

Another sophisticated barometer, which depended on measuring the volume of mercury, was invented, according to Magellan,[273] by the young Count Marsiglio Landriani (1751–1815), though as far as Magellan knew, the idea had not been published. The principle of this barometer is shown in Figure 9.34. *ABDC* forms a barometer, *D* being a special tap, which normally connects *AB* to *C*, but if turned, closes off *AB* and allows all the mer-

---

[272] C.L.C[adet], *J. de Pharmacie*, Vol. 1 (1815), pp. 413–14. This was probably François Antoine Jecker (fl. 1790–1820), the first Paris instrument maker to organize quantity production. See M. Daumas, *Les instruments scientifiques aux XVIIe et XVIIIe siècles* (Paris, 1953), pp. 368–69.

[273] *Description et usages de nouveaux baromètres*, etc. (London, 1779), pp. 152–53. (This document is paged from 87 to 164.)

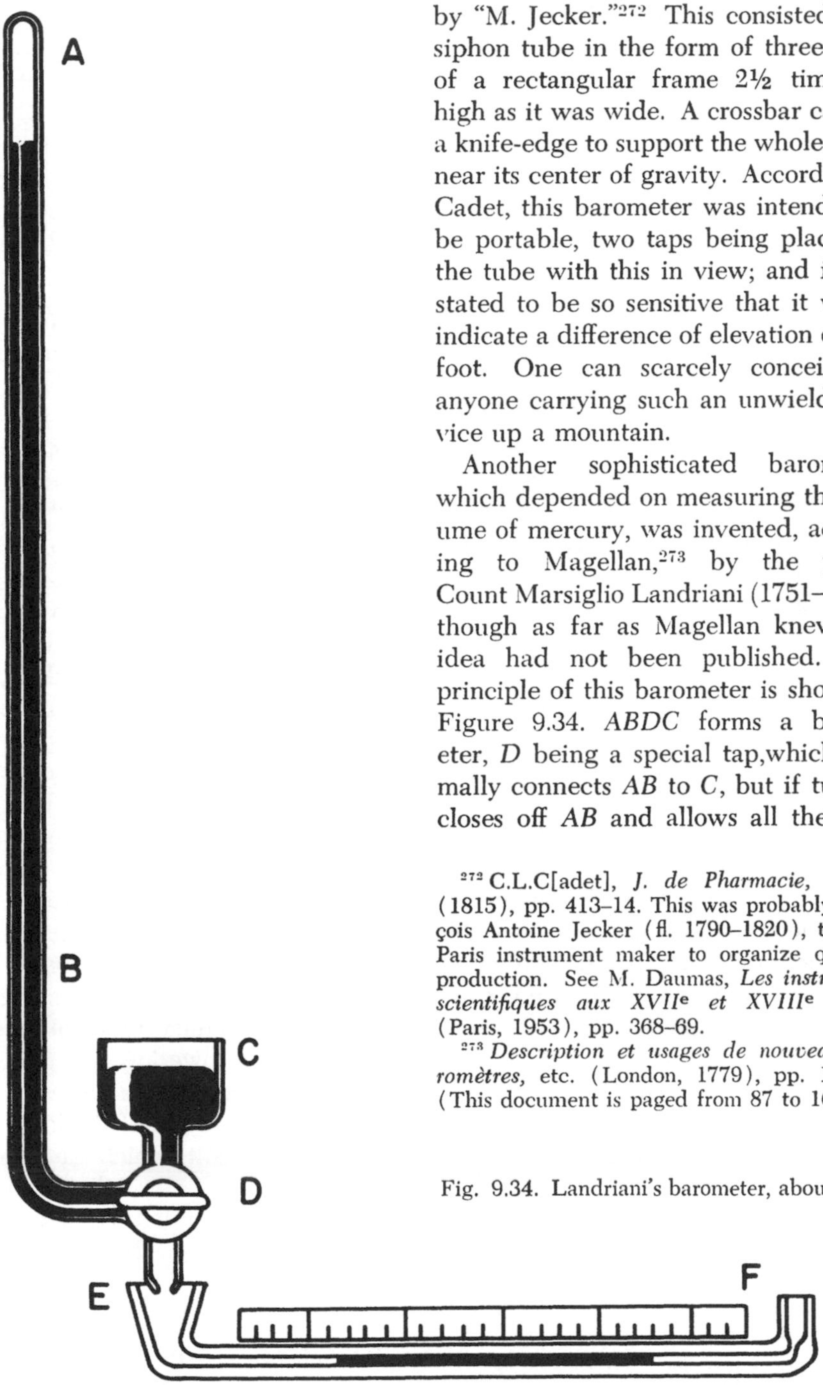

Fig. 9.34. Landriani's barometer, about 1775.

cury in C to fall into the funnel E which forms one end of the graduated tube EF. The length of the resulting mercury thread is measured, and then the mercury is poured back into C and the tap again turned to its normal position.

Magellan suggested as an improvement that the mercury ought to be weighed instead of being measured. This advice may have been taken, for the instrument is referred to in the *Encyclopaedie méthodique* as being used with "a very sensitive balance."[274]

There is such a barometer in Florence,[275] with a glass-lined metal tray under the tap, which would indicate that the mercury was weighed. The tap is relatively enormous, and of glass, in a large glass block. The short tube ends in an ivory cup. The whole barometer is mounted on a flat plate with three leveling screws. It is unsigned and undated, but from the form of the scale and vernier, it is probably early nineteenth-century.

A series of constant-level barometers was suggested, with the necessary theory, by G. Guglielmo.[276] These can have a constant level in the cistern, the chamber, or both. The general principle is that a calibrated burette is connected to the barometer. At some pressure H, determined by another barometer, the level in the chamber (say) is brought to a fiducial point. Now let the pressure change to H', and by ma-

nipulating the burette let us bring back the mercury to the same level. If v and v' are the two readings of the burette, S the cross section of the cistern, then $H' - H = (v - v')/S$. Guglielmo also worked out the theory of the temperature correction of such instruments. It can easily be seen that such an arrangement could be made extremely sensitive.

We may end with two barometers which can be read from a distance. In 1884 a patent was granted to H. B. and H. R. F. Bourne[277] on a device in which the mercury in a siphon barometer short-circuits variable portions of two resistance wires held vertically in the two limbs, and forming part of a Wheatstone bridge. It is rather astonishing that they should get a patent for this, just two years after J. Joly had described a similar but better design.[278] In this two platinum wires were sealed into the top of the vacuum space; to one a stiff iron wire is fastened, to the other a carbon filament (such as was then coming into use in incandescent lamps). The wire and the filament were joined beneath the mercury, below its lowest possible reading. This variable resistor was to be used in a Wheatstone bridge, the balancing resistor to be in a similar tube with means of raising and lowering the mercury, and a scale graduated empirically. To measure the resistance of the long two-wire copper circuit, which would vary with temperature, the coil of a polarized relay was included to short-circuit the barometer when the current was reversed.

[274] *Encyclopaedie Méthodique: Physique,* I (Paris, 1793), 121.

[275] *Museo di Storia della Scienza,* inventory no. 1137. Another, no. 1134, is rather similar and perhaps earlier.

[276] *Atti della r. Accad. dei Lincei,* ser. V, Vol. 2 (1890), pp. 8–17.

[277] British Patent 9116 (1884).

[278] *Nature,* Vol. 25 (1882), pp. 559–61.

Chapter 10

Barometers of High Accuracy

1. INTRODUCTION

THE great majority of barometers are, or ought to be, calibrated by comparison with another barometer which in some way can be made to furnish trustworthy values of atmospheric pressure. While this more reliable barometer may in turn have been calibrated by comparison with a still better one, somewhere along the line we come to a firm requirement for a barometer which can be calibrated from first principles, without reference to any other. This chapter proposes to deal with the history of barometers of this kind.

The difficulty of the problem was scarcely realized before about 1830, when it began to appear that a number of measurements of various sorts (length, temperature, density, the acceleration of gravity) must be made with fair accuracy if the barometric pressure was to be derived from the measurement of the height of a column of mercury, accurately enough to please the surveyors or even the meteorolo-

gists of the time. At about the same period it also became evident that the capillary depression of a mercury surface depended on more factors than the bore of the tube, and the recognition of this awoke instrument makers from the happy dream that any well-constructed siphon barometer might suffice as a standard.[1] The initiative for the provision of something better came from the developing meteorological networks, whose directors were naturally concerned about the comparability of large numbers of barometers, and later, with the advent of telegraphy, became interested in international standardization. Further developments, beginning about 1880, sprang from the necessity for very accurate barometry and manometry in the standardization of temperature scales and in other fields of physics and physical chemistry; these have continued, until at the present time the accuracy of the best instruments is circumscribed rather by the properties of matter than by the construction of the

[1] See Chapter 9, pp. 233 ff.

instruments themselves.[2] I shall not discuss these highly sophisticated instruments at any length, as they were conceived fundamentally as manometers to measure pressures artificially engendered.

In the interests of brevity, I shall denote the barometers to which this chapter is devoted, and which can be calibrated without reference to any other barometer, as *primary* barometers.

2. LARGE CISTERN BAROMETERS

It will be recalled that in 1768 the Cardinal de Luynes succeeded in filling a barometer tube which was about 29 mm. in bore. If this had had suitable mechanical parts, it might well have served as a primary barometer, since the entire capillary depression in such a tube is so much less than the accuracy that was needed in those days that its variation would have been quite negligible. Large cistern barometers were made at intervals throughout the ensuing hundred years. In 1773, relates Assier-Perica, the Prince de Conti had him make a barometer with a very large tube and cistern.[3] He also sets forth his reasons for not liking large barometers, which are very amusing:

they did not indicate pressure changes as well as smaller ones,

all the more because the atmospheric air cannot act freely on a column and a volume of mercury of such a considerable size. Take a privateer of 15 guns, and a ship-of-the-line of 60; at the least movement of the air, the privateer will move more rapidly than the ship.[4]

As an example of false analogy, this can only be surpassed in the history of meteorological instruments by the argument of the director of a famous weather service in favor of the 4-cup anemometer; who ever, he wrote, heard of a 3-cylinder automobile engine?[5] Assier-Perica probably got the inside of his larger tube dirty in filling it.

In 1826 J. G. F. von Bohnenberger, concerned about the variable capillarity in siphon barometers, made a barometer with a tube about 32 mm. in bore, and a very large cistern, in which several smaller tubes could also be stood up.[6] He boiled it out. This barometer was intended for an experimental study of the capillary depression *in vacuo*; and because he found a large central portion of the mercury surface flat, he concluded that the depression in his large tube was negligible, so that it could well be used as a primary barometer.

In 1854 John Welsh was asked by the Kew Committee to construct a large barometer to serve as a standard for Kew Observatory.[7] After a number of false starts,[8] he filled a tube 1.1 inch

---

[2] I have no intention of presenting a treatise on high-accuracy barometry. The reader may be referred to an excellent and up-to-date account by W. G. Brombacher, D. P. Johnson, and J. L. Cross, *National Bur. Stds. Monograph* 8, (Washington, D.C., U.S. Govt. Printing Office, 1960).

[3] *Nouveau traité sur la construction et invention des nouveaux baromètres, thermomètres, hygromètres, aréomètres et autres découvertes de physique expérimentale* (Paris, An X [1802]), p. 15.

[4] *Ibid.*

[5] This is genuine, but to avoid embarrassment to anyone I withhold the reference.

[6] *Annalen der Physik,* Vol. 7 (1826), pp. 378–85.

[7] *Phil. Trans.,* Vol. 146 (1856), pp. 507–13.

[8] See Chapter 9, p. 245 above.

in bore, which was erected in a glass cistern about 5 inches square. Through the metal top of this cistern passed two rods about 4 inches long, one pointed at its lower end, the other flat, either of which could be made to touch the surface of the mercury in the cistern by turning a screw. Each rod had a mark at a known distance from its lower end. Using a cathetometer with two telescopes, the vertical distance from this mark to the level of the upper mercury surface could be measured.[9]

Welsh realized the peculiar problems inherent in reading the level of a large mercury surface by means of a telescope; problems which, as we shall see, have been one of the main concerns of those engaged in precision barometry. Here is his solution:

In order to avoid the inconvenience of light being reflected into the telescope from the surface of the mercury in the tube, a moveable screen is provided, the upper part of which is black and the lower part oiled paper, which is so adjusted as to shut off all light which comes from a higher level than the top of the mercurial column: the surface of the mercury thus presents in the telescope a well-defined dark outline.[10]

The temperature was measured by a thermometer with its bulb dipping into the mercury. "The variations of the temperature of the room are not rapid," wrote Welsh, "so that no sensible error

arises from assuming the temperature of the cathetometer to be the same as that of the mercury."[11] Nothing about vertical temperature gradients; but these must have worried later observers, because at some later date a thermometer was fastened near the middle of the barometer, with its bulb dipping into a glass well of mercury the same diameter as the barometer tube.

This barometer was constructed with great care. F. J. W. Whipple recalled[12] that in 1933 it agreed with the new primary barometer of the National Physical Laboratory, which will be described later.

At the time the Kew Standard barometer was constructed, a vacuum-and-pressure chamber was built at Kew for the verification of marine and other fixed-cistern barometers, particularly for obtaining their "correction for capacity." This enabled three at a time to be compared with a specially-made adjustable-cistern barometer by Adie. The whole thing was mounted on the same wall as the standard barometer and the cathetometer, the latter being also used for reading the instruments in the vacuum-and-pressure chamber.[13]

In 1870 a "nouveau baromètre pour laboratoires" seems to have been made in France according to a "systeme . . . proposé par M. Regnault."[14] In principle it is exactly the same as that of Welsh.

Because it belongs to the relatively small number of primary barometers

---

[9] This has been described partly from the original paper, and partly from the instrument itself, which is now preserved in the hall outside the library of the National Physical Laboratory. Some modifications are apparent. A similar instrument, except for the shape of the cistern and the mounting, hangs beside it. This dates from 1858.

[10] Welsh, *Phil. Trans.*, Vol. 146 (1856), p. 509.

[11] *Ibid.*

[12] *Quart. J. roy. Meteorol. Soc.*, Vol. 63 (1937), p. 132.

[13] Welsh, *Phil. Trans.*, Vol. 146 (1856), pp. 510–12.

[14] C. Mêne, *Rev. hebd. de Chim. scientifique et industrielle* (1869–70), pp. 257–60.

that are of the cistern type, this is the place to describe the new primary barometer of the French *Météorologie Nationale*.[15] This instrument is also on the same principle, but with many refinements. It has a pyrex upper chamber 30 mm. in diameter, and a bent 8 mm. tube dipping into a large rectangular cistern. Vertically beneath the center of the chamber there is a micrometer screw 100 mm. long with a platinum point at its lower end, its upper end being rounded. The platinum point and the mercury are in the grid circuit of an audio-frequency oscillator, and after the contact of the two has been established electrically, one telescope of the cathetometer is set with its cross-hair tangent to the rounded end of the screw. The upper mercury surface is illuminated so as to appear bright against a dark background, and observed directly with the other telescope.

The vacuum above the upper surface is maintained by a two-stage mechanical pump exhausting through solid CO_2 traps, and measured with a Pirani gauge. Distributed along the column there are three mercury thermometers with their bulbs in 8-mm. thimbles filled with mercury. The precision of this ingenious instrument is claimed to be ±0.05 mm. of mercury.

In this barometer the means of observing both ends of the interval are much the same, though the appearance of the two fields of view will be somewhat different. In the 1930's Casella of London designed a cistern barometer in which both mercury levels are set in precisely the same way, by observing (with optical aid) a point and its reflection.[16] As shown in Figure 10.1, the upper point is adjusted to the surface of the mercury (25 mm. in diameter) by moving the whole tube, to which the scale is attached, up or down; while the vernier is rigidly fixed to the lower point. With suitable measurements of length, made once and for all, this could well serve as a primary barometer of moderate accuracy, though for such a purpose it might be well to make the upper surface somewhat larger and to provide for the use of a vacuum pump.

3. LARGE SIPHON BAROMETERS WITH OPTICAL READING

Practically all other primary barometers have been large instruments of the siphon type. The disadvantages of the siphon barometer mentioned in Chapter 9 disappear entirely when the inside diameter of the two chambers is made so large that the capillary depression is negligible to the degree of accuracy envisaged; for then its variations will not be significant at all. Such barometers have always been provided with means of altering the level of the mercury at will.

The first of these which really deserves the name of primary barometer was constructed at St. Petersburg in or about 1830 by A. T. Kupffer.[17] It had a long closed tube and a short open one, both sealed into a cistern with a movable bottom for adjusting the levels of the mercury. These were observed with an elaborate cathetometer having two microscopes which could be re-

---

[15] R. Beving, *La Météorologie* (Paris, 1957), pp. 445–50.

[16] C. F. Casella & Co., Ltd., London, *Catalogue No. 877* (1961), p. 33.

[17] *Annalen der Physik*, Vol. 26 (1832), pp. 446–51.

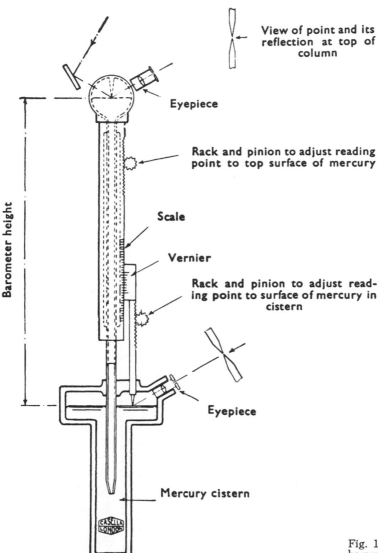

View of point and its reflection at top of column

Eyepiece

Rack and pinion to adjust reading point to top surface of mercury

Scale

Vernier

Rack and pinion to adjust reading point to surface of mercury in cistern

Eyepiece

Mercury cistern

Barometer height

CASELLA LONDON

Fig. 10.1. Casella's large barometer, circa 1930.

versed when the whole thing was rotated through 180 degrees, so as to check the verticality of the frame and the horizontality of the microscopes. Beside the tube a strip of brass was hung; this was engraved with two scratches exactly 750 mm. apart. The lower microscope could be traversed up and down for a short distance and its mounting carried a scale which moved next to the brass strip. The cathetometer could be turned from the tubes to the scale. To make an observation the upper telescope was set to the upper mark, then the cathetometer turned to the tubes. The upper surface of mercury was set to the cross hair in the upper microscope, then the lower microscope was set to the lower mercury surface. The scale attached to it then showed the difference from 750 mm.

Kupffer, as we have seen in Chapter 9, took the rather vague suggestion of Arago and used it to estimate the residual pressure in the vacuum space. Three observations were made, first with the mercury at any arbitrary height, and then after raising it n and $2n$ mm. respectively. Assuming a cylindrical tube, the true barometric height b is found from the three observed heights b', b'', b''' by solving the simultaneous equations

$$b = b' + c/x$$
$$b = b'' + c/(x-n)$$
$$b = b''' + c/(x-2n).$$

The unknown x is the height of the vacuum space in the first observation, taking into account the rounded end of the tube. It will be seen that Kupffer improved on Arago's suggestion a good deal.

It is remarkable how much the development of meteorological instruments in the nineteenth century owes to the directors of famous astronomical observatories, who seem to have felt it to be part of their duties to make meteorological observations of the highest quality. One of the most brilliant of these astronomers was F. G. W. Struve (1793–1864), director of the famous Russian observatory of Pulkowa. In 1845 he was able to publish a sumptuous description of the new observatory and its instruments,[18] on pages 224–26 of which is described a "normal"

[18] F. G. W. Struve, *Déscription de l'observatoire astronomique central de Poulkowa* (St. Petersburg, 1845).

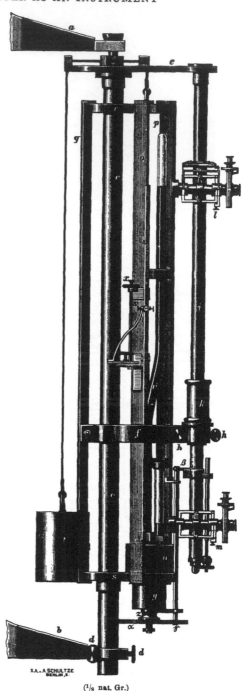

(¹/₈ nat. Gr.)

Fig. 10.2. The Pulkowa primary barometer, 1845. *(Courtesy of the Trustees of the British Museum)*

barometer which in its general arrangement, set the pattern for all such instruments until quite recent times (Figure 10.2).[19] It was made by the St. Petersburg instrument maker T. Girgensohn, who had made Kupffer's barometers a dozen years earlier, and we may take it, I think, that he was largely, if not completely, responsible for the design.

The tubes of this barometer were 0.85 inch in bore, sealed into an iron cistern, the bottom of which was formed by a piston that could be raised and lowered by a convenient handle and appropriate gearing. The longer tube was bent so that the two mercury surfaces could be vertically above one another. It is rather interesting that the frame carrying the microscopes was fixed in height, while that which supported the barometer and the scale was movable, counterpoised by a large weight. The scale itself was of brass, graduated in English inches.[20]

There is little doubt that for a quarter of a century the Pulkowa barometer remained the most accurate in the world, although the tubes were really a little smaller than would have been desirable.

The next improved primary barometer also originated in Russia, where the tremendously energetic German-Swiss Heinrich Wild, whom we have met in Chapter 9, was by 1874 Director of the "Central Physical Observatory" at St. Petersburg, an institution which

dealt with metrology of various kinds as well as meteorology. Wild was much concerned with the results of several intercomparisons made in the years 1866 to 1869 between standard barometers in various countries, which often showed differences of large fractions of a millimeter of mercury.[21] Such inaccuracy was intolerable to Wild, who was thinking in terms of a hundredth of a millimeter, and indeed begins his acount of the new primary barometer with a long theoretical treatment[22] showing the accuracy with which temperature, gravity, density of the mercury, and so forth would have to be known to reach such an absolute accuracy, which is indeed very hard to attain, as he well knew.

There follows an interesting discussion of the various ways of observing the height of the mercury surfaces, and Wild settles for observing them directly with telescopes, and putting screens behind the tubes, their upper halves black, the lower white, so that the mercury appears dark against a brighter background when the screens are properly adjusted. He does, however, emphasize the excellence of the technique of setting on a level halfway between the direct and reflected images of a point just above the surface of the mercury[23]—a scheme which we shall meet later. He also discusses the effect of non-uniformity of the walls of the glass tubes.

The new barometer (Figure 10.3) had tubes with an inside diameter of

---

[19] This barometer was described again in the new *Zeitschrift für Instrumentenkunde*, Vol. 1 (1881), pp. 111–14, by B. Hasselberg, from which paper our figure is reproduced. It was still in operation.

[20] It will be remembered that Kupffer had used the metric system.

[21] *Repertorium für Meteorologie*, St. Petersburg, Vol. 3 (1874), pp. 57–59. The paragraph in which these are referred to is headed "Nothwendigkeit, auf jeder Central-Anstalt ein Normal-Barometer zu haben."

[22] *Ibid.*, pp. 3–18.

[23] *Ibid.*, p. 21.

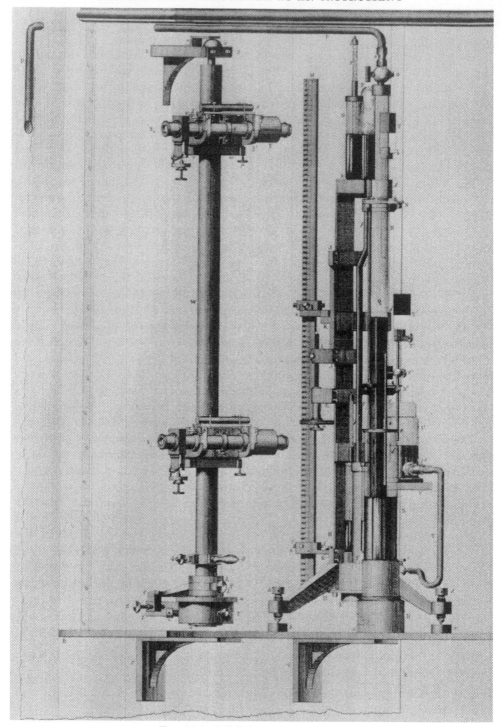

Fig. 10.3. Wild's barometer, 1874.

25.2 mm. It differed from the Pulkowa barometer in two important respects: the barometer and cathetometer were supported separately on a stone wall; and a separate vessel of mercury, which could be raised and lowered and was connected to the barometer by a rubber tube, was used for the adjustment of level. In view of some experiments in gas thermometry which were going on in the Institute, the open limb was extended so that it could be connected to other apparatus in the next room by the pipe shown at the top of the figure, and the instrument thus used as a precise manometer.

Wild measured the temperature of his instrument with two thermometers, one at the top and the other at the bottom, the bulb of each being in a tube of mercury 25.2 mm. in bore. This would make their thermal lag the same as that of the upper and lower chambers, but it seems to have escaped him that the greater part of the tube was much narrower. He was probably saved by a combination of convection and slow temperature changes.

The scale is hung in a Cardan suspension above its center of gravity, to ensure verticality. It was, of course, carefully compared with the standard meter of the Institute, and indeed every possible check was made on all elements of the instrument that could affect its accuracy. In particular, the condition of the vacuum was carefully investigated by the Arago-Kupffer method; and here we find one of the troubles of barometry of high precision:

These investigations, like the more accurate measurements with the standard barometer, can so far be made only in winter, during the period of sleighing, as at any other time the vibrations produced by wagons going past on the street set the mercury in the tube in motion and thus make accurate settings impossible.[24]

After all this, Wild felt that the absolute accuracy of his measurements did really attain ±0.01 mm.[25] and enjoyed crowing over Regnault,[26] who had said that no matter how good the apparatus, one could not expect a barometric measurement to be within less than 0.1 mm. of the truth. Wild was writing at a period when exaggerated hopes were held concerning the accuracy of meteorological observations, and before it was realized that the microstructure of the atmosphere, rather than instrumental error, limits the precision of such measurements. The accuracy that he claimed could be both possible and useful when his instrument was being used as a manometer. He returned to the subject of standard barometers in 1877,[27] maintaining with some reason that the St. Petersburg barometer was the only one in Europe that really fulfilled the conditions which should be met by such an instrument, a conclusion which was confirmed by the careful studies of G. Hellman, published two years later,[28] who found that most of the European observatories of the period had very unsatisfactory barometers.

In 1880 W. J. Marek contributed to the subject a discussion of the best

[24] Ibid., p. 37.
[25] Ibid., p. 52.
[26] Mém. Acad. Sci., Paris, Vol. 21 (1847), p. 69.
[27] Bull. Acad. Imp. des Sci. St. Pétersbourg, Vol. 23 (1877), pp. 86–138. See also Zeits. öst. Ges. Meteorol., Vol. 12 (1877), pp. 417–30.
[28] Repertorium für Meteorol., Vol. 6 (1879), No. 8.

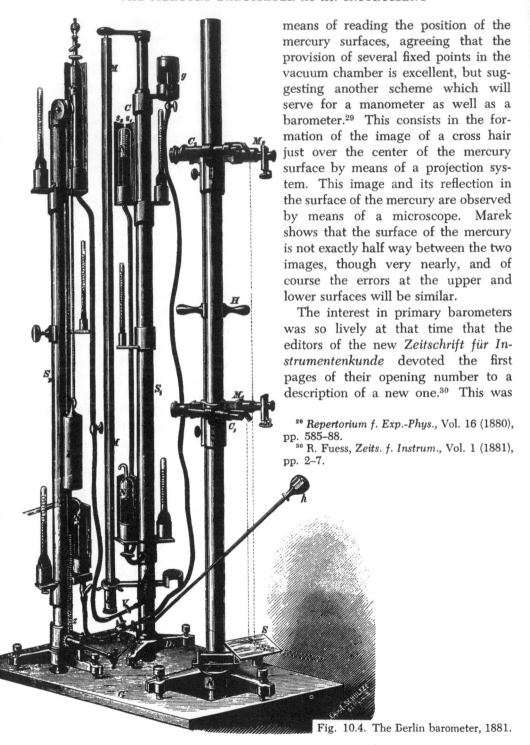

means of reading the position of the mercury surfaces, agreeing that the provision of several fixed points in the vacuum chamber is excellent, but suggesting another scheme which will serve for a manometer as well as a barometer.[29] This consists in the formation of the image of a cross hair just over the center of the mercury surface by means of a projection system. This image and its reflection in the surface of the mercury are observed by means of a microscope. Marek shows that the surface of the mercury is not exactly half way between the two images, though very nearly, and of course the errors at the upper and lower surfaces will be similar.

The interest in primary barometers was so lively at that time that the editors of the new *Zeitschrift für Instrumentenkunde* devoted the first pages of their opening number to a description of a new one.[30] This was

[29] *Repertorium f. Exp.-Phys.*, Vol. 16 (1880), pp. 585–88.
[30] R. Fuess, *Zeits. f. Instrum.*, Vol. 1 (1881), pp. 2–7.

Fig. 10.4. The Berlin barometer, 1881.

made by R. Fuess for the Royal Standardization Commission (K.N.A.K) at Berlin and was the first primary barometer in Germany; it was planned in cooperation with Pernet of Sèvres (the *Bureau International des Poids et Mesures*) and Thiesen of Berlin, and differed in many ways from the one at St. Petersburg.

The actual barometer tube N_1 N_2 (Figure 10.4), and the scale M, are carried on the stand S_1, together with three thermometers and the mercury reservoir g. On the stand S_2 besides three more thermometers, is a manometer formed by the tubes N_3 and N_4, which also serves as an adjusting device for the mercury levels, the vertical position of N_4 being adjustable by means of the rack and pinion Z and the knob h. The internal diameter of the tubes is about 25 mm.

In using the apparatus as a barometer, the mercury in N_3 is brought level with that in N_1, and the connection between them opened. By raising N_3 a little, the levels of the mercury in N_1 and N_2 are also raised, forming a similar meniscus in each—a procedure insisted on by Wild. By using more than one level in N_1, imperfections in the vacuum can be investigated. The method of Marek, using projection systems and microscopes, was adopted, the illumination being provided by an oil lamp outside the double-walled chamber that held the apparatus. There were no miniature electric lamps in 1881! The barometer and manometer tubes were specially investigated as to the uniformity of their walls in the optical sense.

In 1875 the International Bureau of Weights and Measures had been established, with laboratories at Sèvres near Paris. They at first acquired a primary barometer patterned on that of Wild, but in 1878 the International Committee of Weights and Measures decided that a second instrument should be constructed according to an improved design by W. J. Marek, then Director of the Bureau. He reported on the construction and calibration of this beautiful and elaborate instrument in 1884,[31] in a publication from which Figure 10.5 is reproduced. For our purpose we can almost let this figure speak for itself, but it should be recorded that the barometric chambers on the right, and the three other tubes forming the manometer and the leveling vessel, all have inside diameters between 36 and 37 mm., by far the largest up to that time. They are fastened into a steel block arranged so that their axes coincide with an imaginary cylinder having as its axis that of the cathetometer. All the necessary communications between the tubes are taken care of by five stopcocks in this block. The levels of the surfaces were at first read by Marek's method referred to above, but at some time between September 1890 and September 1891, the glass parts of the instrument were remade in order that black glass points could be introduced into the vacuum space of the barometer, which was then filled by boiling under vacuum.[32] They have not been altered since that time.

[31] *Trav. et Mém. B.I.P.M., Sèvres.*, Vol. 3 (1884), D.22–D.53.
[32] Benoît & Chappuis, *Procès-Verbaux Comm. Int. des Poids et Mesures* (1890), p. 20; (1891), pp. 24–26. I am indebted to Dr. J. Terrien, the present director of the Bureau, for this reference and indeed for much information about this instrument.

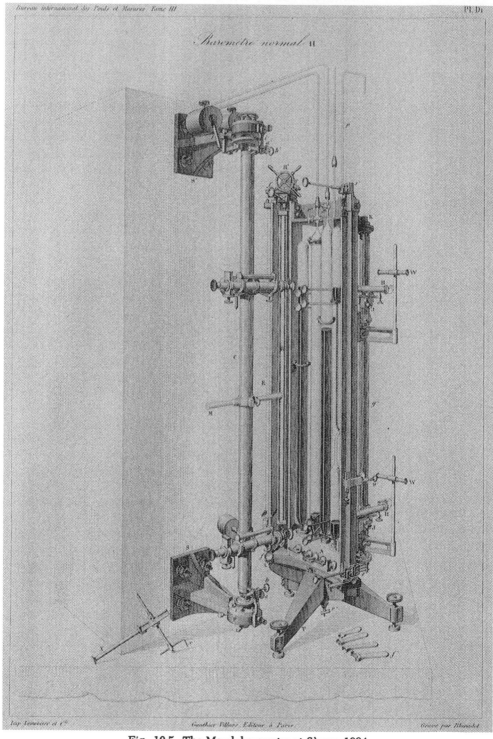

Fig. 10.5. The Marek barometer at Sèvres, 1884.

Probably the greatest difficulty, or at least the main risk of undetected error with any method of reading using a cathetometer, lies in the circumstance that when the telescopes are directed toward the tube, only about half the aperture of their objectives is effective, while it is all used in observing the scale. This poses an acute focusing problem, which in the instrument at Sèvres was solved[33] by providing a slide on each telescope, with which either half of the objective can be obscured, making micrometer readings on a point with each half, and adjusting the focus so that either half gives the same result.

The temperature of this barometer is indicated on three thermometers spaced along the column, read by separate telescopes in suitable positions.

Apart from the alterations mentioned above, and the provision of a screen to intercept thermal radiation from the observer, this splendid instrument was in its original state and in working order in 1961.

Barometry of the precision that such an instrument makes possible requires a consideration of the vapor-pressure of mercury and its dependence on temperature. The first determinations of this quantity at room temperatures were reported by Hagen[34] and by Hertz,[35] both in 1882, and that of Hertz was used by Marek.[36] At 20°C. the value is 0.0013 mm. of mercury. It was redetermined in 1922 by C. F. Hill at the University of Illinois, who ob-

tained somewhat higher values.[37] The latest values adopted by the National Bureau of Standards[38] agree remarkably well with those of Hertz. It will be evident that this quantity can be entirely neglected in all ordinary barometry.

In view of the fact that the Toepler pump was described in 1862, and the Sprengel pump, with which, according to Sylvanus P. Thompson,[39] a vacuum of 0.0005 mm. could be obtained, in 1865,[40] it is remarkable that none of the barometers we have so far mentioned used a mercury pump to produce or renew its vacuum. The first suggestion of this sort seems to have come from L. Grunmach,[41] who not only used a mercury pump to evacuate the space, but a discharge tube to make sure that a good vacuum has been obtained. When the whole tube fluoresces, the pressure is much lower than the accuracy with which the level of the mercury surface can be measured, and therefore more than adequate. Grunmach emphasized that there would still be the mercury vapor in the space. A new primary barometer at the K.N.A.K. in Berlin was made so that it could be evacuated in this way; this was a siphon barometer with 30 mm. chambers and a 12 mm. connecting tube. At the top of the upper chamber was a tee fitting with two stopcocks,

[37] Phys. Rev. (2), Vol. 20 (1922), pp. 259–6G.

[38] Monograph 8, May 20, 1960; Table 3A.

[39] "The development of the mercurial air pump." Telegr. J. and Electr. Rev., London, Vol. 21 (1887), p. 663.

[40] Chem. Soc. J. (2), Vol. 3 (1865), pp. 9–21.

[41] Annalen der Physik, Vol. 21 (1884), pp. 698–710.

[33] Ibid., (1890), p. 18.

[34] Annalen der Physik, Vol. 16 (1882), pp. 610–18.

[35] Ibid., Vol. 17 (1882), pp. 193–200.

[36] Trav. et. Mém., Vol. 3 (1884), D.45.

one leading to a "Geissler tube," the other to a pump.[42]

Very great care was taken in filling this instrument.

After the mercury had been treated for several weeks with ferric chloride and dilute nitric acid, it was washed with benzol and afterwards with boiling distilled water, then dried [he does not say how], and heated to 60-80° C. during the filling, which was done through a double funnel of filter-paper and ledger-paper provided with very fine openings.[43]

Meanwhile, the barometer tube had been carefully cleaned, pumped out, and heated for three days while air which had passed through four drying tubes flowed slowly through it.

Later on Grunmach points out, as Wild had done before, that if we are to measure pressure to ±0.01 mm. of mercury, we must know the mean temperature of the mercury column to ±0.07°C. It will not do to hang one thermometer somewhere on the barometer unless we know that the whole room has the same temperature.

Meanwhile an investigation of methods of reading such barometers was going on in Berlin, where Thiesen,[44] after an investigation of the methods of Pernet and of Marek, came to favor putting the scale directly behind the tube and as close to it as possible, reading the position of the nearest millimeter division and of its reflected image. Although this might, as he said, greatly reduce the effect of the non-

[42] This means of checking the vacuum was adopted again in 1934 by the French weather service. See *La Météorologie, Paris* (1934), pp. 440–44.

[43] *Annalen der Physik*, Vol. 21 (1884), p. 701.

[44] *Zeits. für Instrum.*, Vol. 6 (1886), pp. 89–93.

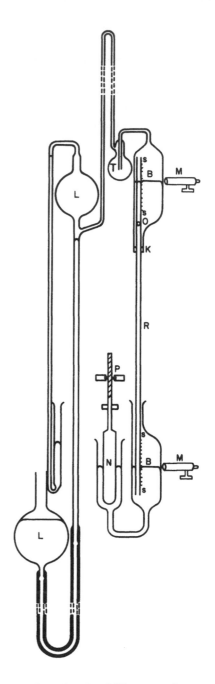

Fig. 10.6. Sundell's primary barometer, 1887. (*Courtesy of the Trustees of the British Museum*)

uniformity of the glass tube, it does not seem to have been widely adopted. But an even more drastic scheme of the same sort was being tried in Finland at the same time.[45] In Sundell's primary barometer the scale was graduated in millimeters directly on the glass tube R (fig. 10.6) which connected the upper and lower chambers, each 40 mm. in diameter. There was an opening in R at O. The system L,L constituted a mercury pump, and T was a drying tube containing phosphorus pentoxide. The microscopes had micrometer eyepieces and were used merely to read the apparent interval between the nearest division and its image in the mercury; but experimental corrections for the curvature of the lines, which could be as much as 0.04 mm., had to be applied.

After this nothing very noteworthy seems to have been reported about barometers of the general type we are considering until 1933, when a highly original design appeared at the National Physical Laboratory, Teddington.[46]

Any doubts about Marek's method of reading the levels of the mercury, thought Sears and Clark, were due to the impossibility of getting optically-worked glass tubing of sufficient accuracy, and so it was decided to construct a barometer of steel, with plane-parallel optical glass windows through which the levels could be observed. Such a construction was made possible by the new simplicity of producing pressures of less than 0.001 mm. by means of diffusion pumps. With such pumps available, an instrument can be

[45] A. F. Sundell, *Acta Soc. Scient. Fenn.*, Vol. 16 (1887), pp. 431–92.
[46] J. E. Sears, Jr. and J. S. Clark, *Proc. Roy. Soc.*, A139 (1933), pp. 130–46.

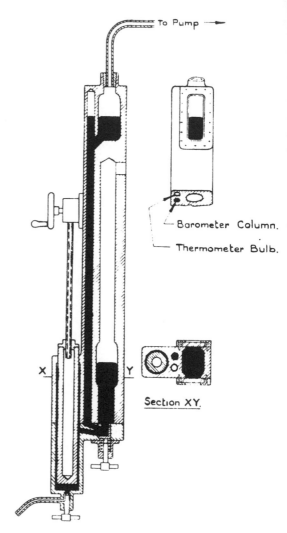

To Pump →

Barometer Column.

Thermometer Bulb.

Section XY.

Fig. 10.7. The primary barometer of Sears and Clark, 1933, diagram. *(National Physical Laboratory—Crown copyright)*

taken down and cleaned at will, and no out-gassing is needed.

The barometer proper, shown diagrammatically in Figure 10.7, consists of a large block of stainless steel having a U-tube bored in it, opened out into two chambers each about 48 mm. square in section and 100 mm. long.

These are covered with plane-parallel glass plates held against flats worked on the steel. The two mercury surfaces are directly above one another. A second hole is bored vertically down the block parallel to the one which forms the barometer tube. This hole contains a special mercury thermometer with a bulb 760 mm. long and 8 mm. internal diameter, and a stem 420 mm. long.

The bend of the barometer tube at the bottom is completed by another block, which also incorporates a cylinder into which a plunger can be moved for the regulation of the levels of the mercury, and a four-way valve to connect or isolate the various parts as needed.

The entire instrument is shown in Figure 10.8 with the standard scale just to the right of the barometer "tube." The microscopes and collimators are mounted on a very heavy stand which can be moved sideways in order to transfer the microscopes from scale to barometer and vice versa. The barometric chambers are so large that it is possible to form the image of a cross wire 18 mm. behind the axis of the barometer, giving the best possible conditions for the application of Marek's method of reading. It is interesting that to obtain the desired accuracy the windows had to be worked parallel to one second of arc. It was concluded from a number of tests that the accuracy of a single observation with this instrument was about ±5.3 microns (±0.0053 mm.) of mercury, and at that time there was an additional uncertainty in the density of mercury corresponding to about 5 microns. For purposes of barometry, as distinct from

Fig. 10.8. The primary barometer of Sears and Clark, general appearance. *(National Physical Laboratory — Crown copyright)*

manometry, no greater accuracy could possibly be of any use.

In 1957 a second barometer of this sort was built, with a simpler vacuum system, using a Pirani gauge to measure the vacuum. The thermometer now consists of five thermocouples in series, placed in the actual barometric column of mercury, and it is estimated that the temperature of the column is known to better than 0.01°C. The means of reading is the same, but the barometer was mounted on anti-vibration mounts, independent of the measuring system.

A similar barometer, modified somewhat because it had to read over a range of pressures from 650 to 780 millimeters of mercury, was later installed at the South African National Physical Laboratory.[47]

In the same year the *Physikalisch-Technische Bundesanstalt* put into service a barometer[48] that was avowedly based on the N.P.L. instrument as far as the barometer chambers, tube, and thermometer are concerned, but although it is claimed to be of the same order of accuracy it is very much simpler in its other arrangements. A vertical part-section is shown in Figure 10.9. The upper surface of the mercury is brought nearly up to a fixed steel point S_1—a phonograph needle!—and a similar but movable point S_2 is then brought nearly down to the lower surface by turning the knob T_1. Attached to this point is a scale M, moving past a stationary mark N. There are three microscopes with eyepiece micrometers, two of which measure the distances between the points and their reflected images, while the third measures the position of the scale relative to the fixed mark. The whole in-

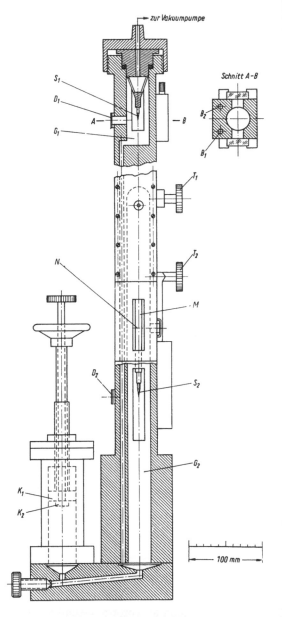

Fig. 10.9. The primary barometer of Gielessen, 1957.

[47] E. C. Halliday & H. J. Richards, *South African J. Sci.*, Vol. 51 (1955), pp. 217–21.

[48] J. Gielessen, *Zeits f. Instrum.*, Vol. 65 (1957), pp. 63–65.

strument is calibrated by measuring the distance between the points for various scale readings, using an independent comparator.

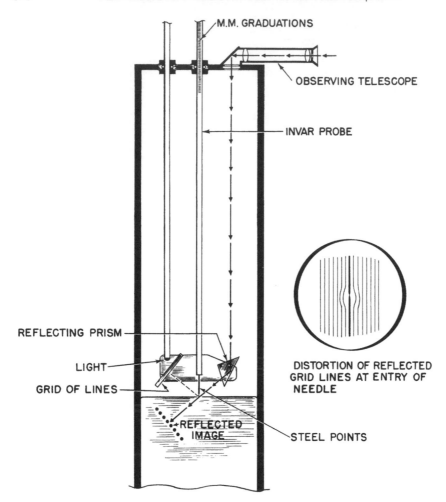

M.M. GRADUATIONS

OBSERVING TELESCOPE

INVAR PROBE

REFLECTING PRISM

LIGHT

GRID OF LINES

REFLECTED IMAGE

STEEL POINTS

DISTORTION OF REFLECTED GRID LINES AT ENTRY OF NEEDLE

Fig. 10.10. The Toronto primary barometer, 1959. (*Dept. of Transport, Meteorological Branch*)

In its use of the principle of setting to points with optical aid, the primary barometer of the Meteorological Service of Canada at Toronto resembles the one just described, though it is much more elaborate. It was designed in the late 1950's by H. H. Bindon and R. E. Vockeroth. An accuracy of 20 microns is claimed in routine use.

The barometer proper is a U-tube of stainless steel, the vertical cylinders being 3 inches in internal diameter and 40 inches long, joined by a 1-inch tube. The most interesting feature of the instrument is that the cylindrical invar scales project into the tubes, a vacuum seal being provided by rubber "O-rings" (Figure 10.10). A small lamp, a graticule, and a reflecting prism are carried along with each scale but on another rod. The contact of the point with the mercury is indicated very sensitively by the sudden distortion of the reflected grid, as shown in the figure.

The scales are observed by means of fixed microscopes with micrometer eye-pieces, and translated parallel to their length by heavy parallel motions derived from commercial lathe parts. The zero of the scales is easily obtained by opening both tubes to atmospheric pressure, after which one is pumped out by a pumping system capable of maintaining a vacuum of 10^{-4} mm. of mercury or better. According to the designers, the greatest single source of error is uncertainty about the verticality of the scale-rods which descend into the chambers.

In 1953 the German meteorological service put into operation at Hamburg a primary barometer of relatively simple construction, but quite accurate enough for the purpose.[49] The bore of the tubes, 32 mm., was chosen because the entire capillary depression is about 0.01 mm. at this diameter. The barometer is a cistern-siphon instrument, the mercury being raised and lowered by means of the adjustable bottom of the cistern. There is no vacuum pump, the Arago-Kupffer method being used to check the vacuum. Between the cistern and the lower wide tube there is a diaphragm with a hole having about the same resistance to the motion of the mercury as has the 3-mm. tube which connects the cistern to the vacuum chamber, an interesting way of preventing overshoot and oscillation when the levels of the mercury are being adjusted. These levels are read by a cathetometer. A grid of black lines 1

mm. wide and 1 mm. apart, sloping at 45° to the horizontal, is placed just behind each mercury surface, and the setting is made on the sharp "break" between these and the "zebra-pattern" formed by their reflection in the surface.[50]

In Finland a primary barometer was described posthumously in 1955 by E. T. Levanto.[51] It was built at the Weights and Measures Office and was expected to give an accuracy of 0.01 mm. In this barometer, which also has 32-mm. chambers, the vacuum is produced by a pump, and measured by a McLeod gauge. The original feature of this instrument is in the method of setting on the surfaces of the mercury. In each chamber floats a loosely-fitting iron ring which supports a point about 0.2 mm. from the mercury surface, the mean of the apparent positions of the point and its reflection being considered the true position of the surface itself. A cathetometer transfers these measurements to a vertical scale. One might wonder whether a change in surface tension in the open limb might have some effect on the height at which the ring floated, but of course any such change would be immediately manifest in the apparent interval between the point and its reflected image.

A somewhat more conventional barometer has recently been described in Japan,[52] and is at the Japan Meteorological Agency in Tokyo. It is a siphon

[49] K. Gödecke, Techn. Mitt. d. Instrumentenwesens d. Dt. Wetterd. No. 26 (1953), pp. 13–20.

[50] This scheme for reading a barometer was described earlier by J. Zeleny, Rev. Sci. Inst., Vol. 7 (1936), p. 289.

[51] Acta Acad. Sci. Fenn. Series A.1, No. 191 (1955).

[52] M. Yoshitake, I. Shimizu, and K. Takeuchi, J. Meteorol. Soc. Japan, Vol. 37 (1959), pp. 104–10.

barometer with 30-mm. chambers and a 10-mm. connecting tube. The menisci are directly observed with the two telescopes of a cathetometer, which is then pointed at a pure nickel scale.

4.   MISCELLANEOUS PRIMARY BAROMETERS

Although the large siphon barometer with optical reading has long been the preferred type of primary barometer, others have been suggested.

Setting a point to a mercury surface is an operation which can be done with ease, and repeated to a precision of at least 0.03 mm. with the naked eye if the mercury is clean and the light good. In 1896 Professor K. Prytz of Copenhagen, after detailing the difficulties of reading the levels of the mercury with a cathetometer, decided on the following construction.[53] In the vacuum space of a siphon barometer (Figure 10.11) there is a glass point, directed downward and on the axis. The mercury can be brought up to this by raising or lowering an auxiliary vessel connected to the bottom of the siphon by a rubber tube. The open limb is lengthened and provided with a cover in which runs a pointed micrometer screw 10 cm. long. This acts on the end of a steel rod, 73 cm. long, pointed at the bottom and polished at the top. Prytz does not say so, but presumably the rod is supported by a spring or counterweight.

The reason for prolonging the open limb is as follows: after the glass point is installed, but before the closed end is sealed, enough mercury is run in to

fill both arms up to the level of the glass point. The micrometer screw is then installed without the steel rod, and its reading noted when its pointed end touches the mercury. If the length of the rod is known, this provides a calibration datum. The vacuum can then be produced, preferably with a good pump, and the rod installed.

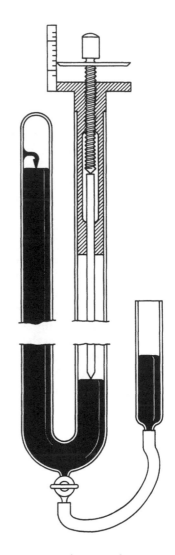

Fig. 10.11. Prytz' primary barometer, 1896.

[53] *Der Mechaniker*, Vol. 4 (1896), pp. 289–91.

About the year 1900, W. Marek and J. Bauer built a barometer at the *Zentralanstalt für Meteorologie und Geodynamik* at Vienna, in which both surfaces were set to points. No description of this seems to have been published,[54] but in 1911 R. Pozděna described a second instrument which was stated to be on the same lines,[55] and this is still at the *Zentralanstalt*, intact though not in use.

This barometer (Figure 10.12) is a siphon barometer with its two tubes K_1 and K_2 let into a steel block with appropriate taps and connections such as might be expected from Marek. The upper chamber has three points, is arranged so that it can be pumped out, and is connected to a discharge tube so that the vacuum can be checked. A removable extension to the open limb permits the two mercury surfaces to be brought to the level of any of the three points.

A slide on a graduated vertical frame C_1C_2 carries a micrometer screw with a pointed end. The zero is obtained by installing the extension tube, letting air into the chamber, bringing the mercury up to one of the three points, setting the index on the slide to the zero of the scale, and reading the micrometer screw after its point has been set to the surface of the mercury in the open limb. The mercury is then lowered, the extension tube removed, the vacuum produced, the slide lowered to [say] the 750 mm. graduation, and the mi-

crometer used on the lower surface while the mercury in the chamber is set to one of the points. Microscopes are used to observe the contacts. An interesting feature is that there are two scales on the frame, one on each column, so that if both are set, an exact parallel motion of the slide is not needed. As a matter of fact, only the zero and 750 mm. graduations are required.

It does not seem to have been considered in this design that it is essential that the glass tube should remain fixed with respect to the frame, and it was soon found that the construction was inadequate in this respect. This was put right by A. Wagner after the 1914-1918 war, and a heat shield installed round the tube.[56]

In 1924 T. H. Laby in Australia described[57] a primary barometer of moderate accuracy "in which the distance between the mercury surfaces is measured by means of the equivalent of a 30-inch steel screw micrometer." The contact of flat steel surfaces with the mercury is used, and the micrometer, which has a heavy steel frame, is standardized with an end standard before its upper end is sealed into the vacuum chamber, which may be exhausted with a pump. The details of the lower end are not quite clear. Laby emphasized the absence of a cathetometer, an instrument which he obviously disliked.

Now there is not very much difference in principle between these barometers and a very much more sophisticated instrument made at Teddington

---

[54] Dr. Untersteiner of the *Zentralanstalt* could find no record of it, after a diligent search during my visit in July 1962.

[55] *Jahrb. k.k. Zentralanst. f. Meteorol. u. Geodynamik*, Wien, n.f. 47 (1911), p. xiii–xxiii.

[56] *Ibid.*, n.f. 60 (1923), p. xviii–xxii.

[57] *J. Sci. Inst.*, Vol. 1 (1924), pp. 342–45.

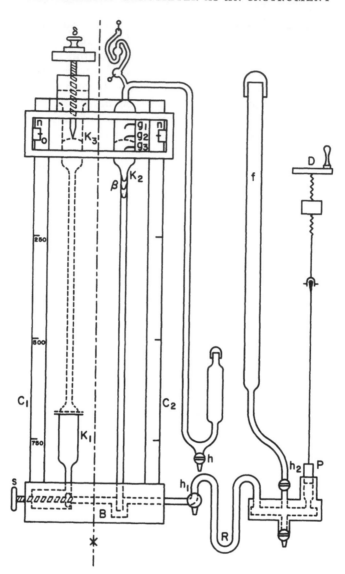

Fig. 10.12. The Vienna primary barometer, 1911.

in 1927.[58] In this instrument (Figure 10.13) the entire mass of mercury is immersed in a water bath. The difference of level between the points A and B is measured by means of a microm-

eter, its value when the micrometer reads zero having previously been determined on a measuring machine. The distance between the platinum point P and the surface of a given weight of mercury in the cup R was also determined before the assembly RST

[58] *N.P.L. Annual Report* for 1927, pp. 156–59.

was sealed on to the main tube. The side-tube *H* takes a stainless-steel plunger *Q* for adjusting the levels. All

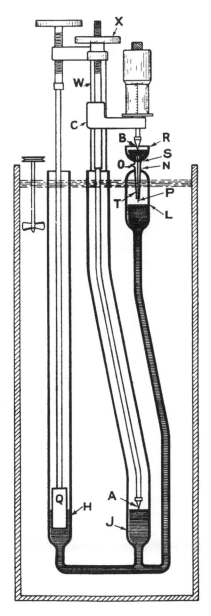

Fig. 10.13. The Teddington primary barometer, 1927. *(National Physical Laboratory — Crown copyright)*

contacts are indicated electrically, and would repeat to ±0.00005 inch (±0.0012 mm.). It was claimed that "the probable error of the final result . . . should not exceed ±0.0005 inch, and is likely to be not more than half that amount." For meteorological purposes this would be quite adequate, but it was evidently not good enough for the National Physical Laboratory, as we have seen.

The third of the barometers envisaged by G. Guglielmo[59] could probably be made into a normal barometer, perhaps by making use of the modern possibility of indicating an electrical contact with an extremely small current, and thus removing earlier objections to contacts with the surface of mercury. Guglielmo's third scheme demands simply that there should be two fixed points in the chamber and cistern (or the two chambers) of a barometer, and that the vertical interval *A* between them should be measured once and for all. The cross-sectional areas *s* and *s′* of the two vessels are also supposedly known. The mercury is gradually raised by letting some flow out of a burette, and the volumes *v* and *v′* shown by the burette are noted at the points where the mercury makes contact with the points in the lower and upper chamber respectively. Then the height *H* of the barometric column will be

$$H = A + (v' - v)/(s + s').$$

This can obviously be made very sensitive. Guglielmo derived formulas for the temperature correction of such a barometer.

[59] See Chapter 9, p. 260 above.

5. MANOMETERS OF EXTREME PRECISION

In recent years the insatiable demands for pressure measurements of increased accuracy, coming from physicists and physical chemists, have led to the development of a number of extraordinarily complex manometers by the national standardization laboratories. While some of these bracket the meteorological range of pressures, they are more accurate than any meteorological measurement needs to be, and a single observation frequently takes so much time that they are almost never usable to anything like their full accuracy when connected to the atmosphere.

Thus it is not quite reasonable to call them barometers, and in this history they will be referred to only briefly.

Probably the most spectacular of these instruments up to this time is the manometer at the National Physical Laboratory in England with a range of 0 to 1200 millibars.[60] This instrument depends on the application of a photoelectric "optical probe" in which the image of an illuminated grating is formed on the mercury surface at an angle of incidence of about 5°, and the reflected light focused on a similar grating system in front of a phototube.[61] The mercury surfaces in this manometer are 11 cm. in diameter. The authors estimate that a single observation of pressure may be considered as accurate to ±0.01 mb. at a confidence level of 99 per cent. This is one one-hundred-thousandth of an atmosphere.

At the National Bureau of Standards at Washington, H. F. Stimson used gauge blocks with which to compare the height of a mercury column.[62] As a detector the electrical capacitances between each mercury surface and a fixed steel plate are compared in a capacitance bridge. This arrangement was later improved by R. J. Berry at the National Research Council in Ottawa.[63]

Finally there is the possibility of using interferometric techniques for measuring the vertical separation of the surfaces.[64] This has been exploited in Japan,[65] and more recently Terrien has suggested an improved method which should yield a manometer of remarkable compactness and stability.[66] The capital difficulty in interferometric methods is the disturbance of the mercury surfaces caused by vibration, but Terrien has also suggested[67] a photoelectric method of overcoming this difficulty.

[60] K. W. T. Elliott, D. C. Wilson, F. C. P. Mason, and P. H. Bigg, *J. Sci. Instrum.*, Vol. 37 (1960), pp. 162–66.

[61] K. W. T. Elliott and D. C. Wilson, *J. Sci. Instrum.*, Vol. 34 (1957), pp. 349–52.

[62] H. F. Stimson, *Comm. Int. des Poids et Mesures, Comité consultatif de thermométrie*, (1952), Annexe T10, p. T82–T103.

[63] *Canadian J. Phys.*, Vol. 36 (1958), pp. 740–60.

[64] Jean Terrien, *Rev. d'Optique*, Vol. 38 (1959), pp. 29–37.

[65] I. Oyama, K. Koizumi, R. Kaneda, S. Sudo, and K. Nishibata, *Report Centr. Inspection Inst. of Weights & Measures, Tokyo*, Vol. 7 (1958), pp. 1–8.

[66] *Rev. d'Optique*, Vol. 38 (1959), pp. 34–36.

[67] *Ibid.*, p. 35.

Chapter 11

Mercury Barographs and Related Apparatus

1. GENERAL REMARKS

THE tendency of human beings to save themselves from routine is very strong, and would alone suffice to account for the more than a hundred designs of mercury barographs, or recording barometers,[1] that appear in the literature. When we add the convenience and general *Übersichtlichkeit* of a graphical presentation of atmospheric pressure against time, the attraction of such instruments is irresistible.

As a step toward a recording instrument, but perhaps less complicated and costly, a barometer which leaves an indication of the maximum or minimum pressure since the last observation might find favor. This class of instruments may fittingly open this chapter. While the many real recording barometers do fall into rough classes according to their mechanism, the arrangement adopted here is necessarily somewhat arbitrary. The barographs will be dealt with in the following order:

Barographs derived from the wheel barometer (non-electrical),

Barographs derived from the balance barometer,

Photographic barographs,

Barographs with servo-systems (except balance barographs),

Barographs using flotation,

Miscellaneous mercury barographs,

Barographs giving a record at a distance.

The chapter will conclude with an account of some instruments designed to record small changes of pressure on a highly magnified scale.

2. MAXIMUM AND MINIMUM BAROMETERS

In 1761 Keane Fitzgerald made a remarkable wheel barometer with a tube more than half an inch in diameter and a three-inch ball at the top, so that

---

[1] The time-honored phrase "self-recording" still seems illogical or at least redundant to me, twenty years after I first said so (*Meteorological Instruments* [Toronto, 1941]). Nor do I like the term "mercurial barometer"; the adjective has overtones of sprightliness and vivacity, the unsuitability of which is scarcely compensated by the delight of writing or reading an extra syllable.

nearly all the movement of the mercury was in the open limb. The axis of the "wheel" was on anti-friction rollers, and the whole thing was beautifully made by Vuillamy, the Queen's watchmaker.[2] In addition to the main pointer there were two extra pointers, very light and well-balanced, one of which was pushed by the main pointer when the barometer was going down, the other when it was rising. This same idea was patented a century later by one John Browning.[3]

In 1798 Alexander Keith described[4] a similar siphon barometer, but instead of the elaborate mechanism of Fitzgerald's instrument, Keith's merely had a bent rod rising from the float in the open limb, and pushing very light indexes up and down a fine stretched wire. It could scarcely have been very sensitive.

Meanwhile Changeux had suggested an entirely different sort of maximum or minimum barometer, which he intended not for a fixed station, but to be let down into a pit or raised to a height "qu'on ne voudrait pas parcourir soimême."[5] This might suggest balloons, except that the description was published just before the first experiments of the Montgolfier brothers. For the maximum-reading barometer, a short graduated tube, inclining downward at a sharp angle and with its end sealed, is added near the top of a siphon barometer tube. Such an addition he

calls an "appendix." At the top of the pit, mercury is added to the open limb until it is just about to flow over into the appendix; then the barometer is lowered and raised again, and the amount which has gone over measured. Changeux saw clearly that a departure from verticality might cause errors, but seems not to have worried about oscillations, so that one may wonder if the idea was ever tried, still more the corresponding idea for a minimum-reading instrument, which had an "appendix" on the short limb.

Marsiglio Landriani of Milan returned to the wheel barometer.[6] He set two of them up, each with the addition of a ratchet with fine teeth on the edge of its very light, balanced five-inch wheel. There was a pawl which let this wheel, and consequently the float, move only in one direction. The maximum barometer had the teeth and the pawl pointing in one direction, the minimum in the other. In the same year Anthony Semple published a "Description of an absence thermometer" [sic],[7] which turns out to be a siphon barometer with an arrangement something like that of Keith, but the wire rising from the float ended in a brass cross-piece, which could push one counterbalanced vernier up a scale, another down. These verniers weighed only 30 grains each, and were hung on silk threads passing over very light pulleys. This might have worked fairly well, at least for a time.

T. S. Traill, probably thinking about the well-known maximum-and-mini-

---

[2] *Phil. Trans.*, Vol. 52 (1761), pp. 146–54. The firm of Vuillamy enjoyed the Royal patronage until 1854.

[3] British Patent 2560 (1861).

[4] *Trans. Roy. Soc. Edin.*, Vol. 4 (1798), pp. 209–12.

[5] *Description de nouveaux baromètres à appendices,* etc., (Paris, 1783), pamphlet, 24 pp.

[6] *Giornale di fisica, chimica, storia naturale, medicina ed arte,* Pavia, Vol. 10 (1817), pp. 54–61.

[7] *Annals of Philosophy,* Vol. 10 (1817), pp. 47–49.

mum thermometer of James Six, proposed[8] the use of a diagonal barometer and a "square" barometer fixed to the same frame, a piece of thick iron wire having been introduced into the diagonal part of the former, to indicate the maximum, and another into the horizontal part of the latter, to show the minimum pressure. These indices (just as in Six's thermometer) can be set back to the ends of the mercury columns by using a magnet. The minimum-barometer part of this was invented all over again by De Celles,[9] and a student at Nancy, Decharme, pleased his professors by inventing not only a maximum barometer and a minimum barometer, but also a combined maximum-and-minimum barometer, using little dumbbell-shaped iron indices.[10] This was simply a tube bent through less than a right angle at the top, as in a diagonal barometer; more than a right angle at the bottom. He also reinvented Changeux's "baromètres à appendices," or as near as makes no matter. Decharme's only real claim to originality was that he pointed out that one ought to know the temperature at the times of maximum and minimum pressure, if the readings are to be corrected. He suggested putting the barometer in a cellar.

It is almost painful to record that in 1903 Louis Anseline Décor of Paris obtained both a Swiss and a British patent for the idea of using a "square" barometer with an iron index.[11]

A similar but more sophisticated idea was patented in 1855 by Negretti & Zambra.[12] To make a minimum barometer, a steel plunger with conical ends is introduced into an ordinary barometer tube. This falls with the mercury, but it is claimed that when the latter rises it flows past the plunger.

Such is the history of the maximum-and-minimum barometer, which the reader may, if he wishes, consider as comic relief, like the porter in *Macbeth*. The one which would have worked best was the earliest, that of Fitzgerald. Let us get on to more serious matters.

3. BAROGRAPHS DERIVED FROM THE WHEEL BAROMETER

The first recording meteorological instrument was undoubtedly the instrument devised by Christopher Wren in 1663, but this did not record atmospheric pressure.[13] It was, however, improved and added to by Hooke some time before 1681, when Nehemiah Grew described the instrument with annoying brevity in his fantastic catalogue of the Royal Society's collections.[14] Among the "Instruments relating to natural philosophy" there was

A Weather Clock. Begun by Sir Chr. Wren . . . to which other motions have since been added, by Mr. Robert Hook Professor of geometry in Gresham-Colledge. Who purposes to publish a description hereof. I shall therefore only take notice, that it hath six or seven

---

[8] *Proc. Roy. Soc. Edin.*, Vol. 1 (1834), pp. 57–58.

[9] *Compt. Rend.*, Vol. 47 (1858), pp. 543–44.

[10] Thesis, Nancy, 1861.

[11] Swiss Patent 27688 (1903); British Patent 3528 (1903). I have not bothered to look for this in other countries.

[12] British Patent 2306 (1855).

[13] See W. E. K. Middleton, "The first meteorographs," *Physis*, Vol. 3 (1961), pp. 213–22.

[14] *Musaeum Regalis Societatis or a catalogue and description of the natural and artificial rareties belonging to the Royal Society*, etc. (London, 1681).

motions; which he supposeth to be here advantagiously made altogether. First a pendulum clock, which goes with ¾ of a 100 llb. weight, and moves the greatest part of the work. With this, a barometre; a thermometre; a rain-measure, such an one as is next describ'd; a weathercock, to which subserves a piece of wheel-work analogous to a way wiser; and a hygroscope. Each of which have their regester, and the weather-cock hath two; one for the points, the other for the strength of the wind. All working upon a paper falling off a rowler which the clock also turns.[15]

Hooke did not publish a description; but in 1726, long after his death, it was published by William Derham in a collection of previously unpublished snippets, mainly by Hooke.[16] One of these begins as follows:

Dr. Hook's Description of his Weather-Wiser; about Dec. 5, 1678.

The Weather-clock consists of two parts. First, that which measures the time, which is a strong and large pendulum clock, which moves a week, with once winding up, and is sufficient to turn a cylinder (upon which the paper is rolled) twice round in a day, and also to lift a hammer for striking the punches, once every quarter of an hour.

Secondly, of several instruments for measuring the degrees of alteration, in the several things to be observed. The first is the barometer, which moves the first punch, an inch and half, serving to shew the difference between the greatest and least pressure of the air. The second is, . . .[17]

For two reasons at least, this instrument is of exceptional interest: first, because it was the very first barograph of any kind, and second, because it introduced the principle of discontinuous recording, generally ascribed to the Comte d'Onsenbray (1678–1754), who invented a famous anemometer.[18] The next use of point-wise recording in a barograph was by Changeux in 1780,[19] and will be referred to again below. Changeux has also been credited with the idea itself, which has the sole (but considerable) advantage of leaving the delicate moving parts of the instrument entirely free to move except during the brief intervals when the marks are being made, so that a recorder need have no more friction than an indicating instrument.

Unfortunately this famous meteorograph of Hooke's has completely disappeared, but there are two reasons for believing that the barograph was an obvious development of the wheel barometer. In the first place, Hooke had invented the wheel barometer; and in the second place, the punch indicating the pressure moved 1½ inches. Now there is much evidence that the extreme range of barometric pressure in England was very early considered to be from 28 to 31 inches of mercury;[20] the reduction of this by half probably means that the float in a siphon barometer was the actuating element. As the paper roller was necessarily horizontal, the paper "falling off" it, we may suppose that a thread passed from the float over a pulley, then horizontally above

---

[15] *Ibid.*, pp. 357–58.

[16] W. Derham, ed., *Philosophical Experiments and Observations of the Late Eminent Dr. Robert Hooke, S.R.S. and Geom. Prof. Gresh.* [am College] *and Other Eminent Virtuoso's in His Time* (London, 1726).

[17] *Ibid.*, p. 41. I have not been able to find the original manuscript of this.

[18] *Mém. Acad. Sci.* (1734), pp. 123–34.

[19] *J. de Phys.*, Vol. 16 (1780), pp. 325–48.

[20] See for example, Oldenburg in *Phil. Trans.*, Vol. 1 (1665/6), p. 153 (quoted on p. 72 above).

the roller, then over another pulley to a counterweight. The "punch" would be carried along by the horizontal portion of the thread.[21]

In 1765 the clockmaker Alexander Cumming (1733–1814)[22] made a magnificent barograph for George III, which is still at Buckingham Palace. This has been described and beautifully illustrated in an article by H. Alan Lloyd.[23] A fine clock movement drives two dials which surround the clock face, the inner dial making a revolution in six months and the outer one in a year. A chart on each dial carries concentric circles corresponding to variations of the barometer, and radial lines at weekly intervals. The barometer is a siphon barometer with a float supporting a light vertical framework carrying two pencils, each of which records on one of the annular dials. There is also a dial and pointer as in the ordinary wheel barometer. George III paid £1178 for this "clock barometer."

In 1766 Cumming made a second barograph for his own use, similar to the royal instrument except that it has only the yearly chart and no dial and pointer and was less splendidly decorated. When Cumming died in 1814

the famous climatologist Luke Howard purchased it and used it in London until 1828, when it was moved to Ackworth, Yorks.[24] This barograph is still in the Howard family, though only the clock was operating in 1962.

The Paris instrument maker Antoine Assier-Perica (or Pericat or Perricat) may have made a somewhat similar instrument; there is an article in the Abbé Rozier's *Journal de Physique* for 1779[25] which is really nothing but an advertisement for Assier-Perica, even giving prices. His most expensive *baromètres* (2000 livres) "qui tiendront un compte exact des révolutions du mercure à chaque instant du jour et de la nuit," may well have been barographs.

Changeux's barograph (Figure 11.1) was thought so well of by its author that he published the article about it[26] separately as a pamphlet with a resounding title page.[27] It was very similar in many respects to that of Cumming, with the important difference that it contained a mechanism for causing the pencil to strike the paper at equal intervals of time, meanwhile leaving the float free to move. On July 22 Le Roy and Brisson made a favorable report about it to the Academy of Sciences, and this was duly reprinted by Changeux.[28]

[21] H. E. Hoff and L. A. Geddes (*Isis,* Vol. 53 [1962], p. 306) give reasons for thinking that it may have been a sort of balance barograph; but I think the wheel barometer a more likely source of this mechanism.

[22] He became an F.R.S.; according to D.N.B. he "wrote largely on the mechanical laws and action of wheels." In 1781 he was made an Honorary Freeman of the Clockmakers Company (S. E. Atkins & W. H. Overall, *Some account of the Worshipful Company of the Clockmakers of the City of London* [London, 1881], p. 186).

[23] *J. Suisse d'Horologerie,* English ed., Vol. 78 (1953), pp. 371–81; Vol. 79 (1954), pp. 46–56. See also "D.C." in *Weather,* Vol. 7 (1952), pp. 252–53.

[24] Luke Howard, *The Climate of London* (2d ed.; 3 vols.; London, 1833), III, 279.

[25] Vol. 14, pp. 327–29.

[26] *J. de Phys.,* Vol. 16 (1780), pp. 325–48.

[27] *Météorographie ou Art d'observer d'une manière commode et utile les phénomènes de l'atmosphère. Contenant la description de deux* Barométrographes *ou baromètres qui tiennent note par des traces sensibles de leurs variations et des tems précis où elles arrivent; avec l'idée de plusieurs autres instrumens météorologiques* (Paris, 1781), 44 pp., 1 plate (p. 44 numbered 42).

[28] *Ibid.,* p. 35–42.

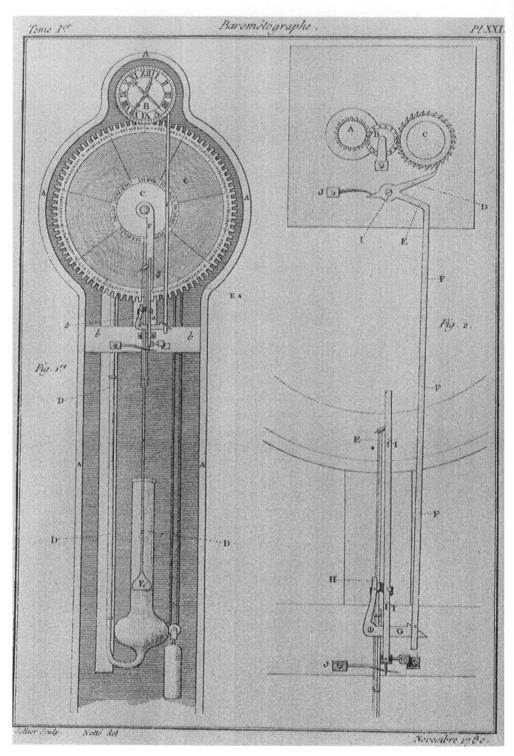

Fig. 1.re

Fig. 2.

Tellier Sculp.   Notté del.   Novembre 1780.

Fig. 11.1. Changeux's "barometrograph," 1780.

Barographs of this pattern were constructed in various places. By 1785 there was one in operation at Mannheim, and there seems to have been another at Munich, but not being used.[29]

There are two recording barometers in the *Deutsches Museum,* Munich,[30] made by Jean Krapp of Mannheim and stated to date from about 1790. They differ in that no. 2061a has a rectangular chart sliding sideways beneath the clock in a frame, and projecting rather a long way outside the case at its extreme position, while 2061b is very like Cumming's barograph except that its chart revolves once a week. The inlaid cases stand about 2 meters high.

The principle of discontinuous recording has been applied to barographs from time to time. The Abbé Felice Fontana (1730–1805) produced an instrument now in the *Museo di storia della scienza* at Florence.[31] This is a siphon barometer with chambers 16 mm. in bore, and a float in the shorter arm, partly supported by a wire from a sector 13 cm. in radius, slightly more than counterbalanced by a weight. The shaft of this, which also carries a second sector, is supported on four rollers 4 cm. in diameter. The second sector is in the form of a cylindrical frame which holds the recording paper, so that the latter moves under a sharp point as the pressure changes. A clock traverses the point in a direction at right angles to the motion of the sector, and at regular intervals depresses the point so that it marks the paper. As far as I know this

barograph is unique in having the chart paper moved by the changes in atmospheric pressure.

In 1843 Karl Kreil of Prague described a siphon barograph with discontinuous recording which was considered excellent for a number of years. A clock draws a chart E (Figure 11.2) to the left at a uniform rate, and also, by means of a cam T on the minute shaft, causes the pencil to make a dot on the chart at uniform intervals of 5 minutes, 11 times an hour, the 12th being blank for the purpose of timing. There is also a delicately balanced mercury thermometer QR which moves a pointer K.[32] There is one of these barographs at the Science Museum, London,[33] made by Dressler of Prague, which was in use at Kew Observatory in 1845.[34] At the *Zentralanstalt für Meteorologie und Geodynamik* in Vienna there is another, somewhat better finished than the Kew instrument.

In 1844, R. Bryson published an account of a barograph[35] in which the float of a siphon barometer had a stiff wire rising from it, carrying at its distal end a small knife-edge. Once an hour this knife-edge was pressed against a drum which revolved on a vertical axis. Japanned black, the drum was covered with whitewash. The resulting hourly readings were read off by placing the drum in a measuring machine constructed for the purpose.

---

[29] G. Hellmann, *Repertorium der deutschen Meteorologie* (Leipzig, 1883), cols. 899–902.

[30] Inventory nos. 2061 a and 2061 b.

[31] Inventory no. 1163.

[32] *Magnet. und meteorol. Beob. zu Prag.,* Vol. 3 (1843), pp. 131–33.

[33] Inventory no. 1876–793.

[34] The sum of £30 was voted by the British Association in 1844 for the acquisition of this barograph (*B.A.A.S. Report* for 1844, p. xxiii).

[35] *Trans. Roy. Soc. Edin.,* Vol. 15 (1844), pp. 503–5.

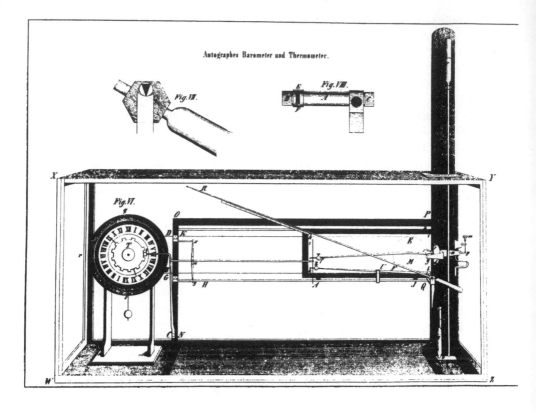

Fig. 11.2. Kreil's barograph, 1843.

Such siphon barographs continued to be described in the most diverse forms, and some people greatly preferred a flat chart to a drum. An instrument which achieved a large flat chart without much sacrifice of compactness was designed by G. A. Schultze,[36] from whose paper Figure 11.3 is reproduced. The tube *ab* is 16 mm. in bore, and has a float *e* in its open limb. By means of a thread and a pulley *D*, this is connected to a wheel *f*, on the axis of which is an arm

[36] *Annalen der Physik*, Vol. 76 (1849), pp. 604–6.

g with a point *h*. Every five minutes a tooth of the wheel *k*, operated by the minute-shaft of a clock, pushes out the lower end of the lever *i* so that its sector-shaped upper end pushes the point into the paper on the chart. This is fastened by clips to a metal plate *c* which is let down by the clockwork at a uniform rate.

This barograph incorporates one novel feature: the reading can be corrected to a standard temperature by the bimetallic construction of the wheel *f*. If it were properly proportioned, such an arrangement could in fact be nearly correct for all pressures, at least over a reasonable range of temperature.

A rather simpler barograph on the same general principles, but without the compensation for temperature and with a vertical drum, was made by "M. Pillischer, Optician, 88 New Bond St., London," about 1850, and is now in the Science Museum,[37] as is that of Admiral Sir Alexander Milne (1806–1896).[38] Of the latter, which resembles that of Kreil, the first is said[39] to have been made by the Admiral himself, and the one in the Science Museum indeed looks "home-made"; but several others were constructed. By 1864 Negretti &

[37] Inventory no. 1939–301.
[38] Inventory no. 1894–119.
[39] Negretti & Zambra, *A Treatise on Meteorological Instruments* (London, 1864), p. 32.

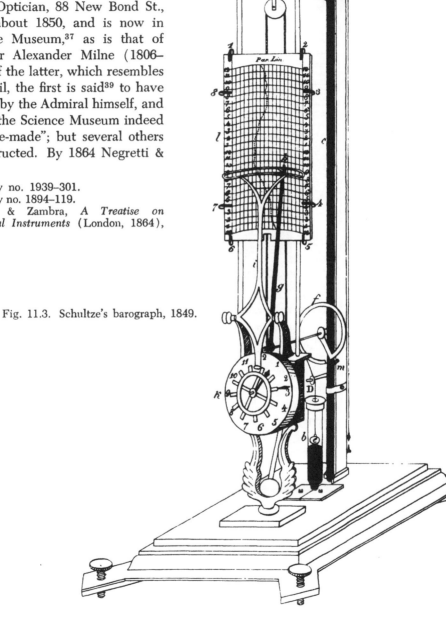

Fig. 11.3. Schultze's barograph, 1849.

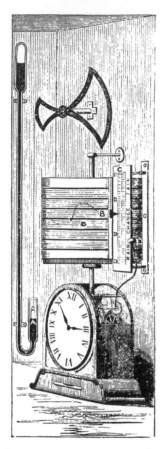

Fig. 11.4. Negretti & Zambra's modification of Milne's barograph.

Zambra were making a mercury barograph which they called Milne's, but which resembles it only in being based on a siphon barometer. Their illustration,[40] which they say is diagrammatic, the barometer tube actually being behind the clock, is reproduced in Figure 11.4.

The director of the Munich Observatory at this time was Johann von Lamont—there is a street named after him near the observatory—and like many other nineteenth-century astron-

omers, he paid a good deal of attention to meteorological observations.[41] One of the recording instruments constructed about 1846 was a barograph of the type we are considering, which made pinpricks in paper on a drum. In this instrument, which seems to have had a tube of about 8 mm. bore, a glass float weighted with mercury was counterbalanced so that it just touched the top of the meniscus in the open limb. A minute before the record was to be made, the clockwork removed part of the counterweight and restored it again, setting the mercury into a brief oscillation which had died down before the pin was caused to pierce the chart. The float was prevented from sticking to the sides of the tube by a thin disk concentric with the float and attached just above it, and just a little smaller than the tube. Temperature compensation was not attempted, though von Lamont suggests that it might have been applied by way of the lever that carried the float. Corrections were made as necessary. There is now no trace whatever of this instrument at Munich Observatory.

For the Great Exhibition of 1851, George Dollond[42] (1774–1852) made an immense meteorograph[43] similar in scope and perhaps in general arrangement to that of Hooke. It was called

[40] Ibid., p. 33.

[41] J. von Lamont, Beschreibung der an der Münchener Sternwarte zu den Beobachtungen verwendeten neuen Instrumente und Apparate (Munich, 1851).

[42] Of the famous firm; the Dollonds were Huguenots who fled from France in 1685. Both John (1706–1761) and George were Fellows of the Royal Society; Peter (1730–1820) joined the American Philosophical Society.

[43] Exhibition of 1851, Reports of the Juries (London, 1852), p. 299.

the "atmospheric recorder." The barograph was based on a large siphon barometer and, like the other instruments, recorded every half hour.[44] As Multhauf writes, "it is not clear that Dollond's instrument was superior to Hooke's, or that its career was longer."[45]

There followed a long period in which the mechanically-recording siphon barometer was entirely neglected in favor of more complicated barographs, until in 1904 William Henry Dines made an excellent instrument with some novel features, which was described in 1929 by his son L. H. G. Dines.[46]

This barograph returned to the system of continuous recording, as used by Alexander Cumming. A diagram of the instrument is shown in Figure 11.5, in which AB is the barometer tube, with large upper and lower chambers equal in size, of which the upper one can be exhausted by a pump and sealed by the U-tube shown at the left. In the lower chamber floats a glass bell C, surrounded by a ring of floating steel balls to keep it central and reduce friction. The glass stem of C fits loosely into a steel tube which supports a balance weight D and a length of steel rod E. The motion of the assembly CDE is transmitted by a pair of platinum wires F to the smaller of two pulleys G and H, fixed on the same shaft which is supported on anti-friction wheels. From the right-hand side of the larger pulley is hung a glass rod J by means of the

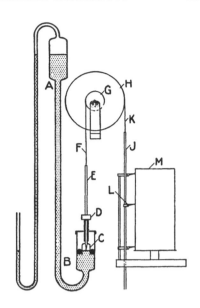

Fig. 11.5. Dines' recording barometer (1904), diagram.

pair of platinum wires K. The pen L, fixed to this rod, records on the drum M the motion of the float, magnified by the ratio of the diameters of the two pulleys, about 4 to 1, and thus records the changes in the barometric height with a magnification of about two. Two fixed pens provide base lines which afford a check on the expansion or contraction of the paper with changes in relative humidity.

The most novel features of this barograph is the means of temperature compensation, which is done by leaving a certain volume of air beneath the glass float C. This volume is subject to calculation.

There is one of these barographs in the Science Museum, London,[47] a beautiful piece of instrument making. The brass label is inscribed "Dines self-

[44] Robert P. Multhauf, *U.S. National Museum,* Bulletin 228 (Washington, 1961), p. 102, has reproduced the illustration from the catalogue of the exhibition.

[45] *Ibid.,* p. 103.

[46] *Quart. J. r. Meteorol. Soc.,* Vol. 55 (1929), pp. 37–53.

[47] Inventory no. 1922–125.

recording barometer J. Hicks. Maker. 8. 9 & 10. Hatton Garden London."[47] There is an earlier version in the showrooms of Heath, Hicks, and Perken at 8 Hatton Garden.[48] Instead of the glass bell surrounded by steel balls, the bell is surmounted by a toothed disk to keep it central, and the rod which rises from it is guided by rollers. Instead of the bifilar platinum suspension there is a flexible watch chain, and the pen is constrained to move vertically by brass guides running on a rectangular glass plate. The barometer tube is a simple siphon with large chambers, but without the additional U-tube which permits the vacuum to be improved.

Dines' barograph was somewhat more sophisticated but probably less durable than the recording barometer of Marvin, also designed about 1904 (Figure 11.6), and instruments of the latter type were in operation in the United States for about forty years.[49] The temperature compensation depended on the use of the correct volume of mercury.[50]

The recording mercury barometer of Richard, Paris, was essentially similar and was made commercially in fair numbers at the turn of the century. The zero adjustment is performed by raising or lowering the entire siphon tube in relation to the mechanism, a screw with

Fig. 11.6. Marvin's barograph, 1904. (*Smithsonian Institution*)

[48] This is described and illustrated in *Meteorol. Mag.*, Vol. 39 (1904), pp. 150–51. James Joseph Hicks started in business at 8 Hatton Garden, London, in 1862, in which year his name first appears in the London Directory. The business, called James J. Hicks after 1874, remained in Hatton Garden until 1952, when it was incorporated in the firm of Heath, Hicks, and Perken, now at New Eltham but with showrooms still at 8 Hatton Garden.
[49] Multhauf, *U.S. National Museum*, Bull. 228, p. 114.
[50] See p. 252 above.

a large knob being provided for the purpose. There is one of these at the C.N.A.M. in Paris.[51]

The last design of a recording siphon barometer that has come to my attention is that of N. Pedicini,[52] in which, to throw nearly all the motion into the open limb, the vacuum chamber was enormously wide, made of two thick glass plates and an iron ring. The record was made on smoked paper with a magnification of 6.

Patents in this field seem not to have been numerous; there was a somewhat fantastic one in 1848,[53] about which nothing further need be said, and another in 1891 in which the principle of discontinuous recording was itself claimed.[54]

4. BAROGRAPHS DERIVED FROM THE BALANCE BAROMETER

With the exception of one design which may never have been constructed, it was nearly two hundred years after the invention of the balance barometer by Sir Samuel Morland that it was first adapted to making a continuous record. There followed a period of intense development which lasted for fifty years, and then the subject was dropped except for one isolated, though highly developed instrument.

There are four fundamental types of balance barograph, which differ not only in appearance but in theory, particularly in the conditions for stability and in the possible means of compensation for temperature changes. The failure to realize that these differences are fundamental led to a great deal of rather bitter and entirely fruitless debate in the 1860's. We may classify these barographs as follows:

a. Balance barographs with bent beams,

b. Balance barographs with straight beams (i.e., with knife-edges exactly, or almost, in a plane),

c. Balance barographs with beams ending in circular arcs,

d. Balance barographs with rolling weights.

Of all these the last type is the most sophisticated.

a. Balance Barographs with Bent Beams

One of the first barographs ever described seems to have been of the type in which the tube is hung on a balance. It is found in the *Theatri statici* of Jacob Leupold, published in 1726,[55] and is illustrated in Figure 11.7. It is questionable whether, on account of friction, such an instrument really would be satisfactory; but its ingenuity is beyond all question. It should be noted that the stability of the system is ensured by the clever use of the spiral

---

[51] Inventory no. 13164.

[52] *Rivista di Meteorologia Aeronautica* (Dec., 1937), pp. 22–28.

[53] British Patent 12,220 (1848), granted to D. Napier and J. M. Napier.

[54] British Patent 7,323 of 1891, dated October 27, 1890, granted to G. Meyer and A. Redier.

[55] Jacob Leupold, *Theatrum machinarum Pars III. Theatri statici universalis, sive theatrum aerostaticum, Oder: Schau-Platz der Machinen zu Abwiegung und Beobachtung aller vornehmsten Eigenschaften der Lufft*, etc. (Leipzig, 1726), p. 305. The spiral cam was also part of a barometer ascribed to Minotto in the *Dizionario tecnologico* (Venice, 1831), II, 376.

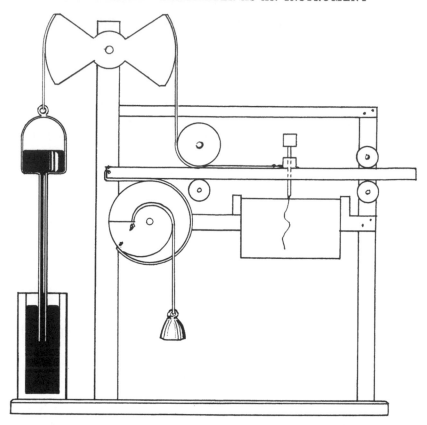

Fig. 11.7. Leupold's design for a barograph, 1726, compiled from the *Theatrum machinarum.*

cam, which has the same effect in sta-bilizing the system as bending the beam; on the other hand, it leaves the deflection linear in pressure, which the bent beam does not. One cannot assume that all the clever devices described and pictured by Leupold actually were built, but of course this might have been tried.[56]

[56] These designs of Leupold's have been discussed at some length by H. E. Hoff and L. A. Geddes, *Isis,* Vol. 53 (1962), pp. 287–310. One of them, in which the cistern and the counterweight were supported from con-centric arcs, would not have been stable.

The earliest balance barograph which certainly worked, and it worked very well, was devised by Father Angelo Secchi (1818–1878), a famous astron-omer and director of the Observatory on Monte Mario at Rome.[57] Figure 11.8 shows its essential features. A large iron barometer tube with an upper chamber about 5 cm. in diameter was suspended from a bent beam *ACB* so as to dip into a cistern *V.* There was an adjustable counterweight at *A.* The

[57] *Nuovo Cimento,* Vol. 5 (1857), pp. 14–17; also *Atti di nuovi Lincei,* Ser. 1, Vol. 10 (1857), pp. 137–45.

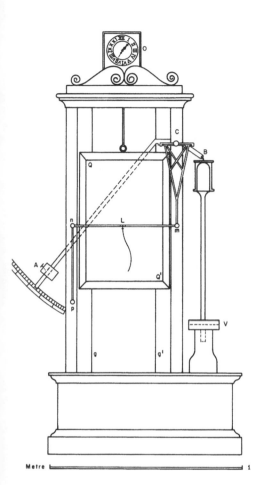

Metre

Fig. 11.8. Secchi's first balance barograph, 1857.*(Courtesy of the Trustees of the British Museum)*

angular deviations of this beam were transformed into a nearly linear motion of the pencil L by a Watt parallel motion hinged at m, n, and p.[58] The pencil drew a curve on paper in a frame QQ', which was let down at a constant rate by a clock O. The mechanism, as will

be seen, was extremely simple. It was also very large, about 1.2 meters wide by 2.1 meters tall, and it ought to be noted that this is the barometric part of a complete meteorograph.

Secchi was quite well aware that the scale of this barograph would not be linear; in fact he had one of his assistants, Father M. Jullien, compute a shape for the upper chamber of the barometer tube so that either the angle turned through by the beam, or the displacement of the pen, might be proportional to the change of pressure.[59] There is no record of such a tube having been made, however. This barograph remained in use, substantially in its original form, for several years, although some of the other parts of the meteorograph were altered as time went on.[60] Secchi then built another and much better barograph into it, which we shall refer to later. He thought so highly of this meteorograph that he sent it, immense as it was, to the Paris Exhibition of 1867.[61]

Over these barographs, and indeed over the theory of the weight barometer in general, there broke out an extraordinary controversy, conducted mostly in the pages of the *Comptes Rendus* and the *Moniteur Scientifique* in the year 1867. The combatants were Father Secchi and a much younger Frenchman called R. Radau (1835–1911). The tone of Radau's articles is such that one is constrained to suppose

[58] A necessary property of this notion is that P,L, and C are in a straight line.

[59] *Ann. di Matemat. pura ed applicata*, Vol. 4 (1861), pp. 337–44.

[60] Angelo Secchi, *Coll. Rom., Bull. Meteorol.*, Vol. 5, No. 4 (30 aprile, 1866), pp. 25–44.

[61] E. Lacroix, *Etudes de l'Exposition de 1867* (Paris, 1867), Vol. 2, pp. 313 ff.

that a pure interest in science was corrupted by some other motive, for instance anti-clericalism; Secchi's replies are more good-natured but not always convincing. There is no doubt that Radau was much the better mathematician.

The controversy seems rather superfluous today, but may have been of some use at the time. It did establish that the balance barometer with the bent beam can be made stable no matter what the shape of the tube, while for the one with the straight beam to be stable, the external cross section of the tube—and anything fastened to it— at the level of the mercury in the cistern must exceed the internal cross section of the vacuum chamber. Radau also showed that the temperature correction for some mean pressure can be zero if the total volume of mercury is rightly chosen.[62]

There seems to be no reason for the controversy to have spilled over into the *Moniteur Scientifique*, except that Radau, who was engaged to write a review of the instruments at the Paris exhibition of 1867 for that journal, seems to have taken a dislike to Secchi and could not stop baiting him.[63] Secchi made a rather plaintive reply in which his best point was that even if his Roman barograph had not a linear scale —as he was aware—it had less friction

than some of the linear ones.[64] Radau[65] replied with a final article which restated several points well known by this time, and was horribly sarcastic at the expense of Secchi. It should not be imagined, however, that Radau's papers were entirely destructive; they ranged over a wide variety of instruments and contained a good deal of useful theory. Two papers[66] by Alois Handl of Lemberg on the theory of the instrument added little or nothing new.

The next bent-beam instrument was made by Hasler & Escher of Berne for Heinrich Wild, then still in Switzerland, who in 1867 published an account of the recording meteorological instruments in the Observatory at Berne.[67] After discarding direct registration with a pencil because of friction, and photographic registration because of its complexity, Wild had settled on pointwise recording. It would seem that the balance barometer was rather a new idea to him, for he ascribes its invention to Secchi and states that the temperature has no influence on the readings of such barometers!

His barometer tube was of 6 mm. bore for most of its length, expanded at the top into a cylindrical chamber 32 mm. in bore. This tube dipped into a wooden cistern provided with two plate-glass sides, with the idea that absolute readings should be taken with a cathetometer. The tube is hung from one end of a balance beam with steel knife-edges, the other end of the beam dipping down at a sharp angle and

---

[62] The sequence of papers in the *Comptes Rendus* is as follows: Radau, *Compt. Rend.*, Vol. 65 (1867), pp. 360–64.
    Secchi, *ibid.*, pp. 443–48.
    Radau, *ibid.*, pp. 502–5.
    Secchi, *ibid.*, pp. 559–62. This contains a calculation of the effect of temperature on the Roman barograph.
[63] R. Radau, *Moniteur Scientifique*, Vol. 9 (1867), pp. 641–46, 705–18, 774–76. (Translated in *Repertorium für phys. Technik*, Vol. 3 [1867], pp. 281–362.)

[64] A. Secchi, *ibid.*, pp. 776–78.
[65] R. Radau, *ibid.*, pp. 881–88.
[66] K. Akad. der Wiss. Wien, II Abt., Math.-Naturw. Kl., Vol. 59 (1869), pp. 7–16; *Rep. für phys. Techn.*, Vol. 6 (1870), pp. 104–12.
[67] *Rep. für phys. Techn.*, Vol. 2 (1867) pp. 161–201.

having an adjustable counterweight on its outer end. The beam also has a long pointer projecting vertically downward, and every 10 minutes a needle on the end of this pointer is pressed against a band of paper moved downward at constant speed by a clock.[68]

Wild claimed an overall error of ±0.2 mm. and ascribed most of it to the sticking of the mercury in the tube, but said that he had not yet adopted the obvious scheme of disturbing the tube shortly before each reading. The barograph was calibrated directly with a cathetometer over a period of months. Wild did not seem to pay any attention to the non-linearity of such a system, but computed a linear calibration by a least-squares solution of 38 sets of observations, the cathetometer measurements being corrected for temperature. It would seem that Radau might have saved his ammunition for Wild. Yet this barograph worked for a long time; in 1877 he reported the addition of a means of disturbing the tube before each mark.[69]

When I enquired about this barograph at Berne in the spring of 1961, I was told that it had been broken up less than two years before. I was even shown some pieces of brass which had formed part of the frame, and were picked out of a box of odds and ends for my enlightenment. This is what happens to such things!

About 1878 or 1879 the firm of R. Fuess in Berlin produced an improved, or at least tidied-up, version of Wild's barograph (Figure 11.9). This was ex-

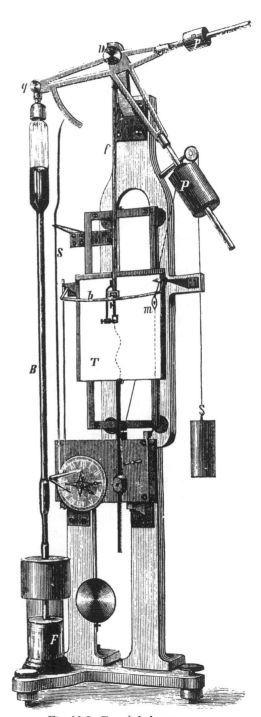

Fig. 11.9. Fuess's balance barograph, about 1879.

[68] There was a later version of this barograph, differing mainly in the form of the chart.

[69] *Bull. Acad. Imp. des Sci. St. Pétersbourg,* Vol. 23 (1877), pp. 492–99.

hibited at the Berlin Industrial Exhibition in 1879, and reported on by A. Sprung,[70] whose name will appear importantly later on in this chapter. The rod S is the member which jogs the beam a short time before each point is recorded.

Meanwhile Wild had revised his ideas a good deal regarding the effect of temperature on the balance barograph[71] and had provided the one at Berne with an ingenious scheme of compensation[72] (Figure 11.10). The effect of temperature is cancelled out by hanging on the left-hand end of the balance beam a glass tube of mercury, into which dips a narrow tube leading to a large bulb partly filled with mercury and partly with absolute alcohol. This bulb is firmly fixed to the frame of the barometer, and a rise in temperature therefore causes some mercury to flow out into the tube hung on the beam. The bar which operates the recording point is now operated by an electromagnet controlled by a clock which makes a contact every 10 minutes. The dimensions of the temperature-compensating *Ausflussthermometer*, and the amount of alcohol, were calculated from first principles and checked by trial. Judging by the excellent results he obtained, it must have been more than adequate. Nor did the departure from linearity seem bothersome.

Similar barographs, with one or two improvements in detailed design, were made for the Technical Museum in Moscow and the Meteorological and Magnetic Observatory at Peking.[73] All in all this relatively simple instrument seems to have been very satisfactory, though it is rather surprising that the Fuess version, without the temperature compensation, should have been as well received as it was.[74]

Wild also suggested[75] that temperature compensation could be obtained by leaving a certain amount of air in the vacuum chamber of the barometer and worked out the theory of this, but, wisely perhaps, did not try out the idea.

We must now turn to the work of Paul Schreiber (1848–1924) of Chemnitz, whose name was remarkably appropriate in view of the reams of paper he devoted to the weight barograph. His first article, which seems to have arisen from a thesis, covers seventy-two large pages,[76] and is mainly on the theory of the bent-beam weight-barometer, which turns out to be very complicated. It is written with Teutonic thoroughness from a much more general standpoint than those of Radau or Handl, but leads to similar results for the main equations. Schreiber made a theoretical investigation of all sorts of errors, such as a variable cross section of the chamber and of the cistern, and so on; and of course, errors due to temperature.

[70] In L. Löwenherz, *Bericht über die wissenschaftliche Instrumente auf der Berliner Gewerbeausstellung im Jahre 1879* (Berlin, 1880), pp. 230–42.

[71] *Bull. Acad. Sci. St. Pétersbourg*, Vol. 15 (1870), cols. 139–47; Vol. 16 (1871), cols. 132–47.

[72] *Repertorium für Meteorologie*, St. Petersburg, Vol. 3 (1874), pp. 120–45.

[73] *Ibid.*, p. 133.

[74] See, e.g., H. Eylert, *Zeits. für Instrum.*, Vol. 6 (1886), pp. 269–72.

[75] *Bull. Acad. Sci. St. Pétersbourg*, Vol. 16 (1871), cols. 143–47.

[76] *Repertorium für exper. Physik.*, Vol. 8 (1872), pp. 245–316.

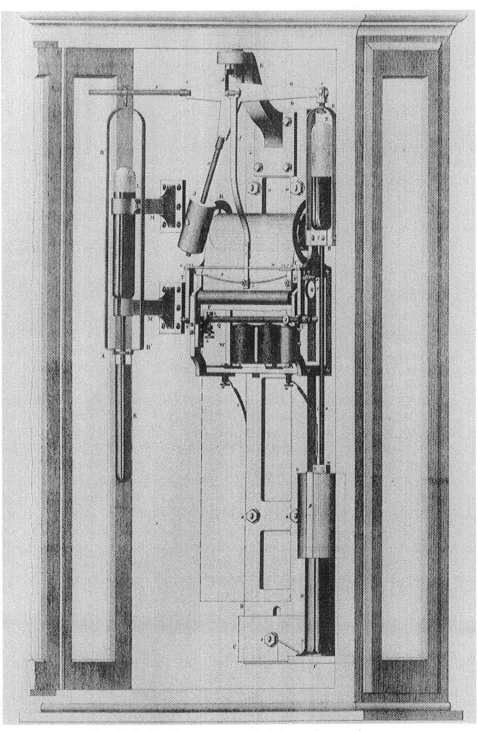

Fig. 11.10. Wild's compensated balance barograph,
1869. *(Courtesy of the Trustees of the British Museum)*

He then applied his theory to the detailed design of such an instrument and built a barograph, of a rather inconvenient shape, which confirmed his conclusions.

Shortly after this Schreiber seems to have joined the staff of the new *Deutsche Seewarte*, the German Naval Observatory, at Hamburg, and in 1878 his new balance barograph was described as part of the description of the observatory.[77] He had forsaken the bent-beam instrument, probably because of its non-linearity, which must have displeased his tidy mind. We shall therefore return to him later.

b. Balance Barographs with Straight Beams

On page 299 we referred to Angelo Secchi's second and improved barograph, which was exhibited at the Paris Exhibition of 1867. This formed part of a huge meteorograph, still preserved at the *Museo Copernicano* on the *Monte Mario* in Rome, which recorded, besides atmospheric pressure, temperature, humidity, rainfall, and wind direction and speed. Figure 11.11, taken from the book by Lacroix,[78] shows a front view of it, in which the barograph, still with its Watt parallel motion, can easily be recognized. The new barograph, however, differs in many ways from the old. In spite of looking bent in the drawing, the beam is straight in the sense that the three knife-edges are nearly coplanar. This fundamental change in design, as we have seen, demanded other changes in order to preserve the stability of the system. To this end the tube is surrounded by a wooden sleeve at the place where it enters the mercury in the cistern. The diameter of this is somewhat greater than that of the vacuum chamber. Such a system will tend to float sideways toward the edge of the cistern, so an arm is provided to prevent this, forming a parallel motion with the beam. It is remarkable that Secchi has duplicated the Watt motion and the pencil, so that the pressure is recorded on both of the diagrams simultaneously produced by the instrument on the two sheets of paper let down by the clockwork in the space of two days.

Secchi published a description of this instrument in 1866[79] and again in 1870.[80] He felt that the new arrangement was a great improvement, especially in respect to the correction for temperature, which, he claimed, was less than 0.1 mm. per 10° C. at a mean pressure.

Another roughly similar balance barometer was described before the *Verein deutscher Ingenieure* by a Professor Bruhns of Leipzig;[81] but it was less highly developed and not really carefully thought out.

Instead of suspending the barometer tube from a balance one may equally well suspend the cistern, leaving the tube fixed. This was the solution adopted by A. Crova at Montpellier in 1874, where a barograph of this sort was said to have functioned without interruption for four years, being com-

---

[77] *Aus dem Archiv d. deutschen Seewarte*, Vol. 1 (1878), No. 1, pp. 18–25.

[78] Lacroix, *Etudes*, Vol. 2.

[79] Secchi, *Coll. Rom., Bull. Meteorol.*, Vol. 5.

[80] *Descrizione del meteorografo dell'Osservatorio del Collegio Romano* (Rome, 1870).

[81] *Zeits. V.D.I.*, Vol. 17 (1873), p. 58.

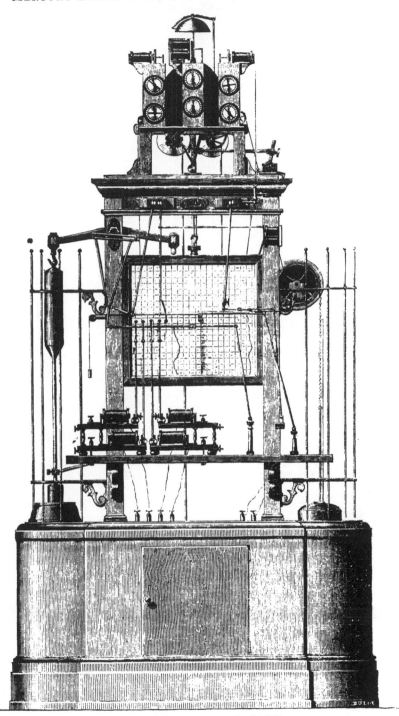

Fig. 11.11. Secchi's meteorograph of 1867.

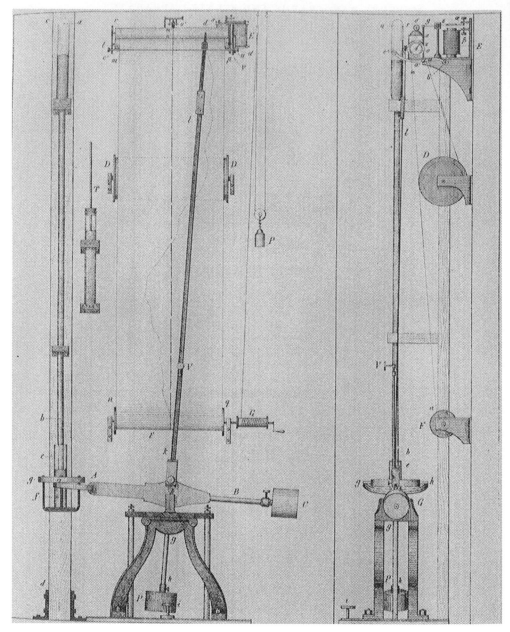

Fig. 11.12. Crova's barograph, 1874. *(Courtesy of the Trustees of the British Museum)*

pared daily with a large standard barometer.[82]

This barograph (Figure 11.12), which registered a point every 10 minutes, had

[82] A. Crova, *Mém. Acad. des Sci. de Montpellier*, Vol. 9 (1878), pp. 153–71.

a fixed tube with an upper chamber 3 cm. in diameter, and a cistern which was suspended on one end of an equal-arm balance. The cistern, 18 cm.² in horizontal cross section, was in a frame carried on knife-edges placed above any possible center of gravity. The beam *AB* was supported by steel knife-edges on agate planes. Crova was quite well aware that it would have been unstable without the weight *P*, "comme le diamètre intérieur de [la chambre] est supérieur au diamètre extérieur de la partie du tube qui émerge du mercure de la cuvette."[83] Moving the adjustable weight *C* changed the zero, moving *P* altered the magnification. In a sense this construction was really equivalent to bending the beam. The pointer was of wood, 80 cm. long. Note that the rollers which drove the paper were moved every 10 minutes by a ratchet, driven by an electromagnet.

Crova gave a simple theory of the instrument without regard to the errors due to change of temperature.[84] Although he knew of the work of others in this direction, he had not tried to compensate for temperature changes.

A rather similar barograph, made by Salleron for the Observatoire de Mont-Souris, was described in 1878 by Marie-Davy.[85]

c. Balance Barographs with the Tube, etc., Supported from Circular Arcs

In Chapter 6 we have already dealt with the weight barometer of Cecchi and Antonelli in which the tube and counterweight were supported from concentric circular arcs. The use of one large wheel is kinematically equivalent to this, and before the date of their barometer Alfred King, an engineer at Liverpool, had designed a barograph on this principle, a small illustrative

[83] *Ibid.*, p. 156.

[84] The first equation (p. 161) seems to be printed incorrectly.

[85] *Annuaire de l'Obs. de Mont-Souris pour l'An 1878*, p. 265.

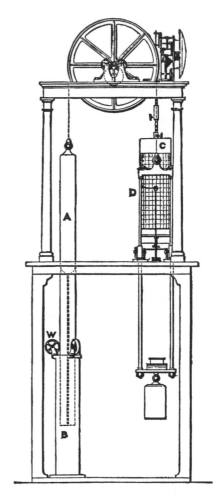

Fig. 11.13. King's barograph, 1863.

model being exhibited to members of the British Association for the Advancement of Science in 1854, and the instrument itself erected at the Liverpool Observatory some years later.[86]

A front view of this instrument is shown in Figure 11.13. The barometer tube *A* had its vacuum chamber 3 inches in internal diameter, and moved up and down in a cistern *B*, constrained

[86] *Report of the Astronomer to the Marine Committee, Mersey Docks and Harbour Board* (Liverpool, 1863), pp. 14–15.

by roller guides and supported by a chain which passed over a large grooved wheel to a counterpoise D in the form of a frame which supported the tracing pencil. This frame was suitably weighted, and guided by more rollers, and the pencil drew a curve on a paper-covered drum geared to a clock. The magnification was 5. In such an instrument everything depends on reducing friction; and King resorted to supporting the large wheel by resting its small axle on two fairly large rollers, keeping it from falling off by two further rollers at each side. He seems to have been fairly successful, as we are told that small oscillations of the barometer were faithfully recorded. The design is an excellent example of nineteenth-century engineering.

This design would probably have remained entirely unknown except that it was illustrated and described in Negretti & Zambra's *A Treatise on Meteorological Instruments*,[87] which received a fairly wide circulation. It was known to Secchi in 1870.[88]

It may also have been known to Paul Schreiber, to whose "thermobarograph" at the *Deutsche Seewarte* we must now return.[89,90] At least the method of supporting the large wheel was a development of King's; but the principle of the barograph was different, and indeed quite original.

It is distinguished from King's by the fact that the part of the tube dipping into the cistern was much smaller than the vacuum chamber. This would have

rendered it unstable, as we have seen, except that part of the counterweight was an iron cylinder dipping into another vessel of mercury.

The entire instrument, a contemporary drawing of which is shown in Figure 11.14, must have been the most tremendous meteorological instrument ever made, at least until such things as radar apparatus began to be put into this category. It really consisted of three instruments: the barograph; a rather similarly constructed gas-thermometer to measure its temperature; and a second gas thermometer with its bulb out of doors. The counterweights of the gas thermometers were simple masses of iron. The marking points were attached to the counterweights and were knocked into the paper once every 20 minutes by hammers operated by the clock which turned the drum. The latter was 30 cm. in diameter and turned once in 200 hours, so that 8 days' record could be obtained before a fresh chart had to be put on.

The mechanical arrangements were of great interest. Schreiber saw clearly that the friction of a wheel on its bearings depends, *caeteris paribus*, on its weight; so he cut the big wheels down to sectors and lightened them as much as possible. The axle of the big wheel rolled on the edges of another pair of wheels and turned the latter through a small angle. Schreiber saw that the friction in this part of the mechanism would be less, the larger the ratio between the radius of these second wheels and that of their axle, so he reduced the second pair of wheels to narrow sectors kept upright by counterweights, and of the same radius as the first wheel. He stopped short of King in

[87] (London, 1864), pp. 34–35.
[88] *Coll. Rom., Bull. Meteorol.*, Vol. 5, p. 27n.
[89] *Aus dem Archiv d. deutschen Seewarte*, Vol. 1 (1878), No. 1, pp. 18–25.
[90] *Repertorium für exper. Physik.*, Vol. 14 (1878), pp. 471–86, 649–702.

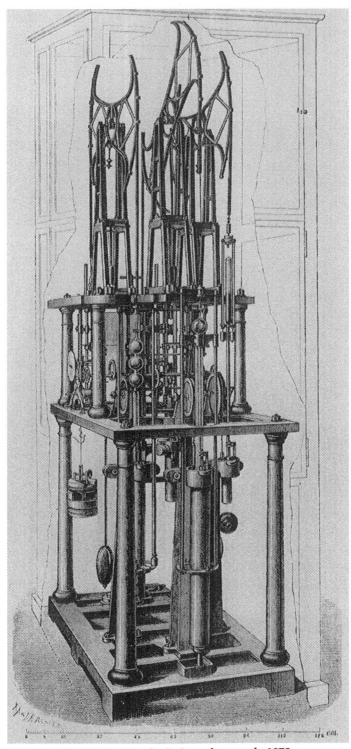

Fig. 11.14. Schreiber's thermobarograph, 1878.

providing polished cheeks, instead of further rollers, to prevent the axle of the first "wheel" falling off the second.

In his third paper[91] he gave a theory of this sort of weight barograph. The constants could be determined by hanging known weights on each side of the balance and by adding and subtracting known amounts of mercury to or from the cistern and the auxiliary vessel. He also calculated the error due to temperature variations, which turned out to be about 0.126 mm. per degree C. at 700 mm. of mercury, and 0.016 mm. per degree C. at 800. He suggested that an automatic correction could be arranged by having an iron cylinder dipped into the cistern by means of the expansion of a zinc rod, multiplied by a system of levers; but possibly the grant he had been given by the Prussian Academy[92] to develop the instrument had run out, or the patience of the Director had been exhausted, for this was not added. This unhappy state is suggested by the rather querulous conclusion to the last mentioned paper, where he complains that while many German journals mention every small technical advance, especially when it is made by some foreigner, they pay no attention to his apparatus.

Schreiber's "thermobarograph," which was really a beautiful piece of machinery, stood for some time among the other meteorological instruments in the *Deutsches Museum* at Munich, but has now been removed. I reproduce a photograph of this display (Figure

Fig. 11.15. Schreiber's instrument in Munich. (*Deutsches Museum*)

11.15) as a memorial to a tradition of instrument design that has vanished forever.[93]

---

[91] *Ibid.*, pp. 649–702.

[92] Schreiber, *Zeits. für Instrum.*, Vol. 6 (1881), p. 257.

[93] A long-term trend in instrument design, which still continues, is to make instruments more compact. The word "miniaturization" has been coined for this process.

d. Balance Barographs Using a Rolling Weight

One of the possible ways of bringing any balance to equilibrium is to traverse a small weight—in chemical practice called a *rider*—along the beam. The application of this simple principle to the recording barometer is especially associated with the name of A. Sprung (1848–1909) of Hamburg.

The first rough idea occurred to Sprung about 1877.[94] It was extremely crude, but embodied the fundamental idea of an electrically-operated clutch, keeping the balance beam oscillating slightly about its correct position, that distinguished all Sprung's barographs. In this the weight did not actually roll on the beam, but acted with a variable leverage on one end of it. It is possible that the idea of the rolling weight came from Paul Schreiber, who, after looking at Sprung's first design, wrote in 1879: "One could have a [*sc.* recording] barometer by hanging the tube on one arm of a two-armed straight lever and letting a running weight engage the other arm."[95] However this may be, Fuess was able to show a rolling-weight barograph of this sort, designed by Sprung, at the Berlin Industrial Exhibition the same year; and Sprung had the pleasure of describing it in his report on the recording instruments at the exhibition.[96]

This barograph (Figure 11.16) had a "steelyard" beam with the tube on

the short arm, the balance being maintained by a weight, driven by a screw, rolling on the top of the longer arm. This weight oscillated continuously about the position of balance. When the electrical contact was open, the weight was moved at a constant speed to the left by the clock, acting through friction wheels. Such a motion would soon tip the beam and close the contact; when this happened, an electromagnet reversed the direction of motion and the weight began to roll to the right, continuing until the contact was broken again. This method of operation was adopted by Sprung with the idea of minimizing errors due to the effects of surface tension in the tube.

As the pen was directly attached to the weight, its motion bore a linear relationship to the atmospheric pressure. Sprung gave an elaborate theory of the temperature error of this barograph, which turned out to be 0.0144 mm. of mercury per °C, varying slightly with the atmospheric pressure. He also suggested that the instrument could be used to make another record at a distance, by duplicating the clock, screw, and gearing, though with Sprung's friction wheels this might well be difficult.

In 1882 he reported on results with this barograph. The mean deviation from an ordinary barometer in seventy-seven comparisons was ±0.11 mm.[97]

The contact in Sprung's instrument was made between a platinum wire and a pool of mercury, and such an arrangement can give a good deal of trouble. In 1884 G. Rung of Copen-

[94] *Repertorium für exper. Physik.*, Vol. 14 (1878), pp. 46–51.
[95] *Ibid.*, Vol. 15 (1879), p. 227.
[96] A. Sprung, in L. Löwenherz, *Bericht über die wissenschaftliche Instrumente auf der Berliner Gewerbeausstellung im Jahre 1879* (Berlin, 1880), pp. 234–42.

[97] *Zeits. Oesterr. Ges. f. Meteorol.*, Vol. 17 (1882), pp. 44–48.

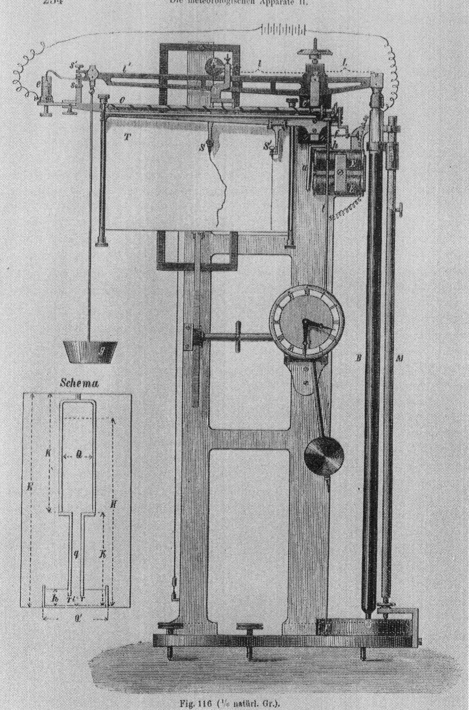

Fig. 116 (¹/₆ natürl. Gr.).

Fig. 11.16. Sprung's rolling-weight barograph, 1879.

hagen, who had one of these baro-graphs, found it preferable to use solid platinum contacts,[98] and in the follow-ing year[99] described a remarkably in-genious variation which dispensed with the use of electricity altogether (Fig-ure 11.17). In this a siphon tube is clamped to the shorter end of the beam, so that the operating force is derived from the displacement of mercury from one limb to the other as the pressure changes. The points *s,s* at the left-hand

end of the beam, which look like elec-trical contacts, are really only stops. A small movement of the end of the beam arrests or releases a fly governor, re-versing the motion of the screw through the action of a differential gearing built into the clock *U*. The clock *W* drives the chart.

[98] *Zeits. für Instrum.*, Vol. 4 (1884), pp. 318–19.
[99] *K. Danske Vidensk. Selskabs Skrifter, Nat. og Math. Afd.*, Vol. 3 (1885), pp. 347–68.

Fig. 11.17. Rung's mechanical rolling-weight barograph, 1885.

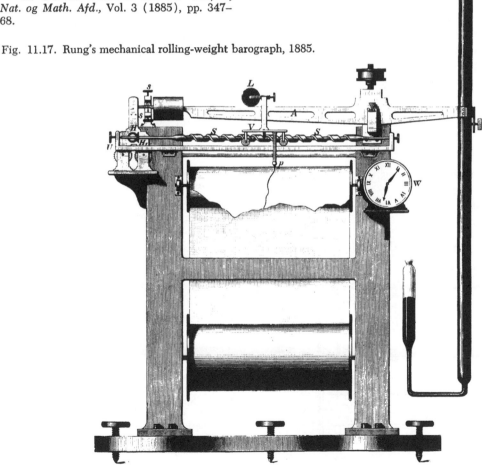

An important and valuable feature of this barograph is that, as Rung showed, it can be completely compensated for temperature changes at a mean value of the pressure by suitably proportioning the tube. He claimed, but not correctly, that his was the first barograph for which this is true.

In 1881 Sprung had also designed a thermograph on the rolling-weight principle, and in 1886 he described a combination thermograph and barograph (Figure 11.18) which not only gave a record of the pressure but also corrected the gas thermometer of the thermograph for the effect of pressure changes.[100] In view of what has gone before, we need only say that P is the barometer tube, $P'P_1'$ the manometer for the thermograph. The mechanical details were very beautifully worked out. In this paper Sprung showed theoretically that for the proper operation of such a barograph the outer diameter

of the part of the tube that dips into the cistern must equal the inner diameter of the upper part of the tube. But it was not until about 1892 that he hit upon a simple scheme for the automatic compensation for temperature of a weight barograph of this type by means of an enlargement in the upper part of the tube, with a definite excess volume (Figure 11.19).[101] The theory of this improvement was done more neatly by H. Koschmieder long afterward.[102]

There was apparently another mechanical arrangement in the 1890's, in which the rolling weight was moved not by a screw, but by a band under tension. This band was driven by two friction wheels controlled by electromagnets in such a way that the beam was always in oscillation. A third electromagnet, put into circuit each minute, dipped a piece of wood into the cistern, disturbing the level of the mercury. Tests at the *Physikalisch-Technische Reichsanstalt*[103] established the excellence of this barograph, but found nothing about the temperature compensation because of the uniform temperature in the room in which the instrument was installed.

Later the screw returned, driven in either direction by a reversing friction gearing, and this barograph seems to have had an exceedingly stable zero.[104] The departures of monthly means at

---

[100] A. Sprung, *Zeits. für Instrum.*, Vol. 6 (1886), pp. 189–98, 232–37.

Fig. 11.18. Sprung's thermobarograph, 1886.

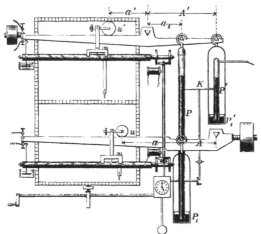

[101] *Deutsches meteorol. Jahrbuch* (1893), Heft 3, p. xi–xv.

[102] *Geophys. Mag.*, Tokyo, Vol. 9 (1935), pp. 23–28. See also W. E. K. Middleton & A. F. Spilhaus, *Meteorological Instruments* (3d ed.; Toronto, 1953), pp. 24–25.

[103] Karl Scheel, *Zeits. für Instrum.*, Vol. 15 (1895), pp. 133–46.

[104] *Zeits. für Instrum.*, Vol. 25 (1905), pp. 37–45, 73–82.

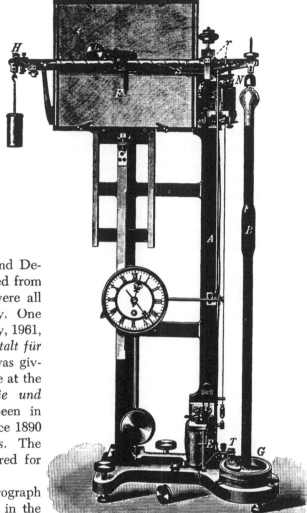

Fig. 11.19. Sprung's final barograph, 1905.

Potsdam between July, 1902, and December, 1904, from those derived from the official station barometer were all less than 0.011 mm. of mercury. One or two are still operating; in May, 1961, I saw the one at the *Zentralanstalt für Meteorologie* in Zurich and it was giving satisfactory service. The one at the *Zentralanstalt für Meteorologie und Geodynamik* in Vienna has been in almost continuous operation since 1890 and still gives excellent results. The instrument was still being offered for sale in 1932.[105]

A variation on Sprung's barograph was designed by C. F. Marvin in the United States, probably about 1890, and described by him in 1894.[106] Ratchet motors, rather than clockwork, were used to move the lead screw. The screw moved only when the pressure was changing.

[105] Fuess's catalog Bg26 (1932), pp. 9–11.
[106] C. F. Marvin, *Barometers and the Measurement of Atmospheric Pressure,* etc., U.S. Dept. Agric., Weather Bureau, Circular F, (Washington, 1894), pp. 29–32.

About 1937 it was suggested to the writer by the late Dr. J. Patterson that we should design a new mercury barograph for the Toronto meteorological observatory. A review of the literature led us to favor the balance barograph; and it was decided to see whether the application of more modern electrical techniques could result in an instrument even better than that of Sprung. The resulting barograph (Figure 11.20) was completed just before the outbreak of World War II.[107] The figure shows the instrument with the chart-mechanism swung out so that the beam can be seen. It is unnecessary to describe this barograph in detail; it differs from Sprung's last design in the use of an equal-arm balance (patterned after a chemical balance), and in the fact that the rolling weight is not in motion except when the pressure is changing. The pressure in millibars and tenths can be read at any time on a counter, and another printing counter prints the figures on the edge of the chart every hour. The form of the tube, however, was precisely that of Sprung.

I should like to see this barograph redesigned using present-day solid state electronic circuits, and possibly a photoelectric device to sense the motion of the beam.

e. Balance Barographs with the Cistern Hung on Springs

In the 1870's and 1880's Daniel Draper was equipping the meteorological observatory in Central Park, New York, with recording instruments of his

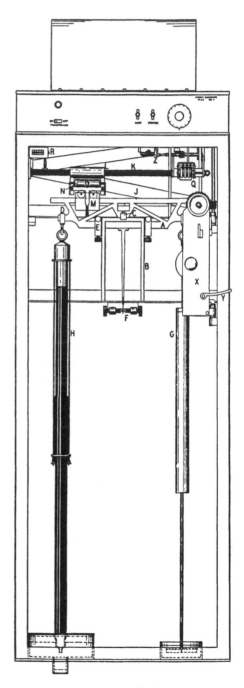

Fig. 11.20. The weight barograph of Patterson and Middleton, 1941.

---

[107] J. Patterson & W. E. K. Middleton, *Quart. J. Roy. Meteorol. Soc.* Vol. 67 (1941), pp. 19–31. See also Middleton & Spilhaus, *Meteorological Instruments*, pp. 25–27.

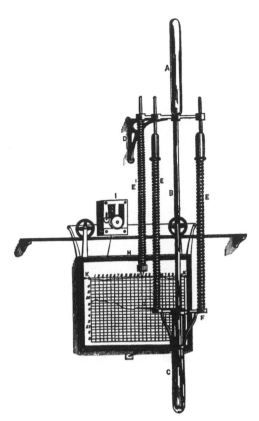

Figure. 11.21. Draper's barograph, 1880.

a mean pressure, and a pen attached to this made a zero line. The fact that the cistern was weighed compensated, to a first approximation, for the effect of temperature on the mercury, though not for that on the glass. This must have been the simplest of mercury barographs. Draper gave it a rugged test by exposing it and an ordinary Fortin-type barometer for a year in the open air (in New York!). In spite of the temperature range, which must have been about 90° F., the greatest difference in their readings was 0.04 inch of mercury.

An illustration in a recent paper by Multhauf[109] indicates that a similar barograph, but apparently without the third spring, was in use at the Lick Observatory in California.

Another barograph with the cistern hung on springs, but mechanically more complicated, was devised sixty years later by S. P. Fergusson.[110]

f. Barographs in which the Center of Gravity of the Tube Moves Horizontally

There is a class of balance barographs in which the displacement of the mercury with changing pressure causes the whole tube to turn about an axis. They are related to the barometers already described in Chapter 6, page 108. The earliest seems to have been built by one Father Baudou of Grenoble.[111] A large barometer tube AB of peculiar shape (Figure 11.22) is

own design. One of these, naturally, was a barograph (Figure 11.21).[108] This was a cistern barometer with the upper chamber ¾ inch in bore. The tube was fixed, and the cistern, in shape like a test tube, was supported by two long helical springs attached to a member carrying a pencil, which wrote on a flat chart moved sideways along a rail by clockwork. As a means of compensating for the effect of temperature on the springs, a third spring carried a weight equal to that of the cistern at

[108] These instruments are concisely described in *Engineering*, Vol. 40 (1885), pp. 535–36, whence Figure 11.21 is taken.

[109] Robert P. Multhauf, "The Introduction of Self-Registering Meteorological Instruments." *U.S. Nat. Museum*, Bull. 228 (Washington 1961), pp. 95–116, Fig. 11.
[110] *Bull. Amer. Meteorol. Soc.*, Vol. 20 (1939), pp. 135–41.
[111] Louis Cotte, *Mémoires sur la météorologie* (Paris, 1788), I, 567–68.

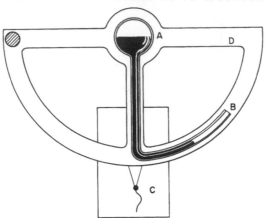

Fig. 11.22. Baudou's barograph, about 1788.

mounted on a cut-out semicircular plate *D*, which must have been nearly a meter in radius. All this is balanced about its center, and tilts as the pressure changes. On an extension of the semicircle is a pencil, which moves over a rectangular chart *C*, let down uniformly by a clock. "Ce barométrographe a été exécuté à Grenoble," writes Father Cotte, "mais je doute qu'il ait toute l'exactitude & la précision de celui de M. Changeux."[112] Nevertheless it was ingenious.

A century later a barograph which falls into this class[113] was installed at the observatory at Lausanne. It had a siphon barometer with a tube of complicated shape, balanced on knife-edges. It does not need the elaborate mathematical analysis given by its authors to convince anyone that its scale was far from linear. Yet it inspired an even more elaborate piece of algebra[114] which

investigated its temperature errors, and purported to show how an alcohol thermometer could be attached which, by shifting its center of gravity as the temperature changes, will compensate for them.

A patent on an instrument of this type was granted in 1923 to S. G. Starling.[115]

5.   PHOTOGRAPHIC BAROGRAPHS

The year 1839 is noteworthy in the history of technology, because in January of that year H. Fox Talbot (1800–1877) made public his process of "photogenic drawing" on paper, and Mandé Daguerre (1789–1851) gave the details of the method of making pictures on metal plates known ever since as Daguerrotype. It was Talbot's process[116] that was first applied to the recording of meteorological observations.

[112] *Ibid.*, p. 568.
[113] H. Dufour and H. Amstein, *Arch. Sci. phys. et nat. de Genève*, Vol. 7 (1882), pp. 19–52. Also in *Bull. Soc. Vaudoise des Sci. Nat.*, Vol. 17 (1882), pp. 549–91.
[114] A. A. Odin, *Bull. Soc. Vaudoise des Sci. Nat.*, Vol. 21 (1885), pp. 163–88.

[115] British Patent 225,055 (1923).
[116] *Phil. Mag.*, Vol. 14 (1839), pp. 196–211; *Proc. Roy. Soc.*, Vol. 4 (1839), pp. 120–21 (Jan. 31); pp. 124–26 (Feb. 21).

This application came with astonishing promptness. The Royal Cornwall Polytechnic Society had a very able and energetic secretary, T. B. Jordan. Within two or three weeks of Talbot's announcement he had clearly seen how the new technique could be used to make a record of the readings of the barometer, and within two months he had done it.[117] (See Figure 11.23.)

The principle of self registration is so obviously advantageous for all instruments used in meteorology [wrote Jordan] that I imagine it is only requisite to show its possibility in order to claim the attention of every admirer of the science. The plan* which I adopt for this purpose, is to furnish each instrument with one or more cylinders, containing scrolls of the photographic paper, prepared according to the directions of H. F. Talbot, Esq. These cylinders are made to revolve slowly by a very simple connection with a clock, so as to give the paper a progressive movement behind the index of the instrument, the place of which is registered by the representation of its own shadows. . . . Its application to the barometer is the most simple. . . . In fig. 1 . . . *ab* is the barometer tube; *C* the cylinder on which the prepared paper is fixed; *M* the height of the mercury in the tube. The scale of the barometer is perforated [*sic*] as usual to admit a portion of the

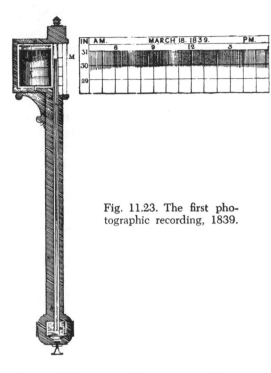

Fig. 11.23. The first photographic recording, 1839.

tube, and the prepared paper is made to revolve as close as possible to the glass, in order to obtain a well defined image.[118]

Hour lines are engraved on the paper, and so may the scale of inches be, or ". . . what is much better, may be printed by the light at the same time, from opaque lines on the tube, which would, of course, leave a light impression on the paper. . . ."[119] This seems an advanced idea to come so soon, and there can be no doubt that Jordan was a most ingenious man. At this time it seemed feasible to use the method only in daylight. It is interesting that the word "photographic" seems to have been current at this early date, at least in Penzance.

[117] T. B. Jordan, *Royal Cornwall Polytechnic Soc., Annual Report for 1838*, pp. 184–89. The date (1838) has caused some confusion; the Society's year seems to have extended into 1839.

* This plan was first named to a committe [*sic*] of the Society, held on the 18th of February; and at a subsequent meeting, on the 21st of March, a paper on the subject was read, and some of the photographic registers were exhibited (Jordan's footnote).

[118] *Ibid.*, p. 184.
[119] *Ibid.*, p. 185.

Later in 1839 Professor J. P. Nichol of Glasgow referred to Jordan's work and made a plea for photographic registration of the readings of meteorological instruments.[120]

It was inevitable that the Patent Office should become involved. James Readman was granted a patent[121] in 1842, of which the fourth claim is a method of recording the readings of a weight barometer (of no particular merit) which forms the subject of an earlier claim. A disc of "photogenic paper" is revolved behind an opaque plate with a radial slit in it. But it is still proposed to make records only in daylight.

The first successful use of artificial light for photographic recording seems to have been made by Mungo Ponton.[122] This was discontinuous, a fresh surface of sensitive paper being moved into position every half hour, and a gas flame being turned up for five minutes. It was done with a thermometer, but Ponton saw that it would be no more difficult with the barometer.

All these schemes had involved merely casting the shadow of the barometer column or other indicator on to the sensitive paper. Francis Ronalds (1788–1873) at Kew Observatory objected to this because of the lack of definition that resulted, and in cooperation with Henry Collen began experiments to improve such records. They used a compound lens system (an early photographic objective) to project an image of a blackened pith ball, floating in the open limb of a siphon barometer, on to a moving piece of calotype paper.[123] A slit between the light and the barometer tube limited the amount of light refracted round the column by the glass tube.[124] An electrometer and a thermometer were also made to record in a similar manner. Ronalds described the apparatus more fully two years later[125] and said that he was giving thought to the problem of compensating for temperature changes, which might be done by a bimetallic system. In the same year Charles Brooke described a scheme in which a float in the open limb of a very large siphon barometer was connected to a lever which moved a very light screen parallel to the axis of a drum covered with sensitive paper.[126] In effect, this was the same idea, except that the lever was mechanical instead of optical. A further description of this instrument, which was installed at Greenwich Observatory, appeared in the annual report of that institution.[127] By that time the source of light was artificial, and a cylindrical lens was used to cast an image of the lamp on to the moving screen. Light passing through a hole in a fixed screen produced a base line on the record.

At Kew, Ronalds continued to improve the instrument, and by 1851 was

---

[120] *The Scotsman*, July 10, 1839; reprinted in *Mechanics' Mag.*, Vol. 31 (1839), pp. 326–27.

[121] British Patent 9280 (1842).

[122] *Trans. Scottish Soc. of Arts*, Vol. 3 (1845), pp. 45–52.

[123] An improved photographic paper invented by H. Fox Talbot.

[124] Henry Collen, *Phil. Mag.*, 3rd ser., Vol. 28 (1846), pp. 73–75.

[125] *Phil. Trans.*, Vol. 137 (1847), pp. 111–17.

[126] *Ibid.*, pp. 59–72.

[127] Royal Observatory, Greenwich, *Magnetic and Meteorological Observations in the Year 1847* (London, 1849), pp. lxxxvi-lxxxvii.

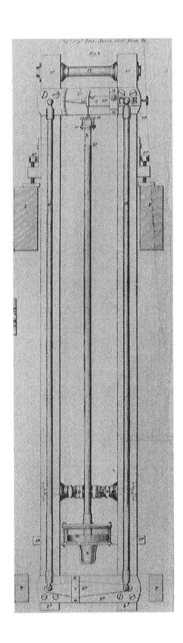

able to describe a very much more advanced design, in which the top of the barometer column was imaged on the sensitive surface by an excellent projection system.[128] In this "Photo-Barometrograph" light from an Argand lamp passed through two condensers which imaged the lamp on the back of a photographic objective. The tube of a large cistern barometer, and a vertical slit ½0 inch by 3 inches just behind it, together acted as a lantern slide, which was projected on to either a Daguerrotype plate or a piece of "Talbotype" paper under glass, behind a slit ⅒ inch wide. This sensitive surface was moved sideways by a clock. The photographic record thus looked very much like that obtained by Jordan. There were no time-marks, but a special plotting device was used to read off the ordinates of the curve at the desired intervals.

The compensation for temperature was done by raising and lowering the entire barometer by the elaborate system of levers shown in Figure 11.24. The expansion of two long zinc rods, arranged so as to be under tension, relative to the timber frame of the instrument lowers the barometer as the temperature rises. One arm of the upper lever is adjustable, and this makes possible a precise compensation at one pressure.

This excellent instrument worked very well indeed. In 1862 Admiral Fitzroy gave his opinion that it was much superior to the barograph devised by Admiral Milne,[129] and while this

Fig. 11.24. Temperature compensation of the Kew barograph, 1851. *(Courtesy of the Trustees of the British Museum)*

[128] Francis Ronalds, *B.A.A.S. Report* (1851), pp. 346–50.
[129] *B.A.A.S. Report* (1862), p. xxxv. For Milne's barograph, see p. 293 above.

was not a very great compliment, it was certainly deserved. In 1863 a very similar instrument, even to the means of compensating for temperature, was made by the Paris instrument maker Jules Salleron for the Lisbon Observatory; in this the same lamp also served as a source of light for a recording psychrometer.[130]

While the idea of moving the entire barometer was fairly satisfactory, it was expensive and complicated and occasionally stuck. Some time about 1867 R. Beckley, a mechanical assistant at Kew who had a hand in a great many instruments, devised[131] two improvements which were subsequently reported on by the Meteorological Committee of the Royal Society.[132] Instead of moving the whole barometer, the expansion of a zinc rod acted on one end of a long magnifying lever. At the other end of this, just in front of the sensitive surface, was a small screen, the shadow of which moved by an amount just equal to the required compensation at a mean pressure, producing an edge which could be used as a baseline in reading off the record. He also arranged a bar which, once every two hours, was moved by the clock so as to interrupt the light and make a time mark. The flat carrier for the paper had by this time been replaced by a drum, and the entire instrument had assumed the form shown in Figure 11.25, taken from the Committee's report. There is one at the Science Museum.[133]

By 1885, when R. H. Scott wrote a history of the famous observatory,[134] nineteen other observatories throughout the world had been supplied with "Kew model" barographs. The one at Toronto, to my personal knowledge, was in daily use until about 1940. The barogram showing its response to the waves of pressure generated by the Krakatoa eruption of 1883 still exists.

The photographic barograph does not seem to have been favorably received on the continent of Europe, though Paolo Volpicelli made one in 1869 at the University of Rome.[135] He began his paper with a long argument in favor of photographic recording, possibly directed against Secchi, who at any rate took it in that way, making caustic "observations,"[136] to which Volpicelli replied, not much less caustically. His barograph was rather elementary, the mercury column casting a shadow on a drum just behind the barometer tube.

6.   BAROGRAPHS WITH SERVO-SYSTEMS

We have already met one or two balance-barographs in which the making and breaking of a small electric current is used to control a relatively large amount of energy for the operation of the recording system. Apart

---

[130] Lisbon, Ann. dell'Osserv. do Infante D. Luiz, Vol. 2 (1863), no. 1, pp. v-vi. But the figure on p. 14 of Salleron's catalog (1858) does not show the temperature compensation.

[131] B.A.A.S. Report (1867), p. lviii.

[132] Meteorol. Comm. Royal Soc., Report for 1867, pp. 40–42.

[133] Inventory no. 1926–937.

[134] Proc. Roy. Soc., Vol. 39 (1885), pp. 37–86.

[135] Atti nuovi Lincei, Vol. 23 (1869), pp. 71–75, 118–22.

[136] Ibid., pp. 76–77.

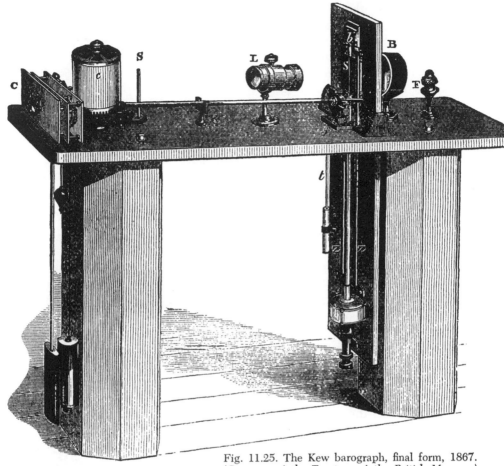

Fig. 11.25. The Kew barograph, final form, 1867.
(*Courtesy of the Trustees of the British Museum*)

from balance-barographs, such servo-systems were used in a number of instruments designed between about 1850 and 1890, almost all based on the siphon barometer with a float in its open limb. Apart from the rather mysterious instrument of Charles Wheatstone, which was part of a meteorograph,[137] this development began at a time when the

discovery of electromagnetism had led to such a flood of "applications of electricity" that it took Du Moncel five volumes, totalling more than 2,000 pages,[138] merely to review them.

The simplest of these barometers was perhaps that of Liais.[139] This was an ordinary siphon barometer fitted with a mechanism which kept one platinum

[137] *B.A.A.S. Report* for 1843, p. xl. See Robert P. Multhauf, *U.S. Nat. Mus.*, Bulletin 228 (Washington, 1961), pp. 105–6, for a discussion of this meteorograph.

[138] Th. Du moncel, *Exposé des applications de l'électricité* (2d. ed.; 5 vols., Paris, 1856–1862).

[139] *Ibid.*, II, 401–2.

point just below the surface of the mercury in the open limb, another just above it. The two points, one slightly longer than the other, were mounted at the end of the same vertical rod, to which the recording pencil was also attached. Each point was in circuit with an electromagnetic clutch, a battery, and the mercury of the barometer; the clutches operated two clockwork drives, which kept the points oscillating slightly.

This had the disadvantage that the mercury was practically undisturbed, and the surface in the open limb must soon have become very bad, with the assistance of the sparking that would take place.

A somewhat better idea, though retaining the mercury contact in air, was that of Regnard,[140] the principle of which is shown diagrammatically in Figure 11.26. The siphon barometer XZ has an additional tube M in which a uniform cylinder BB' can be moved up and down by the screw DD'. A platinum contact f controls a single-pole, double-throw relay, which in turn controls an electromagnetic mechanism which turns the nut R in such a direction as to keep the contact f just touching the mercury in Z; the recording mechanism is geared to the nut R. A rather similar design was made by Montigny.[141]

The "automatic registering and printing barometer" of G. W. Hough,[142]

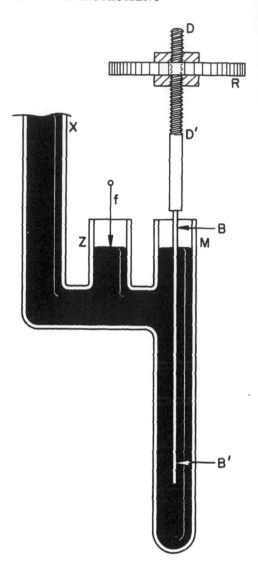

Fig. 11.26. Regnard's barograph, about 1860.

[140] *Ibid.*, IV, 416–18.

[141] *Ibid.*, II, 420.

[142] *Amer. J. Sci.*, 2d ser., Vol. 41 (1866), pp. 43–58. *Annals of the Dudley Observatory*, Albany, N.Y., Vol. 1 (1866), pp. 67–93. Radau (*Moniteur Scientifique*, Vol. 9 [1867], p. 715) was in error in his claim that Hough's barograph was similar to those of Regnard and Montigny.

installed at the Dudley Observatory, Albany, N.Y., removed the electrical contacts from the mercury entirely, though at the expense of some complication. It also provided a printed record of the reading at hourly intervals, a valuable feature which removes all concern about the expansion or contraction of the paper chart. Hough's barograph was based on a siphon barometer with upper and lower chambers about one inch in diameter. An ivory float in the lower chamber carried a wire, at the top of which was a small horizontal platinum disk. Two platinum contacts, one above and one below the disk, were moved as a unit by a screw which was traversed in the appropriate direction to break whichever contact makes. This was done by two clockwork drives, each consisting of a one-tooth wheel released by its appropriate electromagnet, and engaging a 40-tooth wheel attached to the nut which moved the screw. The entire instrument is shown in Figure 11.27. Much of the apparent complication is in the printing mechanism.[143] A later version had two drums with different rates of rotation.[144] This barograph was later modified by H. L. Foreman, at one time Hough's assistant, by having the float supported by a fine wire t (Figure 11.28) from one end of a short lever ℓ, poised on knife-edges at r.[145]

The longer end of this lever was tipped with platinum and placed between two platinum contacts, which were moved by clockwork.

A series of barographs with completely mechanical servo-systems was devised by the Paris constructors Louis and Antonin Redier between 1875 and 1886. The first of these[146] was distinguished by having the entire siphon barometer tube moved up and down by clockwork. It had two clockwork motors; one was constantly in motion and could raise the tube at a constant rate, faster than the mercury in the open limb could ever fall. The other motor, which had no escapement but only a fly governor, started when released by the float and caused the tube to descend twice as fast as the other could raise it.[147] Thus the tube would ascend when this second clock was stopped and descend when it was going. The motion of the tube, magnified, was communicated to the recording pencil. Marie-Davy shows that if the entire tube was made of the same internal diameter, the temperature correction would be very small.[148]

Later the Rediers abandoned the idea of moving the entire tube, but retained the purely mechanical system of recording. In their final barograph[149] (Figure 11.29) the tube is fixed, and a differential clockwork mechanism is used

[143] There are three of these instruments, none quite complete, at the Smithsonian Institution, Washington (Cat. nos. 317,417; 318,283; and 318,284).

[144] Annals of the Dudley Observatory, Vol. 2 (1871), p. xxiv.

[145] C. F. Marvin, Barometers and the Measurement of Atmospheric Pressure, etc., United States Dept. of Agric., Weather Bureau, Circular F (Washington, 1894), pp. 28–32.

[146] Louis Redier, Quart. J. Meteorol. Soc., Vol. 2 (1875), pp. 412–13. Perhaps better described by Marie-Davy, Annuaire de l'Observatoire de Mont-Souris pour l'an 1878, pp. 258–64.

[147] Cf. Rung's barograph, p. 313 above.

[148] Marie-Davy, Annuaire, pp. 261–63.

[149] Antonin Redier, Nouveau baromètre enregistreur à mercure. Pamphlet, 8 pp. (Paris, 1886) (BN 8° R. Pièce. 3377).

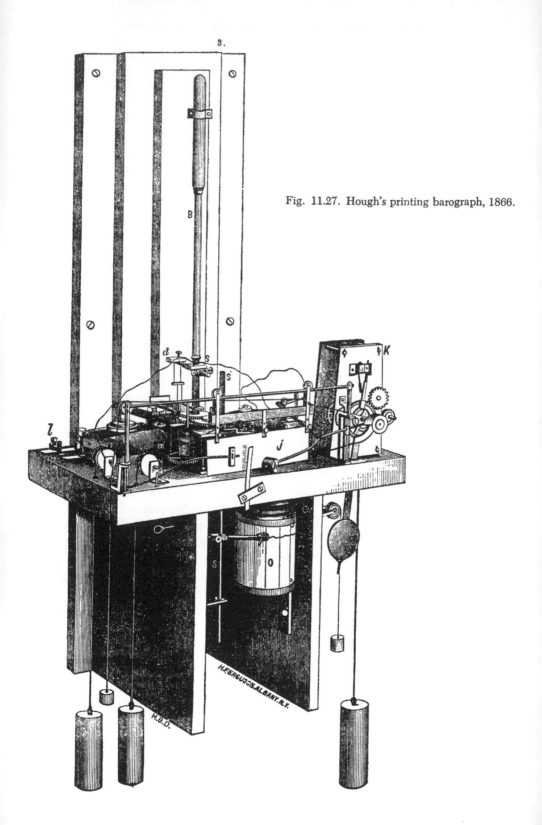

Fig. 11.27. Hough's printing barograph, 1866.

Fig. 11.28. Foreman's modification of Hough's barograph.

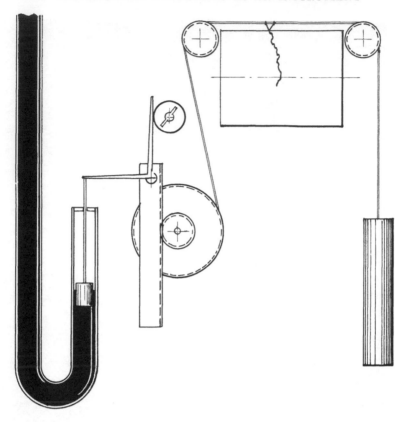

Fig. 11.29. Rediers' last barograph, 1886.

to keep a lever arm just in contact with the stem of the float. A light bell crank has its axis on a sliding piece of brass, which moves vertically in ways and carries a rack. The pinion that moves this rack is on the shaft of a pulley. A cord, wound round this pulley, passes up over two idlers and ends in a weight. The recording pen is moved by this cord, and the pulley is connected to a train of gears and a fly governor, the rotation of which is interrupted when the horizontal arm of the bell crank comes in contact with the float in the barometer. Through a differential gearing, the pulley is also connected to

an ordinary clockwork with an escapement. This latter clock moves the brass slide upward at a constant rate; the former, when it is free, moves the slide downward somewhat faster. The result is that the slide makes small oscillations around the point of contact between the lever and the float. There is one of these barographs, not operating but in beautiful mechanical condition, at the *Zentralanstalt für Meteorologie und Geodynamik* in Vienna.

As the fundamental difficulty with the siphon barometer is the sticking of the mercury in the open limb, it is doubtful whether these ingenious but costly devices really had any advantages over the much simpler constructions described in section 3 of this chapter.

A barograph in which this difficulty is avoided was designed by F. C. G. Müller of Brandenburg in 1874, using the barometer already referred to in Chapter 9,[150] with the difference that there are two platinum points sealed through the top of the vacuum chamber, and differing in length by 0.05 mm. or less.

Circuits connected to the two platinum points control, through relays, a reversing electric motor, consisting of two electromagnets with interrupters and ratchet-and-pawl mechanisms. The card attached to the instrument (Figure 11.30), now in the *Deutsches Museum* at Munich,[151] describes the arrangement: if the mercury does not touch either point, the motor raises the cistern until the mercury touches the longer point; if on the other hand the mercury stands so high that it makes contact with both points, the motor lowers the cistern until the contact with the upper point is interrupted. Figure 11.30 indicates well enough how the motion of the cistern is transmitted to the pen, and the arrangement of the chart. It also shows the two very well-made relays.

The fact that the tube goes right through the cistern simply means that the scale is not contracted. If it did not, the instrument would work just as well, but with a contracted scale. Nothing is said about compensation for

temperature, and it does not seem that this can have been achieved.

Two further applications of modern servo-techniques to the mercury barometer will be mentioned in Chapter 12.

7.   TELEMETERING MERCURY BAROGRAPHS

The desire to be able to transmit the indications of an instrument to a distance was a natural result of the invention of the telegraph, and it is interesting to consider that in little more than a century since a way of doing this was first discovered we have arrived at such almost incomprehensibly complex telemeters as are essential for the success of our space travelers.

In 1843 Professor Charles Wheatstone (1802–1875), one of the inventors of the electric telegraph, described a meteorograph which transmitted to a distance the readings of a wet-and-dry-bulb thermometer and a mercury barometer of the siphon type. It printed the reading of each instrument in figures every half-hour on a paper tape at the receiving station.[152] In the open limb of the barometer, a platinum wire was raised at a uniform rate for 3 inches in 5 minutes by a clock, and then lowered again in one minute. A pair of printing wheels were attached to the train of gears that moved the wire. When the wire emerged from the mercury, a circuit was broken, causing a hammer to fall and print the readings through carbon paper. There was a special device to take care of the occasions when the current is interrupted

[150] *Annalen der Physik,* Vol. 4 (1878), pp. 286–94. See p. 227 above, and Figure 9.20.

[151] Inventory no. 27712.

[152] *B.A.A.S. Report* for 1843, pp. xl–xlii.

Fig. 11.30. The Müller barograph at Munich. *(Deutsches Museum)*

just as a type-wheel was moving. This instrument was made for Kew Observatory, whose management had just been taken over by the British Association.

In 1867 he described another telemeter, applicable to any meteorological instrument with a rotating pointer—for instance a wheel barometer. A magneto at the station where the observations are required causes an electromagnet and ratchet to revolve a contact at the barometer (say) from a zero position, and another similar device to revolve an indicating pointer in the presence of the observer, in synchronism with the contact. When the contact arm touches the pointer, the process is stopped by a relay.[153] There is no record of this having been actually applied to a barometer in this form. But F. van Rysselberghe (or Rijselberghe) of Brussels described the application of a rather similar system,[154] in which a rod was lowered into the open limb of a siphon barometer at a constant speed, and the resulting closure of a circuit stopped a hand or a pen at the other end of a telegraph line. The circuit was broken elsewhere before the rod was withdrawn, to prevent sparking at the surface of the mercury, which was an excellent idea. This "time-cycle" was intended by Van Rysselberghe for application to what we should now call an automatic weather station.

A decade later a very similar telemeter barograph received a medal from the Royal Scottish Society of Arts;[155] it was much like that of van Rysselberghe, except that the wire was withdrawn from the open limb of the siphon barometer in very small steps by an electromagnet in a self-interrupting circuit, operating a ratchet. The number of steps until the circuit is broken by the emergence of the wire is registered by a printer at the other end of the line. When the dipping wire is at the top of its travel, it is brought back by gravity. It is made clear that the instrument was intended for a mountain station.

Inspired by Van Rysselberghe's memoir and by another by E. H. von Baumhauer,[156] the instrument maker Olland of Utrecht produced an elaborate telemeteorograph and sent it to the Philadelphia Exhibition of 1876, where it obtained a medal. This instrument introduced the telemetering device universally known as the "Olland cycle." The principle of this is as follows: A contact is caused to move in a circular path at as uniform an angular velocity as possible and to make contact with a number of fixed contacts as it moves. The completion of a circuit each time one of these contacts is touched establishes at a distant station the momentary angular position of the moving contact. Somewhere on its way it touches an additional contact, also a movable one; this is attached to the pointer of the instrument whose reading is to be transmitted to a distance. By interpolation between the fixed signals the angular position of the pointer is obtained.

---

[153] *B.A.A.S. Report* for 1867, pp. 11–13. *Scientific Papers of Sir Charles Wheatstone* (London, 1879), pp. 208–10.

[154] *Bull. Acad. roy. de Belgique*, 2ᵉ sér., Vol. 36 (1873), pp. 346–74; *Quart. J. Roy. Meteorol. Soc.*, Vol. 2 (1875), pp. 367–74. A barograph on essentially similar lines had been described by J. Morin in 1864 (*Compt. Rend.*, Vol. 59 [1864], pp. 787–88), but it had not the telemetering arrangements.

[155] George R. Primrose, *Primrose's Electric Meteorological Scale Reader Described* (Edinburgh, 1886), leaflet. (BM 8756.ccc.26(3).)

[156] *Archives néerlandaises*, Vol. 9 (1874), pp. 230–58.

Of course there may be several instruments, each with its pointer.

The barometer in Olland's meteorograph (Figure 11.31) was a balance barometer with the tube fixed and the cistern hung from knife-edges at one end of a balance beam, provided with adjusting weights to ensure a center of gravity low enough to make the system stable. A toothed sector, attached to the beam, meshed with a pinion and thus rotated a pointer.[157] Such a barometer would probably be much more durable than that used by Van Rysselberghe.

From about 1880 onward, a large number of patents were granted on more or less practical means of recording at a distance the readings of instruments with a moving pointer, but none of them seem to have been applied to the mercury barograph.

On the other hand, there were a few patents granted in the 1880's for modifications to the barometer tube itself with telemetry in view. Of these the most likely is that given to J. Enright,[158] who suggested sealing platinum wires into the tube at regular intervals, and connecting them to the junctions between a number of resistances in series. With an indicating or recording galvanometer in the circuit, the readings of the barometer could be displayed at a distance.

In the previous year Daniel William Kemp of Leith had described to the Royal Scottish Society of Arts—which seemed to get more than its share of fanciful barometers—an arrangement consisting of twenty-five barometer tubes, each with a platinum contact in the vacuum chamber, varying systematically in length by tenths of an inch. A 26-wire cable would connect this with an observing station.[159] My only reason for mentioning this *tour de force* is to be able to put on record the following paragraph, which excites a certain sympathy:

Here the recording instruments would be placed. These may be as simple or as complicated as desirable, merely indicating the height of the mercurial column when wanted, or doing so continuously. I confess to not favouring the latter plan, as I shrink from the responsibility of advocating any method which would add so materially to the already countless millions of observations which meteorologists have accumulated from the sources now at their command.[160]

8. BAROGRAPHS USING FLOTATION

The first vague idea for a mercury barograph on the principle of flotation was that of the Rev. Arthur M'Gwire,[161] in which the tube, sustained by a large piece of wood, was to have a pencil attached to it. It would certainly never work. A patent along somewhat similar lines was granted to Thomas T. MacNeill seventy years later.[162] Meanwhile John Stevelly, a professor of natural philosophy at Belfast, had suggested a design in which the tube of a barometer was fixed, and the cistern suspended from a hydrometer contained in a fixed vessel of some appropriate liquid.[163]

---

[157] Snellen, *ibid.*, Vol. 14 (1879), pp. 180–208.

[158] British Patent 4163 (1883).

[159] *Trans. Roy. Scottish Soc. Arts*, Vol. 10 (1882), pp. 537–46.

[160] *Ibid.*, p. 542.

[161] *Trans. Roy. Irish Acad.*, Vol. 4 (1791), pp. 141–43.

[162] British Patent 1733 (1861).

[163] *Trans. Roy. Irish Acad.*, Vol. 17 (1837), pp. 471–80.

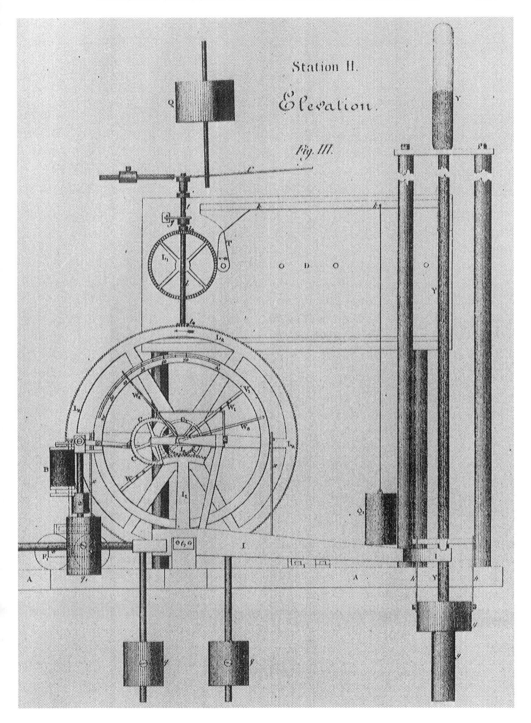

Fig. 11.31. Olland's telebarograph, 1876.

Stevelly showed the conditions for stability, and how to compute the magnification and the corrections for temperature. The instrument would undoubtedly have worked, but it was not the sort of thing that anyone would build.

Guiseppe Agolini of Florence did, on the other hand, build[164] and patent[165] a barograph of this type which, if not world-shaking, merits a figure (11.32) and a brief description because of its ingenuity and rather elegant design. It differs from all others in that the float is in the upper chamber, and connected directly to the pen arm by a completely simple and rigid system of steel parts. Another float in the large cistern provides a baseline. The weights and buoyancies of the moving parts are so adjusted that the upper float follows the motion of the mercury mainly on account of surface tension. The chamber has to be exhausted by a vacuum pump. Unfortunately this instrument would have the temperature errors of any ordinary cistern barometer.[166]

Another idea is to float the cistern in a larger bath of mercury, and measure its vertical motion as the pressure changes. Such a scheme was patented by James Readman in 1842;[167] and in 1872 H. C. Russell of Sydney demonstrated a rather similar barograph to the Meteorological Society in London,[168]

which, to judge by the discussion, does not seem to have received it with any enthusiasm. The movements of the cistern were communicated to a recording pencil by rather complicated electrical means, not at all clear in the brief description.

9.   MISCELLANEOUS MERCURY BAROGRAPHS

There are a few recording barometers which do not seem to fit into any scheme of classification. Among these are sev-

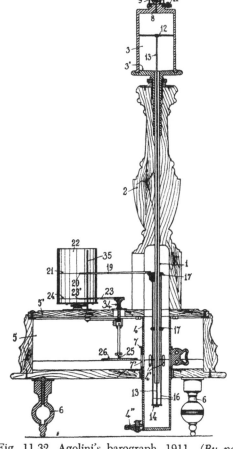

Fig. 11.32. Agolini's barograph, 1911. *(By permission of the Controller, H. M. Stationery Office)*

---

[164] G. Alfani, *L'Osservatorio Ximeniano ed il suo materiale scientifico. VI. Il barometrografo Agolini* (Firenze, 1913).

[165] D.R.P. 249149 (1911); British Patent 20,800 (1911); U.S. Patent 1,068,726 (1913).

[166] In the interests of clearness a simplified drawing has been reproduced. The actual instrument had a number of small mechanical refinements.

[167] British Patent 9280 (1842).

[168] *Quart. J. Meteorol. Soc.*, Vol. 1 (1872), p. 122 (discussion, p. 124).

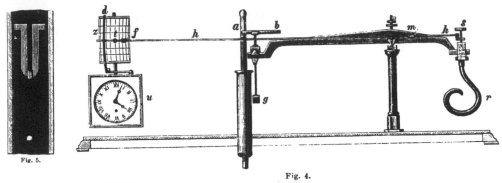

Fig. 5. Fig. 4.

Fig. 11.33. Fuess's magnetic barograph, 1883.

Fig. 6.

eral schemes for making hourly readings by setting up a number of barometer tubes and closing one off at each
hour.[169],[170],[171] The simplest of these is
the one due to M'Farlane, which has
twelve tubes, and an ingenious and
simple system of closing their ends
under the mercury. This is done by a
simple disk, covered with leather, which
is pressed against the flat end of the
tube. It is correctly noted that the only
observation of temperature required is
one taken when all the tubes are read.

An instrument *sui generis* was the
complicated barograph invented by A.
G. Theorell of Upsala. This was first

described in 1868[172] as part of a meteorograph, quite frankly on the principle of Wheatstone,[173] the time-scale
system being used not for telemetering,
but merely to furnish a record on the
spot. It was probably chosen because
of the possibility of recording temperatures in the same way. Three years
later he fitted the instrument with a
printing mechanism which recorded the
pressure hourly in figures.[174]

A meteorograph on exactly similar
principles, but less neatly designed, was
constructed just a little later by Hough,
for the newly established meteorological
office of the Signal Corps, United States
Army.[175] Like that of Theorell it recorded wet- and dry-bulb temperatures
as well as barometric pressure.

Finally there is the ingenious barograph devised by R. Fuess, which had
a magnet, enclosed in a rubber cylinder,
floating in the vacuum space of a cistern barometer.[176] To keep the magnet

[169] H. H. Blackadder, *Trans. Roy. Soc.
Edin.*, Vol. 10 (1825), pp. 337–47.
[170] J. von Lamont, *Beschreibung der an
der Münchener Sternwarte zu den Beobachtungen verwendeten neuen Instrumente und
Apparate* (Munich, 1851), pp. 7–9.
[171] P. M'Farlane, *Quart. J. Meteorol. &
Phys. Sci.*, Vol. 1 (1842), pp. 286–88.

[172] *Nova Acta Reg. Soc. Sci. Upsala*, Ser.
III, Vol. 7 (1868), No. 2.
[173] See p. 329 above.
[174] *Kongl. Svenska Vetenskap-Akademiens
Handl.*, Vol. 10 (1871), pp. 3–10.
[175] *Annals of the Dudley Observatory, Albany, N.Y.*, Vol. 2 (1871), pp. xxxvi–xlii.
[176] *Zeits für Instrum.*, Vol. 3 (1883), pp.
194–97.

horizontal the assembly was weighted with a small platinum ball at the end of a short platinum wire. A larger U-shaped magnet *b* (see Figure 11.33) was supported on the left-hand knife-edge of an unequal-arm balance of 2 to 1 ratio; on the right-hand knife-edge was hung a curved tube *r* filled with mercury, and about half as long as the barometric column. A second, lighter beam *hh* was supported on knife-edges collinear with the central knife-edges of the main beam. The right end of *hh* was fitted with a hardened screw *s* which rested on the hardened point *q*, part of a float which swam on the mercury in the curved tube *r*. With a rise of temperature *q* was raised, and hence the small beam, which carried the pencil *t* at its left end, failed to follow the large beam by just the right amount to compensate for the expansion of the mercury in the barometer tube. The screw *s* acted as a fine adjustment for the pen.

Fuess made tests to find out whether there was friction between the magnetic float and the tube. If the space around the float was narrow, the mercury always lubricated the float and it was frictionless. The narrow space (¼ mm.) also made surface tension unimportant, and the outer magnet was found to follow the inner one to within about 0.04 mm. The temperature compensation was of course exact only for one value of pressure.

10. MERCURY MICROBAROGRAPHS

It is a matter of definition whether a recording barometer with a given magnification should be called a microbarograph or just a barograph. Let us put the dividing line at a magnification of 10. On this definition most microbarographs depend either on light liquids or on the elastic properties of solids; these will be dealt with elsewhere. Here we shall merely mention two instruments which are really sophisticated descendants of Hooke's "double barometer" described in Chapter 6. In the instrument of Streit,[177] shown in Figure 11.34, a bulb in the lower limb of a siphon barometer has toluol above the mercury and is connected by a narrow tube as shown to a second bulb of equal diameter containing some more toluol, the surfaces at *B* and *E* being at about the same height. There is an air bubble *A* in the horizontal part of the thin tube. A magnification of 10 is practical, and the instrument is well suited to photographic registration.

Fig. 11.34. Streit's microbarograph, 1959.

[177] W. Streit, *Mitt. Naturf. Ges. Bern*, n.F., Vol. 17 (1959), pp. 19–25.

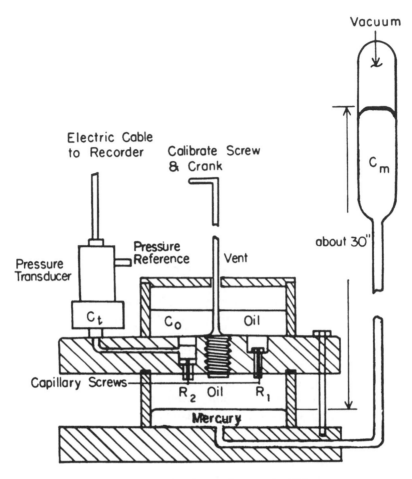

Fig. 11.35. Van Dorn's microbarograph, 1960.

A very advanced microbarograph has recently been described by W. G. Van Dorn.[178] This instrument, shown in Figure 11.35, is a siphon barometer in which the mercury in the lower chamber is covered by silicone oil in two chambers connected only by the capillary tube R_1, the upper one being open to the atmosphere. The changes in pressure in the lower chamber are transmitted through a second capillary R_2 to an electrical transducer. The dimensions of the capillaries are so chosen in relation to the volumes of the various parts that atmospheric waves are fully recorded over a desired range of periods, in this case about 30 to 3,000 seconds.

[178] *J. Geophys. Res.*, Vol. 65 (1960), pp. 3693–98.

The Mercury Barometer
in North America

A NUMBER of the barometers and barographs mentioned up to this point have had their origin in Canada or the United States; so that it might be questioned whether this chapter is needed at all. It seems desirable, however, to review the early history of the instrument and note some North American constructions in later times which seem to have been specially conceived for use under North American conditions. We must keep in mind the purpose of this book, which is to trace the development of the barometer as an instrument, not to record the growth to meteorological observations made with it, or of the explanation of barometric changes.

Within this limit the barometer has almost no history in North America in the seventeenth and eighteenth centuries. It is necessary, however, to document the almost complete dependence of America on European instrument makers throughout this period as far as the barometer is concerned.[1] "Yankee ingenuity" applied itself to other and quite different problems during this time.

The first record of a barometer in North America seems to date from the year 1727, when Thomas Hollis, a London merchant who was the greatest of the early benefactors of Harvard, gave the College a remarkably complete set of the physical apparatus of the day, as well as endowing Professorships of Divinity and of "Mathematics and Natural Philosophy," the latter of which, Cohen notes,[2] was the first endowed scientific professorship in the New World. As to the apparatus, it was an excellent set, and it came from England. We know exactly what was in it, for a contemporary inventory is preserved at Harvard.[3] It is headed "A Catalogue of the Mathematical & Philosophical Instruments belonging to ye Apparatus given to Harvard College, by mr Thomas Hollis of London, Merchant, with price Sterling." The 55 items are classified under "mechanicks,"

---

[1] For a general view of American science in Colonial times and somewhat later, see Brooke Hindle, *The Pursuit of Science in Revolutionary America, 1735–1789* (Chapel Hill, N.C., 1956).

[2] I. Bernard Cohen, *Some Early Tools of American Science*, etc. (Cambridge, Mass., 1950), p. 3.

[3] Harvard Archives, *College Book No. 6* (ms.), pp. 20–22.

"opticks," "hydrostaticks," "pneumaticks," and "miscellanies." Under "pneumaticks" we read

1. Two setts of Tubes for Torricellian experiments   01-05-00 [i.e., £1:5:0]
2. A Frame for supporting them 00-07-06
3. Apparatus for Monsr Auzouts Experiment   01-05-00
4. A large Double Air Pump with it's Apparatus   26-05-00

and so on, including

12. A portable Baromator [sic][4]

This inventory is followed by the following receipt in the hand of Isaac Greenwood, the first Hollis Professor of Mathematics and Natural Philosophy:

Cambridge, September 6, 1731.
  The Particulars of y: foregoing Catalogue, the Generous Benefaction of Mr Hollis to Harvard College I acknowledge to be now in my sole Custody at M^r. Hollis's Chambers, for y^e. use of such as are his Students, and Subscribers to the Hollitian Lectures.
ISAAC GREENWOOD[5]

The portable "Baromator" may well have been the first such instrument to be sent to North America. The inclusion of the apparatus for Auzout's experiment[6] is interesting; this, and indeed the choice of instruments throughout, was probably based on the prospectus for the course of lectures given in 1712 by William Whiston and Francis Hauksbee, already referred to in Chapter 4.[7] For there is a second inventory dated April 19, 1738,[8] in which the instruments are the same, but references are given to the prospectus of

1712.[9] However, items 1 and 2 under "pneumaticks" now become

1. Two Setts of Tubes for Torricellian & other Experim^{ts} viz. thirty one in number & most of them whole.
2. The frame for supporting them. Unknown. [make?]

Item 12 has finally been spelled in the conventional manner.

In the "Miscelanies," besides 12 pounds of mercury, we find "Twelve glass Tubes of different bores taken notice of in No. 1. Pneumaticks."[10] Glass tubes were a most valuable commodity in America áll through the seventeenth and eighteenth centuries, and indeed well into the nineteenth. In 1779 the fate of a small glass tube was decided by a resolution of the President and Fellows of Harvard College. On July 21 of that year it was voted

That Mess^s. Gannett [one of the Fellows] and [Professor] Winthrop be authorized to exchange a small tube belonging to the Apparatus, from 3 to 4 feet in length, with Mr. John Prince, for a barometrical tube & frame offered to this body for that Purpose.[11]

I have been unable to determine the date at which glass tubing suitable for scientific purposes was first made in North America, but it was probably after 1850. According to Davis,[12] very few glassworks were successful in the eighteenth century.

[4] *Ibid.*, p. 21.
[5] *Ibid.*, p. 22.
[6] See p. 49 above.
[7] P. 81 above.
[8] *College Book No. 6*, pp. 35–38.

[9] These have been collated by Cohen, *Some Early Tools*, pp. 133 ff.
[10] *College Book No. 6*, p. 37. Cohen, p. 142.
[11] Harvard Archives, *Corporation Records*, Vol. 3P., p. 46.
[12] Pearce Davis, *The Development of the American Glass Industry* (Cambridge, Mass., 1949).

To return to the Hollis instruments, they were all consumed in the disastrous fire of 1764. Meanwhile the meteorological instruments had been used by John Winthrop (1714–1779), then Hollis Professor of Mathematics and Natural Philosophy, for an almost continuous series of observations dating from 1742, which were resumed after the fire and continued until the year of his death.[13] At the end of the diary for 1743 there are two unnumbered leaves which show that he knew at least some of the difficulties:

Y 2d column contains y hight of y barometer in english inches & decimals [of an inch].[14] Agreeable to Dr. *Jurin's* admonition in Philosoph. Transact. N°. 379,[15] I made use of a common or open barometer
very
wch I filld & inverted [as] carefully [as
            clear it of   as
possible] so as to [exclude all] air [from effectly as I cd
y tube]   . Y diameter of y bore of
        is 27⁄100
my tube [was] ⅓ of an inch, & y diam-
            to   is
eter of y cistern in   wch it [was] im-
        is
mersd [was] 3 inches; so yt y variatⁿ of
                        is
y hight of the mercury is y cistern [was]
    not y 1⁄100 th   is
but] y [1⁄100] th part of wt it [was] in y tube. [at y same time].

This passage is of interest because it is not credible that a complete barometer with a three-inch cistern could

have been transported intact from England at that time, and because by his own account Winthrop at least filled the tube himself. The tube, incidentally, was nearly 7 mm. in bore, a very respectable diameter at that period. Unfortunately we have no description of the scale or of the general appearance of the instrument; but it is at least possible that Winthrop's barometer may have been the first such instrument made in North America.

Meanwhile the teaching of "Natural Philosophy" was beginning at New Haven. In 1734 Yale College obtained some apparatus, including a barometer, from "Mr. Scarlett" of London.[16] These instruments were probably presented by some well-wisher, for in a list of benefactions for the year 1734 we read "A reflecting telescope, a microscope, barometer, and sundry other mathematical instruments."[17]   These were probably not very well cared for, as in 1744 we find the then President of Yale, Thomas Clap (1703–1767) writing rather plaintively to the Secretary of the Royal Society, Cromwell Mortimer:

We have a very good Borometer [*sic*] and Thermometer both together, but the Borometer Tube is broke, and I believe no Man in this Country can put in another. I have had Thôts of Sending it to London, if I was Acquainted with any Gentleman

---

[13] Harvard Archives, ms. HUG. 1879.207 (3 volumes).

[14] The square brackets, etc., are in the original, as if the Professor was editing the ms. for publication.

[15] James Jurin, *Phil. Trans.*, Vol. 32 (1723), pp. 422–27.

[16] Henry M. Fuller, in *Papers in Honor of Andrew Keogh* (New Haven, 1938), p. 166. This was probably Edward Scarlett, either the father or the son, of Dean Street, Soho. See E. G. R. Taylor, *The Mathematical Practitioners of Tudor and Stuart England* (Cambridge, 1954), p. 287.

[17] Thomas Clap, *The Annals or History of Yale College in New Haven* (New Haven, 1766), p. 97.

there who understood those things, and would be pleas'd to take the Care of it.[18]

If repaired it did not last, for in his appendix on the "Present State" of Yale College, dated 1765, Clap writes:

We have a good air-pump, set of globes, telescope, small astronomical quadrant, microscope, thermometer, theodolite, and an electrical machine: but no other apparatus, of any great consequence.[19]

Fuller states that in 1747 there were still a barometer and thermometer at Yale; but in a list dated June 23, 1779, they no longer appear.[20] In 1789 it was still necessary to send to London for instruments;[21] the shipment (quite a large one) included "Fahrenheit's thermometer verified—graduated from 40 below cypher to 220 above. In the same case an Hygrometer & a Barometer with nonius division."[22]

The complete dependence on Europe is underlined in a letter from Benjamin Franklin to Peter Collinson, dated May 28, 1754. He writes that one of three barometers was broken in transport, and that

This Loss added to the former, leaves me but 2 out of Nine. . . . Two others lately brought over here for two of my acquaintance, are broke just in the same manner . . . if they cannot be made to bear Carriage, 'tis to no purpose to send any more.[23]

Even after the Revolution the situation improved very slowly. There was one instrument maker of special ability in Massachusetts, the Rev. John Prince, rector of the First Church in Salem.[24] Naturally he was an amateur; but it was he who in 1779 negotiated for the precious glass tube with the Corporation of Harvard College, as above noted. Instrument makers seem to have come to Philadelphia in the 1780's, though these do not appear to have been of a high order, and it is not clear to what extent they were actual makers and to what extent merely importers. Gillingham[25] has found some details regarding John Denegan, Alloysius Ketterer, and Martin Fisher, who seem to have had the same business in rapid succession between 1785 and 1791, and has established that in 1796 a Joseph Gatty from Italy set himself up at 79 South Front Street. In view of Gatty's claim to make any form of glass, and the fact that he also dealt in fireworks, it is at least doubtful whether he made any barometers of scientific importance.

Things were little better in Cambridge. There is an inventory, made in 1779, on some unnumbered pages at the end of the "Hollis book" (College Book No. 6), which includes the item "13. A compound Barometer void of Quicksilver & the Case unglued in part"; but in a further inventory "taken January 1790," there is no mention of a barometer. However, barometers were being used at Harvard and elsewhere in New England. Edward Wigglesworth, the Hollis Professor of Divinity, was mak-

[18] Royal Society, Letters & papers, Decade I, no. 296, verso. The letter is undated, but was received on May 31, 1744.
[19] Clap, The Annals, pp. 86–87.
[20] Fuller, Papers, pp. 170 and 172.
[21] Ibid., p. 173.
[22] Ibid., p. 176.
[23] L. W. Labarree & W. J. Bell, Jr., eds., The Papers of Benjamin Franklin (New Haven, Yale University Press, 1962), Vol. 5, pp. 331–32.

[24] Cohen, Some Early Tools, p. 63.
[25] Harold E. Gillingham, Penna. Mag. of Hist. & Biog., Vol. 51 (1927), pp. 291–308.

ing observations in 1784 with a barometer and thermometer by Champney, London.[26] At the same time Samuel Williams, the Professor of Natural Philosophy, was using instruments by Nairne, until the arrival of a new set presented to Harvard by the Elector Palatine, on September 8, 1785.[27] The barometer was of course the high quality "bottle barometer" designed by Hemmer,[28] and Williams compared the old and new instruments, assumed the Mannheim ones correct, and reduced his earlier observations to the new basis.[29] Observations from Cambridge continued to be published in the annual Mannheim volumes until 1787.

I have found only one record of a barometer having been made in the United States during this time, and this was by another Winthrop, James, who went on a "tour to Lake Champlain" in August 1786. The barometer used by James Winthrop was "prepared each time; the mercury being discharged after every observation, for the convenience of transportation."[30] The height of the column was actually measured from the surface of the mercury in the cup, which was 2.5 inches in diameter, the bore of the tube being 0.3 inch.

The Harvard observations were continued by Webber, the President, from 1790 until 1807, and then by John Farrar, at that time Hollis Professor of Mathematics and Natural Philosophy.[31]

We have a list of the barometers used in this extensive program. Until July 1802 the Mannheim barometer continued in use; this was followed by "a barometer made by Champney," probably the one used by Wigglesworth; in 1810 this was in turn replaced by "one made by W. & S. Jones, London, provided with a floating gauge and a scale of correction."[32] It is interesting that the barometric heights were reduced to a temperature of 55° F.

It is probable that some, at least, of these barometers had suffered in coming across the Atlantic. The mean pressure for the years 1805 to 1810, derived from observations with the Champney barometer, differed by more than five times the standard deviation of either from the mean, measured with the Mannheim instrument, for the years 1790 to 1795.

In the country as a whole, barometers were scarce for a long time. In 1817 Josiah Meigs, then Commissioner of the General Land Office in Washington, promoted a plan to establish a regular series of daily meteorological observations at various places. Congress failed to approve the scheme, and Meigs tried to persuade the "Registers" [registrars] at the local land offices to observe on a voluntary basis. But, Meigs' great-grandson pointed out:

. . . of course it was not possible to obtain barometric observations; Meigs himself does not seem to have had one of these so important, but at that time expensive and still rather rare instruments; and it is very unlikely that any of the Registers had them.[33]

---

[26] *Ephemerides Societatis Meteorologicae Palatinae anni 1785* (Mannheim, 1786), p. 661.
[27] *Ibid.*, p. 637.
[28] See p. 134 above.
[29] *Ephemerides, ann. cit.*, p. 638.
[30] *Mem. Amer. Acad. Arts & Sci.*, Vol. 2 (1793), p. 127.
[31] John Farrar, *Mem. Amer. Acad. Arts & Sci.*, Vol. 3, part II (1815), pp. 361–412.

[32] *Ibid.*, p. 361.
[33] W. M. Meigs, *Life of Josiah Meigs*, etc. (Philadelphia, 1887), p. 83.

The blank forms sent out contained no column for atmospheric pressure.

It will be seen that up to this time the barometer as an instrument had almost no history in North America. In the 1830's the situation was changed by James Green.

This remarkable instrument maker, born in England, established a business in Baltimore in 1832, naturally enough as an importer as well as an artisan. He moved to New York, most probably in 1849, and in 1864 took as an apprentice his 16-year-old nephew, Henry J. Green, who became his partner in 1879. When James retired in 1885, the name of the firm was changed to Henry J. Green, which it still bears.[34] It is safe to say that most of the barometers of scientific interest made in the United States between 1840 and 1940 were manufactured by this company.

Let us go back to 1830 and 1831, when James Green made a trip to London and Paris

. . . for the purpose of availing myself of the great advantages which those cities afford for improvement in mechanical skill . . . I had frequent intercourse with one of the most deservedly distinguished instrument-makers of the former city, who I found entertained a more favorable opinion of the extent of the demand for good instruments in the United States than my experience authorized; that opinion doubtless arises from the frequent orders he received to furnish such for this country.[35]

Green devoted the rest of his life to satisfying this demand. By 1835 he is able to say that it has been his "uniform practice" to boil the mercury "in all instances where such accuracy was desirable," and also to use redistilled mercury.[36]

He became a member of the Maryland Academy of Science and Literature and made a "standard barometer" for this body.[37] This was of a highly original design, with a cistern which may be described as the inverse of that of Eizenbroek (or Prins),[38] the mercury being allowed to spread out at a constant level *underneath* a glass plate 3 inches in diameter. This covered a mahogany dish with a shallow depression ⅛ inch deep and 3 inches across and a deeper hole 1.1 inch in diameter into which the tube dipped.

The upper end of the instrument was equally ingenious. The scale in fiftieths of an inch was engraved on a short brass tube attached to and enclosing the barometer tube and extending two inches above it; there were "windows" front and back. Fastened to the top of the brass tube there was a nut in which worked a micrometer screw attached to a larger tube which fitted around the scale tube; to read the barometer, the plane of the lower edge of this outer tube was brought into coincidence with the top of the meniscus, the micrometer screw being used to interpolate between scale divisions. The tube was furnished with "Professor Daniell's platinum guard,"[39] an illustration of James Green's willingness to use the latest ideas, though not really an improve-

---

[34] For this information I am indebted to Mr. Richard Whatham, the present General Manager, who has preserved many interesting documents relating to the firm.

[35] James Green, *Amer. J. Science*, Vol. 27 (1835), p. 293.

[36] *Ibid.*, p. 292.

[37] James Green, *Trans. Maryland Acad. Sci. & Lit.*, Vol. 1 (1837), pp. 138–42.

[38] See p. 222 above.

[39] See p. 249 above.

ment. His regard for accuracy is shown in the stress he lays on the determination of the density of the mercury used.

In the Library of Congress there is a *Catalogue of Mathematical and Optical Instruments, Philosophical and Chemical Apparatus, constructed and sold by James Green, No. 43 South Street, Baltimore, and No. 175 Broadway, New York, 1844.* This presents somewhat of a bibliographical problem. The above title is on the colored wrapper—in no less than eight type faces!—but on the actual title page the last word is *Baltimore,* and there is no date. We cannot accept the date 1844 without question, however, for inside the wrapper there is an extensive notice dated "New York, 1849." This note is interesting; after telling his customers that he has opened in New York, but that the business in Baltimore will be continued under the care of his brother, he goes on:

In addition to his own manufactured goods, his personal acquaintance with the principal makers in Europe, gives every facility for the importation of instruments; and at the same time, secures an early knowledge of all improvements made there. . . . Especial attention will be still given to the making of the more accurate class of instruments required for refined observation—particularly in *Meteorology*—both for stationary standards and portable instruments.

I think the date 1849 is most likely to be the correct one, and that the date on the front of the wrapper is wrong, probably copied thoughtlessly from a similar catalog which may be supposed to have been issued at Baltimore five years before.

Let us examine the catalog itself. It shows clearly that people in the United States were beginning to think about the design of barometers, and it will appear that some of them were greatly helped by James Green. The section "Meteorology" begins:

Barometer, as designed by
F. R. Hassler   [$] 35 00

The same, with rack work motion to vernier, and Green's screw adjustment for avoiding the necessity of putting the fingers in the mercury of the cistern in setting up the instrument. Tube five-tenths bore   45 00

Barometer, as made for J. H. Alexander. In this instrument the graduation is made on the glass tube itself, and read by a vernier tube sliding over it: the tube is closed by an elastic iron cistern, preventing the entrance of air when out of use, at the same time allowing of expansion and contraction from changes of temperature with safety to the tube,
     25 00 to 35 00

A detailed description of this barometer is given in Silliman's Journal, vol. xlv. No. 2, Art. 1. Green's barometer with invariable line of level of the mercury in the cistern   50 00

A detailed account of this barometer is given in the transactions of the Maryland Academy of Science and Literature, vol. 1.

Troughton's Tripod Barometer, altered so that the attached thermometer bulb is immersed in the mercury of the cistern   60 00

Smaller size, without the tripod   35 00
Syphon barometer (Bunten's)
with two vernier readings. This
is the lightest and most portable of all the forms of the
mountain barometer, and sufficiently accurate for most
service   35 00
Common mountain barometer   25 00
All the above barometers are
accurately graduated, and read
by Vernier to .01.002.001 [sic]
of the inch, as may be desired:
and the tubes well boiled
throughout their length.

The catalog continues with some
cheaper barometers.

I have quoted this price list at length
because it seems to me to give an insight into James Green's philosophy of
instrument making, and also because it
introduces us to a couple of nineteenth-
century American barometers.

Ferdinand Rudolph Hassler (1770–
1843) was a Swiss immigrant who became, for a time, superintendent of the
United States Coast and Geodetic Survey, an able but somewhat pompous
man who managed without much difficulty to fall foul of the United States
Congress. His "Description of a new
form of transportable original barometer" is to be found in a Congressional
document.[40] The barometer displayed
a good deal of originality. It was of the
type in which the filled tube, a flask of
mercury, and a basin are carried separately. To the bottom of the tube was
applied a sort of iron clevis, and an
iron wedge which held a padded plate
over the mouth of the tube. In setting

it up, the tube was inverted into a basin
of mercury and the wedge pulled out
with the fingers under the mercury, a
defect which James Green was at pains
to remedy. The scale was of brass and
attached to an iron rod. Round the bottom of this rod was an ivory float, very
light, and there was a scratch round the
rod which was made to coincide with
the top of the float. A vernier was engraved on a thin split tube of brass,
and in Hassler's own design this slid
friction-tight up and down the tube.
Green added a rack and pinion. It may
be assumed that after Hassler's death,
Green made another change; for Hassler
was strongly against boiling the mercury in the tube.

Alexander's barometer was somewhat
akin to Hassler's, but the graduations,
including the fiducial mark, were engraved on the glass tube itself.[41] To
make it portable, a shallow inverted
iron dish with a central boss was cemented to the bottom of the glass tube.
A flat and somewhat flexible sheet of
iron was held to this with a threaded
collar, and the deformation of the sheet
would allow changes in the volume of
the mercury with changes of temperature. A special tool was provided for
removing the bottom of the cistern
under the mercury. Alexander hand-
somely acknowledged that this way of
terminating the tube was suggested "by
Mr. Green, the artist who constructed
the various portions of the instrument."[42]

At about this time Green must have
realized that such freakish constructions

---

[40] Doc. 176 H.R., 27th Congress, 2d sess.,
pp. 15–21 [1842].

[41] Amer. J. Science, Vol. 45 (1843), pp.
233–42.

[42] Ibid., p. 240.

had no future, and set himself to produce a satisfactory portable instrument
of more conventional design. His first
tentative solution is represented by an
instrument in the museum of the Smithsonian Institution,[43] which looks like a
Fortin barometer without its ivory
point, but having two large rectangular
windows in the brass case surrounding
the glass portion of the cistern. The
level in the cistern was set by lining up
the surface of the mercury with the tops
of these windows.[44] This instrument is
inscribed "Smithsonian Inst^n Ja^s Green,"
and in much bigger and more deeply-
engraved letters, "N.Y.," from which I
surmise that it may have been under
construction when Green moved to
New York, probably in 1849. It is evident that a number of these must have
been made for the Smithsonian Institution, and it is possible that these were
the barometers referred to by Joseph
Henry in 1856, when he wrote:

A system of meteorological observations
was established by the Smithsonian Institution, in 1849. . . . The institution . . .
procured standard barometers and thermometers from London and Paris, and,
with the aid of Professor Guyot, a distinguished meteorologist, copies of these were
made, with improvements, by Mr. James
Green, a scientific artist of New York.[45]

If so, one can conclude only that Guyot
was something less than distinguished
as a designer of instruments. But however this may be, it was not long after
this that Green designed the famous

Fortin-pattern barometer which, for
portability, durability, and ease of
cleaning, has probably never been surpassed. It was described by Green in
1856;[46] we should like to know whether
Guyot helped in the design of this one.

The essential improvement introduced by Green was the complete
elimination of cement or glue as a
means of assembly.

In Fortin's barometer, he writes, and
also Delcro's [sc. Delcros'] modification of
it, a cement is used to secure the mercury
against leakage at the joints. This, sooner
or later, is sure to give way; and tested
under the extremes of the thermometrical
and hygrometrical range of this climate
especially, has made the defect more evident. This was removed by the substitu-
tion of iron in the place of wood; but it
was soon found impracticable, in this
form of cistern, to prevent damage from
rust. These objections led to the present
plan of construction, which effectually
secures the joints without the use of any
cement. The surfaces concerned are all
made of a true figure, and simply
clamped together by the screws, a very
thin leather washer being interposed at
the joints. This would not be permanent,
however, but for the special care taken in
preparing the box-wood. The box-wood
rings are all made from the centres of the
wood and concentric with its growth.
They are worked thin and then toughened,
as well as made impervious to moisture
by complete saturation with shellac. This
is effected by immersing them in a suitable
solution in vacuo.[47]

The construction of Green's cistern
is shown in Figure 12.1. It will be observed that the long screws hold together the boxwood top of the cistern,

---

[43] Inventory no. 224, 424.
[44] See Smithsonian Institution, 10th Annual
Report of the Board of Regents (Washington,
1856), p. 222. Such instruments are there
referred to as "Green's first barometers."
[45] Report of the Commissioner of Patents for
the Year 1855 (Washington, 1856), I, 370.

[46] Smithsonian Institution, 10th Annual Report (Washington, 1856), pp. 251–58.
[47] Ibid., p. 253.

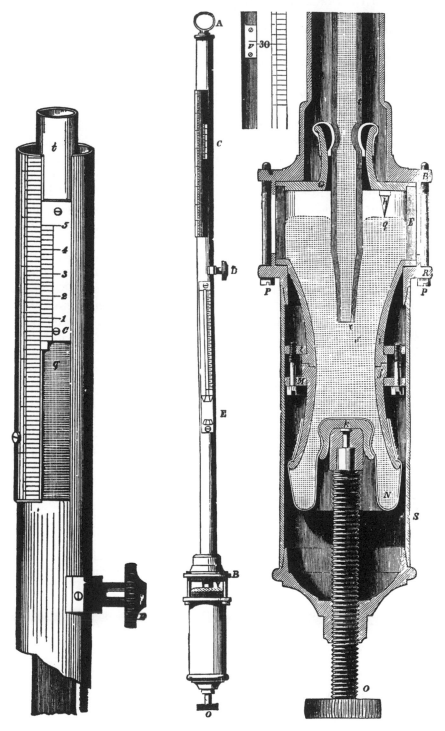

Fig. 12.1. Green's "Fortin-type" barometer.

the glass cylinder, and the upper of the two large boxwood pieces, and at the same time fasten all these to the brass frame. The two larger pieces of boxwood are held together, with a thin gasket between, by two brass rings, each split along a diameter, but on diameters at right angles, so that the loosening of the four screws, and the removal of one of them, allows the rings to be taken off.

This is a most ingenious idea which entirely dispenses with reliance on threads cut in the boxwood; and the whole design has been shown in the course of time to be most successful. Barometers of this type were carried on many mountain expeditions, notably the tremendous survey made by Major R. S. Williamson of the U.S. Corps of Engineers, who contributed so greatly to our knowledge of the limitations of barometric hypsometry.[48] Numerous examples made in the nineteenth century are still in use.

The cistern is very easy to disassemble for cleaning, thus removing one of the serious objections to barometers of this type. When the barometer is upside down, the mercury comes just to the level of the junction between the boxwood parts. With the lower one removed, the end of the tube can be firmly closed with the finger and the rest of the mercury in the cistern poured out into a cup. The remainder of the cistern can then be taken apart and cleaned thoroughly while the tube and its attachments are stood upside down in a corner. This facility of clean-

ing undoubtedly contributed greatly to its long innings as a station barometer at the stations of the U.S. Weather Bureau, where it was retained after many other services had adopted the fixed-cistern barometer.

The scale of the Green barometer is adjustable over a short distance with respect to the protecting tube. It appears from the account referred to above that Green adjusted his barometers by actual measurement from the ivory point, allowing for the calculated effect of capillarity.

While Green was making barometers for the Smithsonian Institution and others, the United States Navy was obtaining its barometers elsewhere. In 1854 we are told in the report of the Kew Committee of the British Association that ". . . fifty barometers were ordered from Mr. [Patrick] Adie . . . for the use of the United States Navy, all of which are to be verified at the [Kew] observatory."[49] The man responsible for this order was the famous M. F. Maury (1806–1873), then a lieutenant. But apparently the Navy also dealt with Green, for there is a letter signed by Maury in the possession of Henry J. Green Instruments, Inc., which reads in part as follows:

National Observatory,
Washington,
Sept. 26, 1854

SIR:

Enclosed please find bill of lading for one box containing 14 Marine barometers and 4 Mountain ditto shipped to your address.

Can the Marine barometers be so repaired as to be made equal in all respects to those lately purchased from you? If

[48] Robert Stockton Williamson, *On the Use of the Barometer on Surveys and Reconnaissances* (New York, 1868). (U.S. Army, Corps of Engineers *Professional paper no. 15.*)

[49] *B.A.A.S. Report,* 1854, p. xxxi.

so, what will be the cost of repairs for each? . . .

Respectfully, &c
M. F. MAURY

Mr. James Green
422 Broadway
New York

We can only suppose that some of Adie's barometers had come to grief on their way across the Atlantic. If so, this must have aroused Maury's interest in a local source of supply, quite apart from considerations of patriotism. In any event, Green continued to have excellent relations with the departments of government, as may be gathered from the scattered letters which have been preserved. In 1867, for example, he delivered some "Standard Barometers" to "the Several Navy Yards."[50] We may even note a tendency to abandon other barometers in favor of Green's, as in the following letter:[51]

Navy Department
Bureau of Navigation,
Hydrographic Office,
Washington, May 6th, 1869.

. . . .

I have on hand 46 Barometers which I wish to dispose of at Sale as I prefer not to issue such to the Navy even if repaired, as your pattern has been adopted.

Very Respectfully
E. SIMPSON
Commander
In Charge Hydro. Office.

James Green Esq.
175 Grand St
New York

If these were the Adie barometers that Maury had bought, patriotism must have entered in, for such instruments have usually remained in service many times thirteen years.

Not only the Navy acquired his barometers. There is a letter dated May 23, 1870, written on behalf of the Chief Signal Officer to James Green, directing

the ten (10) Barometers ordered by this Office to be made with the improvement contained in that furnished as the Standard Barometer of the U.S. Naval Observatory in this city—that is, the lower edge of the back part of the sliding or vernier scale should be serrated or cut into saw teeth so that the light may appear between the teeth and assist in the adjustment of the scale to the top of the mercury.[52]

In the course of the next two decades most of the stations of the Signal Service were equipped with Green's Fortin-type barometers, and as early as 1872 complete instructions for cleaning their cisterns were supplied to the observers.[53] The Weather Bureau, organized in 1891, took over these instruments and the excellent habit of acquiring them. Nevertheless, the Signal Service had not hesitated to import other barometers. The Kew Committee of the British Association reported in 1871:

Tubes for the construction of a Welsh's Standard Barometer[54] on the Kew pattern, together with the necessary metal mountings, and a cathetometer, have been made under the superintendence of the Committee for the Chief Signal Office, [sic] Washington.[55]

---

[50] Fragment of a letter from the Bureau of Navigation, in the possession of Henry J. Green Instruments, Inc.

[51] Henry J. Green Instruments, Inc., collection.

[52] Same collection.

[53] Instructions to Observer Sergeants, Signal Service, U.S.A. on Duty at Stations (Washington, 1872), pp. 15–16.

[54] See p. 262 above.

[55] British Assn. Report (Edinburgh, 1871), p. li.

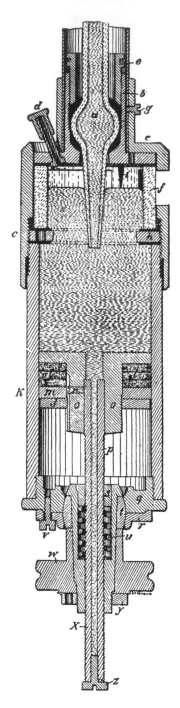

Fig. 12.2. Tuch's barometer cistern.

In 1878 the Signal Service imported ten large Fortin-type barometers by Adie, with tubes about half an inch in bore. Four of these were still being used as standards by the U.S. Weather Bureau in 1934, maintained by international comparison, as the Bureau had no primary barometer.[56] Two of them, of which no. 1710 is famous for the constancy of its corrections, were being used as working standards in 1962.

Henry J. Green continued his uncle's solicitude for accuracy after the latter retired in 1885. In 1887 he carried a barometer to England and back to New York after having it compared with the standard at Kew.[57] The reputation of the firm remained high; in 1894 there were negotiations for a copy, or near-copy, of the Pulkowa normal barometer.[58]

A certain amount of trouble was sometimes experienced with the leather bag of the Fortin-type barometer, and sometime after 1880 a cistern which did without a leather bag was devised by Charles B. Tuch of the instrument division of the U.S. Weather Bureau.[59] The principle of this will be sufficiently evident from Figure 12.2; it should perhaps be noted that this use of a piston

[56] S. P. Fergusson, *Monthly Weather Rev., Washington,* Vol. 62 (1934), p. 364. See also R. N. Covert, *J. Opt. Soc. Amer.,* Vol. 10 (1925), pp. 313–15.

[57] Letter dated December, 1887, from G. M. Whipple, the Superintendent of Kew Observatory, to Henry J. Green. Collection of Henry J. Green Instruments, Inc.

[58] Letter of Dec. 13, 1894, from Prof. Wm. Harkness of the U.S. Naval Observatory (Green collection). For the Pulkowa barometer see p. 266 above.

[59] C. F. Marvin, *Barometers and the Measurement of Atmospheric Pressure, etc.,* U.S. Weather Bureau, Circular F (Washington, 1894), p. 11.

was made by Horner about 1800, but with different materials.[60] Tuch's barometer contained a very large number of parts, and must have been expensive to construct, but it appears to have functioned very well; and there is one in the Smithsonian Museum[61] inscribed "J. Green N.Y.," the piston of which still operated smoothly in 1962.

Let us now go back to 1835, when Charles F. Durant described the first barometer cistern, as far as I know, invented in the United States.[62] In this a small hollow steel sphere was cemented on to the tube with the end of the tube, drawn down to a point, inside, and a small hole near the top of the sphere. All this was under the mercury and was intended to damp its motion during transport. In addition there was a gauge glass, like the water gauge of a steam boiler, in which the level of the mercury could be observed!

The first United States Patent on a barometer was No. 1,951, issued to W. R. Hopkins of Geneva, N.Y., on January 27, 1841. It must be confessed that except for its having been the earliest it would merit little attention. Hopkins' barometer consists of an iron or steel tube and a cistern. Knife-edges are fixed to the middle of the tube and at its upper end, and there is a valve in the cistern to close the lower end of the tube. To use the barometer, it is hung up and the valve opened; when the mercury has come into equilibrium with the atmospheric pressure, the valve is shut, and then the tube is placed horizontally with the central knife-edge resting on a plane. Weights added to

a scale-pan hung on the terminal knife-edge indicate the atmospheric pressure. As it is expressly stated that the tube should be large, one wonders how the mercury would distribute itself in the tube.

There were no more patents on barometers until 1857, when no. 18,560 was issued to T. R. Timby. This was for a "bottle-barometer" with a tap just below the enlargement and a short length of rubber tubing below it, to prevent breakage in transport if the mercury should expand. There is one of these in the Peabody Museum at Salem, Mass.[63]

Timby got another patent[64] in 1862, this time for a barometer cistern with a screw-operated stopper in the form of a small elastic bag in the middle of an elastic cistern bottom. This was to be pressed around the end of the tube when the instrument was being carried about, and left some room for expansion of the mercury in the tube.

In the Smithsonian Museum there is a rather elementary barometer with an iron cistern,[65] patented[66] in 1860 by L. Woodruff of Ann Arbor, Michigan. It has not much to recommend it. A score of other nineteenth-century patents are mainly reinventions. Occasionally something really ingenious and original appears, as in the patent[67] granted to C. H. Stoelting in 1904 for a portable barometer with a turned-up tube ending in a cistern closed by a loose leather "hat," into the "crown" of which is tied a plunger. A point, forming the zero of

[60] See p. 214 above.
[61] Inventory no. 314, 712.
[62] Amer. J. Sci., Vol. 27 (1835), pp. 97–104.

[63] Inventory no. M.1723.
[64] U.S. Patent 36,872.
[65] Inventory no. 314,608.
[66] U.S. Patent 28,626 (1860).
[67] U.S. Patent 756,905.

a movable scale, projects through the center of the bag. In use, this point is set to the level of the mercury in the cistern, but for transport the plunger is moved down until tube and cistern are full of mercury. This is an example of the sort of idea that would be very good if there were not many better ones.

A characteristic invention is the "tornado alarm," consisting of a siphon barometer with an additional short tube connected to the ordinary one by a capillary.[68] A sudden change in pressure will produce a difference in the level of the mercury in the two tubes, whereas normal rates of change will result in a negligible difference. By the use of two floats and an ingenious mechanism, the differential motion closes an electrical circuit. One's only reservation might be that by the time the sudden change of pressure had manifested itself, it might be otherwise obvious that a catastrophe was in progress.

The twentieth century produced two related patents on portable barometers, both assigned to the Central Scientific Co. of Chicago, and both intended to make possible the production of a fairly accurate instrument in quantity at a reasonable price, for school laboratories and similar applications.[69] This is a barometer with an adjustable cistern and a fiducial point. We shall illustrate this by Figure 12.3, taken from the later of the two patents.

Apart from a gasket and the fiducial

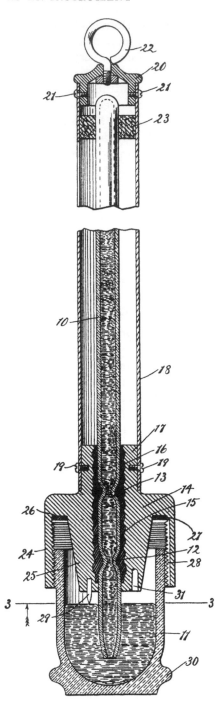

Fig. 12.3. The barometer of Klopsteg and Sachse (1927).

---

[68] U.S. Patent 572,536 (1896), granted to William T. Flournoy of Marionville, Miss.

[69] A. M. Krause, U.S. Patent 1,353,482 (1920); Paul E. Klopsteg and H. P. Sachse, U.S. Patent 1,632,084 (1927).

point, the entire cistern contains only two parts, both designed to be molded from plastic materials, the lower one, of course, transparent. It is intended that enough mercury should be put in the cistern to make it necessary for a little to escape through a groove provided for that purpose, just before the lower part of the cistern closes against the gasket. To provide for expansion of the mercury during transport, a recess in the form of an incomplete annular groove, or else a number of small holes, is provided in the lower end of the upper portion of the cistern. A little air, trapped in this recess, can be compressed should the mercury expand.

It was inevitable, of course, that modern servo-techniques should be applied to the barometer, especially in view of the requirement for automatic weather stations. In 1942 Waldo H. Kliever patented such a device on behalf of the Minneapolis-Honeywell Regulator Co.[70] On the mercury in a large barometer tube floats a short iron cylinder, kept away from the walls by a sharp-edged ring. Around the tube is a sleeve containing two inductances separated by a short interval, which form adjacent arms of an inductance bridge. The output of this bridge is fed to a phase-sensitive amplifying system which causes a two-phase motor to move the sleeve so as to keep it at a constant height in relation to that of the float, i.e., to follow the float. If the sleeve is driven by a lead-screw, for example, a dial or counter, from which the pressure may be read, may be geared to this. Developments of this idea, using a differential transformer rather than an inductance bridge, were under way in 1960.

Another idea is to follow the movement of the surface of the mercury photoelectrically, a procedure which can be given almost any desired sensitivity. We have referred to a barometer on these lines, described in 1948.[71]

The American genius naturally produced a number of barographs, but these have been dealt with in the preceding chapter.

[70] U.S. Patent 2,357,745.

[71] See Chap. 9, p. 203 above.

Chapter 13

Luminescence in the Barometer

O N an evening in 1676 Jean Picard (1620-1682), one of the astronomers in the new Paris Observatory, chanced to carry a barometer into a dark room and noticed that as the mercury oscillated up and down in the tube a light was produced. The explanation of this phenomenon took a long time.

In the official manuscript *Registres* of the *Académie royale des Sciences* this interesting observation received only the following brief mention: "Mr Picard nous a proposé un autre phanomine [sic] sur le barometre lequel estant agité durant la Nuict fait paroistre de la Lumiere."[1] In the *Journal des Sçavans* there appeared a much fuller account:

It is well known that the simple barometer is nothing but a glass tube, hermetically sealed at the top and open at the bottom, in which there is quicksilver, standing ordinarily at a certain height, the space above being empty. M. Picard has one at the Observatory which, when in the dark it is moved enough to make the quicksilver oscillate, seems to produce sparks and

throws off a certain intermittent light which fills all that part of the tube in which the vacuum is produced. This happens at each oscillation only when a vacuum is being produced, when the quicksilver is descending. The same experiment has been tried on several other barometers of the same construction, but up to now it has succeeded with only one. As it has been resolved to examine the matter in every way possible, we shall give in more detail all the circumstances that are found out about it.[2]

The date of the observation is often given as 1675, probably because the date of the above extract from the *Journal des Sçavans* is erroneously so stated in a marginal note in the more easily consulted *Mémoires* published in 1730.[3]

The observation does not seem to have led to much written discussion for about twenty years. In 1694 Philippe de la Hire (1640–1718) contributed a short note to the records of the Acad-

[1] *Acad. R. des Sci., Registres,* 1676, fol. 82v.

[2] *Journal des Sçavans,* May 25, 1676, p. 112.
[3] *Mémoires de l'Academie royale des sciences depuis 1666 jusqu'à 1699* (Paris, 1730), X, 566. The marginal note reads "1675.P.112." The transcription of the text is exact.

emy, which was not published till much later.[4] De la Hire recorded that when in 1676 Picard found his barometer to be capable of producing light, he examined a number of instruments belonging to other people, but found none of them with this property. He gave some of his mercury, "qui avoit été revivifié de Cinabre," to De la Hire, but the barometer the latter made from it would not shine. After Picard's death in 1682, De la Hire took his barometer apart and refilled it with the same mercury; but it was no longer luminous. At about the same time Cassini,[5] the Director of the Observatory, had noticed that his own barometer had begun to be luminous, and it had remained so. Finally Picard's barometer, which De la Hire had repaired, again began to give light when moved; but some years later it lost the property again. De la Hire thought at first that whatever made the mercury luminous must have been dissipated or used up, but then he remade it once again and got a very luminous barometer. There was a curious difference between this barometer and that of Cassini; in the latter the light seemed attached to the surface of the mercury, but in his instrument it seemed to fill all the vacuum.

The phenomenon had also been observed in England, as is clear from a pamphlet published by the barometer maker John Patrick some time shortly before or after 1700.[6] Patrick made,

and possibly also independently invented, the conical barometer[7] usually ascribed to Amontons.

It hath also been observed in some of these pendent barometers, [he writes], if it be carry'd into a dark room, and the quicksilver put into a vibration, by lifting it up and down, that it will throw forth flashes of lightning from the top of the quicksilver, to the top of the tube in the vacium [sic]: which noble experiment requires the solution of the most profound naturalist.

It got the attention of a great mathematician, Johann Bernouilli the elder (1667–1748), who had read about the phenomenon in a little book by Dalencé.[8] He described his own experiments in a letter from Groningen to Pierre Varignon (1654–1722) of the Académie des Sciences.[9] This was later printed in the Mémoires.[10] He had formed quite clear ideas about the explanation of the phenomenon, which he defended at intervals during the next two decades.

He found that his own barometer gave a little light when disturbed, and as it had been constructed only four weeks earlier, this disposed of Dalencé's suggestion that perhaps only an old barometer would show the effect. He then made a barometer very carefully with very pure mercury, but got no light, so he concluded that neither very pure mercury nor absence of air was sufficient, though they were both essen-

---

[4] *Hist. Acad. r. des Sci.* (Paris, 1733), Vol. 2, pp. 202–3.

[5] Jean Dominique Cassini (1625–1712), the first of four generations of Cassinis who, one after the other, directed the affairs of the Observatory until the Revolution.

[6] *A New Improvement of the Quicksilver Barometer, Made by John Patrick, in Ship-Court in the Old-Baily, London.* n.d., 1 leaf.

Re Patrick, see p. 120 above.

[7] See p. 119 above.

[8] *Traittez des barométres, thermométres, et notiométres, ou hygrométres.* Par M^r. D°°° (Amsterdam, 1688).

[9] *Registres,* Vol. 19, fols. 243r–253r. (30 June 1700).

[10] *Mém. Acad. r. des Sci.,* 1700, pp.178–90.

tial. He devised two methods of filling a barometer without having to pour the mercury through the air: first, by suction from the top with the tube inclined, then sealing it off; next with the aid of an air pump and "receiver." In both cases he obtained luminous barometers, and he reasoned that a tarnish on the mercury, formed when the barometer is filled in the ordinary way, prevents the luminosity. He also noted that the light was produced when the mercury is descending in the tube.

Bernouilli therefore concluded that a very subtle substance (la matière du premier élément of Descartes) must come out of the mercury when it descends, while at the same time a somewhat less delicate substance (Descartes' globules celestes) enters through the walls of the tube; these combine to cause light. Noting that a very little water or alcohol in the barometer completely spoils the effect, Bernouilli concludes that this prevents the premier élément coming out of the mercury.

These conclusions were of course only one way of interpreting the experiments, and depended on completely ad hoc hypotheses that could not be tested experimentally. They were, in fact, a splendid example of Cartesianism in action.

The Academy, mainly disciples of Descartes, took them very seriously. However, when they tried to make a luminous barometer on July 3, 1700, they failed, and a committee of three, Cassini, Couplet, and Varignon, were appointed to begin again "à la manière de Mr Bernoulli [sic]."[11] They were obliged to inform Bernouilli of their non-success, and Bernouilli wrote

[11] Registres, Vol. 19, fol. 237r.

another letter,[12] obviously with a chip on his shoulder, suggesting that perhaps they had not done it carefully enough, and "peut-être que le tuyau dont on se servit, n'étoit pas assés sec ny assés net; car la moindre humidité ou graisse empêche l'apparition de la lumiere."[13] It is noteworthy that the argument was not about the theory, but only about the recipe for making a luminous barometer.

In this letter Bernouilli also described his "mercurial phosphor," simply a phial partly filled with clean mercury and evacuated. When this is shaken a bright luminescence results. This experiment impressed Francis Hauksbee (d. 1713) in England, who between 1705 and 1711 made experiments on what we should now call electroluminescence.[14] Hauksbee ascribed these and other luminous effects to friction and connected them with electrification.[15] He had decided against the idea that some property of the glass was responsible, after an earlier admission that

one might with some probability conclude, that the light produc'd proceeds from a quality in the glass, upon such a friction or motion given it; and not from the mercury, any other ways than as a proper body, which falling or rubbing on the glass, produces the light. And that which would seem farther to corroborate such a conclusion is, that some time ago I took a mercurial barometer, and rubb'd the

[12] Mém. Acad. r. des Sci. (1701), pp. 1–9.
[13] Ibid., p. 3.
[14] For a full discussion see E. Newton Harvey, A History of Luminescence (Philadelphia, 1957), pp. 275–81. I cannot agree with Harvey (p. 273) that the effect of impurities in preventing the luminescence of the barometer is an example of quenching, at least in the modern sense of the term (ibid., pp. 405–6).
[15] Phil. Trans., Vol. 25 (1706–7), pp. 2372–77.

upper or deserted part of the tube between my fingers, and a light ensu'd,
without the motion of the quicksilver. Yet
for all this the conclusion is doubtful. . . .[16]

and he described an experiment in
which mercury, contained in a heavily
varnished pot, was shaken in a vacuum,
and light appeared. He also referred,
in this connection, to the light observed
when loaf sugar is broken.

There is no doubt whatever that Bernouilli was the first to produce the
"mercurial Phosphor," but that is no
reason for him to accuse Hauksbee of
plagiarism, as he did in 1719.[17] Hauksbee in his experimental work went far
beyond Bernouilli. It seems that the
latter had taken over from Descartes
more than the primary matter and the
celestial globules; he had inherited the
feeling that what his reason told him
could not be wrong. The fact that
Hauksbee and others did not accept his
explanation seemed to worry him to an
inordinate extent. His book is directly
derived from a thesis read by an astute
student[18] before Bernouilli, who is of
course referred to throughout as *Celeberr.* [etc.] *Praeses.* In this remarkable
example of apple polishing, Nebel calls
Hauksbee a clever workman *(peritus
artifex)*, and claims that all the experiments of the latter are derived from
those of Bernouilli. This effusion ends
in no less than seven sets of Latin
verses and is really rather sickening. It
is interesting that on page 61 Nebel

admits that he has not seen Hauksbee's
book,[19] only large extracts[20] in the *Acta
Eruditorum* of Leipzig. I do not think,
however, that Hauksbee was misrepre-
sented in that review.

A number of essays on the subject
appeared in the years after 1709, most
of them negligible. A fresh breeze blew
in with one J. M. Barthius,[21] who had
obviously read widely in the subject,
and pointed out the awkward fact that
no one knew what light really was. In
1717 J. J. Dortous de Mairan (1678–
1771), in an essay[22] which received a
prize from the Academy, followed Bernouilli's explanation for the barometer
light, with the variation that instead of
the "first element" coming out of the
mercury it was the "sulphur" of the
mercury which emerged; and that this
could not get out unless the mercury
was retreating from the "autre air plus
subtil, qui tient un espece de milieu
entre l'air proprement dit, & la matiere
étherée, & qui entre & sort par les pores
du verre . . ."[23] This was merely a different terminology for substances that
could not be demonstrated to exist.

J. M. Heusinger[24] advanced the subject by pointing out that a barometer

---

[16] *Ibid.,* p. 2279.

[17] *De mercurio lucente in vacuo* (Basle,
1719). Also in *Johannis Bernouilli opera
omnia,* ed. Cramer (Lausanne & Geneva,
1742), II, 323–92.

[18] W. B. Nebel, *Dissertatio physica de mercurio lucente in vacuo* (Basle, 1719).

[19] *Physico-mechanical Experiments on Various Subjects, Containing an Account of Several Surprising Phenomena Touching Light
and Electricity, Producible on the Attrition of
Bodies* (London, 1709) (2nd ed.; London,
1719).

[20] *Acta Eruditorum,* 1709, p. 238.

[21] *De luce barometrorum* (Leipzig, 1716).

[22] *Dissertation sur la cause de la lumière
des phosphores et des noctiluques* (Bordeaux,
1717).

[23] *Ibid.,* p. 39.

[24] *De noctiluca mercuriali sive de luce quam
argentum vivum in tenebris fundit* (Gissae,
1716). Quoted by Harvey, *A History,* p. 274;
also by Du Fay, *Mém Acad. r. des Sci.*
(1723), p. 297.

which is filled so well that no space is apparent when it is inclined is likely to be luminous; but a *small* bubble will not always prevent it. The mercury, too, need not be very pure; he had obtained luminescence with an amalgam of 23 parts of mercury and 5 [!] of lead. But the mercury must be very free from air and especially from moisture.

The next step was taken by C. F. de Cisternay du Fay in 1723,[25] who had learned from a German glassworker how to make barometers luminous "à coup sûr." This process has already been described on page 243, but we must again note that only two-thirds of the mercury was boiled; and this limitation may have been what made Du Fay so sure of getting a luminous barometer in this way. As far as theory was concerned he did not make much progress, but he felt that the luminosity was an important test of the excellence of a barometer:

. . . apart from their singular property, luminous barometers are much better than others in ordinary use, because, being perfectly empty of common air, they are free from the fault common to all others, which is . . . that they act like thermometers and necessarily vary with heat and cold, which never happens to luminous barometers, and in consequence this makes them preferable to all other kinds of barometer.[26]

That Du Fay's glassworker may not have been entirely up-to-date is suggested by an interesting book published only two years later by an obscure clergyman of Dabrun, Johann Georg

Leutmann,[27] seventy-one pages of which are devoted to the luminescence in the barometer. Leutmann records, and for the first time as far as I am aware, that a barometer with a *very* good vacuum may not be luminous, but can be made so by introducing a little air. This, of course, is not news to a twentieth-century physicist familiar with the phenomena of electrical discharge through gases; nevertheless it shows that a very good technique of filling barometers was available to Leutmann. But he was still in the dark about the theory. "The light in baroscopes, therefore," he says, "is a tremulous motion made by volatile salts coming out of the mercury, and impinging on bubbles of air *(in bullulas aëreas)*, against which these salts strike, not in a completely absolute vacuum, but with a few particles of rarefied air remaining there."[28] When the mercury descends, the volatile salts come out of the top and strike the air particles; but when the mercury ascends it sweeps them up again. Hence the light is seen only when the mercury is descending.

This is less Cartesian, as we have come from "primary matter" to "volatile salts"; but it is no nearer an explanation.

The failure of a really excellent barometer to luminesce was again insisted on by Petrus van Musschenbroek (1692–1761),[29] who recommended heating the tube strongly and filling it with boiling mercury. This advice was repeated in the well-known textbook of

---

[25] *Mém. Acad. r. des Sci.* (1723), pp. 295–306.

[26] *Ibid.*, pp. 305–6.

[27] *Instrumenta meteorognosiae inservientia* (Wittenberg, 1725).

[28] *Ibid.*, p. 59.

[29] *Essai de Physique* (Leiden, 1739) (2d ed.; 1751), p. 656. This first appeared in Dutch as *Beginselen der Natuurkunde* (Leiden, 1736).

J. T. Desaguliers (1683–1744),[30] who may or may not have had any personal experience in the matter. "When you would know whether the tube be well fill'd," he writes, translating Musschenbroek, "you must shake it a little in the dark, so as to make the mercury rise and fall: if then you see no light upon the surface of the mercury, it is a sign that the barometer is perfect; but if it gives light, it is not as it should be: for then you may be sure that there is a little air above, to which the light adheres."[31]

It was hard for some people to believe that a barometer which did not luminesce could be a good one. This opinion was certainly not commonly accepted; for instance, in a general article on the barometer,[32] written in 1763, the astronomer J. J. Le François de Lalande says that "The celebrated London physicist Mr. Wilson[33] showed me some experiments which clearly prove, contrary to the common opinion, that an excellent barometer which is not luminous may become so, if a very small portion of air is introduced into it."[34] De Luc could not believe it, and attacks Musschenbroek at some length for what he considers an egregious error, which has done much harm,[35] as indeed it may have done, if it were taken to mean that every non-luminous

barometer is a good one. But it is evident that De Luc believed that the luminescence was an evidence of a perfect vacuum.

It is worth noting that when Bohnenberger was building his primary barometer many years later,[36] he remarked that the wide tube would give light *before* being boiled out, but not afterwards.

As Harvey remarks,[37] it is surprising that Du Fay, who later became famous for his researches on electricity, should not have realized the true nature of the barometer light, already suggested by Hauksbee. The same remark might apply to Desaguliers. The first unequivocal demonstration of the electrical nature of the luminescence, however, had to wait until 1745, when C. F. Ludolff[38] enclosed the upper part of a barometer tube in another glass vessel, which could be exhausted by means of a pump. Ludolff caused the mercury in the barometer to move up and down by blowing and sucking air into and out of the cistern, without moving the barometer itself. When this was done, threads and small bits of paper which had been left inside the outer tube were attracted to the barometer tube. The vacuum in the outer tube disposed of the possibility that currents of air might have moved these light objects. Ludolff found that only barometers which luminesce behave in this way. Very similar experiments must have been made at about the same time by J. N. S. Allamand (1713–1787), a Professor

[30] *A Course of Experimental Philosophy* (2 vols.; London, 1734 & 1744), Vol. 2, pp. 272–74.
[31] *Ibid.*, Vol. 2, p. 273.
[32] *Connoissance du temps pour l'année 1763* (Paris, 1765), pp. 199–221.
[33] Benjamin Wilson (1721–1788), an extraordinarily versatile man, was a professional portrait painter as well as a physicist.
[34] Lalande, *Connoissance,* pp. 202-3.
[35] *Recherches sur les modifications de l'atmosphère,* I, 47–48.

[36] See p. 262 above.
[37] *A History,* p. 275.
[38] *Mém. Acad. r. des Sci. et des Belles-Lettres de Berlin (Classe Physique)* (1745), pp. 3–7.

at Leiden, according to Abraham Trembley of Geneva, who informed the Royal Society of them.[39] They did not, however, include the use of the evacuated outer jacket.

This phenomenon naturally interested Benjamin Franklin. Writing to James Bowdoin from Philadelphia on December 13, 1753, he says:

My barometer will not shew the luminous appearance by agitating the mercury in the dark, but I think yours does. Please to try whether it will, when agitated, attract a fine thread hung near the top of the tube.[40]

Not even in recent times has a really clear idea of the mechanism of the luminescence been attained. In 1957 W. R. Harper reviewed the matter[41] and examined a number of hypotheses

which might account for the charging of mercury, but without being very partial to any of them. It cannot, he thought, depend on friction in the ordinary sense of the word, but must be, in some sense, contact charging. Proof is at hand that "mere contact is capable of generating *static charges of sufficient magnitude* to account for triboelectrification."[42]

Nor is the barometer light entirely innocuous. According to a National Bureau of Standards document,[43]

This [electrical discharge] seems to accelerate the formation of gaseous products or promote outgassing of the glass. In any event painting the outside wall of the glass tubing, but not where it would interfere with visibility of the meniscus, with a metallic electrically conducting paint greatly reduces the rate at which the vacuum deteriorates.

[39] *Phil. Trans.*, Vol. 44, Part I (1746), pp. 58–60.

[40] A. H. Smyth, ed., *The Writings of Benjamin Franklin* (10 vols.; New York, 1905–1907). III (1905), 192.

[41] *Advances in Physics*, Vol. 6 (1957), pp. 365–417.

[42] *Ibid.*, p. 390 (Italics Harper's).

[43] W. G. Brombacher, D. P. Johnson, and J. L. Cross, "Mercury barometers and manometers," *N.B.S. Monograph* 8 (Washington, 1960), p. 17.

Part III

BAROMETERS OTHER THAN MERCURY BAROMETERS

"Utrumque fortasse inuentum non modo perfectius, sed ad usum etiam accommodatius reddi posset, si quis rem, de qua agitur, denuo suscipere velit."—E. Zeiher, *Nov. Comm. Acad. Imp. Sci. Petrop.*, Vol. 8 (1763), p. 274.

Chapter 14

Other Liquid Barometers

1. GENERAL REMARKS

IN spite of the greater compactness of the mercury barometer, it has been thought worth while from time to time to construct barometers using other liquids. This task has been undertaken from various motives. The earliest water barometers set up by Berti and Pascal, which have been referred to in Chapters 1 and 3, were of course the result of scientific curiosity, as were indeed nearly all those erected in the seventeenth century. In later times, however, such barometers have been erected either for public display, or privately as a hobby, with one nineteenth-century exception that we shall deal with later.

The liquids used have been water, glycerine, linseed oil, sulphuric acid, and olive oil, all of which have their disadvantages. Nearly all these instruments have been fixed-cistern barometers, except for two which were arranged as barographs. Only the early experimental instruments are of much scientific interest, and the rest need not detain us long.

2. SEVENTEENTH-CENTURY WATER BAROMETERS

After Berti and Pascal, the earliest water barometer seems to have been erected by Otto von Guericke, the mayor of Magdeburg, some of whose other experiments have been referred to in Chapter 9. The date was about 1654. Figure 14.1 is taken from the famous and much-reproduced engraving that illustrates his description.[1] The various sections of the lead tube each had a cup-shaped enlargement at the top into which water could be poured as a seal. The top section was of glass, and in it, on the surface of the water, floated a manikin standing on a cork. The "cistern" was a tub resting on the ground. Guericke called the manikin his *semper vivum* and noted that "it is not perpetually in motion, but only as the weather changes in the whole of an extended region round about. It is not a thermoscope, altered by heat and

[1] Guericke, *Experimenta nova (ut vocantur) Magdeburgica*, etc. (Amsterdam, 1672), p. 100.

Fig. 14.1. The water barometer of von Guericke, circa 1654.

cold."[2] We may be a little surprised that such an acute observer as Guericke failed to notice the effect of temperature on the water barometer, but perhaps he did not observe it for long enough.

Another instrument with a manikin was set up in 1683 by Comiers d'Ambrun,[3] who called it *L'homme artificiel anemoscope, ou prophète physique du changement des temps.* This was not a water barometer as is commonly supposed, but a sort of wheel-barometer with a magnifying system of pulleys.

But before this there had been a good deal of interest in the water barometer in England. In 1661 such an instrument was erected at Townley Hall by Richard Townley, Henry Power, and others. The tube was of tin and had a glass section at the top.[4] Soon the Royal Society came to feel that they should have one at Gresham College, and we read that on June 10, 1663, "Dr. Goddard, Mr. Pell, Dr. Pope, and Dr. Croune, were appointed to have the care of setting up the long glass tube for the Torricellian experiment with water."[5]

The committee reported on June 22, recommending that two tubes should be put up, and this seems to have been accomplished in time for the meeting of July 16. Between then and July 29 it was observed that the water rose and fell according to the weather. Hooke evidently undertook to make systematic observations, for there exists in manuscript "An Account of y[e] Observ.ons of y[e] Glass-Cane of 40 foot long, given in by y[e] operator, Nov. 15. 1663."[6] This is a table of observations covering the period July 23 to August 29, 1663.

In the following year the Society returned to the question, and on August 17, 1664, the Torricellian experiment was made with spirits of wine (rather dilute), water, and brine; the results were as we should expect. Many of the Fellows were certainly acquainted with what Pascal had done at Rouen,[7] but in the true spirit of the times insisted on seeing for themselves.

Some time later, but in any event before 1669, Robert Boyle built a water barometer outside his house, putting his improved vacuum pump on the roof in order to pump it out. For an illustration of this barometer the reader may be referred to Figure 4.6; most of the tube was of tinned iron with a glass section near the top.[8] It is questionable whether this should really be called a barometer, the purpose of the experiment—*nullius in verba* again?—being to find out how high water could be raised by suction.

Under the Observatory of Paris had been dug a shaft 28 meters deep, with a winding staircase around the walls. About 1683 Edme Mariotte put a water barometer in the stairwell thus created; and because of the extreme uniformity of the temperature in such a place, no better site for such an instrument could be imagined. The tube, 40 feet long,

[2] *Ibid.*
[3] *Mercure Galant*, March, 1683, pp. 164–214. Reviewed in *Acta Erudit.*, Vol. 3 (1684), pp. 26–28.
[4] Henry Power, *Experimental Philosophy in Three Books* (London, 1664), p. 131.
[5] Birch, *History of the Royal Society*, (London 1756–57), I, 255. See also p. 74 above.

[6] Royal Society, *Classified Papers*, iv (1), no. 8.
[7] See p. 41 above.
[8] *A Continuation of New Experiments Physico-Mechanical*, etc. (Oxford, 1669), pp. 42–44. *Works*, (1772 ed.), III, 205–9.

was made of 5 to 6 pieces of glass, fused together in the great hall of the Observatory by the *émailleur* Hubin.[9] There seems to be little further information about this instrument; but it must have been completed toward the end of 1682, because early in 1683 he reported to the Academy about it.[10]

3.   LATER LIQUID BAROMETERS

There seems to be almost no information about water barometers in the eighteenth century. A. T. Nolthenius, who made a special study of these large barometers,[11] found only one reference; it seems that C. A. Hausen, a professor at Leipzig, made one at some date earlier than 1759, the tube being constructed of sections of brass tubing screwed together, with leather gaskets at the joints and a glass jar at the top. This was described later in a German textbook.[12] It is worthy of note that De Luc, in his historical review of the barometer,[13] mentions no water barometers at all.

In 1801 Luke Howard, the famous London meteorologist, constructed a

linseed-oil barometer outside his house and compared it frequently with a mercury barometer.[14] He found the temperature effects very large and gave up the project.

The exceptional water barometer referred to at the beginning of this chapter, which was intended as a serious piece of research equipment, will now be described. It was erected in the Royal Society's quarters by John Frederic Daniell (1790–1845), better known as the inventor of the hygrometer and the electric battery which bear his name. Daniell, whose scientific judgment was not always sound, thought that observations of a water barometer might throw light on atmospheric tides, as well as on the variation of the vapor pressure of water with temperature. As for the latter, this seems a highly indirect method of experiment, while the former purpose can be carried out only by a statistical study of a series of excellent hourly observations made over a period of time, long enough to enable the variations of pressure associated with weather changes to be averaged out. Nevertheless an elaborate water barometer was built in 1830, and reported on two years later at great length.[15]

It was a fine example of physics in the age of steam. (See Figure 14.2.) The cistern was in the form of a small steam boiler, so that the water, previously distilled, could be freed from air by boiling and immediately put into the tube. To make this possible, a stopcock was provided at the top of

---

[9] E. Mariotte, *Traité du mouvement des eaux* (Paris, 1686), pp. 90–92. *Oeuvres* (nouv. ed.; La Haye, 1740), I, 362. See also Wolf, *Histoire de l'Observatoire de Paris* (Paris, 1902), p. 103, who ascribes this barometer to J. D. Cassini.

[10] Acad. r. des Sci., Paris, *Registres* (ms.) 1683, fol. 9ᵛ: "M. Mariotte a fait son rapport des expériences qu'il a faittes auec le grand Barometre d'eau." The report does not seem to have been copied.

[11] *Faraday* (Groningen), Vol. 18 (1949), pp. 49–64; Vol. 22 (1953), pp. 61–66.

[12] A. G. Kästner, *Anfangsgrunde der angewandte Mathematik* (3rd. ed.; Leiden, 1780), pp. 1610–11. Quoted by Nolthenius (1949), p. 58. I cannot find this in Kästner's book.

[13] *Recherches sur les modifications de l'atmosphère* (Geneva, 1772), Vol. I.

[14] Luke Howard, *The Climate of London* (London, 1818), p. xi.

[15] J. F. Daniell, *Phil. Trans.*, Vol. 122 (1832), pp. 539–74.

Fig. 14.2. Daniell's water barometer, 1832.

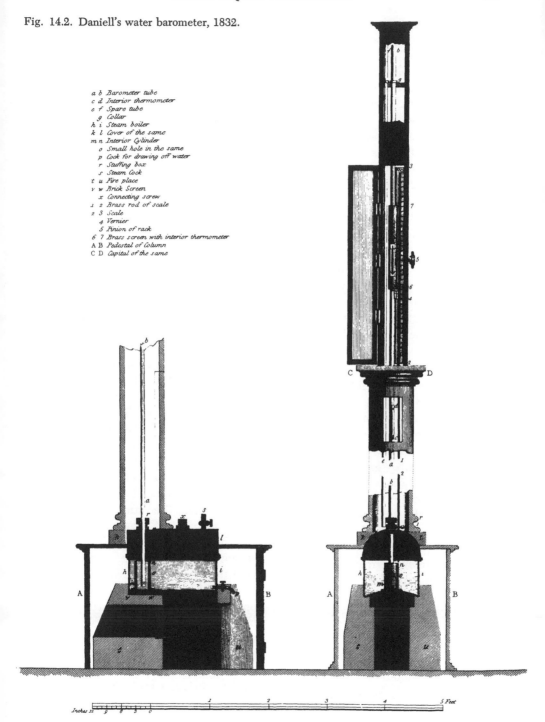

a b Barometer tube
c d Interior thermometer
e f Spare tube
 g Collar
h i Steam boiler
k l Cover of the same
m n Interior Cylinder
 o Small hole in the same
 p Cock for drawing off water
 r Stuffing box
 s Steam Cock
t u Fire place
v w Brick Screen
 x Connecting screw
1 2 Brass rod of scale
2 3 Scale
 4 Vernier
 5 Pinion of rack
6 7 Brass screen with interior thermometer
A B Pedestal of Column
C D Capital of the same

Inches 12 9 6 3 0   1   2   3   4   5 Feet

the tube and another on the closed cistern. First both were opened and the water boiled for some time; then the lower stopcock was partly closed so that the steam pressure could raise the water in the tube. After some had been allowed to flow out of the top, the upper cock was closed, the fire removed, and the lower tap opened, when the water fell to its proper level in the tube. The top of the tube was then sealed off. A layer of castor oil was placed on the water in the cistern to keep air from being redissolved.

It is interesting that Daniell was able to get his glass tube, one inch internal diameter and 40 feet long, manufactured in one piece. The description of the drawing of this tube, which he personally supervised, makes interesting reading.[16]

Daniell's water barometer was in use for two years, the heights being reduced to 39.38° F., the temperature at which water has its maximum density. The difference between the readings of the water and mercury barometers, in mercury units, gradually became greater, and it was suspected that air must have got in. Daniell's explanation was that the layer of castor oil had suffered some chemical change which rendered it pervious to air. Even if it had proved more permanent, however, this barometer could scarcely have justified its construction. One serious limiting factor would be the uncertainty in the mean temperature of such a tall column of liquid, and of course the oscillations in anything but calm weather were notable.

In 1844, nevertheless, the Royal Society asked Daniell to put the in-

strument back into working order, but he died suddenly at a meeting of the Society before the work was finished. It was later transferred to the Crystal Palace at Sydenham, where it was refilled, but did not work properly for very long, and was finally destroyed by fire in 1866.

Meanwhile the American physicist Joseph Henry had become interested in the erection of a large barometer at the Hall of the Smithsonian Institution in Washington, but was somewhat daunted by Daniell's difficulties.[17] The use of sulphuric acid was suggested to him by Professor Schaeffer of the Patent Office and independently by Professor Ellet of New York. The instrument maker James Green (see Chapter 12) undertook to set up the instrument. Because of the high density of the acid, the tube had to be only 20 feet long. A tube of that length and ¾ of an inch in bore was inverted in a glass bottle 4½ inches in diameter with two mouths, through one of which the tube passed, while to the other was connected a drying tube containing calcium chloride, as of course it was necessary to prevent water being absorbed from the atmosphere by the sulphuric acid. We may wonder whether the calcium chloride was a strong enough drying agent to prevent this.

Two thermometers were attached, one near the top, one near the bottom of the tube. The large correction for temperature was investigated by measuring, in the laboratory of the Smithsonian Institution, the density of the acid at various temperatures.

16 Ibid., pp. 540–41.

17 Joseph Henry, Proc. Amer. Assn. for the Adv. of Science, 10th meeting, 1856, pp. 135–38.

In 1865 Alfred Bird of Birmingham, described as an "experimental chemist," published an account of a water barometer which he had erected in his house in 1859.[18] This was ingeniously made of half-inch compo tubing and one-inch glass tubing. The process of filling it with distilled water was carefully planned so that after distillation the water never came in contact with air, and a thick layer of olive oil was left over the water in the cistern. An interesting feature of this barometer was that a large portion of the vacuum space consisted of a coil of compo tubing in a tank which could be filled with ice, so that the correction for the vapor pressure of water could be obtained directly.

Bird did this as a hobby; but we now come to James B. Jordan, the only man who ever made a business side line of erecting large barometers.[19] Jordan was employed at the Mining Record Office, and had obtained some experience in constructing large barometers with water as a liquid by building one for the Crystal Palace Company in 1867, the year after the fire. This was still in operation in 1881.[20] He also built one at about the same time for "the Museum in Jermyn Street."[21] Later he turned to glycerine, and he tells us why:

Although the water barometer is an instrument of much interest as a weather glass, it is found to be of little value for indicating with certainty the variations in atmospheric pressure, owing to the effect of temperature on the aqueous vapour above the column; this fact led to the examination of other fluids than water for a barometric column, and among those tried glycerine appears to be best suited for the purpose.[22]

Against this can be argued its high viscosity, making the removal of air more difficult, and its strong tendency to dilute itself by absorbing water from the atmosphere. The latter could be prevented by putting a layer of mineral oil over the glycerine in the cistern.

He must have had a persuasive manner, for he was granted £30 by the Committee which administered the government's fund for aiding scientific research, to set up such a barometer at Kew Observatory. The upper, glass part of the tube, with its scale, is still preserved there,[23] together with the cistern (Figure 14.3). The rest of the tube was made of "gas pipe." There was nothing special about this barometer, except perhaps the simple and ingenious way in which it was filled. To facilitate this there was a valve at the bottom of the tube under the liquid in the cistern, and a stopper at the top. Some glycerine which had been warmed for some time was first put in the cistern, and the valve closed. The tube was then filled to the top and stoppered, and the valve reopened. The resulting Torricellian vacuum was allowed to draw out most of the air

[18] Phil. Mag., Vol. 30 (1865), pp. 349–55.

[19] James B. Jordan, The Glycerine Barometer (London, 1881). Pamphlet, 19 pp. + colored plate. Also Proc. Royal Soc., Vol. 30 (1880), pp. 105–8.

[20] Jordan, The Glycerine Barometer, p. 5.

[21] Ibid. This museum was "The Museum of Practical Geology . . . devoted to the exhibition of all the mineral products of Great Britain and the colonies." London (Illustrated). A Complete Guide to Places of Amusement, etc. (London, 1877), p. 57.

[22] Jordan, The Glycerine Barometer, p. 5.

[23] Inventory no. 160.

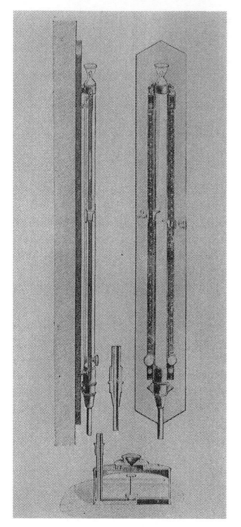

Fig. 14.3. Jordan's glyc-
erine barometer, 1880.

in order to avoid misconception, that it is
not pretended that the glycerine barometer can take the place of the standard
mercurial barometer as an instrument of
precision. All that can be claimed for the
new instrument, is the obvious advantage
that it possesses of indicating accurately
the smallest changes of atmospheric pressure by wide oscillations of a large fluid
column, the simplicity of its construction
enabling it to be used in situations where
it is likely to be of more value to the
observer than the ordinary barometer.[25]

His profession led him to think of
coal mines, and he suggests another
use for the glycerine barometer:

It has been observed in our fiery collieries
that a sudden depression of the barometric
column indicates an atmospheric change
which may probably be the first cause of
an explosion. The firedamp, which had
accumulated in places out of the range of
the ventilating system, is set free by the
reduced pressure, and thus streams forth
into the workings of the colliery, where
it forms, with the atmosphere, an explosive
mixture. There can be no doubt that in
such places the glycerine barometer would
prove of much value to those who are
responsible for the safety of the workings,
for the instrument indicates, far more
clearly than the mercurial barometer, all
those minute fluctuations in atmospheric
pressure which are constantly occurring.[26]

Although it is difficult to see exactly
what the Research Committee had in
mind, this was a workman-like job,
and various others were built by Jordan. Of these the most famous was
that at *The Times* office in Printing
House Square.[27] Storms on the days
following its installation led to a some-

from the glycerine, and then the valve
was shut again and the tube again
filled with warm glycerine. This process could be repeated.[24]

Jordan was modest in his claims. "It
may be well to remark," he writes,

[24] But the account in the *Scientific American* (*Supplement,* June 20, 1885, p. 7894)
says that a vacuum pump was used.

[25] *The Glycerine Barometer,* pp. 10–11.
[26] *Ibid.,* p. 12.
[27] *The Times,* London, Oct. 25, 1880, p. 10.

what over-enthusiastic leading article on October 29 ("It takes an observer of some practical skill to read a mercurial barometer in tenths," etc.), and readings at two-hour intervals were published in the paper for some years. Other similar instruments were set up at the South Kensington Museum in London, the Royal Scottish Museum in Edinburgh, and the *Deutsche Seewarte* in Hamburg.[28] The Edinburgh barometer, erected in 1882, is still complete eighty years later, though its vacuum is defective.[29] And on November 23, 1880, the Board of Directors of the South Eastern Railway Co. held a meeting at which "it was decided to order a Glycerine Barometer for Folkestone Harbour at a cost of £25."[30] This barometer was erected on the wall of a building at the docks so that it could be seen by prospective cross-channel passengers, a good way, one would imagine, of ensuring that the more nervous ones would be seasick. It has long since been removed, but is still remembered by one or two very old people at Folkestone.

Whether or not the inspiration was in Jordan's success I do not know, but in the succeeding decade there were at least three liquid barometers erected at widely separated places. In 1884 Charles Caron reported on a glycerine barograph erected at Beauvais,[31] the first such instrument to be made to give

a record. This barograph was part of a meteorograph, made for the *Commission météorologique de l'Oise*, which recorded rainfall, pressure, temperature, humidity, wind direction, and wind speed. To avoid the use of electricity Caron made all his instruments in large sizes. The barometer was in the form of a siphon with copper chambers 15 cm. in diameter joined by a 2.5 cm. lead tube. A float in the lower chamber was made to actuate the pen through a reducing lever system, so that one centimeter on the chart corresponded to exactly one centimeter of mercury. The force available for actuating the pen works out at about a kilogram per centimeter deflection, a figure that the makers of smaller barographs might envy.

This instrument, which was filled by a process similar to that used by Jordan, had worked well for two years when the report was issued, having been compared regularly with a Fortin barometer.

The next glycerine barometer in point of time was across the Atlantic in New York, erected by Zophar Mills, a merchant, in an office building at 146 Front Street.[32] The glass tube was in one piece and had to be lowered through holes made in the roof and two floors. There was a large copper cistern, and some trouble was taken to have this exactly at mean sea level, but otherwise the barometer was of no special interest.

A second barograph, this time using water as a liquid, was built in 1889 by

[28] Nolthenius, *Faraday*, Vol. 18 (1953), p. 62.

[29] Private communication from Mr. Robert W. Plenderleith of the Royal Scottish Museum, May 14, 1962.

[30] I am obliged to the Archivist of the British Transport Commission for this quotation.

[31] *Baromètre à glycerine* (Beauvais, 1884) (BN: 8°R. pièce. 2880.)

[32] Anon., *Scientific American*, Vol. 55 (1886), p. 403; see also C. Maze, *Cosmos*, Vol. 6 (1887), pp. 211-13.

J. Jaubert, director of the *Laboratoire d'études physiques*, in the *Tour St. Jacques* at Paris.[33] This incorporated a siphon barometer having a straight one-piece glass tube, 2 cm. in diameter and 12.65 meters long. A curved copper tube joined this to another glass tube 2 meters long which formed the short limb in which the barometric variations were at first read. Then in 1890 the recording mechanism was installed, consisting of a hollow copper float connected to the stylus by a silk thread passing over two small pulleys. The fluctuations of level were recorded on a large drum, an electromagnet bringing the pencil into contact with the paper every 90 seconds.

This instrument seems much less workman-like than the earlier one of Caron, which had continuous registration. In particular the long, uniform glass tube seems unnecessary, although it is fascinating to read that it was brought from the glassworks at St. Denis, 9 km. away, on the shoulders of six men.

In the twentieth century, large barometers have occasionally been constructed in connection with exhibitions. The most appropriate of these was certainly the olive-oil barometer designed by Father Guido Alfani for the exhibition held in 1908 at Faenza to commemorate the tercentenary of the birth of Torricelli.[34] This had an upper chamber of glass, 20 cm. in diameter and 100 cm. long, a lead tube, and a cistern 50 cm. in diameter and 25 cm. high which could be floated on water for adjustment. It could be, and indeed had to be read at a distance, being attached to an ornamental pylon surrounded by fountains and a pool. It was found difficult to get and to keep the olive oil free from air.

Finally we may refer to a water barometer constructed by A. Andant of the Sorbonne at the *Palais de la découverte* in Paris in connection with an exhibition illustrating the scientific work of Pascal.[35] This was in 1950, and by this time Pyrex tubes were available as a standard product for the chemical industry. Five 2.2 meter lengths were used in this barometer, 5 cm. in diameter, and they are joined by short lengths of rubber tubing, clamped on by metal bands of the sort familiar to the user of a garden hose. The construction is evidently quite satisfactory, for the instrument was still in operation in 1962. The cistern is a large rectangular battery jar.

With this we may leave these large barometers, never very useful as scientific instruments, but possessing a certain appeal to the imagination of the public, which the mercury barometer undoubtedly lacks.

[33] *Astronomie*, Vol. 9 (1890), pp. 216–19.
[34] B. Latour, *Cosmos*, Vol. 59 (1908), pp. 456–59.

[35] *Palais de la découverte. Catalogue de l'exposition de l'oeuvre scientifique de Pascal* (Paris, 1950), pp. 27–28.

Chapter 15

Air Barometers
and Related Apparatus

1. INTRODUCTION

THE mercury barometer, while certainly more accurate than any other, suffers from a number of drawbacks which from time to time have appeared very serious to certain people. Its shape and length, for example, seem to be against it, as is the difficulty of using it at sea. Unless specially made for portability, it is fragile in transport, and we have seen in Chapter 7 what efforts were required to surmount that particular difficulty.

The great majority of the instruments to be discussed in this chapter were devised either for use at sea, or with the idea of making the barometer shorter and hence easier to carry. This gives most of them a family resemblance, enhanced by the choice of air or some other gas as the "working substance," which has led me to choose the term "air barometer" as a convenient name for this whole class of instruments. Nothing more is intended by this word than to express the fact that they all make use of the relations between the pressure and the volume,

or the pressure and the density, of some gas or other.

Some of these instruments indicate, and others record, small fluctuations in pressure. These will be called microbarometers or microbarographs. Still others record the variation in the rate of change of pressure with time, and will be referred to as variographs for dp/dt.

2. THE "STATICAL BAROSCOPE" OF BOYLE AND GUERICKE

The "statical baroscope"—Boyle's term—described both by him[1] and by Guericke,[2] consists quite simply of a large glass bubble (Boyle) or a thin exhausted metal ball (Guericke, see Figure 14.1) hung on one arm of a

[1] In a letter to Oldenburg, *Phil. Trans.*, Vol. 1 (1666), pp. 231–39 (No. 14, July 2, 1666). In Boyle's posthumous *The General History of the Air* (London, 1692) (Fulton no. 194), this letter is reprinted (pp. 90–99) and given the date March 24, 1665. See also *Works* (1772 ed.), V, 648–53.
[2] *Ottonis de Guericke experimenta nova (ut vocantur) Magdeburgica de vacuo spatio*, etc. (Amsterdam, 1672), p. 101.

balance, and counterpoised by a small weight. Guericke wrote about it in a letter to the Jesuit Kaspar Schott dated Dec. 30, 1661, in which he shows that he had observed that the weight of the air varies.[3] As far as Boyle is concerned, the invention of this device presumably dates from about 1661 or 1662, as it is mentioned in Boyle's *New Experiments and Observations Touching Cold*, which, as we have already shown, was most probably written in 1663. We read:

I might mention . . . an experiment I thought on, and also attempted last winter to show ev'n upon a ballance the varying gravity of the atmosphaere in one and the same place, by hanging a small metalline weight at one end of a pair of scales so strangely exact, that they would turn with far less than the 500. part of a grain; and counterpoising it at the other end with a hermetically seal'd glass bubble, which being blown as large and as thin as could possibly be procur'd of so small a weight might by its great disproportion in bulk to the metalline body lose more of its weight than that would upon the ambient airs growing more heavy. But the particular account of this attempt belonging to another place, the trial ought not to be more than hinted here . . .[4]

These scales are hard to believe in, and by the time Boyle wrote to Oldenburg on March 24, 1665, they responded to only $\frac{1}{30}$ grain, which is more reasonable. With this balance the instrument was sensitive to a change in pressure of about $\frac{1}{8}$ inch of mercury.

Boyle saw some advantages in this device, in spite of its lack of sensitivity. The chief of its good qualities was in its clear demonstration of the weight of the air:

At first it confirms *ad oculum*, our former doctrine, that the falling and rising of the mercury depended upon the varying weight of the atmosphere, since in this baroscope it cannot be pretended that a *fuga vacui*, or a *funiculus*, is the cause of the changes we observe.[5]

Franciscus Linus' theory thus had another nail to its coffin. Boyle also thought the instrument easier to construct than the mercury barometer.

The "statical baroscope" has long been used in elementary physics lectures in connection with the air pump, to demonstrate the weight of the air. As a meteorological instrument it has found few advocates, but Sir Christopher Wren is stated to have devised "balances which are useful to other purposes, that shew the weight of the air by their spontaneous inclination."[6] Two centuries later A. Heller[7] attempted to revive it, using an optical lever to measure the deflection of the beam. It could indeed be made very sensitive by using micro-weighing techniques, but would then become complex because of the necessity of protecting it from all convection.

[3] Schott, *Technica curiosa*, etc. (Nürnberg, 1664), pp. 52–53.

[4] Boyle, *New Experiments and Observations Touching Cold*, etc. (London, 1665) (Fulton no. 70), pp. 19–20. *Works* (1772 ed.), II, 485.

[5] *The General History of the Air*, p. 96. (*Works*, *ed. cit.*, V, 651.) This was also the use Guericke saw for it.

[6] Thomas Sprat (1635–1713), *The History of the Royal Society of London, for the Improving of Natural Knowledge* (London, 1667), p. 313. See also p. 102 above.

[7] *Annalen der Physik*, ser. 5, Vol. 22 (1871), pp. 311–14.

3. HOOKE'S "MARINE BAROMETER" AND RELATED INSTRUMENTS

We have seen in Chapter 7 how Robert Hooke discovered or imagined the difficulties of reading a mercury barometer at sea and suggested the constricted tube. But he had another and, he believed, better idea:

I Bethought [me] of some other ways of Discovering the pressure of the air where no such motion is to be feard. And for this I think nothing can be better then a weather glasse made wth pure air and quicksilver wherein the quicksilver is left open to the pressure of ye air and soe becomes agitated by a double principle of motiõ. to witt by heat and the pressure of the air. In this Instrument I make use of cleaned Mercury, first that It may not by the great heats of the air under the torrid Zone be evaporated & diminished, 2dly that It may not be frozen by the colds of the frigid, And 3dly that the included air may not be augmented by any steames that may arise from almost any other subjacent liquor. For this I take to be very secure against any mutation of the air as to heat & Cold. To Distinguish therefore what part of its motiõ Is ascribable to the heat of the air I would have an other thermometer made wth Spirit of wine and Seald as I have elsewhere Described Now both these thermometers ought to be graduated together in a stove or some such place at once, whereby the corresponding degrees of expansion arising from heat & cold may be visible, And If by making observations wth them at Land the Discrepancy of them to one another be compard to the mercuriall standerd it will be very easy to calculate a table of Differences which shall exhibit all the various degrees of pressure answering to all the various Degrees of heat and cold. And soe only by the help of Two such weather glasses all the various pressure of the air

may be very practicably discovered at Sea and consequently the mutations of the weather may perhaps in great measure be timely enough Discoverd by the inquisitive & Diligent Mariner.[8]

This was read to the Royal Society on January 2, 1667/8, and Hooke was asked to make such an apparatus, which he brought to the meeting of January 16. It consisted of an air thermometer, using mercury as the indicating liquid, and an ordinary spirit thermometer with which to determine the temperature independently. It is rather astonishing that nothing whatever about this was published at the time, and one can only surmise that Oldenburg was feeling even more annoyed with Hooke than usual. It had to wait until 1701 for a laudatory notice[9] by the celebrated astronomer Edmond Halley (1656–1742), who had taken one on the famous voyage of exploration in the Atlantic in 1699–1701 during which he charted the variation of the compass. After describing the instrument and referring to his voyage, he goes on to say that

. . . it never failed to prognostick and give early notice of all the bad weather we had, so that I depended thereon, and made provision accordingly; and from my own experience I conclude that a more useful contrivance hath not for this long time been offered for the benefit of navigation.[10]

---

[8] Royal Society, *Classified Papers*, Vol. XX, item 48, autograph.

[9] *Phil. Trans.*, Vol. 22 (1701), pp. 791, 788, 785, 794. (The pagination is extremely erratic. This begins with the third of three pages numbered 791.)

[10] *Ibid.*, p. 794.

The inconstancy of the zero of this instrument was known to Halley but did not worry him:

It has been observed by some, that in long keeping this instrument, the air included either finds a means to escape, or deposit some vapours mixt with it, or else for some other cause becomes less elastick, whereby in process of time it gives the height of the mercury somewhat greater than it ought; but this, if it should happen in some of them, hinders not the usefulness thereof, for that it may at any time very easily be corrected by experiment, and the rising and falling thereof are the things chiefly remarkable in it, the just height being barely a curiosity.[11]

A long time elapsed before anyone bothered about the absolute accuracy of a *marine* barometer. At the History of Science Museum, Oxford, there is a marine barometer of Hooke's type, ascribed to John Patrick.

Guillaume Amontons, whom we have

met before as the inventor of the conical barometer, reinvented Hooke's barometer in 1705, except that he used a lighter liquid instead of mercury.[12] A tube 5 feet long, open at the top, had a 180° bend near the bottom, and the end was blown into a two-inch bulb. He hung a thermometer near it and, like Hooke, made an empirical correction; but he added a sliding scale that could be shifted according to the temperature indicated by the thermometer.

The instrument was greatly developed by G. W. Richmann of St. Petersburg,[13] who described several versions, of which one is shown in Figure 15.1, and which all had the glass parts in water baths. In the one illustrated, the vessel *ABCDE* is filled with mercury from *X* to *Y* and is attached to an inclined plane hinged at *F*, which can be leveled by means of the screw *HK*.

[11] *Ibid.*, p. 785 [793].

[12] *Mém. Acad. r. des Sci.*, 1705, pp. 49–54.
[13] *Nov. Comm. Petrop.*, Vol. 2 (1749), pp. 181–209.

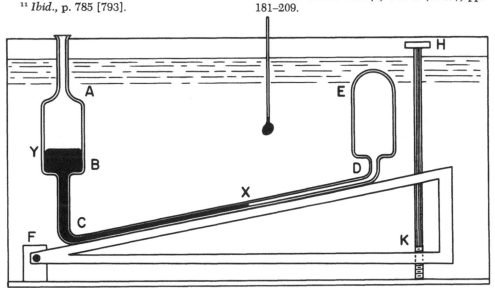

Fig. 15.1. One of Richmann's air barometers, 1749.

A thermometer is in the water bath with the rest of the apparatus, which was certainly not designed for marine use, but rather to obtain an expanded scale.

J. H. de Magellan says a very good word for the "marine barometer" of Hooke and describes an elaborate one with two taps and an adjustable leather bag, which, one would imagine, effectively prevented it retaining a calibration.[14] This complication, of course, is in the name of portability, an end which obviously led to the design described by Daniel Wilson in 1817,[15] which had a closed iron chamber half-filled with mercury, half with air (Figure 15.2). The thermometer was in the sample of air, a considerable advantage, and there was a stopcock to make the instrument more easily portable. This barometer was calibrated at 75° F. with the help of an air pump. When the instrument was to be used by a traveler, the chamber was put under the armpit until the thermometer read 75° F., then the manometer tube placed vertical and the observation made. The idea has a sort of wild reflection of practicality, but presumably it is too hot to travel when the temperature is higher than this.

The use of the observer as a source of heat reached its apex, one would think, in the "mouth barometer" made by P. Grützner about 1895 for mountaineering.[16] This was simply an air thermometer with a small flat bulb

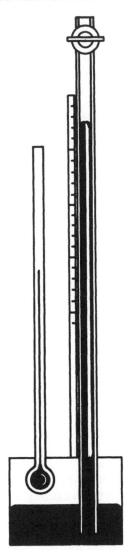

Fig. 15.2. Wilson's air barometer, 1817.

which could be put under the tongue! In 1902 its inventor defended this instrument by quoting a list of comparisons with an aneroid barometer on mountaineering expeditions;[17] he had also taken a clinical thermometer with

[14] J. H. de Magellan, *Description et usages des nouveaux baromètres,* etc. (London, 1779), pp. 146–49.

[15] *Annals of Philosophy,* Vol. 9 (1817), pp. 313–22.

[16] *Mitt. d. deutsch. u. österr. Alpenvereins* (1895), p. 157.

[17] *Annalen der Physik,* Vol. 9 (1902), pp. 238–42.

him, so that we are really back with
Hooke. In extenuation of this astonishing idea it may be pleaded that at
this particular period the aneroid was
somewhat under a cloud because of
its elastic errors.[18]

In 1818 Hooke's, or rather Amontons' barometer was invented all over
again, being given the learned name
Sympiesometer (συμπιεζω, compress,
μετρον, measure). Alexander Adie of
Edinburgh patented it in that year,[19]
and in 1819 published a description.[20]
The patent is entitled "An improvement on the air barometer." The improvement was a sliding scale, which
Amontons' also had, but at any rate the
Sympiesometer was much smaller and
neater than Amontons' barometer.
Most, but not all of the discussions of
the air barometer after 1818 refer to
the instrument as the sympiesometer.

Adie's instrument (Figure 15.3) has
a bulb A, filled with hydrogen, and
another bulb C which, with part of the
connecting tube, contains colored
almond oil. A thermometer is mounted
on the same board. The scale of pressures MN is made to slide against a
fixed scale of temperatures OP, both
being graduated empirically. To use
the instrument, the thermometer is
first read, and an index mark on MN
set opposite the reading of the thermometer on OP. The pressure is then
read from MN opposite the level of
the oil in the tube.

The sympiesometer was calibrated
by comparison with an ordinary barometer in a pressure-vacuum chamber,

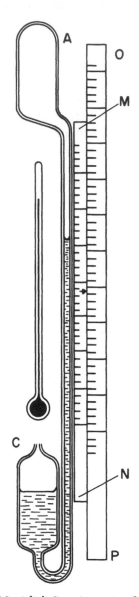

Fig. 15.3. Adie's Sympiesometer, 1818.

[18] See p. 437 below.
[19] British Patent 4323 (1818).
[20] *Edinburgh Phil. J.*, Vol. 1 (1819), pp.
54–60.

and the scale OP was calibrated by
varying the temperature with the pressure constant at some mean value. The
correction for temperature will, of
course, be exact only at this pressure.

Before making the instrument public, Adie had it tested on ships in the Tropics, in the Arctic, and near the coast of Scotland. All the reports seem to have been most enthusiastic. Adie quotes a letter from the Commander of the *Isabella*, one of the ships of the Ross expedition to the Arctic, in which it is stated that

The Sympiesometer is a most excellent instrument, and shews the weather far better than the marine barometer. In short, the barometer is of no use compared to it . . . in my opinion it surpasses the mercurial barometer as much as the barometer is superior to having none at all.[21]

There were still no synoptic maps of the oceans, of course; and as for the ordinary marine barometer, the construction of the tube had not yet been standardized.[22]

There is a sympiesometer by Adie in the Science Museum, London.[23] There is also another "thermobarometer,"[24] dated 1839, the invention of which is ascribed to Ronketti.[25] This differs from the sympiesometer in two ways: first, mercury is used as the working fluid; and second, the bulb of the thermometer is sealed into the bulb holding the sample of gas whose volume is measured. This should be a decided improvement; but it is found in Wilson's instrument and also in one

described in 1823 by J. J. Prechtl[26] in a paper of almost unbelievable verbosity. It was used with the measuring tubes in a horizontal plane, two level bubbles being provided for that purpose. This instrument was intended not as a marine barometer, but for the measurement of heights.

There is a world of difference between the two uses. In 1829 the eminent Scottish scientist James David Forbes (1809–1868), while praising the sympiesometer as a marine barometer[27] —"as a marine barometer, its superiority in accuracy and utility, as well as convenience, seems fully established"—practically shot it dead as an instrument for mountain work, on the basis of his own experiments with it. Surprisingly, its apparently greater portability than the mercury barometer was an illusion, for it had to be carried right way up, and while concussions might not break it, they would separate the column of oil. But apart from portability altogether, it had two fundamental faults: the oil gradually absorbed the hydrogen, and the thermal lag of the bulb of hydrogen was very small, much less than that of a mercury thermometer. Forbes also suggested that it would be better if the bulb of the thermometer were placed inside the hydrogen.

In spite of Forbes, patents on similar instruments were granted at intervals; in 1840 to Charles Cummins[28] for a sympiesometer of somewhat different shape to Adie's, and using sulphuric

[21] *Ibid.*, p. 59.

[22] See p. 165 above.

[23] Inventory no. 1918–14. There are a number of sympiesometers by various makers in several museums.

[24] Inventory no. 1929–962.

[25] John George Ronketti was for a time a partner of Enrico Negretti in Cornhill, London. But the instrument in the Science Museum was made by Carpenter & Westley.

[26] *Jahrb. des k.k. polytechn. Instituts in Wien*, Vol. 4 (1823), pp. 284–327.

[27] *Edinburgh J. Sci.*, Vol. 10 (1829), pp. 334–46.

[28] British Patent 8462 (1840).

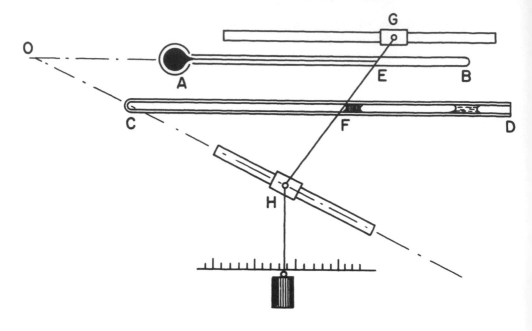

Fig. 15.4. Hans and Hermary's barometer, 1873 (diagram).

acid instead of almond oil; in 1850 to A. O. Harris,[29] who proposed to use carbon dioxide as the gas, and mercury as the liquid. In 1886 S. A. de Normanville patented[30] the idea of proportioning the bulbs of a sympiesometer and a thermometer so that the movement of the liquids in each, at constant pressure, is the same for any given change in temperature. The pressure scale was attached to the thermometer and moved with it, being raised or lowered until the two liquid columns were at the same level. The pressure was then to be read against a fixed index.

The further development of the sympiesometer is not very interesting except for one rather astonishing display of ingenuity by two Captains in the French artillery, M. L. P. Hans and H. A. H. Hermary,[31] which actually was given a silver medal at the *Exposition Universelle*, Paris, 1878, and a gold medal at the *Exposition des Sciences* in 1879. We must state at once that the juries at these exhibitions were misguided, as it is doubtful whether Hans and Hermary's barometer was ever more than a conversation piece or a decoration for a front hall; but it could in principle be calibrated theoretically, which distinguishes it from earlier instruments of this type. The principle is stated as follows: "The pressure being constant, the intersection of a straight line passing through the absolute zeros of a liquid and [an] air thermometer with another straight line passing through the reading points of the two ther-

[29] British Patent 13,422 (1850).
[30] British Patent 3625 (1886).

[31] British Patent 2147 (1875).

mometers will always be exactly the same whatever may be the temperature."[32] This is true if the scales of both thermometers are linear in the region in which they are used, and if "absolute zero" is extrapolated on this basis.

This being admitted, let us try to show by a diagram (Figure 15.4) just what they did. Suppose the tubes of a liquid-in-glass thermometer *AB* and an air thermometer *CD* are mounted parallel to one another. Let the extrapolated absolute zero of the former be at *O*; that of the air thermometer will be at *C* if it is made of uniform tubing. *E* is the end of the column of liquid; *F* the inner end of the reading index of the air thermometer (a short column of sulphuric acid). The principle states that the intersection *H* of *OC* and *EF* will be invariable with temperature, provided the pressure does not change. This property is used by having a thread attached to a slider at *G*, which moves along a horizontal rod, and passing another slider at *H*, which slides along the direction *HCO*. The thread ends in a weight after passing in front of a scale on which the pressure is read. In the actual instrument both thermometers are folded round the back of the panel which bears the scales. There is also a short column of non-drying oil near the open end of the air thermometer to protect the sulphuric acid from the atmosphere, and this is also patented.

If I seem to have devoted an undue amount of space to this curious barometer, it is because there are three of them at the *Conservatoire National des Arts et Metiers*,[33] one of them recording its successes on the exhibition circuit, and because the principle of their operation is so entirely obscure

[33] Inventory nos. 9136, 9524, and 10916.

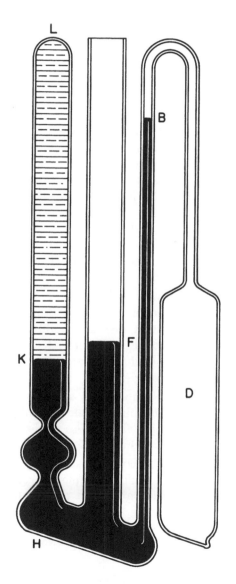

Fig. 15.5. Calantarients' compensated barometer, 1870.

when one is confronted with the instruments themselves.

There were several suggestions for making the air barometer direct reading, that is to say self-compensating for temperature changes. A vague idea for doing this was patented by D. S. Brown in 1853,[34] but an instrument was built and described in 1870 by J. A. Calantarients which at least showed how such an end might be attained.[35] In this (Figure 15.5) the bulb D is filled with slightly rarefied air, KL with liquid sulphuric ether. They are so proportioned that the fall which would take place in the scale tube B with a rise in temperature is exactly compensated by the rise in F due to the expansion of the mercury and ether in the system LHF. The two narrowings at the lower end of LH are intended to prevent ether escaping when the instrument is laid on its side. How the mercury stays in is not clear.

The difficulty with this sort of arrangement lies mainly in the time required to attain thermal equilibrium, and it is difficult to see how it could have been thought adequate for observations *en voyage*.

Somewhat simpler and slightly different in principle, the air barometer of the Abbé L. Jeannon also made use of two liquids (Figure 15.6).[36]

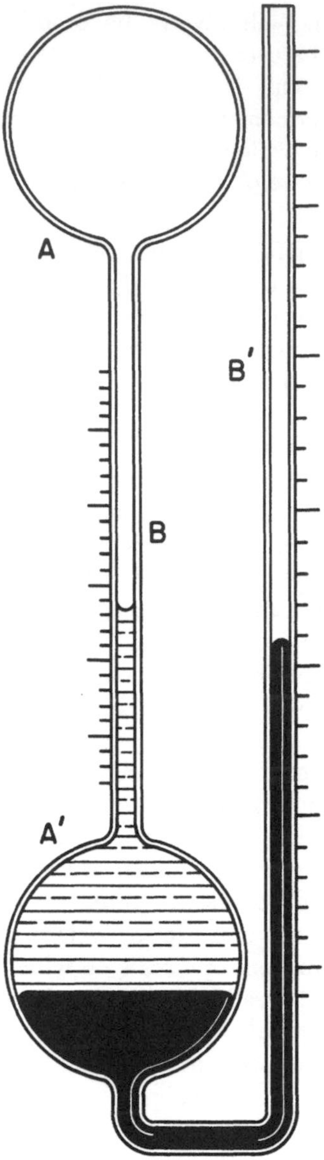

The bulb A contains air. Suppose first that the bulb A' contained only mercury, which also filled part of the tubes B and B'. Then it ought to be possible to proportion the volumes so that at constant external pressure the

Fig. 15.6. Jeannon's barometer, 1863.

expansion of the mercury with temperature, and its consequent rise in B', would just compensate for the increased pressure in A, and keep the level constant in B. A scale of pressure could then be placed on B and a scale of

[34] British Patent 243 (1853).
[35] *Phil. Mag.*, Vol. 39 (1870), pp. 371–74.
[36] *Les Mondes*, Vol. 1 (1863), pp. 581–83.

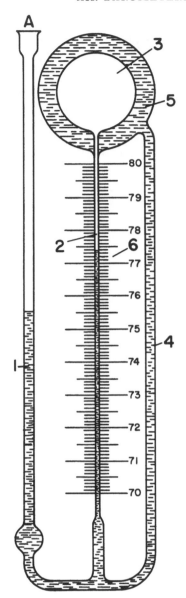

Fig. 15.7. Borgesius' compensated barometer, 1912.

by making B smaller than B'. This construction was patented in 1877 by R. M. Lowne.[37]

Probably the best thought out of these compensated air barometers was that of Dr. A. H. Borgesius of The Hague.[38] As shown in Figure 15.7, the gas is almost entirely enclosed in the liquid, making it likely that their temperature will be the same. When the temperature rises, the expansion of the liquid will raise the level in the tube 1 just enough, if the instrument is correctly proportioned, to compensate for the increased pressure in the gas 3. One of these instruments, with some sort of oil as a working liquid, is in the History of Science Museum at Leiden.[39]

Apart from a number of patents of no particular value, this seems to complete the variations on Hooke's and Amontons' air barometers.

We now turn to a unique instrument patented in several countries by A. S. Davis of Leeds,[40] of which there is an example in the Science Museum, London.[41] This depends on taking a certain volume of atmospheric air, compressing it with a certain head of mercury, and measuring its new volume. The air chamber 2 (Figure 15.8) and the measuring tube 1 are immersed in water 8 in a tube 7, in order to prevent rapid changes of temperature. A certain amount of mercury 4 is held in the upper chamber 3; the remainder of the fittings at the top constitute a drying tube which air must pass to

temperature on B'. Now if the upper part of A' and also part of B were filled with a lighter, non-volatile liquid, the scale could be magnified, and a further magnification could be obtained

[37] British Patent 88 (1877).
[38] D.R. Patent 261,090 (1912).
[39] Inventory no. Th30.
[40] British Patent 16,285 (1900). D.R. Patent 132,349 (1901). U.S. Patent 670,178 (1901).
[41] Inventory no. 1905-74.

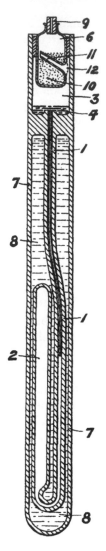

Fig. 15.8. Davis' barometer, 1900.
*(By permission of the Control-
ler, H. M. Stationery Office)*

reach the inside. This apparatus is
normally horizontal, so that the chamber 2 is open to the atmosphere. To
read the pressure it is rotated into a
vertical position, so that the mercury
closes the mouth of the tube 1 and
flows down into it until the pressure in

2 is balanced by the pressure of the
atmosphere plus the head of mercury
in the tube 1 and the chamber 3. The
position of the end of the column in 1
is read on a scale. The instrument is
then laid horizontal again ready for
the next reading. There is actually an
axis, stand, and counterweight to ensure that the opening 12 remains on top
when the tube is horizontal.

It remains to describe an extremely
ingenious constant pressure air barometer invented by Anton Steinhauser,[42]
and shown in Figure 15.9.

[42] *Repertorium der Phys.*, Vol. 23 (1887),
pp. 411–25.

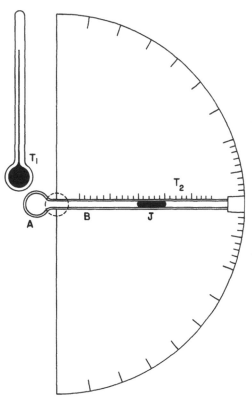

Fig. 15.9. Steinhauser's con-
stant-pressure barometer, 1887.

A bulb A and a narrow tube B, open at the end, are mounted so that they can rotate about the center of a circular scale. In B is a column of mercury J, which must be rather more than half as long as the extreme range of pressure which it is desired to measure, expressed in mercury units. The bulb contains air. At a mean pressure b_0 the tube B is placed horizontal and the scale T_2 established by varying the temperature of the whole apparatus, in comparison with a mercury thermometer T_1. The pressure scale is then computed from the fact that if s is the length of the index, and the tube B is inclined at an angle θ to the horizontal, the pressure in the bulb A is $b_0 + s. \sin \theta$. To read the pressure at any time, the tube is inclined until the left-hand end of its index shows the same reading on T_2 as the temperature indicated by the mercury thermometer. This instrument would no doubt be very good until the outer end of the tube became dirty.

4. VARIABLE-CISTERN AIR BAROMETERS

There are two air barometers depending on the pressure-volume relation that seem to fall into a class of their own. The earlier is due to Henry Meikle,[43] and is shown diagramatically in Figure 15.10. A constant mass of air is enclosed in the bulb A. By varying the volume of the cistern B, the mercury in the tubes C and D is brought to the same height, so that the pressure in A is the same as that

[43] *Phil. Mag.*, Vol. 62 (1823), pp. 214–18.

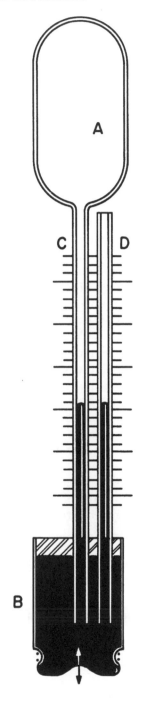

Fig. 15.10. Meikle's air barometer, 1823.

outside. The latter is then a function of the volume, indicated by the reading of the scale on C, and the temperature, indicated by a separate thermometer. Meikle intended this barometer for the measurement of heights, and it ought to have been more reliable for this than the sympiesometer, but the question of thermal lag would still be awkward.

The second variable-cistern air barometer, in which mercury is brought to a constant level in a bulb and the height of a column of mercury measured, was patented half a century later.[44] Its principle is exactly that of the well-known McLeod gauge. A considerable improvement was suggested in 1937 by the well-known Finnish meteorologist Vilho Väisälä.[45] He enclosed the volume of gas in a double-walled vessel, the space between the walls forming the "bulb" of a mercury thermometer which presumably measured the temperature of the gas. In this the instrument resembles that of Borgesius already referred to, though it is most unlikely that Väisälä had heard of the latter. One may still wonder about the effect of a changing ambient temperature.

5. THE DIFFERENTIAL BAROMETER

The difficulty with the barometers described in the two preceding sections lies in the necessity of measuring the temperature of an enclosed volume of

[44] T. T. MacNeill, British Patent 3280 (1870).
[45] *Ann. Acad. Scient. Fennicae,* Ser. A, Vol. 48 (1937), No. 7, 6 pp.

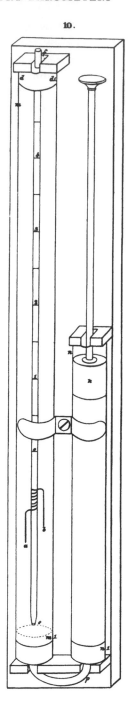

Fig. 15.11. The differential barometer, 1842.

air. We now come to a class of instruments in which this requirement is replaced by a much simpler one, namely that this temperature should stay constant while a measurement of atmospheric pressure is being made.

The first idea of such a barometer seems to have been due to E. F. August,[46] whose name is better known in connection with the wet- and dry-bulb psychrometer. A rather similar instrument was described by Hermann Kopp in 1840,[47] but as all these barometers rest on the same principle, we shall confine ourselves to the improved construction described by Kopp in 1842.[48]

This instrument, a diagram of which appears in Figure 15.11, depends on a measurement of the excess pressure necessary to compress air at a known volume to a known smaller volume. On depressing the piston in the cylinder k, the mercury rises in the cylinder m until, when it reaches e, it cuts off a certain volume of air in m. As it rises farther, it arrives at either of the points a and b. The volume in m above each of these points is known, and the excess pressure needed to compress the original volume to either of these new volumes is read on the scale in the tube ef, which has its zero at e. If the two volumes are v' and v'', and the excess pressures p' and p'', it can be shown that the atmospheric pressure is

$$p = (p'v' - p''v'')/(v''-v')$$

provided the compression is isothermal. Kopp found this instrument, only 25 cm. high, very useful on his travels, and claimed that its error was less than half a Paris line (0.6 mm.) of mercury.[49]

In 1854, when the need for such things was rapidly disappearing, Karl Kreil described another sort of differential barometer which could be made half as long as an ordinary mercury instrument.[50] It can be understood from Figure 15.12, in which $ABCD$ is a closed vessel, like a barometer cistern, fitted with a piston FG, a valve L, and two points M and N which define two volumes above their levels, one twice the other. The vessel is partly full of mercury. With the valve L open, the piston is adjusted until the mercury touches M. The entire barometer tube HP will then be full of mercury. After shutting the valve, the mercury is then adjusted to touch N. Doubling the volume of the enclosed air, this halves the pressure; and if there is a scale on HP with the level of N as its zero and graduated in half-units, the pressure of the atmosphere can be read on it as on an ordinary barometer. The correction for temperature would seem to be similar to that for an ordinary instrument with an adjustable cistern.

Finally in 1876 E. Dufour of Nantes described an air barometer[51] which

[46] *Annalen der Physik*, Vol. 3 (1825), pp. 329–40.

[47] *Annalen der Physik*, Vol. 40 (1840), pp. 62–67.

[48] *Annalen der Physik*, Vol. 56 (1842), pp. 513–41, Abs. in *Ann. de Chim. et de Phys.* 3d. sér., Vol. 6 (1842), p. 383.

[49] Note that the old French measurements were still in use. See Chapter 8.

[50] *K. Akad. d. Wiss., Wien, nath-naturw. Kl., Sitzungsber*, Vol. 14 (1854), pp. 397–98. This was reinvented by Blondeau *(Compt. Rend.*, Vol. 46 [1858], pp. 939–41).

[51] *Assoc. française pour l'Avanc. des Sci., Congrès de Clermont-Ferrand 1876*, pp. 300–2.

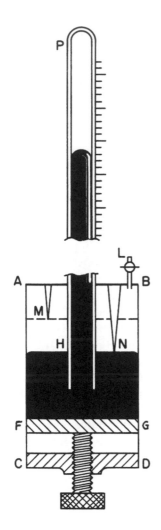

Fig. 15.12. Diagram of Kreil's traveling barometer, 1854.

6. THE FLOATING BELL

In 1704, when, as we saw in Chapter 6, the magnification of the scale of the barometer was a major concern, a Mr. Caswell of Oxford wrote to Flamsteed giving an account of a new baroscope he had invented. This was eventually communicated to the Royal Society.[52] It consisted of a ballasted copper box, terminating at the bottom in an open tube, and having at the top a closed cylinder of uniform cross section which emerged from the water in which the device floated. Fine wires were stretched diagonally from this cylinder, in order to make small changes in the level of immersion of the float more evident. The theory given by Caswell is not adequate; the magnification depends, to a first approximation, on the relative cross-sectional areas of the tube at the bottom and of the cylinder which emerges from the water; this ratio can be made extremely large, and Caswell claimed a magnification of 1200 with reference to the ordinary mercury barometer. He noted that every gust of wind, and every cloud passing overhead, set it in motion.

More than a century later an essentially similar device was described, also in the *Philosophical Transactions*, by John Thomas Cooper.[53] Such instruments can give almost any desired magnification, and we shall find the principle applied with success when we come to consider variometers. Temperature has a large effect on them,

was extremely compact, on the principle of Kopp's instrument except that the increment of pressure could be measured by a micrometer screw. The chief difficulty with such an instrument might be leveling it.

[52] *Phil. Trans.*, Vol. 24 (1704), pp. 1597–1603. (There are two pages numbered 1597.)
[53] *Phil. Trans.*, Vol. 129 (1839), pp. 425–31.

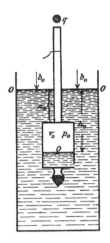

Fig. 15.13. The *Luftdruckära-ometer* of Fischer, 1900.

and for any accuracy special precautions are necessary.

Such precautions were taken, toward the end of the nineteenth century by K. T. Fischer, in an instrument which he called the *Luftdruckäraometer* (air pressure hydrometer).[54] It was indeed rather like a hydrometer, except that there was a hole near the bottom into which water could enter (Figure 15.13). Fischer first found by experiment that by means of melting ice a chamber could be kept constant in temperature to within ±0.002°C. for an entire month. The instrument was immersed in a vessel kept in such a thermostat. He then showed how a change of barometric pressure from b_0 to b was related to the relative movement of the float $(x-x_0)$ in terms of the dimensions of the instrument and the initial pressure. Being inde-

pendent of accelerations, it was very suitable for use in the gondola of a balloon, and one was constructed in 1896 and performed very well in such service. In considering these experiments, we must remember that the elastic aftereffect of the aneroid barometer was very troublesome at the time, especially when large and rapid changes of pressure were involved.

7. THE "REPEATING BAROMETER"

We now come to an instrument designed to be extremely easily portable; just a glass tube with a scale engraved on it, and a valve at each end.[55] It contains a column of mercury, shorter than the tube, and originally in contact with the top valve. To use it, the tube being held vertical, the bottom valve is opened, then closed; the top valve being opened, the mercury falls a little, compressing the air beneath it. The top valve is then closed again, and the bottom one opened; the mercury again falls, rarefying the air above it. After n such cycles the total fall is measured, and the pressure calculated from a formula derived with the aid of a good deal of mathematics. To help in the use of his *baromètre répétiteur*, the Baron d'Avout published some convenient tables.[56] A good deal of thought seems to have gone into the design of the valves in this instrument.

[54] *Annalen der Physik*, Vol. 3 (1900), pp. 428–37. Also in *Meteorol. Zeits.*, Vol. 17 (1900), pp. 257–74.

[55] Baron d'Avout, *Compt. Rend.*, Vol. 44 (1857), pp. 658–61.

[56] Baron d'Avout, *Notice et tables destinées à accompagner le baromètre répétiteur*. Pamphlet (Paris, 1857).

8. THE "CAPILLARY BAROMETER"

Finally we arrive at the simplest device that can be used to measure, even roughly, the presence of the atmosphere; nothing more than a straight glass tube, open at both ends, and a thread of mercury.

The possibility of measuring pressure with such an apparatus seems to have occurred first to Franz Emil Melde (1832–1901), a Professor at Marburg, who had the whimsical notion of finding out what kinds of experiment you could do with narrow glass tubes.[57] The tube he used for this purpose had an internal diameter of about 2 mm. It was provided with graduations, and a calibration of its volume per unit length was established by sliding a mercury thread along it, and making measurements in the usual way.

Now suppose we have a thread of mercury, say half the length of the tube, somewhere in the middle. We stop one end firmly with a finger, and bring the tube vertical, the stopped end down. Then if the thread of mercury is h mm. long, a volume of air v will be held under a pressure $B + h$, where B is the atmospheric pressure in millimeters of mercury. Still keeping the tube closed at the same end, we invert it, and the same air, now under a pressure $B - h'$, now occupies a volume v'. Applying Boyle's law, $B = (vh + v'h')/(v' - v)$. The beauty of this is that it does not assume that the bore of the tube is uniform, or even conical, as long as it has been calibrated.

[57] *Annalen der Physik,* Vol. 32 (1887), pp. 659–70.

A similar instrument was described in the following year by T. H. Blakesley,[58] and was apparently made commercially by Watson Brothers at 4 Pall Mall, London. The tube was now permanently closed at one end and was assumed to be cylindrical. C. Fischer rightly pointed out that Melde had the priority.[59]

Actually the instrument is quite insensitive unless it is very long, and if it is very long its advantages disappear.

9. MICROBAROMETERS, AND A MICROBAROGRAPH

The first application of the air barometer to the indication of minute variations of pressure seems to have been made by the celebrated physicist James Prescott Joule (1818–1889), whose name will always be associated with the determination of the mechanical equivalent of heat. In his own words[60] the instrument "consists of a large carboy connected by a glass tube with a miniature gasometer[61] formed by inverting a small platinum crucible over a small vessel of water. The crucible is attached to the short end of a finely suspended lever multiplying its motion six times." Presumably there would have been a tap which

[58] *Phil. Mag.,* ser. 5, Vol. 26 (1888), pp. 458–60.

[59] *Zeits. für Instrum.,* Vol. 10 (1890), pp. 65–67.

[60] *Manchester Lit. & Phil. Soc., Proc.,* Vol. 3 (1863), p. 47. *(Scientific Papers* I, 534.)

[61] British English for the large variable-volume gas-holders which disfigure our cities. A misleading word.

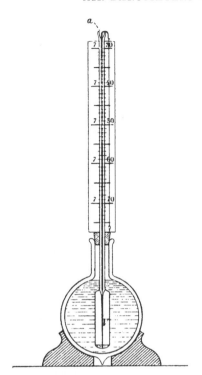

Fig. 15.14. The air barometer of Behn and Kiebitz, 1903.

constant, but Behn and Kiebitz give an exact theory. As to the correction for ambient temperature, this involves only the small column of glycerine. It is shown that a magnification of 10 or more in mercury units is entirely practicable, and apart from the choice of a liquid this would seem to be a useful instrument for some special purposes, especially as in a good vacuum flask the temperature can be maintained constant for at least a month.

could be opened momentarily to equalize the pressure inside and outside the apparatus when the lever had gone off scale.

The invention of the vacuum bottle by Sir James Dewar in 1892 gave rise to many ingenious applications. An excellent example is the "short glycerine barometer" of U. Behn and F. Kiebitz.[62] In this (Figure 15.14) a volume of air V is kept at a constant temperature of 0°C in a good-sized vacuum flask filled with chipped ice and water, the pressure variations being observed on a column of colored glycerine about 1 mm. in diameter. The volume of the sample of air is, of course, very nearly

[62] *Phys. Zeits.*, Vol. 4 (1903), pp. 543–45.

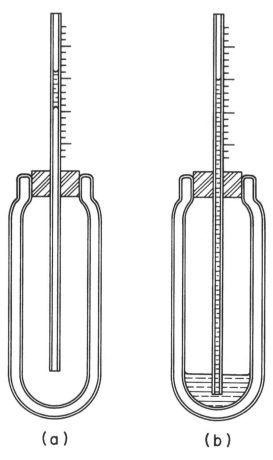

(a)   (b)

Fig. 15.15. Montrichard's microbarometers, 1909.

The suggestion which appears to have been made about 1909 by the Marquis de Montrichard[63] seems less happy, as a known constant temperature could not be attained. The two possible arrangements are shown in Figure 15.15; at *a* we have an isolated column of liquid as an index, at *b* a column dipping into some liquid in the bottle. In the first arrangement there is no theoretical limit to the magnification, but in the second it is limited by the density of the liquid. Mercury was found to be unsuitable; it sticks. There are two of these instruments in the C.N.A.M.,[64] perhaps the prototypes.

If this sort of instrument is to be made into a microbarograph, photographic recording is the simplest form.

In 1911 G. J. Gibbs obtained a patent[65] on a device of this kind, consisting of a large insulated vessel connected to a horizontal glass tube which is turned up at the end into a small bulb, open to the air. A small drop of non-volatile liquid in the tube forms an index, the image of which can be projected on to a photosensitive surface. There is a tap to bring the vessel to atmospheric pressure, and thus center the index, at any time.

A further development of this general idea was patented in 1952 by G. E. Conover of Garland, Texas;[66] but this was a null instrument, having a chamber of fixed volume joined by a capillary containing a drop of liquid to a second chamber, the volume of which could be altered in a known

[63] *Der Mechaniker*, Vol. 17 (1909), pp. 253–54.
[64] Inventory nos. 14388[1] and 14388[2].

[65] British Patent 28,625 (1911).
[66] U.S. Patent 2,617,304 (1952).

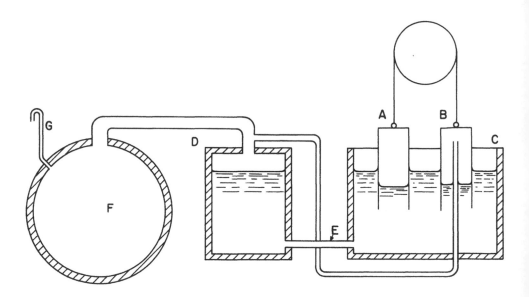

Fig. 15.16. Whitehouse's variograph, 1871.

way. Either chamber could be put in communication with the atmosphere or closed at will, and the whole thing was contained in a thermostat. Changes in pressure were measured in terms of the change in volume of the adjustable chamber required to bring the drop back to its original position.

10. VARIOGRAPHS FOR dp/dt

It would be impossible to leave any of the microbarometers described above running for any length of time unattended without the index going off the scale.

The cure for this trouble is very simple, and seems to have been discovered in 1871 by Wildman Whitehouse.[67] It is to connect the large vessel of air with the atmosphere by a very narrow passage, for example, a length of capillary tubing, so that the major changes of atmospheric pressure, which take place slowly, will find their way into the large vessel, while rapid fluctuations are recorded as before. There is, of course, no sharp boundary between fast and slow oscillations; the slower the variations, the less completely they will be recorded.[68]

In Whitehouse's instrument (Figure 15.16) there are two bell-shaped floats A and B in the same open cylinder C partly full of water and connected by a tube E to a closed cylinder D, also partly full of water. The air space of

the latter is in communication with a large vessel F buried in the ground to protect it from changes of temperature, and communicating with the air through a capillary tube G. This large vessel also communicates with the space under the right-hand float, which therefore descends when $dp/dt > 0$, and vice versa. The two floats are connected by a thread passing over a pulley to which a recording pen may be connected.

The second cylinder with the water in it does not seem to be essential, and a simpler solution was adopted by the famous English meteorologists Sir Napier Shaw and William Henry Dines in their "microbarograph."[69] In this the measuring element is a small bell floating in mercury, a tube from the large insulated vessel rising into the space under the bell, the motion of which is magnified by a system of levers.

An entirely different variograph was described by H. Benndorf and W. Zimmermann.[70] A large vessel 50 liters in volume, of which the top portion appears in Figure 15.17, is fitted with a chimney Z about 5 cm. in diameter, supported separately from the vessel but joined to it by a water seal at R. In the chimney hangs a very light aluminum fan wheel, held up by a fine wire and carrying a mirror like that of a galvanometer. The deflection of this mirror is proportional to the air speed in the chimney, which in turn is proportional to dp/dt. To calibrate the instrument the cylinder G

---

[67] *Proc. Roy. Soc.*, Vol. 19 (1871), pp. 491–93.

[68] The theory of instruments of this sort is given, for example, in W. E. K. Middleton & A. F. Spilhaus, *Meteorological Instruments* (3d. ed.; Toronto, 1953), p. 54.

[69] *Quart. J. Roy. Meteorol. Soc.*, Vol. 31 (1905), pp. 39–52.

[70] *Meteorol. Zeits.*, Vol. 55 (1938), pp. 273–83.

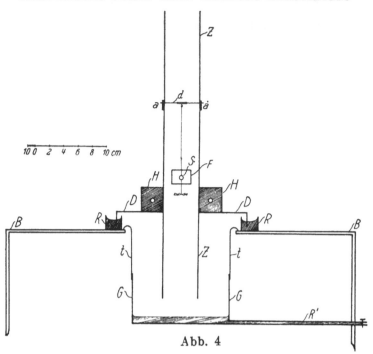

Fig. 15.17. The variograph of Benndorf and Zimmermann, 1938.

is provided, hung from the top by wires t. This can be filled with water, or emptied, through the tube R', and after the water level has been raised enough to seal the bottom of the chimney, further addition of water produces an artificial wind which can readily be calculated from the rate of flow of the water and the dimensions of the apparatus.

The response of this instrument to variations of short period is very small, in contrast to the instruments described before. But it can be made to have almost constant sensitivity to variations of longer period.

An interesting variograph with the property of being insensitive to very high or very low frequency fluctua-

tions, but fairly uniformly responsive to those with a period of about one to 30 minutes, was developed in 1918 by Professor Toshi Shida of Kyoto, but described only in 1936.[71] It is shown schematically in Figure 15.18, where A and B are two glass vessels, A containing water, and B a mineral oil of density 0.87 as far as the level C, where a float is arranged to follow the motion of the interface between the two liquids. The ratio of the cross sections in each vessel was about 11, and the magnification was about 40 in mercury units. The float was partly supported by a platinum wire passing over a pulley G to a counterweight F.

[71] *Kyoto, University, Mem. Coll. Sci.,* Vol. A19 (1936), pp. 237–44.

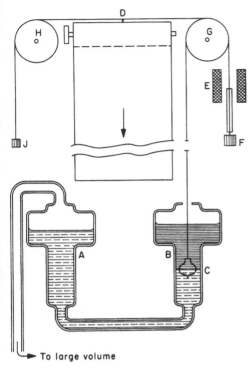

Fig. 15.18. Shida's variograph, 1918.

marks on the band of smoked paper which is moved past D by a clock. The vessel A is connected to an underground chamber with a volume of about 1 cubic meter, and provided with a capillary leak.

In the "microbarovariograph" of Donn[72] a very uniform sensitivity is attained over a range of periods from 1 to 200 seconds by the use of two leaks, a slow one into the reference volume and a faster one into a second volume called the "compliance chamber." The motions of a cork float are recorded electrically.

Finally a very simple variometer may be mentioned, due to F. von Hefner-Alteneck.[73] This consists merely of an insulated one-liter flask, with a capillary leak, and a slightly curved and nearly horizontal graduated tube holding a drop of colored liquid. The curve is concave upward. It is suggested that this instrument might be used in a boat for the determination of wave heights.

Another wire passing over G and H carried a stylus D and terminated at one end in a counterweight J and at the other in an iron armature which could be disturbed by energizing the solenoid E in order to make time-

[72] W. G. Donn, *Trans. Amer. Geophys. Union*, Vol. 39 (1958), pp. 366–68.
[73] *Annalen der Physik*, Vol. 57 (1896), pp. 468–71.

Chapter 16

Dead-Weight and Elastic Barometers

1. INTRODUCTION

WE NOW come to two classes of barometers which differ from all those which we have dealt with up to this point, in that they depend neither on hydrostatic equilibrium nor on the relations between the pressure, volume, and density of gases. They depend instead on the measurement of the force which results from the pressure of the air on some area in a mechanical system.

If we are weighing some solid object, we may compare it with known weights on a pair of scales, or we may, if the law allows, make use of a spring balance. The same two methods have been applied to the measurement of atmospheric pressure, and even a combination of the two. We may thus call our two classes dead-weight barometers and elastic barometers.

2. DEAD-WEIGHT BAROMETERS

Blaise Pascal, that complex man who came to the end of his short life as a religious philosopher, began it with a great fund of mechanical ingenuity which, one may suspect, was scarcely tapped. Not content with inventing a calculating machine, he tossed off the following idea in a fragment (all that was ever published) of a proposed treatise on the vacuum.[1] Take a bellows—which must evidently be thought of as something like a concertina—three inches in diameter. Close it and then seal it. Fasten one side to a roof beam, and from the other hang a heavy chain reaching the floor. Then as the atmospheric pressure changes, various amounts of the chain will be lifted off the floor, so that the weight of the chain suspended from the bellows at any moment will equal the force of the air on its lower surface. This way of using a chain is somewhat similar to that employed nowadays in certain balances for rapid weighing, except that in these the chain hangs in a loop, which is better. Nevertheless, in principle Pascal invented a new kind of

[1] *In Traitez de l'équilibre des liqueurs et de la pesanteur de la masse de l'air* (Paris, 1698), p. 144. *Oeuvres,* ed. Brunschvicg, II. 515–17.

barometer, even though at that date it would probably have been impossible to keep such a bellows completely airtight. The precise date of this invention must remain in doubt, but it was certainly earlier than 1654, and most probably about 1650.

Ten years later Robert Boyle realized that the pressure of the air might be measured by hanging a dead weight on the piston of his air pump, which had first been pushed to the end of the cylinder, and the valve closed.[2] In reading his discussion of this experiment we receive one of the mental shocks that Boyle sometimes provides. It has appeared that he knows exactly what he is doing, and then we come to the following passage:

It might also be tried with cylinders of several diameters, exquisitely fitted with suckers, that we might know, what proportion several pillars of the atmosphere bear to the weights they are able to sustain or lift up; and consequently whether the increase or decrease of the resistance of the ambient air can be reduced to any regular proportion of the diameters of the suckers.[3]

We realize that in spite of all his interesting and beautiful experiments—this is the thirty-third—Boyle had not clearly understood that the force on the piston was equal to the pressure times the area.

It would, of course, be possible to construct an instrument of this sort by having a piston in the neck of a flask full of air. A small weight could then try to pull the piston out, the degree of its success being a measure of the atmospheric pressure. Also, alas! of the temperature. A suggestion of this kind was made in 1686 by a learned Jesuit.[4]

From that day almost to the present the dead-weight barometer has slumbered, a sleeping princess waiting to be awakened by two suitors from the U.S.S.R. in 1959. One of these, P. V. Indrik, described[5] a highly precise piston barometer in which nearly all the upward force is balanced by weights, the remainder by a weak spring. Dibutyl phthalate, a liquid with a vapor pressure of only about 10^{-5} mm. of mercury at room temperature, and having some lubricating properties, is used as a seal. When a measurement is to be made, the cylinder is evacuated by means of a high-vacuum pump, and the piston and weight rotated slowly by hand to be sure there is no sticking. In comparison with a primary mercury barometer the precision is stated to be 0.005 to 0.006 mm. of mercury, and this, it may be pointed out, is about equal to the absolute accuracy of the best mercury barometers. The correction for temperature is stated to be smaller by a factor of 18 than that for a mercury barometer.

The instrument of K. I. Khansuvarov[6] is on the same principles, but about one quarter as precise. It is found desirable to rotate the piston 1 (Figure 16.1) in the cylinder 3 by

[2] New Experiments Physico-Mechanicall, etc. (London, 1660), pp. 237–38. (Works [1772 ed.], I, 71–73.)
[3] Works, I, 73.

[4] Francesco Lana-Terzi, Magisterium naturae et artis (Brescia, 1684–92). Vol 2 (1686), p. 286 and Plate 15, fig. 8.
[5] Measurement Techniques (Izmeritel'naya Tekhnika) 1959, February, p. 23 [I.S.A. Translation (1959), p. 108].
[6] Ibid., 1959, February, p. 24 [I.S.A. Translation (1959), pp. 109–13].

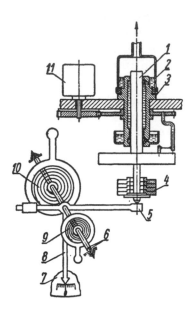

Fig. 16.1. Khansuvarov's piston barometer, 1959.

3. THE INVENTION OF THE "ANEROID" BAROMETER

The *idea* of the aneroid barometer, though not its name (α, without; $\nu\eta\rho\sigma$, liquid) nor its actual construction, goes back to about 1700, and occurred to no less a man than Gottfried Wilhelm Leibniz (1646–1716). In the voluminous correspondence between Leibniz and Johann Bernouilli[7] there are several letters in 1698 and again in 1702 dealing with this matter. In the first of these (June 7, 1698), Leibniz wondered about "a little closed bellows which would be compressed and dilate by itself, as the weight of the air increases or diminishes."[8] On July 29 he returned to the subject again, saying that he would very much like a bellows made of durable material, for a portable barometer, and many other uses. This is all very vague. But in 1702 he is much more definite; writing on February 3, he says that he has thought "about a portable barometer which could be put in the pocket, like a watch; but it is without mercury, whose office the bellows performs, which the weight of the air tries to compress against the resistance of the steel spring."[9]

means of a motor 11 while an observation is in progress. The lower end of the piston bears the weights 4 and operates the lever 5 on shaft 6, which has spiral springs 9, 10, and an index 7. The article includes a scheme for a barograph, in which an inductive transducer and amplifier cause an electric motor to twist a spring as the pressure changes, so as to maintain equilibrium.

It should be noted that such barometers are necessarily secondary instruments, requiring calibration by comparison with a mercury barometer. An absolute determination of pressure on such principles, to such accuracy, would involve the measurement to the effective area of the piston to an impossible degree of exactitude. Yet as secondary instruments they would appear to have many advantages.

This is clearly an approach to the construction which was successfully carried out nearly a century and a half later, at least in so far as the pressure of the air is counterbalanced by the elasticity of a metal. Bernouilli's reply

[7] *Virorum Celeberr. Got. Gul. Leibnitii et Johan. Bernouillii Commercium Philosophicum et Mathematicum* (2 vols.; Lausanne and Geneva, 1745).

[8] *Ibid.*, I, 368.

[9] *Ibid.*, II, 70.

is not in the *Commercium,* but according to Hellmann[10] he wrote to Leibniz pointing out that all the materials the latter had suggested for a bellows would suffer from the effects of humidity. On June 24, 1702, Leibniz replied, "I should like to use a metallic bellows, in which the folds will be furnished with strips of steel (*chalybeis laminis*). In this way the effects you fear will be nullified."[11] We may regret with Hellmann that Leibniz probably had no instrument maker available who could carry out this idea.

The next man to suggest a spring for this purpose was I. E. Zeiher,[12]

[10] G. Hellmann, *Meteorol. Zeits.,* Vol. 8 (1891), pp. 158–59, quoting from *Leibniz'ens mathematische schriften,* ed. C. J. Gerhardt, (Halle, 1856).

[11] *Commercium,* II, 78.

[12] *Nov. Comm. Acad. Imp. Sci. Petrop.,* Vol. 8 (1763), pp. 274–78.

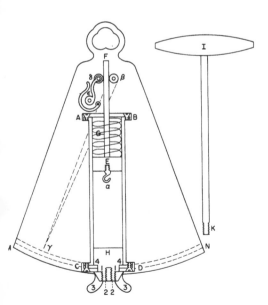

Fig. 16.2. Zeiher's marine barometer, 1763.

whose "marine barometer" was actually constructed. It consisted of a cylinder *ABCD* (Figure 16.2) with two pistons *E* and *H* carefully fitted into it. The upper one was attached to the top of the cylinder by a spring *G*. The lower one, at first in contact with the upper, could be pulled down by means of a special tool so as to evacuate the space, and clamped by a wing-nut 3 on to a leather washer 4. After this had been done the piston *E*, and the rod *EF* attached to it, would move with changes in atmospheric pressure. This motion was communicated to a pointer γ by a friction wheel β, against which the rod was pressed by another wheel δ, moved by a spring ϵ. In case this "friction gearing" might slip, there was a scratch on the rod which coincided with the surface of the cover *AB* at a known pressure, permitting readjustment. Should the vacuum be lost, it could easily be re-established by operating *H*.

It is indeed unlikely that a vacuum could have been preserved for very long in this construction, and the idea seems to have been quickly forgotten.

The next elastic barometer dates from 1797, and was thought of by Nicolas Jacques Conté, the director of the ballooning school at Meudon, whose iron barometer was referred to in Chapter 7. The description[13] is not very clear, nor is the figure (Figure 16.3), but it seems to have been a chamber about the size and shape of a watch, with a solid iron or brass back, and a thin flexible steel top, held up

[13] *Bulletin des Sciences, par la Société philomathique,* 11 Floréal an VI (April, 1798), p. 106.

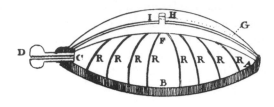

Fig. 16.3. Conté's elastic barometer, 1798.

by springs inside. This could be evacuated, and there were means, not described, of converting its deflections into the motion of a pointer. Conté was not satisfied with this because it was too sensitive to changes of temperature, and it would never have been heard of again if it had not been referred to in a famous legal struggle fifty years later.

So we come to Lucien Vidie (1805–1866), who gave the aneroid barometer both its name and its usual form. We know a great deal about the circumstances surrounding the development of this instrument, thanks to a sympathetic, if highly partisan, biography by Vidie's friend Auguste Laurant.[14]

Lucien Vidie,[15] born at Nantes (Loire-Inférieure), was destined for the bar; but he was of an extremely sensitive and retiring disposition, and left the legal profession in 1830 and went in for steam engineering. About 1838 he began to think about the improvement of steam-pressure gauges, for which mercury columns were then universally employed; this led him to the barometer. In spite of his final technical success, he was so poor a businessman that others reaped the financial reward.

One of the most remarkable things about the aneroid barometer is that its successful construction was believed at that time to be impossible because of the properties of metals. In the first place, metals were believed to be slightly porous.[16] Vidie, however, succeeded in making thin metal structures which held a vacuum. In the second place, and more serious, the *Académie des Sciences,* reporting in 1844 on a study of elasticity by G. Wertheim, set down an official finding that there is no real elastic limit in metals.[17] In other words even a small force, acting for sufficient time, will produce a deflection. In still blunter words, Vidie was on a wild-goose chase. Nevertheless, in 1843 he had succeeded in producing some satisfactory metallic barometers.

Vidie did not take out his basic patent of 1844 in his own name, but in that of Pierre Armand Lecomte de Fontainemoreau, the head of a patent agency in both London and Paris, who seems to have treated him very fairly; in the specification Vidie appears only as "a certain foreigner residing abroad."[18] Before applying for the

---

[14] *Histoire des baromètres et manomètres anéroides. Biographie de Lucien Vidie* (Paris, 1867).

[15] There has been some confusion about his name, often spelled Vidi. Laurant (*Histoire des baromètres,* p. 74) tells us that Vidie got tired of receiving letters addressed to Vidié or Vidier, and began to sign his name without the final e.

[16] Pouillet, *Eléments de physique expérimentale* (3d. ed.; Paris, 1837), I, 28 (quoted by Laurant, *Histoire des baromètres,* pp. 196–97).

[17] *Comptes Rendus,* Vol. 18 (1844), pp. 921–32.

[18] British Patent 10,157 (1844), p. 1.

patent, however, he had submitted the barometer, through Fontainemoreau, to one Andrew Pritchard in London, a retired manufacturing optician, for an opinion. On July 27, 1843, Pritchard wrote that it was, as far as he could see, new, and potentially of great value, "however, I would not advise the expenses of a patent until other experiments are made to ascertain the range of the instrument."[19] Pritchard must have been much impressed, for he later certified that "in or about the month of August 1843, accompanied by . . . de Fontaine-Moreau I ascended to the dome of St. Paul's cathedral London, with an instrument made according to the specification and drawing so submitted to me and herein described for the purpose of testing its fitness for measuring heights and . . . the experiment was successful."[20]

It is extremely improbable that Vidie knew of the existence of Conté's attempt to make a metallic barometer; indeed, he denied it under oath in a court of law, saying that if he had known that such a man as Conté had made the attempt and failed, he would have abandoned the idea.[21] Nevertheless there is an extraordinary resemblance between the main features of Conté's design, as far as it can be determined, and of the design shown in Figure 1 of the basic patent.

There is the same rigid box with a flexible top and internal springs. But in Vidie's instrument the dimensions are more reasonable for an experimental apparatus, and there is a mechanism which compensates for the effect of temperature on the elasticity of the diaphragm.

Patents were worded differently in 1844; but let us consider the passage in the patent which would nowadays follow the words "What I claim is:"

The "New Mode of Constructing Barometers and other Pneumatic Instruments" consists more especially in the application of thin sheets or diaphragms of metals, glass, india rubber, or other flexible air-tight substances, to certain apparatus employed for measuring the pressure and elasticity of the air and other fluids, in such manner as to form a kind of elastic cushion or buffer, susceptible of the slightest variation of the pressure of the atmosphere or fluid with which it is in contact, and, consequently, indicating the amount of the same by the greater or less depression of the said yielding substances; and the Invention consists, generally, in the application of the above principle to all those pneumatic instruments in which any reciprocating motion and oscillation takes place upon a variation in the pressure or weight of the superincumbent column of the atmosphere, or in the pressure of the liquid in which it is immersed, or in the elasticity of a gascious [sic] body, and which instruments are commonly called barometers and manometers, etc.[22]

Thus Vidie's patent seems to cover the use of "thin sheets of diaphragms" of all "flexible air-tight substances." His original barometer, in the form used to illustrate his claims, is shown

---

[19] Quoted by Laurant, *Histoire des baromètres*, p. 385.

[20] *Ibid.*

[21] *Ibid.*, p. 138. Conté had greatly impressed Napoleon, who took him on his expedition to Egypt.

[22] British Patent 10,157 (1844), p. 2. A French Patent, no. 12,473 (1844) was also taken out, and in 1846 a U.S. Patent, no. 4,702.

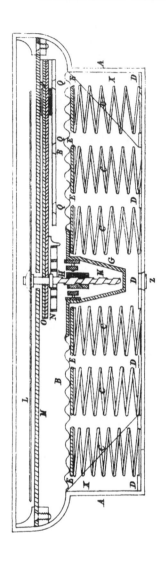

Fig. 16.4. Vidie's aneroid, first form, 1844.

cal springs C (the plan view, not reproduced, shows 33 of them), each fitted with a flat cap E and based in a shallow recess in the bottom of the box A. It is noted that "steel plates folded in a zigzag manner" can be used instead. A cup G is soldered into a hole in the center of the diaphragm; in this cup, supported on gimbals, is a nut H which engages a steeply-threaded screw K to which the pointer L is attached. Before the upper part of the instrument is assembled, the chamber is exhausted through D and the plug Z ingeniously soldered in. Temperature compensation is cleverly provided by a bimetallic strip O which supports the bearing in which K turns, and backlash is taken up by a spiral spring N. About five other possible constructions are illustrated, three of them using a metal bellows.

It was all very well to make a brilliant invention, but it was another thing to get the instruments produced. In 1845 Vidie ordered 100 of them from Redier, a well-known watchmaker; but they were unsatisfactory and he refused to take delivery. Redier hauled him into court, and called Bunten, whom we met in Chapter 7, as a witness. Bunten was "barometer-maker to the Institute," and his opinion that the instruments were satisfactory carried great weight, though of course it really had little value. The great Arago, whose powers of sarcasm were only matched by his boundless self-confidence, was also called as an expert. Vidie was a fool to persist, he said, when the Institute had declared absolute elasticity a chimera. Naturally the court awarded damages to Redier. There was nothing for Vidie

in cross section in Figure 16.4. The lower part is a strongly-made brass box A, strengthened by brackets X. The diaphragm B is corrugated "so as to enable it to be depressed or elevated to the greatest extent requisite without rupture." It is supported by helito do but to hire workmen and manufacture the barometers himself.

He succeeded in making them to his satisfaction, but selling them was another matter. No doubt because of Arago's opinion, they were practically unsalable in Paris. In despair, he went to England, where E. J. Dent (1790–1853) took an interest in them.[23] They appealed to the English, especially to mountaineers and sailors, and within a few years, according to Laurant, Vidie had delivered more than 5,000 aneroids to England, but had sold less than 100 in France.[24]

In spite of this, Vidie presented a barometer to the *Académie,* and the not very distinguished committee to which it was referred made an entirely non-committal report.[25] On September 16, 1848, he sent one to Sir G. B. Airy, the Astronomer Royal, at Greenwich. Airy had comparisons made with the mercury barometer over a period of six months, and on May 8, 1849, reported very favorably on the aneroid.[26] "I do not perceive," he wrote, "that the differences follow any law depending on the height of the barometer, or on the height of the thermometer, or on their changes. I think that, upon the whole, the reading of the Aneroid Barometer has diminished a very little with time; but the apparent

diminution is so small that its existence is very uncertain."

Meanwhile competition, or as Vidie felt, infringement, was in preparation. In 1846 a German railway engineer called Schinz had discovered that a curved tube of elliptical cross section would change its curvature when subjected to internal pressure; and by 1848 steam-pressure gauges on this principle were in use on locomotives in Germany.[27] It seems that Schinz took out a Prussian Patent, no. 3 of 1849.[28] This is, of course, the familiar pressure gauge universally known as the Bourdon gauge. Bourdon, a Paris instrument maker, patented precisely the same construction on June 19, 1849,[29] three months after the first of the articles in the *Eisenbahn-Zeitung;* but it is entirely possible that he invented it independently. There were additions on September 3 and October 17, 1849. Both Vidie's barometer and Bourdon's, as well as the latter's steam gauge and thermometer, were exhibited at the Great Exhibition of 1851 in London, and each was awarded a Council Medal. On July 9 of that year, Vidie had a number of Bourdon's instruments seized and prosecuted him for infringement.

The following March 19 the case was decided in Bourdon's favor, apparently on the grounds that Vidie's barometer was in principle nothing else but the one Conté had described

---

[23] Dent was a chronometer maker of great distinction. In 1948 he published in London *A Treatise on the Aneroid, a Newly Invented Portable Barometer. With a short historical notice on barometers in general, their construction and use.* Pamphlet, 34 pp., reprinted 1850.

[24] Laurant, *Histoire des baromètres,* p. 84.

[25] *Compt. Rend.,* Vol. 24 (1847), p. 975.

[26] Laurant, *Histoire des baromètres,* pp. 391–92. Airy's letter is quoted in full.

[27] *Eisenbahn-Zeitung,* Vol. 7 (1849), 5 March and 2 April.

[28] L. B. Hunt, *J. Sci. Instrum.,* Vol. 21 (1944), p. 39.

[29] French patent 4408. Charles Cowper also patented a similar device (British Patent 12,889, 1849).

in 1798 and was therefore in the public domain. One must admit—in spite of Laurant, who makes Bourdon out to be a sort of commercial Machiavelli and general ogre—that a perusal of Vidie's patent might well lead to that conclusion. Vidie appealed and, on July 13, 1852, lost; he took it to the highest court available, and on January 7, 1853, lost again.

Vidie, who seems to have been convinced of the justice of his cause, as litigious people must be, found a way to reopen the matter in 1858. This time, through the brilliant efforts of an advocate called Senard, he was awarded damages of 25,000 francs, which after two appeals was reduced to 10,000 francs on July 9, 1861, exactly ten years to the day from his original seizure of Bourdon's instruments. It was a Pyrrhic victory; the basic patent had expired, and a renewal was refused.

Let us get back to the history of the instrument. The French Patent 12,473 of 1844, although described as a *brevet d'importation*, has very little resemblance to the British one. It emphasizes the use of a *forme d'inégale resistance* and suggests a flattened hollow sphere, which, flattened a good deal more, became the well-known aneroid capsule. In a *brevet d'addition* of October 31, 1844, one of the improvements is the leaving of a definite quantity of air inside the capsule for temperature compensation. In 1862, when competition was widespread, Vidie took out a British Patent[30] covering twelve different mechanical improvements to the aneroid barometer.

There are no drawings, and most of these devices seem to be of minor interest.

Early Vidie barometers are evidently rare. There is one in the C.N.A.M. at Paris,[31] graduated from 63 to 80 cm. in mm.; unfortunately the mechanism is not visible.

4. THE BOURDON BAROMETER

It is a reflection of the characters of the two rival inventors that while the curved, flattened tube (originally due to Schinz) is universally called the "Bourdon tube," the memory of Vidie is preserved only in the German language, in which the flattened evacuated capsule is called the *Vidiedose*. Bourdon made a fortune out of his tube, which was and is eminently suitable for use in steam-pressure gauges, and could also be applied directly to thermometry by filling it completely with some liquid. While the barometer was Vidie's first concern, the pressure gauge being an afterthought, Bourdon, with his commercial acumen, realized that the money was on the engineering side.

Bourdon's 1849 patent and its two additions constitute an extraordinary document, illustrated with over fifty figures. He patented almost every conceivable construction of the flattened-tube manometer, curved in one plane, coiled into a helix, or twisted. He extended it to what he called "metallic lenses," clearly infringing Vidie's patent, though I do not know whether he ever made any of these. He even described a steam engine in which,

[30] No. 682(1862).

[31] Inventory no. 17540.

instead of a piston rod, the free end of the curved tube was attached to the connecting rod and crankshaft!

The steam engines, and even the pressure gauges, are outside our orbit. As to the Bourdon barometer, it can be made extremely simple, because of the comparatively large motion of the free end of the tube. There is a beautiful early example in the *Istituto Galileo Galilei* at Florence,[32] having a Bourdon tube about 9 cm. in radius and extending round about three-quarters of a circle, which, with the very simple link, toothed sector and pinion which move the pointer, can be clearly seen behind the glass front of the instrument. One would think that it would have been at least competitive with Vidie's instrument, but experience has decided otherwise. Apart from a type of barograph manufactured in Paris up to the early years of this century, and those in some early balloon-meteorographs, almost all elastic barometers have been derived from those of Vidie. As a matter of fact their superiority soon became evident; Laurant quotes with relish a letter dated November 5, 1859, from Father Secchi, the Director of the *Osservatorio del Collegio Romano*, whom we met in Chapter 11. After describing a special aneroid that he wished to obtain, Secchi adds, "Il va sans dire que je veux la construction de Vidie, et non celle de Bourdon, qui ne *vaut rien*."[33] One senses that Laurant supplied the italics.

There is an interesting Bourdon barometer preserved at the C.N.A.M. in Paris,[34] in which one end of a large Bourdon tube is rigidly attached to the short arm of a steelyard, the free end of the tube carrying a weight. The system may be balanced by sliding another weight along the longer arm of the steelyard, which is graduated. As the pressure changes, the change in shape of the Bourdon tube moves its center of gravity.

Apart from one very recent design (see p. 425), the later patents on the Bourdon barometer do not seem to have covered anything that Bourdon did not think of in 1849.[35] There were occasional attempts at special applications. The great experimenter F. Kohlrausch made one turn a small mirror supported by a steel spring, so that a scale and telescope at 3-meters distance gave a sensitivity of about 25 scale divisions per mm. of mercury. He called the instrument a *Variationsbarometer*.[36]

An attempt at a theory, quite inadequate, was made in 1872 by the Rev. E. Hill.[37] The complete elastic theory, which is complicated, was given much later by H. Lorenz.[38]

5. EARLY DEVELOPMENT OF THE ANEROID

On July 28, 1845, Vidie took out another French patent,[39] this time in his own name, in which he covered the use of the familiar curved spring, as

---

[32] Inventory no. A196.
[33] Laurant, *Histoire des baromètres*, p. 402. The recipient of the letter is not identified.
[34] Inventory no. 16409.

[35] E.g., W. E. Newton, British Patent 1185 (1860); R. W. Thomson, British Patent 1006 (1866).
[36] *Annalen der Physik*, Vol. 150 (1873), pp. 423–26.
[37] *Quart. J. Meteorol. Soc.*, Vol. 1 (1872), pp. 50–55.
[38] *Zeits. des V.D.I.*, Vol. 54 (1910), pp. 1865–67.
[39] No. 1149.

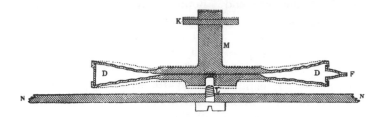

Fig. 16.5. Vidie's vacuum box, 1849, after Dent.

well as various devices for temperature-compensation. There seems to have been no English patent on this; but barometers constructed in this manner were being sold in England by E. J. Dent at least by the end of 1847, and the construction of these instruments is clearly shown in Figures 16.5 and 16.6. They must have been much easier to make than the form shown in the original English patent.

Figure 16.5 shows the "vacuum-vase," as Dent calls it,[40] "in a compressed state, after the air has been exhausted by the air-pump through the tube F. The dotted lines running nearly even with the corrugated surface are intended to show the position which that surface will assume after the introduction of the gas, which effects a compensation for the results of varying temperature."

This instrument was therefore compensated for temperature not by a bimetal, as suggested in the British patent of 1844, but by leaving a certain amount of gas in the capsule, as described in the French *brevet d'addition* of October 31 in that year. From the fact that Dent refers to "a gas" and does not state what it is, I infer that the capsules at least, in the instruments sold by Dent, were made in France by Vidie; and

from Laurant's rather vague account it seems likely that he made the complete barometers.

This method of temperature-compensation, which is exact only at one pressure,[41] seems to have been adequate, to judge by two tables published by Dent[42] showing comparisons of two of these instruments with mercury barometers by purchasers in England. The earliest observation in these tables is dated January 6, 1848, which indicates that Dent had begun to sell them before this date.

The rest of the mechanism is shown in Figure 16.6, from which it will be seen that the external spring was helical at this time, the large member CC merely being a lever, pivoted on posts B. For setting the hand to agree with a mercury barometer, a screw A at the back of the case raised or lowered the base of the spring.

It is interesting that a very similar mechanism was described in 1859, the external helical spring being considered to be a new invention made by R. Deutschbein of Hamburg.[43] It was felt

[40] *A Treatise*, (1849), p. 16.

[41] W. E. K. Middleton & A. F. Spilhaus, *Meteorological Instruments* (3d. ed.; Toronto, 1953), p. 48.
[42] *A Treatise*, pp. 18–19.
[43] F. Petersen, *Zeits. des V.D.I.*, Vol. 3 (1859), p. 28.

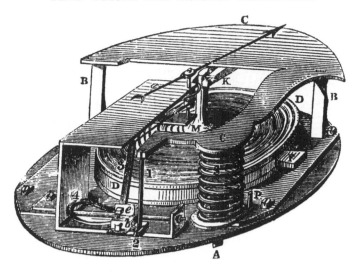

Fig. 16.6. The aneroid mechanism, 1849, after Dent.

to be much better than the French aneroids; and the fact that the helical spring was thought to be new suggests that Vidie had abandoned it for the flat spring before any great number of his instruments had reached Germany. Actually the flat spring ought to be better, for a helical spring tends to twist as it is compressed.

The popularity of the aneroid in England is further attested to by a pamphlet published in 1849 by J. H. Belleville of the Royal Observatory, Greenwich.[44] "Much has been made against its variations from temperature"; he wrote, "the writer has made experiments with his own instrument and with many sent him for comparison . . . in a range of temperature from 28° to 80°, the variations have seldom exceeded a tenth of an inch."[45] Belleville's requirements for accuracy, and indeed those of most users, were extremely liberal.

I have not found much information about the history of the aneroid between 1850 and 1860. Vidie (under the name Vidi) sued "Smith and Another" for infringement in 1854,[46] but they were making Bourdon barometers. From Laurant's biography one gets the impression that there were many *contrefaiseurs*, but there are no details. At any rate there were several makers soon after the patent expired in 1859; the most successful being Naudet, Hulot, & Cie. According to Le Roux[47] they made 20,000 aneroid barometers between 1861 and 1866. They called them *baromètres holostériques*, which ought to mean that they were entirely solid; but the name stuck, and references occur in the continental literature to Naudet barometers and to holosteric barometers for the rest of the nineteenth century. They acquired a great reputation and were widely imitated. The temperature com-

[44] *A Manual of the Barometer*, etc. (London, 1849). iv+53 pp.
[45] *Ibid.*, p. 44.

[46] 23 L. J., Q. B. 342 (1854).
[47] *Soc. d'Encouragement pour l'industrie nationale* no. 165 (1866).

Fig. 16.7. Negretti & Zambra's
aneroid mechanism, 1864.

pensation was bimetallic, as in the
instruments manufactured in England
by Negretti & Zambra, whose design is
shown in Figure 16.7.[48] These makers
must have made rapid progress in this
line, for by 1862 they were able to supply two aneroids to James Glaisher
(1809–1903) for use in his famous scientific balloon ascents of that year, the
highest of which nearly resulted in his
death. He compared an aneroid with
a siphon barometer at pressures down
to 7 inches of mercury and found it "to
read correctly," but he did not seem
concerned about differences of the order
of 0.1 inch.[49]

At the instigation of Admiral Fitz-
Roy, Negretti & Zambra embarked on
a development which would now probably be called miniaturization, and by
1861 produced an instrument of the
shape and size of a pocket watch, with
a setting pointer operated by a milled

rim.[50] The Astronomer Royal, Sir George
Airy, suggested that an altitude scale
should be added and presented them
with a computation based on an atmosphere at a temperature of 50° F., presumably at all elevations.[51] These became extremely popular with travelers;
I remember that my father used to carry
one at the other end of his watch
chain. Scientific men differed among
themselves in their opinions; thus C. V.
Walker was highly critical,[52] while G.
J. Symons[53] reported two years later
that he had carried one 20,000 miles
and had been perfectly satisfied with
the results, contrary to Walker's experience. One might well imagine that the
chance of getting a "dud" would be
fairly large.

The smallest and most elegant aneroid I have ever seen is in the National
Maritime Museum, Greenwich.[54] It has
a chased gold case only ⅞ of an inch
in diameter; but it is scarcely a scientific instrument.

We see that by about 1864 the aneroid had been more or less standardized
as far as its ordinary use is concerned.
It was excellent as a domestic "weatherglass," and highly useful as a marine
barometer under the conditions pre-
vailing at the time. As a scientific instrument of precision, especially for the
measurement of heights, it had a long
way to go.

It had had a bad start, really. Vidie
was neither a scientist nor an instru-

[48] Negretti & Zambra, A Treatise on Meteorological Instruments (London, 1864), p. 52.
[49] B.A.A.S. Report (Cambridge, 1862), pp.
481–82.

[50] Proc. Brit. Meteorol. Soc., Vol. 1 (1861),
p. 84.
[51] Negretti & Zambra, A Treatise, pp. 54–
55.
[52] Proc. Brit. Meteorol. Soc., Vol. 1 (1863),
pp. 380–82.
[53] Ibid., Vol. 2 (1865), pp. 418–22.
[54] Inventory no. B.4.

ment maker, and the original arrangement of the parts, as in the 1844 patent, was too neat to be practical. The first form which actually emerged looks as if it has received the attention of a bewildered watchmaker trying to make the best of a strange new device. Nothing was even statically balanced except perhaps the pointer, the opportunities for friction were numerous, and the details of construction seemed bound to lead to small uncertainties in the magnification. That it worked as well as it did was due to the fairly large operating forces available. Quite apart from mechanical considerations, the elastic properties of materials were not properly understood, as the *Académie des Sciences* had so convincingly demonstrated in 1844.

The improvement of the aneroid barometer therefore followed two paths: first, a complete departure from the original construction, and later, great metallurgical advances. While this was going on, a great deal was discovered about the elastic behavior of metals in general and the aneroid in particular.

For many purposes, aneroids continued to be made—and are indeed still made—of a form very like that arrived at by Naudet, Hulot, & Cie about 1860. The bearings were provided with jewels, for which application a patent was obtained in 1861,[55] and which was suggested all over again by Ralph Abercromby in 1876.[56] The illustration of the aneroid in Negretti & Zambra's thick catalogue of 1886 was extremely like our Figure 16.7, except that it was bet-

ter drawn.[57] Similar aneroids today are —or at least might be—superior mainly from a metallurgical standpoint.

6. IMPROVEMENTS IN THE VACUUM CHAMBER

As far as the vacuum chamber itself is concerned, it seems to have changed little in the nineteenth century, at least in any fundamental way. Various metals were used, mainly copper, brass, and later nickel alloys such as "German silver" *(Neusilber)*. Not much seems to have been published in either the scientific journals or the patent literature, the various makers having their own shapes and their own ways of joining the edges, etc., always keeping their own counsel. About 1867 the firm of Breguet was using a flat external spring with the multiple chambers of their barographs, which will be referred to below. A little later, Negretti & Zambra[58] put internal leaf springs in each chamber of the pile.

Things began to change, or at any rate the changes began to transpire, at the beginning of the new century. In 1903 J. Y. Fulton patented the use of a metal bellows;[59] Vidie had of course done so too, but the time was at hand when such a construction would be technically feasible. In 1914 R. Fuess

---

[55] James Pitkin, British Provisional Patent 2947 (1861).

[56] *Quart. J. Meteorol. Soc.*, Vol. 3 (1876), pp. 87–89.

[57] Negretti & Zambra, *Encyclopaedic Illustrated and Descriptive Reference Catalogue of Optical, Mathematical, Physical, Photographic, and Standard Meteorological Instruments*, etc. London (1886), p. 22. I am indebted to Messrs. Negretti and Zambra for the privilege of examining this most fascinating book, and also for dating it.

[58] *Ibid.*, p. 27.

[59] British Patent 11,548 (1903).

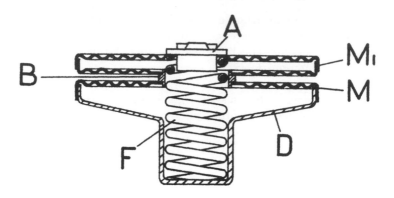

Fig. 16.8. Fuess's patent capsule, 1914.

obtained a patent on the use of an internal helical spring[60] and showed how one spring could be used for more than one capsule (Figure 16.8). In 1922 we hear of a capsule with flanged edges and "substantially inflexible or inoperative portions joining the local regions with the flexible areas."[61] The idea was to keep solder from spreading to the operating parts of the capsule. Capsules generally had been lapped as in Figure 16.8, but from this time on, more and more of them were flanged.

Springs are troublesome, and in 1931 P. Kollsman of New York patented an aneroid chamber with no springs, which "supports pressures solely by the strength and elasticity of its walls."[62] Such a chamber would be more especially advantageous for altimeters, and Kollsman undoubtedly had this in mind.

Without anticipating section 7 of this chapter, it may be stated here that the greater part of the effect of temperature on an aneroid is due, not to changes in its dimensions, but to alterations in the elastic constants of the materials of the capsule and springs. It had been pointed out by E. Kleinschmidt in 1928[63] that some nickel steels with a zero temperature coefficient of dilatation have a positive temperature coefficient of Young's modulus. In 1937 the Bendix Aviation Corporation patented a construction in which the two sides of an aneroid chamber are made of different metals, for example beryllium-copper, for which Young's modulus decreases with increasing temperature, and "modulvar," a nickel alloy for which it increases.[64] R. Fuess[65] in Germany took out a similar patent, and Wilhelm Lambrecht in Göttingen, not to be outdone, patented the use of different shapes as well as different materials for the two sides of the capsule.[66] Finally the end of the road in this direction was reached by the metallurgists when an alloy known

[60] D.R.P. 288,537 (1914). See also D.R.P. 309,578 (1918).

[61] J. P. Brown and V. H. Gregory, British Patent 198,838 (1922).

[62] U.S. Patent 1,930,899 (1933); British Patent 393,272 (1931).

[63] *Beitr. Phys. freien Atm.*, Vol. 14 (1928), p. 190, note.

[64] U.S. Patent 2,162,308 (1939); British Patent 501,337 (1937).

[65] D.R. Patent 717,707 (1939).

[66] D.R. Patent 735,918 (1940).

as "Ni-Span C" was put on the market, the modulus of elasticity of which does not vary with temperature. The construction of aneroid capsules of this material has recently been patented.[67]

Let us go back to 1938. In order to avoid some of the difficulties due to the elastic behavior of metals, it was proposed to make capsules of non-porous ceramic material, mounted in metal frames.[68] These do not seem to have won acceptance. A less unorthodox idea came from the firm of Negretti & Zambra,[69] in their flat capsules of hardened and tempered steel, nickel-plated, and soldered together at the flanged edges. They required no additional springs, and had good elastic properties. The patent describes an elaborate shape, the reason for which is not given; but the deflections are stated to be linear with pressure over a considerable range.

Meanwhile the metal bellows had not been forgotten, and had become a reliable article of commerce. For barographs they became widely used, supported by an internal helical spring. This construction was first adopted by the firm of Julien Friez, Baltimore, in 1920.[70] Now when a helical spring is compressed, one end tends to rotate with respect to the other, and about 1942, the writer, in designing a barograph for the Canadian Meteorological Service, incorporated both radial and thrust ball bearings in the bellows.[71]

We have come a long way from the little brass box. It has to be admitted that the most spectacular improvements in these capsules have resulted from the requirements of the aviation industry for the indication of large ranges of pressure with high accuracy. The users of barometers are just beginning to reap the benefits of this.

7. COMPENSATION FOR CHANGES OF TEMPERATURE

The elastic assembly of spring and chamber which reacts to changes in atmospheric pressure may be thought of as a compound spring. A change in the temperature of the instrument will have two effects: to change the dimensions of the apparatus and to change the elastic constants of this spring. Since the temperature coefficient of the elastic constants is usually more than an order of magnitude greater than the coefficient of dilatation, the greater part of the effect is due to the change in the elastic constants. However, at least as far as the capsule is concerned, the two effects are usually additive.

Until recently there were two possible ways of compensating an aneroid barometer for temperature: by the use of a bimetallic link in the mechanism, and by leaving the correct amount of air or other gas in the aneroid chamber. Both of these were known to, and

[67] P. W. Harland, U.S. Patent 2,698,633 (1955); The Bristol Co., British Patent 874,333 (1961).

[68] The General Electric Co., Ltd., and H. C. Turner, British Patent 515,750 (1938). F. B. von Rautenfeld of Riga had experimented with fused quartz long before (Dissertation, Leipzig, 1916).

[69] H. N. Negretti, P. E. Negretti, and H. W. Ibbot, British Patent 528,124 (1939).

[70] Private communication from Mr. W. Boettinger. See also p. 429 below for further information.

[71] See W. E. K. Middleton and A. F. Spilhaus, *Meteorological Instruments*, p. 45.

indeed patented by, Vidie. Now there is a third way: to make the capsule of a material with no temperature coefficient of Young's modulus, such as "Ni-Span C."[72] With the bimetal, exact compensation is only obtainable at one pressure. With the inclusion of gas, it is possible, if the internal volume is small, to have the compensation exact at two pressures, and fairly good between them.[73] With the special capsule, it can be almost perfect at all pressures, provided that the mechanical parts of the barometer are correctly designed so as to balance out the effects of their dilatation.

The aneroids made in the last quarter of the nineteenth century usually had the bimetallic compensation. The Bourdon barometers used in early balloon meteorographs were compensated by inclusion of gas,[74] but these hardly come within our province. In 1910 H. Daniels and A. J. Daniels patented a construction in which the bimetal, instead of being itself one of the levers, moves the fulcrum of one.[75] Later J. W. Barnes proposed making the bimetal in the form of a helix, one end being fastened to the pointer and the other to its axis.[76] Finally, an extremely sophisticated design was evolved by the Kollsman Instrument Corporation,[77] in which the bimetal 14 (Figure 16.9)

applies a force to the front of the aneroid chambers through the pins 15, 16, the compensation being adjustable by turning the screw 45.

The second method of compensation, by leaving the correct pressure of gas in the chamber, is of course easier to use in the sort of instrument which is operated by a metallic bellows. It is rather hard to mass-produce such things because of the variability of the individual components, and in 1921 R. Fuess patented the scheme of putting two capsules or metal bellows in series, each with its own inner spring. Half the capsules would be pumped out before assembly, the other half being left with various amounts of gas. A uniform result would be obtained by selection and combination.[78]

In 1932 J. Jaumotte pointed out that if the internal volume of an aneroid capsule is very small, it will be possible to compensate for temperature at two pressures, and very nearly at all pressures between.[79] The steel aneroids referred to above[80] can be compensated in this way; and in 1939 Paul Kollsman patented, among other things, the idea of an aneroid chamber with the front and back corrugated in a complementary way, so that the volume could be kept extremely small.[81]

The development of alloys which become stiffer as they get warmer opened up interesting possibilities to the makers of aneroids. In 1939 C. J. Jenny patented[82] the use of dissimilar metals

---

[72] The construction due to Scammell (see p. 417 below) really offers another method of temperature compensation.

[73] See Middleton and Spilhaus, *Meteorological Instruments*, pp. 48–49. See also ref. 79 below.

[74] H. Hergesell and E. Kleinschmidt, *Beitr. Phys. freien Atm.*, Vol. 1 (1905), pp. 109–11.

[75] British Patent 3055 (1910).

[76] British Patent 200,649 (1922).

[77] U.S. Patent 2,628,501 (1953); British Patent 670,913 (1950).

[78] D.R. Patent 355,499 (1921).

[79] *Assoc. française pour l'Avanc. des Sci.*, Congrès de Bruxelles 1932, *Compt. Rend.*, pp. 167–70.

[80] See note 69 above.

[81] U.S. Patent 2,150,771 (1939).

[82] U.S. Patent 2,162,308 (1939).

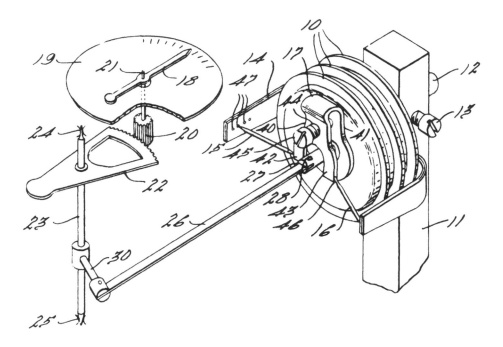

Fig. 16.9. Kollsman's bimetallic compensator, circa 1950.

for the two sides of the chamber, and four years later the use of two springs of different metals within a metal bellows was patented by A. T. Newell.[83]

Finally we come to the construction of capsules from new materials which do not alter their elastic properties with change of temperature—the ideal solution. The dilatation of the rest of the instrument now becomes important, and is minimized by careful design.[84]

8. THE FURTHER IMPROVEMENT OF ANEROID MECHANISMS

On page 409 we hinted that entirely new designs came to supplement the

usual construction of the aneroid barometer. These designs are very numerous, and some attempt at classifying them must be made. The most fundamental dichotomy is between aneroid barometers which can be read by merely looking at a dial and pointer—tapping the instrument first is allowed!—and those in which the instrument must be *set* before a reading can be obtained. To put it in more technical language, we can divide them into direct-reading barometers and null-barometers. We shall consider the direct-reading instruments first.

One small but fundamental improvement was possible in the old construction using a leaf spring. In its original form, with the bottom of the capsule fastened to the base of the instrument,

[83] U.S. Patent 2,311,900 (1943).
[84] See note 67 above.

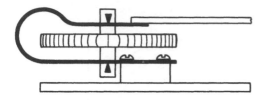

Fig. 16.10. Capsule supported on knife edges.

the pull of the spring is axial only at one particular pressure, and at any other pressure there is a lateral component to the force of the spring. In 1923 V. H. Gregory of the Royal Aircraft Establishment, Farnborough, patented a construction in which "the blade spring . . . is supported by small and constant areas or abutment surfaces which are arranged so that the spring exerts a force along the axis of the cap-

sule."[85] One way of arranging this is to use knife edges passing through studs fastened to the capsule, as in Figure 16.10. But even an aneroid constructed in this way is somewhat affected by its attitude, and two years later[86] Gregory patented a "balanced" construction—balanced as to its heavier parts, at least —the main points of which are indicated in Figure 16.11. The frame of the instrument is indicated by the number 1. The number 13 indicates a bimetal used in an interesting way.

This brings us to the use of several capsules in series, an idea dating from at least 1862, when T. E. Blackwell patented it.[87] If we have n capsules

[85] British Patent 215,844 (1923).
[86] British Patent 258,991 (1925).
[87] British Patent 3264 (1862).

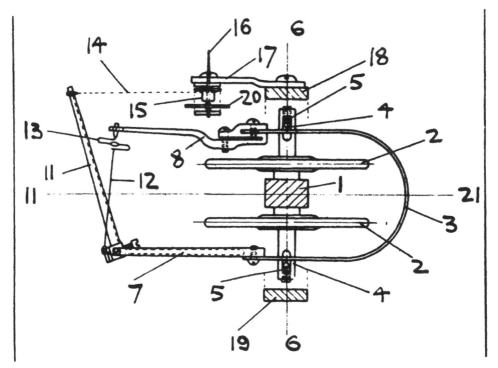

Fig. 16.11. Balanced aneroid, 1925.

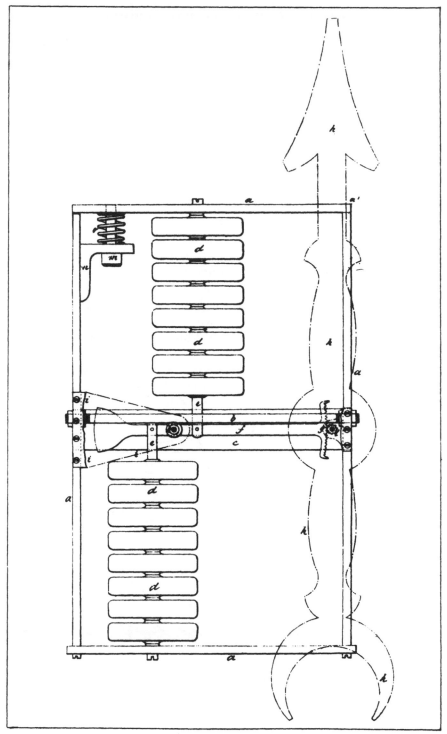

Fig. 16.12. The display barometer of Guichard and Co., 1881.

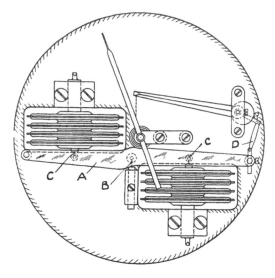

Fig. 16.13. Precision aneroid movement, about 1939. (*Negretti & Zambra Ltd.*)

coming friction. Between about 1865 and 1920, this construction was universal for aneroid barographs. In 1881 the Paris firm of S. Guichard & Co. patented a combination of multiple capsules and balanced construction, having in view "the production of a barometer, so arranged as to give precise indications sufficiently large to be read upon a dial similar to those of public clocks."[88] This simple and elegant scheme is shown in Figure 16.12, in which the spring-loaded screw at the upper left-hand corner serves to adjust the zero. One must assume that the capsules have internal springs. A similar use of opposed groups of capsules was patented by the well-known optical firm of C. P. Goerz in 1915,[89] but with the very different aim

instead of one, the available motion is multiplied by n, leaving the actuating force the same, a valuable aid in over-

[88] British Patent 2020 (1881). This is in the name of B. J. B. Mills.

[89] D.R. Patent 299,667 (1915).

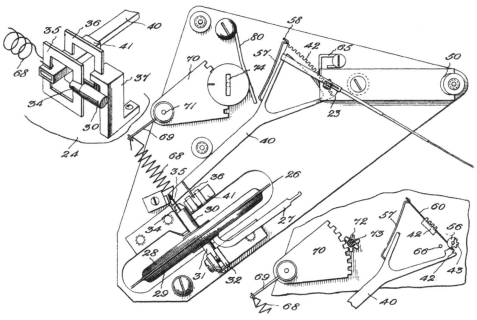

Fig. 16.14. Wallace and Tiernan's aneroid, 1945. (*By permission of the Controller, H. M. Stationery Office*)

of rendering the instrument insensitive to attitude or to shock—the requirements that inspired Gregory a decade later.

The next advance in purely mechanical direct-reading aneroids was made by Negretti & Zambra about 1939, in applying the steel capsules referred to above. They evolved the very superior mechanism shown in Figure 16.13, in which it should be noted that the main lever A is carried not on a pivot but on a flexible strip of stainless steel B. Similar strips C,C connect the lever with the two sets of chambers. All the parts except the chain and the very light link D are balanced.

The use of elastic links instead of pivots was carried a step farther by C. F. Wallace in the United States,[90] leaving no bearings except those of the pointer. The main details are reproduced from the later patent in Figure 16.14. The aim was not to achieve insensitiveness to attitude or vibration, but to make as good an aneroid as possible for use in a fixed position.

A scheme as different from this as it could well be was patented in 1948 by E. W. Scammell of Birmingham (Figure 16.15).[91] Scammell used a bellows and a spring, each of low spring-rate,[92] pulling against each other. The relatively large motion of the end of the bellows makes possible a simple mechanism, but unfortunately the entire instrument would also be rather large.

All these aneroid barometers are, in effect, spring balances. Why not use

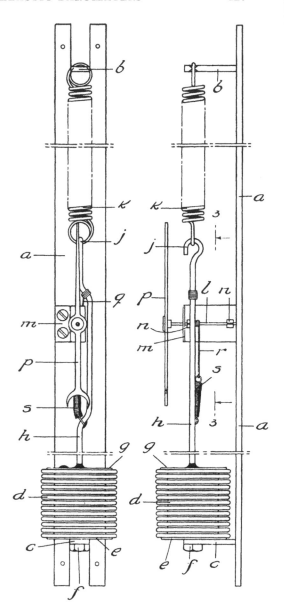

Fig. 16.15. Scammell's aneroid, 1948. *(By permission of the Controller, H. M. Stationery Office)*

[90] U.S. Patents 2,368,905 (1945), and 2,431,098 (1947); British Patents 601,589 (1943), and 617,063 (1945).

[91] British Patent 614,746 (1948).

[92] Spring-rate is the force required to produce unit deflection.

weights? This occurred to J. C. Mewburn, who in 1878 patented the idea of fixing an aneroid capsule on to the end of a balance arm in such a way that the center of gravity might change with the pressure.[93] Many years later this idea was ingeniously developed by K. B. Walz in Germany,[94] in whose barometer, clearly intended for a fixed installation, the change in the center of gravity of the aneroid capsules and their attachments causes the whole mechanism to rotate until equilibrium is established by the motion of a rolling ball.

Let us now consider the very few direct-reading aneroids with optical indication. The first of these was due to F. H. Reitz[95] and was very simple. Intended for the measurement of heights, it consisted only of a vacuum chamber, a spring, and a scale divided into hundredths of a millimeter. This was observed through a microscope and could be read to 0.001 mm., which corresponds to a height difference of about 25 cm. at sea level. It was made commercially by R. Deutschbein at Hamburg.

[93] British Patent 4454 (1878).
[94] D.B. Patent 887,275 (1951); U.S. Patent 2,953,024 (1960); British Patent 819,014 (1956).
[95] Zeits. für Vermessungswesen, Vol. 2 (1873), pp. 363–64.

In 1950 the idea of using an optical lever with a Bourdon or aneroid barometer was patented by I. E. McCabe.[96] In his design, no pivot bearings were used, these being replaced by elastic links. In the following year R. Fuess patented a rather similar device except that the scale, now made on glass by microphotography, is projected on to a screen greatly magnified. Figure 16.16 is a diagram taken from the patent specification.[97] The scale is effectively 3 meters long between the pressures 495 and 925 mm. of mercury, and the instrument is stated to be accurate to 0.1 mm. of mercury.

We shall now leave direct-reading aneroids, and go back to just before the time that Vidie's basic patent expired.

In 1857 the famous German meteorologist Ludwig Friedrich Kämtz (1801–1867) was in Zurich, where the instrument maker J. Goldschmid showed him an entirely novel type of aneroid barometer. Kämtz returned in 1858 and

[96] U.S. Patent 2,530,068 (1950).
[97] D.B. Patent 857,691 (1951). See also Fuess Catalog 112.1 (1952).

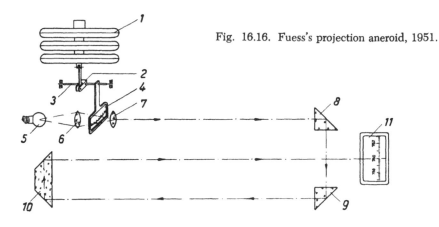

Fig. 16.16. Fuess's projection aneroid, 1951.

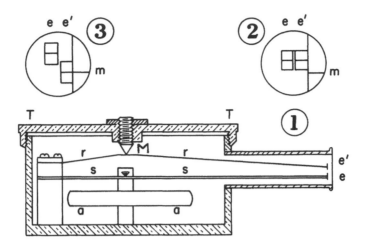

Fig. 16.17. The Goldschmid aneroid, about 1870.

bought one. In 1861 he wrote a paper about it,[98] in which he praised its stability and freedom from friction, but noted that it had a rather large temperature correction and that the scale was not quite linear. Strangely enough Kämtz did not describe the instrument, and indeed no complete description seems to have been printed before 1870.[99] A more concise and clearer description, however, is contained in the cover of the leather case of one of these instruments made by Hottinger & Cie, Zurich, and now in the museum of the C.N.A.M. of Paris.[100] On a circular card are three drawings as in Figure 16.17, the larger one being an axial cross section through the main body of the instrument (9 cm. in diameter and 6 cm. high) in a direction chosen to include the tube which accommodates the reading indexes. In the actual instrument

there is also an eyepiece, and a short thermometer with its bulb inside the main body and its scale projecting beneath the index tube. Under the drawings there is a text in French, of which the following is a free translation:

Referring to fig. 1, when the graduated head TT turns in a clockwise direction the screw M presses the spring rr at the side of the tension spring ss,—which latter is set in motion by the evacuated box aa below—until the marks on the heads ee' of the springs are in line, as in fig. 2. The reading A opposite the index line should be corrected according to the table on the divided head. The reference mark m (fig. 2) is fixed; consequently if the screw is turned to bring the spring rr down until the mark on the head e' is aligned with m, the same reading should always be obtained, if the instrument is not out of order. If the figure obtained is a higher number than the correct one, it should be reduced, and the reading A by the same amount; and *vice versa*. Each adjustment should be made in a downward direction. Small differences indicated by a comparison with a normal

[98] L. F. Kämtz, *Repert. für Meteorol.*, Vol. 2 (1861), pp. 241–45.
[99] J. Goldschmid, *Zeits. der. österr. Ges. für Meteorol.*, Vol. 5 (1870), pp. 177–86.
[100] Inventory no. 8917.

mercury barometer are corrected by moving the circle which carries the index mark. Then a reading is again made on the fixed mark m. The instrument should always be kept in its case. It gets better every year. Reading on the fixed mark, $m = $ [in ink] 776.5.

The table of corrections referred to is really two tables, one for index correction, the other for temperature. The scale corrections vary from $+3.5$ to -3.1 mm., fairly smoothly. The temperature correction is about -0.8 mm. for 10° rise.

The great advantage of the Goldschmid aneroid was that the capsule was under no constraints except the resistance of the spring. It was a remarkable first attempt at a null instrument of this sort, although it sacrificed something to compactness.

It was more practical than the next idea of this kind, patented by E. T. Loseby;[101] a portable aneroid in which a micrometer screw holds a steel "drop-piece" between its end and a stud on top of the aneroid capsule. As the screw is slowly backed off, a point is reached where the piece drops out. This indicates the atmospheric pressure.

In 1878 Anton Schell described an elaborate *Stand-Aneroidbarometer*, i.e., an aneroid barometer intended for fixed stations, invented by Arzberger and Starke.[102] As shown in Figure 16.18, it was a large instrument, set on a base that could be leveled. The motion of the upper end of the two chambers rotated a second level about an axis x,

and it was restored to level by rotating it about an axis y by means of a micrometer screw with a graduated dial. Air seems to have been left in the chambers; there is no sign of controlling springs; but by elaborate experiments the scale-reading s was related to the pressure b and the temperature t, and also to the time z in days after calibration (!) by a formula of the type

$$b = A + Bs + Cs^2 + Dt + Ez + Fz^2$$

It was apparently assumed that the instrument would have a "rate" like a chronometer, and that this would be calculable for a long time. This analogy between a chronometer and an aneroid occurs more than once in the literature. It is part of a general feeling which seems to have been strong in the last half of the nineteenth century, especially in Germany, that it was more interesting to improve the calibration procedures than to improve the instruments.

The next year, W. C. Röntgen devised an equally stationary aneroid barometer[103] which was rather like Goldschmid's in principle, except that the correct position of the chamber was read by means of a telescope, observing a mark by reflection in a small mirror which was turned by the motion of the chamber. The sensitivity was $\frac{1}{200}$ mm. of mercury, which was, of course, unusable except for the observation of small fluctuations of pressure.

Another development of the Goldschmid aneroid was patented in 1893 by Constant Dutoit of Montreux,[104] and it is ingenious enough to be of interest.

---

[101] British Patent 3454 (1863). It was also described in *Quart. J. Meteorol. Soc.*, Vol. 5 (1879), pp. 191–92.
[102] *Repert. für exper. Phys.*, Vol. 14 (1878), pp. 730–61.

[103] *Repert. für exper. Phys.*, Vol. 15 (1879), pp. 44–49.
[104] Swiss Patent 7232 (1893).

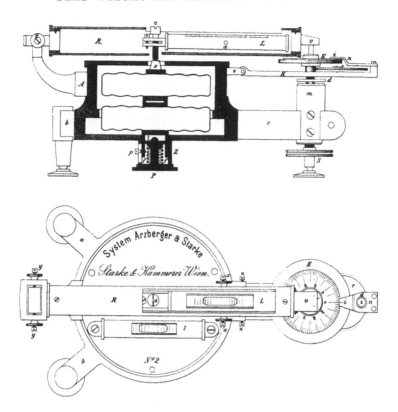

Fig. 16.18. The aneroid barometer of Arzberger and Starke, 1878.

Imagine a metal box with a lid that screws on with a fine-pitched screw. Fastened to the bottom of the box is an aneroid capsule, the top of which carries a piece of black glass with a flat upper surface. In a hole in the lid is a plano-convex lens, the convex side, of large radius, facing the black glass. Newton's rings, formed by interference between the two, serve as a setting index, and the reading is made on a scale engraved around the edge of the lid, as in the Goldschmid instrument. The difficulty with this elegant scheme might well be that the black glass would tilt.

G. Forbes and J. M. Gorham pat-ented a null aneroid for stationary use in which the pressure is measured by the dead weight that has to be placed on the diaphragm to return it to a standard position, observed optically.[105] This brings us naturally to the series of patents taken out by Josua Gabriel Paulin of Stockholm, beginning in 1916, covering the design of a portable aner-oid on the null principle which for two or three decades was probably the best-known instrument of this class.[106] It would take too long to describe in

[105] British Patent 12,713 (1899).
[106] D.R. Patents 303,189 (1916); 492,613 (1926); 492,614 (1927); 558,329 (1929); also corresponding patents in other countries.

detail the various stages of Paulin's invention; but the general principle is as follows: the motion of the free face of an aneroid capsule is restricted by stops to a very small interval, and its exact position in that interval is indicated by a pointer having a very large magnification. A spring has one end attached to the top of the capsule, the other being movable. This spring balances the force due to the pressure of the atmosphere, and the position of its outer end when the pointer is brought to a zero mark serves as a measure of the atmospheric pressure.

The design gradually evolved in the direction of greater simplicity and the complete elimination of sliding parts. In the 1916 design there were still two pivot bearings, and also two springs, a large fixed one carrying most of the load, and a weaker one whose extension was measured. By 1926 all pivots had been eliminated by the use of elastic suspensions. As the patent drawings appear rather complicated, a schematic reduction is offered as Figure 16.19. A

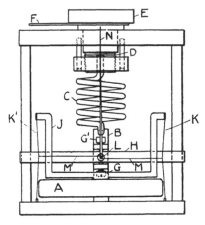

Fig. 16.19. The Paulin aneroid, about 1926 (schematic)

is the aneroid capsule, on which is mounted a lug B, to which is attached a strong spring C. The upper end of the spring can be moved vertically, without twisting, by the screw D and the knob E, to which is attached the pointer F, moving over a dial graduated in units of pressure or altitude.

The motion of the lug B, and consequently of the upper face of A, is restricted to about 0.025 mm. by the stops G,G', which are attached to B and work against a crossbeam H. To B is also attached a cradle J, and to the ends of this are fastened slightly bowed phosphor-bronze strips K,K', having their lower ends attached to the base plate of the instrument.

A shaft L, really a wire held in tension by two helical springs (not shown), passes through holes in the beam H and the lug B. The horizontal wires M,M' from the middle of K,K' are attached, one to the bottom of L, one to the top, so that an upward movement of J moves M to the right and M' to the left, rotating L counterclockwise and moving the pointer N to the left. The movement of N is very large compared to that of J. To read the instrument, the knob E is turned until N is opposite a fixed mark, and the pressure then read from the position of the pointer F. A sensitivity of 0.02 mb. is easily attained.

It is interesting that in 1862 Lucien Vidie patented the elastic suspension. The fifth claim in this patent[107] is for "a bow for holding a wire or metallic blade stretched, when I employ the said wire or blade for giving the rotative motion to the axis of the hand or other

[107] British Patent 682 (1862).

piece of the mechanism."[108] By 1862 Vidie had practically retired from business, and made no use of his patent, but it is impossible not to admire the prescience of this remarkable man.

The most recent designs of precision aneroid have used the capsule in an unconstrained condition, or nearly so, attention being concentrated on the means of determining the exact position of some member attached to it. In 1953, August Lang described an optical

means of reading.[109] A steel point on the end of a group of aneroid chambers bears on an agate plane, so that motion of the point tilts a mirror, all the other constraints of which are elastic. The folded beam of light from a straight-filament lamp, reflected several times by this mirror, is adjusted to zero deflection by moving the whole aneroid system with a micrometer screw, on whose drum the pressure is read. The

[108] *Ibid.* p. 5.

[109] Hamburg, *Deutsche Wetterdienst, techn. Mitt., Instr.* no. 21 (1953), pp. 7–9.

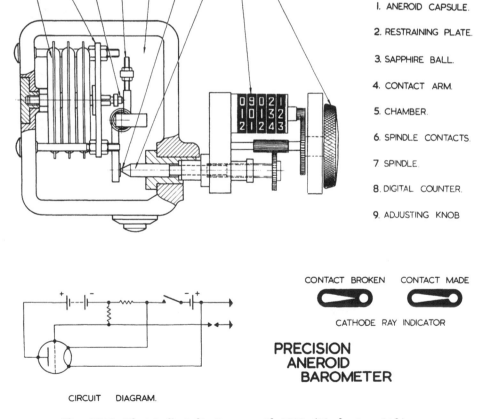

1. ANEROID CAPSULE.

2. RESTRAINING PLATE.

3. SAPPHIRE BALL.

4. CONTACT ARM.

5. CHAMBER.

6. SPINDLE CONTACTS.

7. SPINDLE.

8. DIGITAL COUNTER.

9. ADJUSTING KNOB

CONTACT BROKEN   CONTACT MADE

CATHODE RAY INDICATOR

PRECISION ANEROID BAROMETER

CIRCUIT DIAGRAM.

Fig. 16.20. Electrically-indicating aneroid, 1960. *(Mechanism, Ltd.)*

magnification is high enough to give a rapid indication of the direction in which the pressure is changing, and this is undoubtedly one of the great advantages of such a system for marine work.

Finally, an electrical means of determining the position of the free end of the aneroid capsule has been used by Mechanism Ltd. of Croydon, England, in an instrument designed for the Air Ministry.[110]   The working parts are

[110] British Patent 816,073 (1957).

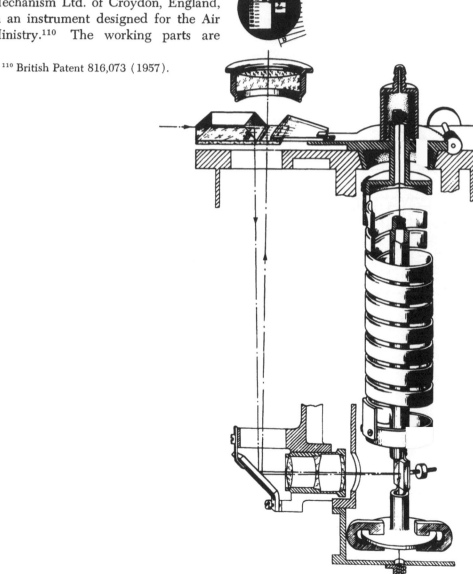

Fig. 16.21. The "microbarometer." (*Askania-Werke*)

shown in Figure 16.20, which is self-explanatory with the information that the electrical contacts (6), when they touch, alter the appearance of a small cathode-ray indicator, the circuit being designed so that the current made or broken by the contacts is extremely small. Things are arranged so that each digit on the counting mechanism represents 0.1 millibar, and temperature compensation is approximated by leaving an appropriate amount of dry air in the capsule.

9. THE BOURDON BAROMETER AGAIN

After several decades of neglect, the Bourdon barometer received a most interesting application in a "microbarometer" designed by the Askania-Werke, Berlin-Mariendorf,[111] which appears to have a relative accuracy of about 0.02 mm. of mercury over reasonable periods of time. In this instrument (Figure 16.21), one end of a helical Bourdon tube is attached to a very precisely-made rotating mechanism, the other—apparently by way of a bimetal—to a front-surface mirror carried on a torsion wire, which merely provides a frictionless bearing.

A series of 30 overlapping partial ranges of pressure is provided by the rotation of the top end of the Bourdon tube, and an index observed by autocollimation in the mirror permits readings in each range to 0.01 mm. of mercury. Oscillations are critically damped by an eddy-current device. The instrument is cleverly designed for portability and ease of leveling in the field, and

[111] Catalog Sheet E1439a, 1956.

seems to be technically of a high order.[112]

10. ANEROID BAROGRAPHS

The aneroid barograph in its ordinary form is such a familiar instrument, and displays its mechanism so clearly in its glass case, that a general description seems superfluous. Instead, I shall reproduce the earliest illustration I have been able to find (Figure 16.22), which dates from 1867, when the Parisian maker Breguet exhibited one at the Paris International Exhibition.[113] The general design of this instrument is surprisingly like that of modern barographs, but note three differences: there was one external leaf spring over the pile of capsules; the clock, separate from the drum, also told the time; and the record was made by a point on smoked paper, relatively frictionless but also very messy.

By 1878 an electromagnet had been added, which made hour marks on the base line. The barograph was compensated for changes of temperature by leaving some air in the capsules.[114]

By 1886 other manufacturers were making such instruments, for instance, Negretti & Zambra.[115] They used seven independent capsules, each with its internal leaf spring, soldered together with bushings between, as shown in

[112] Cf. H. U. Sandig, *Veröff. deutsche geodät. Komm., Reihe A*, Heft 10b (1955), pp. 49–56; C. A. Heiland, *Mech. Engng.*, Vol. 73 (1951), pp. 971–74.
[113] See R. Radau, *Repert. für phys. Techn.*, Vol. 3 (1867), pp. 330–31.
[114] *Annuaire de l'Obs. de Mont-Souris pour l'an 1878*, pp. 256–57.
[115] *Encyclopaedic Illustrated . . . Catalogue*, etc., London (1886), p. 27.

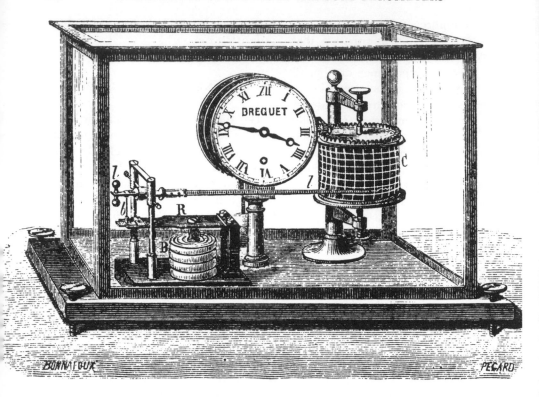

Fig. 16.22. Breguet's original aneroid barograph, 1867.
(Courtesy of the Trustees of the British Museum)

Figure 16.23; a construction which was common for another half century. Variations were introduced into the mechanism, as for example by Fuess, who patented the idea of carrying the vacuum chambers on one end of a heavy bimetallic strip, for temperature compensation, instead of fastening them to the base plate of the instrument.[116]

In all such instruments the hour lines on the chart are curved, a circumstance that is perhaps untidy but does not really cause much inconvenience to the user. In 1919 Negretti & Zambra pat-

[116] R. Fuess, British Patent 13,761 (1913).

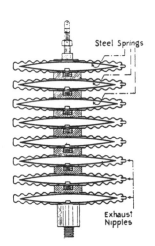

Fig. 16.23. Aneroid barograph. (*Negretti & Zambra*)

ented a barograph which, through a system of levers, records in rectangular co-ordinates.[117] This appeared in their catalogue,[118] but does not seem to have found much favor, partly, no doubt, because it was more costly, but also because of increased friction. In the *Deutsches Museum* at Munich[119] there is an elaborate barograph with a rectangular chart, signed "C. P. Goerz Berlin Nr. 3560," in which an aneroid mechanism with nine chambers in series is arranged to revolve one of two similar pulleys round which a fine chain passes, kept stretched by a light spring and carrying a pen. The chain passes just in front of a paper band driven upward by a clock and apparently making one excursion in three weeks.

There were, of course, minor variations; in 1904, for instance, Short & Mason of London patented an aneroid barograph in which the vacuum chamber and part of the first link are beneath the base plate.[120] But the main advance since about 1920 has been the construction by several firms of aneroid barographs with a magnified scale in terms of mercury units, the so-called *microbarographs*. The firm of Richard pioneered in the field of open-scale barographs with a large instrument having sixteen separate aneroid chambers and giving a magnification of about 3.[121] More recent instruments have been developed by, or with the

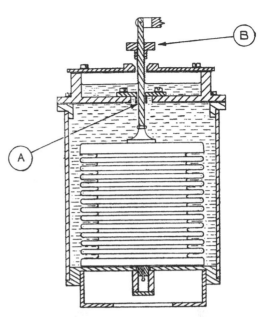

Fig. 16.24. Marine barograph. *(By permission of the Controller, H. M. Stationery Office)*

encouragement of, national meteorological services, and it would be invidious to mention any one in particular. They all make use of metal bellows, usually with internal springs, and a simple lever system. For marine use, the bellows may be immersed in a cylinder of silicone oil, the first link emerging through a small hole in the bottom of an upper chamber partly filled with the oil (Figure 16.24).[122] The response to rapid fluctuations of pressure is thus damped, and an anti-vibration mounting is also used.[123]

A return to the external spring and bimetallic compensation for tempera-

[117] British Patent 8722 (1919).

[118] List B3 (1920), p. 125.

[119] Inventory no. 50730.

[120] British Patent 22,556 (1904). The firm of Short and Mason is now part of the Taylor Instrument Co. of Rochester, N.Y.

[121] There is one at the Smithsonian Institution, Washington, Catalog no. 308,190.

[122] Great Britain, Minister of Supply, British Patent 641,127 (1948). The plug B is screwed down for transport.

[123] See Great Britain, Air Ministry, *Handbook of Meteorological Instruments, Part I*, (London, 1956), pp. 73–75.

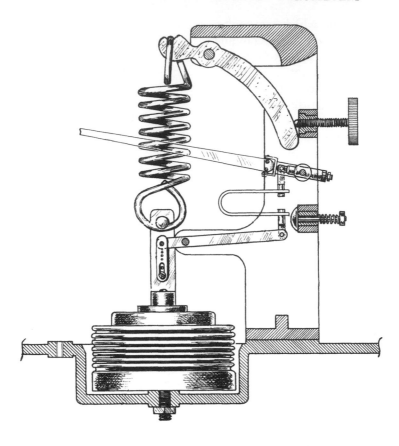

Fig. 16.25. Canadian barograph. *(Dept. of Transport, Meteorological Branch)*

ture has recently been made in the Canadian Meteorological Service, with ease and cheapness of construction in mind (Figure 16.25).[124] The construction affords a convenient zero adjustment and a means of making time marks without touching the pen arm. A slotted link permits transport by air without danger of straining the parts.

In the United States the development of the aneroid barograph was largely

associated for about twenty years with the firm of Julien P. Friez of Baltimore. Julien Pierre Friez, a Frenchman from Belfort, founded the firm, and in 1893 issued an *Illustrated Catalogue of Meteorological Instruments and Apparatus, with Special Instructions on the Equipment of Meteorological Stations; the Installation and Care of Instruments, and the Compilation of Records.* This was issued from 107 E. German Street; but in 1896 Friez moved to larger premises on Central Avenue and set up a meteorological observatory which is still called the Belfort Observatory. In

[124] Canada, Dept. of Transport, Meteorol. Br., *Instrument Manual 11* (Toronto, 19 July 1957).

1899 William Boettinger joined the firm as an apprentice; and it is to Mr. Boettinger, now retired and an octogenarian, that I owe most of my information about the firm.

In 1918 the meteorologist S. P. Fergusson showed a metal bellows to Mr. Boettinger, and the first barograph using such a device was made by the firm in 1920 for use on an aircraft at McCook field, Dayton, Ohio. A station barograph using such a bellows was put on the market in 1925; and in 1928 a larger instrument with a magnification of 3—optimistically called a microbarograph—was produced. In 1939 Boettinger patented[125] an extremely ingenious use of a bimetal for the shaft of the first lever of a barograph. This bends with change of temperature and so alters the ratio of the arms of the lever. It is intended to be used in addition to the usual device of leaving some gas in the chamber, which by itself compensates for changes of temperature only at one pressure. This was applied to a "dual-traverse" barograph first produced in 1925, in which, by using the ingenious lever system of the Fergusson recording rain gauge,[126] the pen is caused to traverse the drum twice, first up and then down, as the pressure rises through its range.

In 1942 Boettinger designed a barograph with two opposed metal bellows in which most of the usual bearings were replaced by elastic links. In 1945 the Friez firm became part of the Bendix Corporation.

The barographs which we have been considering so far, except perhaps that of Goerz, are in the main line of development and are indeed simple and elegant instruments, well suited to the use made of them by countless thousands of meteorologists, amateur as well as professional. But aneroid barographs have also been devised on quite different lines, and it is these to which we shall now direct our attention.

The earliest of these of which I have found a description was devised at the Observatory of Modena by D. Ragona, and described in the Supplement no. 1 to the 1867 report of that institution. This was very elementary, a sharp stylus being attached to the end of the pointer of a Naudet aneroid barometer, and a mechanism built round it by which this stylus was caused to prick a hole in a moving band of paper at equal intervals of time.

A much more sophisticated development of this idea is that of Hipp,[127] who called it a *Barometrograph*. This is an elaborate instrument (Figure 16.26) in which a point is pricked every ten minutes in a band of paper pulled off a supply roll. The stylus, moved horizontally by a rather complex lever system, is pressed into the paper as an electromagnet draws up an armature, and on its release the paper is advanced by a ratchet and a spring-loaded pawl. There is one of these at Munich,[128] and also a second one[129] incorporating a bank of switches which could be used to transmit the indication to a distance with a precision of about 2 mm.

[125] U.S. Patent 2,165,744 (1939).

[126] S. P. Fergusson, *Monthly Weather Rev.*, *Washington*, Vol. 49 (1921), pp. 379–86.

[127] *Zeits. österr. Gesell. für Meteorol.*, Vol. 6 (1871), p. 104. Hipp was an instrument maker at Neuchâtel.

[128] *Deutsches Museum*, Inv. no. 7244.

[129] Inventory no. 5581.

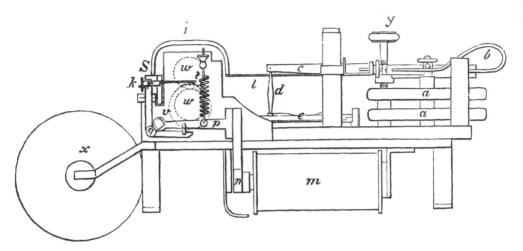

Fig. 16.26. Hipp's *Barometrograph*, 1871.

The beautifully constructed barograph shown in Figure 16.27,[130] made by J. Goldschmid of Zurich, belongs to about the same period. Apart from having its stylus moving in a vertical plane, this differs from that of Hipp in being entirely mechanical. It also makes only one reading an hour, the part that looks like a combination clock face and circular saw being the cam that gradually swings the semi-cylindrical hammer away from the stylus, letting it fall exactly on the hour. The stylus is quite free nearly all the time. In this instrument the motion of the paper is continuous, not intermittent as in that of Hipp.

This form of barograph was later made by Hottinger & Co. of Zurich, according to a description published in 1878.[131] By 1881 they had made the

stylus move in a horizontal plane, as in Hipp's barograph; but it was still entirely mechanical.[132] A second fixed stylus gave a base line.

In 1894 J. Bartlett patented a barograph which recorded on a vertical circular chart instead of on a drum.[133] It is peculiar that this form of record, which appeals so much to engineers, has never found favor with meteorologists, who also seem to prefer instruments which stand on a shelf to those that hang on a wall. Finally, shortly after the war of 1914–1918, the interests of mountaineers were considered by the Paris firm of Richard. In 1962 the present administrator of the company, M. Claude Bruneau, showed me a beautiful little pocket barograph about the size of a package of cigarettes, which records point-wise at regular intervals on a moving band of chart paper. If this is the smallest of baro-

[130] This is at the Meteorological Office, Bracknell. I am indebted to Mr. C. H. Hinkel for this photograph.
[131] *Zeits. österr. Gesell. für Meteorol.*, Vol. 13 (1878), pp. 174–76.

[132] *Ibid.*, Vol. 16 (1881), pp. 273–81.
[133] British Patent 1900 (1894).

Fig. 16.27. Goldschmid's barograph, about 1870. *(Meteorological Office)*

graphs, the largest aneroid barograph must surely be the one which stands in the entrance hall of the firm at 25 rue Melingue, and has most of the pressure of the air on a pile of enormous aneroid capsules balanced by a heavy weight hung from the lower end of this assembly. The record on the immense drum is very detailed.[134]

[134] The firm of Richard Frères was founded in 1858, and greatly extended by Jules Richard (1848–1930), the son of the founder. The first Richard barograph, made in 1878, depended on a Bourdon tube and had a Watt parallel motion to permit a rectangular chart. But by 1882 the firm was making an aneroid barograph with 8 capsules, and of the usual shape (see *Bull. Mensuel de la Soc. d'Encouragement pour l'Ind. Nat.*, sér. 3, Vol. 9 [1882], pp. 531–43).

11. ANEROIDS WITH SERVO SYSTEMS

Either for indicating at a distance, or for making a record, mechanisms have been applied to the aneroid in which the small motions of the capsule or bellows control, through electrical or mechanical apparatus, much larger motions of an index or a pen.

The first such barometer of which I have found any trace was made by Redier of Paris about 1875, and was at one time in the Science Museum, London. This had a dial 1.50 meters in diameter, the index "actuated by a small aneroid connected through a lever with a clock having two trains which

Fig. 16.28. The tower barometer at Munich, transmitter. *(Deutsches Museum)*

work in reverse directions. When the air pressure is constant, the fans of the two trains are held fast by a cross piece of the levers, but with a rise or fall of the pressure one or other of the fans is released, and its train thus freed to turn the index round the dial."[135] It would revolve, of course, until the index has moved far enough for the "fan" to be arrested again. Such a fan is more usually called a fly governor.

Thirty years later, the mechanism of the Sprung rolling-weight barograph[136] was applied to an aneroid by J. Maurer of Zurich.[137] The aneroid capsule used was a twenty-year-old one by Usteri-Reinach. In 1961 I could find no trace of this instrument at Zurich.

About 1925 a tremendous distant-indicating barometer, still in operation, was constructed for the tower of the Deutsches Museum at Munich by the firm of G. Lufft, Stuttgart. The dial of

[135] Great Britain, Board of Education, *Catalogue of the Collections in the Science Museum, South Kensington,* etc., *Meteorology* (London, 1922), p. 31. The inventory number of this instrument was 1876–774.

[136] See p. 311 above.
[137] *Meteorol. Zeits.,* Vol. 25 (1908), pp. 367–69.

this instrument is no less than 6 meters in diameter. The transmitting apparatus, installed in the room where meteorological instruments are on exhibition, is shown in Figure 16.28. The dimensions of this machine can be visualized from the datum that the diameter of each aneroid capsule is about 20 cm. Through relays, installed below, the contacts at the end of the skeleton pointer operate either of two direct-current motors which rotate the entire aneroid mechanism in the appropriate direction to keep the contacts centered. The actual rotation is done step-wise through a maltese-cross mechanism, so that two larger motors in the tower can rotate the huge pointer in exact agreement with the rotation of the aneroid.

In 1926 J. G. Paulin patented the application of what amounts to a mechanical servo-system to his null reading barometer.[138] The purpose of this extremely complicated system was, however, to increase the sensitivity of the barometer and do away with the necessity of observing the null indicator. It had over a hundred parts, and does not seem to have become well known. It was developed into a "precision barograph" in the United States thirty years later.[139]

In Germany A. Graf has made the Askania "microbarometer" into a very sensitive barograph by using differential photocells and an electrical recording system similar to that used in the self-balancing potentiometers so common in industry.[140] A good case has been made for the use of such an instrument in the measurement of heights with the barometer.[141]

Electronic circuitry could not fail to be applied to the barometer. One of the most ingenious of elastic barometers consists essentially of a wire held in tension by an aneroid chamber acting as a spring; the tension in the wire will thus vary with the pressure of the atmosphere. The wire is in the field of a permanent magnet, so that if an alternating current is passed along it, it will vibrate. A vacuum-tube oscillator in which the wire forms part of the resonant circuit is used to drive it, and the frequency of the resulting current, which is a function of the pressure, is measured or recorded by an electronic frequency meter.[142]

We may close this section with a reference to a remarkably simple device patented by A. Ovtschinnikoff in 1953,[143] which might well form the basis of large barometers for public display, though its author intended it only as a differential instrument to measure small changes in pressure. This is a self-setting null instrument, and has a metal bellows attached to a frame at one end, the other end being fastened to a rod which can be moved axially in either direction by a reversing motor. The end of the bellows which is attached to the frame is formed by a flexible diaphragm carrying a contact which is kept centered between two fixed contacts as the other end of the bellows is moved in and out by the

[138] D.R. Patent 492,613 (1926).
[139] D. E. Copple, U.S. Patent 2,712,238 (1955).
[140] *Zeits, angew. Phys.*, Vol. 3 (1951), pp. 107–10.

[141] A. Graf, *Veröff deutsche geodät. Komm., Reihe A*, Heft 10b (1955), pp. 65–79.
[142] Frank Rieber, U.S. Patent 2,473,610 (1944).
[143] U.S. Patent 2,646,682.

reversing motor, operated through these contacts and a suitable relay system. A counter, a dial, or a recording pen could be geared to the motor shaft, and of course the readings could be transmitted to a distance by any of the usual electrical devices.

12.   ELASTIC MICROBAROGRAPHS AND VARIOGRAPHS

If the reader will recall the discussion of microbarographs and variographs in Chapter 11, he will see that many of the devices used could easily be replaced by elastic components, giving a record of p or of dp/dt depending on the particular arrangements. In the last decades of the nineteenth century the firm of Richard in Paris were in fact manufacturing a *statoscope*, which was nothing more than a large aneroid chamber with a low spring-rate, placed inside a closed insulated box which could be opened to the atmosphere through a tap from time to time. The rest of the mechanism was much like that of an ordinary barograph. In 1910 E. A. T. Hue patented a clockwork mechanism to bring the reservoir into communication with the outside air at regular intervals.[144]

A better idea would be to perform this operation automatically whenever it is needed, that is to say whenever the pen approaches the edge of the chart. G. A. Suckstorff of Göttingen has described such a microbarograph,[145] in which the interior of a very flexible capsule with a very weak internal spring

is connected by a narrow lead tube to an empty 100-liter flask buried 2 meters in the earth. The magnification is 10. When the pen approaches either edge of the chart a magnetic mercury valve is operated through a time-delay system giving fairly quick operation but a release time of 3 minutes. This valve connects the system to the open air and leaves time for the pressures inside and outside to become equal, bringing the pen back to zero.

This is all right as long as the actual value of the pressure is unimportant. To D. Sonntag of Collmberg Observatory it was of importance, and he described a microbarograph of special construction which made it possible to measure it, even with a magnification of 50 in mercury units.[146] In this instrument a small mirror is turned by the motion of the aneroid chambers. This mirror is at one focal line of an elliptical cylinder, around part of which 31 plane mirrors are suitably located, so that any one of them will reflect light to the other focal line, where a cylindrical lens gathers the light to a point on a drum covered with photographic paper. The 31 mirrors are arranged so that each one takes over just as its neighbor reflects the light to one edge of the drum. In default of special apparatus at Collmberg, it was calibrated as well as possible by comparison with two mercury barometers.

Elastic variographs for dp/dt seem to have been mainly electrical, for example that of H. Benioff and B. Gutenberg,[147] which consists essentially

[144] D.R. Patent 229,504 (1910).
[145] *Zeits. für Geophys.*, Vol. 12 (1936), pp. 245–49.

[146] *Zeits. für Meteorol.*, Vol. 6 (1952), pp. 68–74.
[147] *Bull. Amer. Meteorol. Soc.*, Vol. 20 (1939), pp. 421–26.

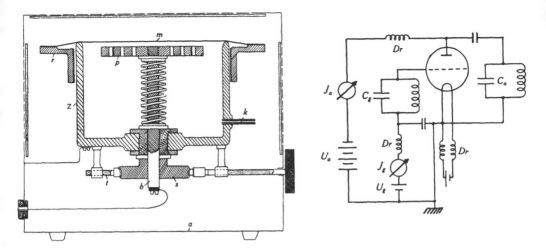

Fig. 16.29. Saxer's dp/dt variograph.

of a permanent-magnet loudspeaker mounted in the wall of a vessel holding 200 liters. The E.M.F. induced in the coil, which is proportional to dp/dt, is recorded photographically.

The dp/dt variograph of Saxer, also electrical, uses a condenser microphone of special construction.[148] The capacitance between the disk p (Figure 16.29) and the tinfoil membrane m forms part of the plate-tank circuit of a tuned-grid, tuned-plate oscillator, and over a certain range the grid current of such an oscillator is linear in this capacitance. Slow changes of pressure are reduced by having a capillary leak, and the response is of course more pronounced the higher the frequency of the fluctuations of pressure.

Finally, for the investigation of very rapid changes in pressure, A. C. Crehore and G. O. Squier developed an inter-ferometric barograph.[149] A stack of eight aneroid chambers, with a front-surface mirror attached to its upper end, was supported in a strong frame. To the frame was attached, adjustably, a half-silvered optical flat. Fringes produced in the light from a mercury arc were imaged in the focal plane of a camera which had a slit in front of a moving film; thus the record was in the form of a series of wavy lines, the result of a time-series of central sections through a constantly expanding or contracting system of circular fringes, so that the number of fringes that appear or disappear may be counted. The record looks exactly like a picture of a tangential section through a somewhat irregular fir tree. The magnification was about 50,000. Nothing is said about the effect of wind on the room in which the instrument was contained, but it may

[148] Leonhard Saxer, *Helv. Phys. Acta*, Vol. 18 (1945), pp. 527–50.

[149] *Bull. Mt. Weather Obsy.*, Vol. 4 (1912), pp. 115–20.

be suspected that many of the "waves of pressure" could be traced to this cause.

13. ELASTIC BAROMETERS WITH LIQUID INDICATORS

In 1852 Enrico Negretti and Joseph Zambra patented, among other things, the idea of joining a flexible capsule to a vertical glass tube and filling them with liquid up to an appropriate mark.[150] The increasing pressure of the air, for example, would tend to compress the capsule and force the liquid up the tube until hydrostatic equilibrium was once more attained. This arrangement is thus a sort of hybrid between the elastic and the hydrostatic barometer. Eleven years later, on January 7, 1863, Negretti & Zambra took the unusual course of issuing a formal disclaimer, through the Patent Office, of all the parts of this patent dealing with barometers. In April of the same year a patent[151] was issued to W. H. Mitchel covering this sort of barometer. The tube now ended in a closed chamber filled with gas, the expansion of which was to provide compensation for changes in temperature. There seems to be no good reason why this would not work, but even less reason why anyone would want to make such an instrument. Almost exactly the same construction was, nevertheless, patented all over again ninety years later.[152]

It was also perhaps inevitable that someone should think of combining the mercury barometer with the aneroid, and in 1872 A. H. Emery and John Johnson took out a patent for such an instrument.[153] In their barometer the pressure of the atmosphere was balanced by that exerted by a short column of mercury, plus the pull of a set of tension springs. The latter were attached to a spider which in turn was fastened to the center of a flexible metal diaphragm forming the bottom of a chamber filled with mercury, from which rose a narrow tube, closed at its upper end and containing the column of mercury used as a measuring device. The springs were stretched until a vacuum space appeared at the top of the tube, and the device presumably calibrated by comparison with an ordinary barometer. Such an instrument would have no advantages over the ordinary aneroid, and is mentioned merely as a curiosity.

14. INDICATORS OF DIRECTION OF PRESSURE CHANGE

The "setting hand" which is such a familiar feature of the household barometer was applied to the aneroid at least as early as 1849.[154] By using this hand, one can tell whether the barometer has gone up or down since it was last looked at and "set." But perhaps one has forgotten to set the hand; or perhaps the barometer has been down and is on the way up again; what then? One gives it a smart tap, of course, and watches which way the pointer jumps. Unfortunately the success of this maneuver is less, the better the quality of

[150] British Patent 14,002 (1852).
[151] British Patent 889 (1863).
[152] British Patent 742,020 (1954).

[153] U.S. Patent 127,752 (1872).
[154] E. J. Dent, A Treatise (1849), p. 22.

the instrument; and a good deal of ingenuity has from time to time been devoted to devising means of indicating at a glance whether the atmospheric pressure is rising or falling. We may as well state at once that a barograph of the most moderate quality is preferable to any of them.

In 1861 the instrument maker John Browning patented[155] "two loose hands or indicators which will yield without offering appreciable resistance to the barometer needle, and be pushed thereby, and remain stationary (until reset) at the points where so pushed. . . ." As we have seen,[156] Keane Fitzgerald had applied precisely the same device to a wheel barometer exactly a century before. This was, of course, more an indicator of maximum and minimum pressure than of "tendency."

A decade later, a public barometer at Chartres was fitted with a real indicator of the direction of the pressure change.[157] This was a loose, counterbalanced pointer concentric with the main pointer of the barometer. At the outer end, the extra pointer carries a small rectangular plate with two "ears," either of which can be pushed by the main pointer, so that the position of the latter with respect to the little plate shows at a glance whether the pressure is rising or falling. That is, if the motion in one direction has been appreciable.

Between about 1880 and 1935 or later, the patent files accumulated a number of specifications, not very various in conception, for such devices. Most of them require a good deal of pressure change to produce a clear indication. One very simple device that did not is the *elettro-barometro* of Alberto Miglioretti,[158] in which the pointer of an aneroid barometer pushes around two light electrical contacts, fastened together with a space between them and insulated from each other and from the frame of the instrument. Two electric lamps or other signals can thus be installed to indicate whether the pressure is rising or falling.

15. ELASTIC PROPERTIES AND ACCURACY

At the birth of the aneroid, as we have seen, the attending doctors doubted whether the properties of materials would even permit its safe delivery. Their fears proved unfounded; nevertheless the infant that received such tender care did turn out to be afflicted with two, or perhaps three, serious constitutional defects, associated with the elastic properties of metals.

The first serious investigation of an aneroid barometer from this point of view was made in the United States by J. Lovering,[159] who took a barometer, no. 1265, made by Lerebours & Secretan of Paris, and put it in a chamber in which the air could be compressed or rarefied. He noticed that if, for example, the pressure was reduced and then brought back to its original value, the

---

[155] British Patent 2560 (1861).
[156] Chapter 11, p. 285 above.
[157] De Vésian, *Congrès scientifique de France, 36ᵉ session* (Chartres, 1870), pp. 131–34.

[158] Italian Patent 54,851 (1900); British Patent 15,965 (1900).
[159] *Amer. J. of Sci. & the Arts*, Vol. 9 (1851), pp. 249–55.

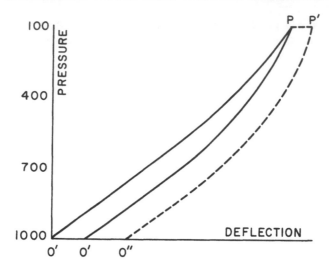

Fig. 16.30. Illustrating hysteresis and creep.

aneroid did not return all the way, but remained at a somewhat lower reading and only gradually regained its original deflection. Lovering's research was of course only a first attempt, but it served to point out the existence of the difficulty.

It will be best to acquaint the reader with the phenomena as they are now known. Referring to Figure 16.30, suppose an aneroid to be calibrated by taking it through a cycle of pressure variation from about 1,000 mb. to some low pressure, for example 100 mb., and back again to 1,000 mb. Let us assume that the whole procedure takes an hour or so. As the pressure decreases, a certain relation between pressure and deflection will be observed, leading to a calibration curve such as OP. As the pressure increases, it will be found that the deflection no longer follows the original curve, but the curve PO' instead, and when the pressure has come

back to its original value, a small deflection OO' will be indicated. The magnitude of this has been greatly exaggerated in the figure, for clarity. This phenomenon has been given the name of *elastic hysteresis*, probably by analogy with the magnetic properties of materials.

If now the pressure be maintained at its maximum value, the deflection will gradually return from O' toward O. Furthermore, if the pressure had been kept at its minimum value, the deflection would have gradually increased, approaching some limit P'; if then it were brought to the maximum pressure again, the new deflection would be O''.

The gradual change in deflection at constant pressure, PP' or O'O, is known as *creep*, or in German as *elastische Nachwirkung*. It proceeds rapidly at first, reaching a maximum value asymptotically after many hours. There may be a further gradual alteration in the calibration of an aneroid, even if no air leaks into the chamber, and this is ascribed to molecular adjustments in the material of the capsule, which has, after all, been subjected to severe deformation in manufacture. This is known as *secular change* and is nowadays reduced to manageable proportions by suitable heat treatment.

Let us go back to 1860 or thereabouts. If Lovering had not noticed these phenomena in the laboratory, the mountaineers, who had bought hundreds of aneroids, would have done so on the mountains. Representations from this quarter led to extensive tests at Kew in 1868 and 1869, reported by Balfour Stewart (1828–1887), then Superintendent of the famous Observatory.[160] Aneroid barometers were borrowed from several manufacturers, and tested for the effects of changes of temperature and pressure. Most of them were found to be well compensated for changes of temperature, but showed considerable hysteresis and creep.

At an international meteorological conference held at Leipzig in August 1872, it was decided that the aneroid barometer should not be used at meteorological stations except as an interpolation instrument. The Meteorological Society in London met to discuss this question and agreed with the decision.[161] The following year Paul Schreiber, in a paper of his usual length, demurred.[162] He thought that it was time that some real investigation was done;

but nevertheless he believed that the aneroid could be substituted for the mercury barometer after it had been carefully tested at a central station for index correction, temperature correction, and rate *(Gang)*. He obviously thought that these instruments changed their correction uniformly with time and compared them with chronometers in this respect. He dismissed the tests of Lovering and Stewart, the last on the fairly good grounds that his pressure changes were stepwise. He made numerous comparisons among six aneroid barometers over a period of several months, but the range of pressure was small, and they did not really determine anything about the elastic errors.

C. M. Bauernfeind, the Professor of Surveying at Munich, investigated the zero-, scale-, and temperature-corrections of three Naudet aneroids over a moderate range of pressures, but without coming upon the elastic aftereffect or hysteresis.[163] He checked his calibrations by going on a train journey in Bavaria, with the aneroids on the seat beside him, and a thermometer which could be hung outside the window. The seat, we are told, was upholstered. Readings were made at every station, the elevation of which had, of course, been found by ordinary leveling when the railway was built.

Meanwhile comparisons were being made between aneroid barometers of different patterns, in particular between the usual construction typified by those of Naudet and the newer types due to Goldschmid and to Rietz. One of each type was used by the professor of sur-

[160] *Proc. Roy. Soc.*, Vol. 16 (1869), pp. 472–80.

[161] *Quart. J. Meteorol. Soc.*, Vol. 1 (1872), pp. 223–24.

[162] *Repert. für exp.-Phys.*, Vol. 9 (1873), pp. 193–241.

[163] *Akad. Wiss. München, Abhandlungen, math.-phys. Klasse, 3te Abt.*, Vol. 11 (1874), pp. 27–80.

veying Wilhelm Jordan,[164] whose results were really not at all conclusive; Joseph Höltschl of Vienna compared Naudet's and Goldschmid's aneroids and wrote a book[165] about his tests, which was severely and probably justly criticized by both Jordan and Schreiber. He found the Goldschmid much inferior, but he tested over 100 Naudet instruments and only one by Goldschmid! The sequel was an astonishing volume,[166] privately printed, which for its picturesque language and its abundance of quotations from Schopenhauer, deserves a prominent place even among the turgid polemics of German professors.

Meanwhile the elastic errors were being investigated by Grassi in Italy,[167] and a little later by Carl Reinhertz in Germany,[168] who pointed out that the phenomenon of elastic "creep" had been discovered and named by W. E. Weber in 1835.[169] Reinhertz correctly identified the two aspects of the problem, the effects which manifest themselves during the change in pressure, and those which show up after the change in pressure is complete. In extensive laboratory experiments he used (1) a constant *rate* of change of pressure with various total changes, and (2) a constant total pressure change and various rates. He found things very complicated, the error at any time being a function of all the history of the instrument for some time before the observation. He also found that some aneroids were much better than others in this respect.

The next contribution was from the celebrated mountaineer Edward Whymper (1840–1912),[170] who tested over seventy aneroids in the laboratory, and used eight in the Andes. He showed that the usual method of "pointing" the scales of these instruments, in which the pressure is fairly rapidly changed and the indication noted at once, has the result that the altitudes determined from them in high mountains must be almost always too great, sometimes by several hundred feet. Travelers, he noted, often feel that recomparing their aneroids with a mercury barometer after returning to near sea level ensures the accuracy of their readings at great elevations; but this may be far from the truth. Whymper's book was written from the very practical standpoint of an immensely experienced mountaineer.

It was evident to Whymper that in practice it was impossible to correct the readings for the elastic errors. The only way out was to make better instruments; and in the following two years he took out two patents in partnership with the instrument maker J. J. Hicks.[171] In the specification of the first we read,

[164] *Zeits. für Vermessungswesen,* Vol. 2 (1873), pp. 364–73.

[165] J. Höltschl, *Die Aneroide von Naudet und von Goldschmid* (Wien, 1872). Quoted by Schreiber, see note 162 above.

[166] J. Höltschl, *Stultitia et mala fides oder; Die Weisheit und Biederkeit der Aneroïd-Gelehrten in Süd- und Mitteldeutschland* (Vienna, 1877).

[167] *Ricerche sperimentale* etc. (Rome, 1875 and 1877); quoted by Reinhertz, *Zeits. für Instrum.,* Vol. 7 (1887), p. 154.

[168] *Zeits. für Instrum.,* Vol. 7 (1887), pp. 153–70, 189–207.

[169] *Annalen der Physik,* Vol. 34 (1835), pp. 247–57. Weber named the phenomenon *Nachwirkung* (aftereffect).

[170] Edward Whymper, *How to Use the Aneroid Barometer* (London, 1891).

[171] British Patents 11,008 (1892); 4387 (1893).

"To prevent errors in aneroid barometers due to their being exposed for some time to low pressures, the parts when fitted together, or some of the parts independently, are exposed to an annealing process at a high temperature for a long time."

Whymper's book made a great impression and led to demands that the method of testing aneroids at Kew should be reviewed. A very extensive investigation was therefore undertaken at the observatory, and reported on by its Superintendent Charles Chree in 1898.[172]

Part of this long paper is devoted to a statistical analysis of tests which had already been made at Kew on about 300 aneroid barometers. In these tests the pressure had been changed at a rate of 1 inch of mercury in 3 or 4 minutes. Chree found that whatever the total range of pressure, all the observations lay nearly on the same curve if the differences between the ascending and descending readings were plotted against the pressure change as a fraction of the total pressure change. In addition to this analysis, special tests were made on four aneroids bought from J. J. Hicks for the purpose. With these, the above conclusion was confirmed, and it may be worth noting that for a pressure range of 9 inches of mercury the *mean* difference between the readings ascending

[172] Charles Chree, *Phil. Trans.*, Vol. A191 (1898), pp. 441–99.

and descending was over 0.1 inch in three out of the four instruments.

When the pressure was held at its lowest point, the fall of the reading in a given time was nearly proportional to the range of pressure and was an exponential function of time. A similar behavior was found on return to the higher pressure. Temperature, in the range 50° to 80° F., had a very minor effect on all these phenomena.

There were more conclusions, but in general the result was to show the great difficulty of predicting the behavior of an aneroid, for example, when being carried up a mountain, unless tests had been made at a very similar rate of change of pressure, including stops. In this respect the mountaineer's problem is more serious than the problem faced by the aneroids carried up by sounding balloons, which ascend at an approximately uniform, and in any event predictable, rate. These latter can easily be calibrated on a time schedule, both of pressure and temperature, similar to that in which they will be used. But this is outside the province of this book.

The subsequent history of the elastic errors of the aneroid, insofar as it is written down at all, lies in the Patent Office files, and is largely a story of advances in metallurgy. It is now possible, thanks to the development of aircraft instruments, to make aneroid barometers reliable enough to supplant mercury barometers at many meteorological stations.

APPENDIX

Appendix

LIST OF THE MORE IMPORTANT OR
INTERESTING BAROMETERS EXAMINED

THE following entries are derived from a combination of actual notes made in the museums, etc. referred to, catalog entries, and information extracted, generally afterward, from the literature. One of the most difficult decisions that had to be made in the preparation of this book was whether to go first to the museums or to the libraries. On the idea that it would give me the "feel" of the subject, I decided to see the museums, especially on the Continent of Europe, first. Very few of these have catalogs; almost none have printed catalogs that are still "in print."

Not all the barometers in any museum are dealt with here; only those with features which make them in some way especially interesting, unique, or illustrative of a principle of construction, and, in museums with printed catalogs, those regarding which there is a difference of opinion between the catalog and the writer. Nor are they completely described in the way that they ought to be (but seldom are) in a catalog. What I have tried to do is to isolate the interesting features, referring to the main text where fuller explanation or description is there available.

There must be many places where barometers are on exhibition which I have not visited. A few which I have visited have not been mentioned, chiefly for the reason that this book is not concerned in any way with barometers considered only as decorative objects. In all the places where I have examined barometers, I have had nothing but the most delightful courtesy and helpfulness from Directors and staff.

The list is arranged alphabetically by cities, and in each institution by inventory numbers, followed by any barometers which have no number.

BOLOGNA, ITALY. UNIVERSITA. ISTITUTO DI FISICA.

(Not a museum, but a University Physics Department which has preserved many of its old instruments.)

1715 (?) Cistern barometer with an (ivory?) float now lost; the ivory guide has a horizontal mark. Scale plate in-scribed *"Bellani fece;"* scales in Paris inches and lines, and in cm. and mm. No verniers.

—— A handsome barometer in a wooden frame, with the inscription "Giuseppe Brusca, 1806." Adjustable cistern with a large plug through which it could be filled. Zero evidently established by overflow.

445

Scales & verniers in English inches (marked "London") and French inches (marked "Paris"). (*See p. 174.*)

—— A neat siphon barometer carrying the inscription "T. Salleron 24 rue Pavée (du Marais) Paris" and the engraved number "180." Gay-Lussac type with two scales in mm., verniers to 0.1 mm.

BRACKNELL, ENGLAND. METEOROLOGICAL OFFICE

(The official museum of M.O. 16, removed from Harrow to Bracknell in 1961.)

1 One of the beautiful barometers "Invented and Made By Dan Quare London." Scales in inches and 20ths on each side of the tube; no verniers but two indexes moved by screws operated by decorative knobs at the top of the brass box with a glass front which encloses the scales. For portability, the leather bottom of the cistern may be raised; but there is no zero adjustment. (*See p. 150.*)

10 A mountain barometer in a cylindrical mahogany case. Fixed cistern, but scale not contracted. On an ivory band there is the inscription "Newmans improved portable iron cistern 109 Regent Street London Correction for capacities 1/50 Neutral Point 30.194 Capillary Action +.040 Temperature 62°." The scale is graduated in inches and 20ths from 18.3 to 32.0 in., with a vernier to 0.002 in. (*See p. 149.*)

60 "Kew pattern" station barometer, inscribed "Adiè London No. 950" and made ca. 1903. Inches and 20ths, vernier to 0.002 in. Very like modern Kew pattern barometers in general appearance.

112 French station barometer, inscribed "Tonnelot Fab^{nt} rue Massillon, 3, Paris." (*See p. 152.*)

4101B Fixed-cistern barometer in a wooden case; scale on ivory, 27 to 31 in., inscribed "West, 41 Strand, London." Vernier to 0.01 in.

4102B1 ⎰ Two almost identical barometers
4102B2 ⎱ by Newman. The first is marked "⋏M 35 J. Newman 122 Regent Str London." Fixed-cistern made portable by closing the end of the tube with a screw-operated plunger. The screw has a squared outer end, which is hidden by a brass cap after the barometer has been installed. Tubes about 5 mm. inside diameter. Ca. 1825. (*See p. 157.*)

4103B A barometer in a mahogany frame, inscribed "W & S Jones Holborn London." This is incomplete, but seems to have had a leather bag which could be compressed for transport. In the little door which covers the scale when not in use there is a thermometer in degrees F. with "fever heat" at 112. These makers were in business at 30 Lower Holborn from about 1793 to after 1850.

4104B A barometer in a rectangular mahogany case, signed "Bate, Poultry, London ⋏No. 6," and having an ivory scale from 27 to 31 inches in tenths, with a vernier to 0.01 in. The index is clearly designed to read the edge of the mercury column. The cistern is enclosed in a turned brass body; it is made of wood with a leather bottom which can be compressed by a screw with a plate on the upper end, but only after the bottom section of the brass cover has been removed.

4107N This barometer is inscribed on its silvered scale "H. Olland. Utrecht. 178." It is provided with a Cardan suspension and is of metal construction. Metric scale. Index designed to read the top of the column, but providing no way of avoiding the effects of parallax. The box that encloses the scale is a trapezium in horizontal section, the front being of glass about 5 cm. wide, the sides of silvered brass, and the rear, about 1 cm. wide, of ground glass; it is very easy to illuminate the tube and scale.

4108B One of Newman's mountain barometers is a square brass tube with the scale on a smaller brass tube which slides inside the other, the vernier being fixed and the scale therefore reversed. Windows at front and rear near the top of the scale-tube through which the top of the mercury column can be sighted. On the back of the scale tube is "J. Newman 122 Regent Street London. Capacities 1/52." The scale is graduated from 19.3 to 32.0 in. in 20ths; vernier to 0.002 in. The cistern is Newman's design with a divided iron box. (See p. 148.)

4109F   A large barometer in a wooden case, inscribed "Depot de la Marine N°. 14 Salleron 24, Rue Pavée au Marais Paris." The tube about 8 mm. bore. Simple scale in cm. and mm., surrounded by a considerable treatise on single-observer forecasting. A relatively enormous thermometer. No vernier or index. The cistern, encased in a wooden block, is made of glass in the form of two approximately spherical chambers, the upper one about 30 mm. in diameter, communicating by an aperture of about 15 mm. with the lower chamber, which is about 22 mm. across. About three-quarters of the way up the upper sphere there is an air hole.

4110B   Similar to no. 10 in many respects. The inscription is "Newman's improved portable iron cistern/122 Regent Street London/Correction for capacity 1/58 /Neutral point 30.205 / Capillary action .042/Temperature 59°." The vernier slide can be moved up and down by hand and set by turning a knurled nut at the top of the instrument, thus avoiding any increase in the diameter (about 35 mm.) of the cylindrical mahogany case. There is a narrow slit at the back through which light can come, but the index only goes round the front of the barometer tube.

4111B   Similar to the ordinary "Kew pattern" barometer, but having a dial reading to 0.1 mb., geared to the setting slide, in place of a vernier. The main scale is divided only in tens of millibars. Made by the Cambridge Scientific Instrument Co., and presented by them to the Meteorological Office in 1921. (See p. 201.)

4113F   This is a cistern-siphon barometer of the type developed by H. Wild about 1875. While it bears no signature, it was probably made by Turretini in Geneva. (See Fig. 9.22 and p. 230.)

4116F   A marine barometer bearing the Crown and Anchor device of the *Marine Impériale* and the signature "Ernst à Paris N° 202." It has a sturdy mahogany frame and an ivory scale. The thermometer is °C. and °R. The cistern is cleverly arranged so that it can be made portable by reversing an unsymmetrical piece at the bottom. (See Fig. 7.11 and p. 155.)

4117B   A "Fitzroy" barometer with the usual curved tube and cylindrical bulb and valve. The large Fahrenheit thermometer is inscribed "S. Maw & Son London." Fitzroy's weather remarks and a diagram of the vertical structure of the atmosphere, tinted in blue. (See p. 127.)

4144B   Marked "Casella London M.O. No. 2205." The upper part of this is like the ordinary M.O. "Kew-pattern" barometer, but about two-thirds of the way down, the tube goes into a circuit and comes up under the iron cistern. The outer brass tube is attached to a plate, which is supported by three very heavy iron posts from a lower plate, holding the cistern. There are leveling screws and two bubble levels. The advantage of this construction is not clear.

BRUGES, BELGIUM. A HOTEL.

——   A cistern barometer, probably early nineteenth-century, with a very large glass cistern and a tube 5 or 6 mm. in bore. It is on a board 25 cm. wide, graduated in cm. from 5 to 28, cm. and ½ cm. from 28 to 65, and cm. and (very roughly) mm. from 65 to 87 cm. The interest is in the large number of geographical indications. (See p. 131.)

CAMBRIDGE, ENGLAND. UNIVERSITY. WHIPPLE MUSEUM.

(The small but highly interesting History of Science Museum of Cambridge University.)

281   A barometer with an adjustable cistern for portability, but no way of setting the zero. Interesting chiefly because it is inscribed "J. Sisson London." An extremely elementary index, and a vernier with 10 divisions covering 1.1 inch.

747   A barometer signed "Nairne & Blunt London," and hence dating from before 1800, having an adjustable cistern but no apparent way of adjusting the zero. On a heavy mahogany plank almost entirely faced with silvered brass. The scale is in inches and 10ths from 26.1 to 31.0 in.; the vernier to 0.01 in. A very large thermometer, graduated 0-128° F. The tube of this barometer has a sharp kink forward at about 25.5 in., to bring it out of the board and into the plane of the scale.

1090   A barometer inscribed "Bennett London" of no special interest except for the index, which is flat on top except for a semi-circular notch about 2 mm. wide, presumably so that either the summit or the edge of the meniscus may be read. The index is hinged to the vernier slide, which has 10 divisions covering 2.1 in., also half-divisions. There is also a clampable "reminder," which can be brought up to support the vernier slide at the level of the last reading. A John Bennett was working in London about 1770.

1130   A "double-diagonal" barometer, inscribed "Samuel Lainton Maker Halifax." Each tube has a "bottle" cistern, concealed by a wooden cover, and is very heavy, about 12 mm. outside diameter and perhaps 2 mm. bore. Thermometer in Fahrenheit and Centigrade; "Fever heat" at 113° F.

FAENZA, ITALY. SOCIETÀ TORRICELLIANA.

(This little museum is housed in the Biblioteca Communale. There are three or four barometers, of which only one seemed of particular interest.)

——   A barometer inscribed "Lenvie Paris," probably early nineteenth-century. This has a float with an ivory stem and index projecting out of the cistern; the instrument is on a heavy walnut board with four brass leveling screws projecting backward toward the wall. A scale in Paris inches and half-lines, and a vernier with 20 divisions reading to 1/40 line or 1/480 in. Also a millimeter scale without a vernier. Contained no mercury in 1961.

FLORENCE, ITALY. ISTITUTO GALILEO GALILEI.

(The gabinetto di fisica of a university, excellently arranged and cared for. There is a manuscript catalog, which must, however, be used with caution.)

A87   An aneroid barograph inscribed "Simplex barograph R.d N.o 468701 Pietro Pannini Firenze no. 283." (Note that this is in English.) The spring is external and the temperature compensation by a bimetallic lever. Stated to be made in 1915.

A196   An early Bourdon barometer with a very simple mechanism, clearly visible through the glass front of the cylindrical case.

A197   A mountain barometer of the Fortin type, graduated from 47 to 81 cm., with its own brass tripod and a Cardan suspension about half way up the tube. Notable for a very neat drive for the vernier slide, saving space and weight; the rack is cut partway through the brass tube, and the pinion and its knob are mounted on the slide and move with it. Ascribed by the catalog to Deleuile of Paris.

A239   A beautifully-made aneroid of the Goldschmid pattern, by Usteri-Reinach of Zurich. (See p. 418.)

FLORENCE, ITALY. ISTITUTO GEOGRAFICO MILITARE.

(The museum of the Survey Corps of the Italian Army. Well arranged and cared for. There is an extensive printed catalog. The museum contains about 20 mercury barometers and a number of aneroids; most of them are of standard modern types. We shall mention one or two that are not in this category.)

157 A "Normalbarometer" inscribed "R. Fuess Berlin No. 89." This is the large type developed according to the ideas of H. Wild. It dates from 1882, and according to the catalog was shipped empty and filled by the Officina Galileo. (*See p. 231.*)

6241 A Fortin-type barometer made by Fuess, in which the vernier slide is moved up and down by a nut on a long screw, operated by a knurled collar. (*See p. 201.*)

6242 Cistern barometer with an ivory float, apparently adjusted by pouring. A mark on the stem of the float is brought level with two marks on a small frame that guides it. Italian, probably early nineteenth-century.

6243 A siphon barometer something like that of De Luc (*see p. 137*) but with brass scales and verniers reading to 0.1 mm. The tube about 4 or 5 mm. in bore. No maker's name, but on a rather similar barometer without a number, there appears "Barbanti in Torino No. 2." The verniers have to be slid up and down with the fingers. In 1829 Carlo Barbanti was listed in the *Guida di Torino* as machinist to the Royal Observatory at Turin, and in 1837 he was appointed machinist to the *Reale Accademia delle Scienze di Torino.*[*]

FLORENCE, ITALY. MUSEO DI STORIA DELLA SCIENZA.

(One of the few museums devoted to the history of science alone. There is an extensive catalog, published in 1954; certain doubts concerning some of the entries will appear below.)

[*] I am indebted to Professor Mario Gliozzi of Turin for this information.

1131 A siphon barometer, the mounting entirely of brass, signed "Galgano Gori fece in Firenze l'Anno 1846." Stated in the catalog to be according to the ideas of G. B. Amici, it has a movable scale, the zero of which can be set to coincide with the lower mercury surface. It must have been intended for portability, as there is a threaded collar obviously designed to receive an outer protecting tube for the upper half of the mounting, which is square. At the bottom is a low folding tripod, but no obvious means of ensuring that the barometer is vertical. Dottoressa Bonelli, the Director of the Museum, informs me that the reference to Amici arose from her early researches in the *Archivio Statale*.

1132 An early cistern-siphon barometer with a cistern of variable volume, enabling the level of the mercury in the shorter tube to be brought to the zero of the scale. It is wrongly stated in the catalog that this is analogous to no. 1131. Mounting entirely of brass. Low tripod with three leveling screws. The scale, index, and vernier are rather modern in appearance; the top of the mercury column can be seen against the light. This barometer is interesting because, in addition to a scale in mm. from 500 to 840, it has three other scales and verniers which have been identified in hand-written inscriptions as "Pollici francesi," "Pollici inglesi," and "Soldi di braccio." (*See p. 172.*)

1134 A slightly simpler version of the no. 1137 (*q.v.*).

1135 A beautiful barometer signed "Faits Portatifs par Dan Quare A. Londres." (See above, Bracknell, no. 1.)

1136 Another, similar to no. 1135.

1137 This is the "stereometric" barometer ascribed by Magellan to Landriani. The tube terminates below in a glass block with an enormous ground glass tap, which can (1) connect the tube to an ivory cup, (2) connect the cup to a glass-lined metal tray beneath, closing off the tube. A side tube goes to a small steel tap and an ivory cylinder with a screw cap, apparently for filling. The whole barometer is mounted on a brass plate with three leveling screws. The large glass tap has "Canna aditto" [addetto?] on the ends, "Casca in mercurio" on the flat faces. The scale is in inches. This instrument probably dates from about 1800. (*See p. 259.*)

1146   A siphon barometer in a cylindrical mahogany case, with narrow windows at the back, closed by hinged shutters, so that the tops of the mercury columns can be seen against the light. Scales in inches. Twenty-one geographical notations engraved beside the upper scale. (*See p. 129.*) Considering the Bruges barometer mentioned above, one might disagree with the statement in the catalog that these notations indicate that "the instrument was chiefly intended for the measurement of altitudes in the high mountains." They rather seem to have been pure whimsy.

1149   A portable siphon barometer in a rectangular mahogany stick. This is described in the catalog as of the Gay-Lussac type (*see p. 140*). But it really has a movable scale, as is also noted in the catalog.

1150   This is De Luc's barometer (*see p. 136*) and seems complete, except that it is without mercury. As the catalog says, the scale may have been drawn by De Luc himself.

1152   Not a Fortin barometer (as the catalog states) in the sense now understood, but a simple cistern barometer with a boxwood cistern and a brass scale, and no index or vernier.

1153   A barometer with a glass cistern that can be moved up and down, while the tube remains fixed to a wooden board. An ivory point attached to this board indicates the zero of the scale, which is engraved on the tube itself. The scale is in mm., and a brass index and vernier can be slid up and down the tube by hand.

Surrounding the vacuum space is a glass vessel open at the top, presumably to be filled with ice in order to reduce the effect of vapors. A thermometer, now lost, dipped in the cistern. Hard to date, but at any rate before 1832, when it appeared in an inventory.

1156   A very large balance barometer of the type in which the lower end of the tube is horizontal. (*See p. 107.*) There is also a crude wooden model of this type of instrument; this is stated in the catalog to have been listed in an inventory of 1776.

1160   A barometer signed "Dollond London," with the tube in a rectangular mahogany frame, suspended by a neat four-legged stand. At the bottom a heavy lead weight encloses the cistern, the end of which is made of leather, exposed to view because the adjusting screw and its plate have been lost, so that it is rather remarkable that the instrument is still full of mercury.

1162   A barometer, erroneously referred to in the catalog as "Tipo Fortin," chiefly distinguished by the possession of three thermometers, one with a very short scale, dipping into the cistern; one about 40 cm. long in the air near the middle of the tube; and a short thermometer sealed into the vacuum space. It would, of course, have been impossible to boil the mercury in the tube.

1163   Felice Fontana's recording barometer. A siphon barometer with upper and lower chambers 16 mm. in bore has a float in the shorter arm, partly supported by a sector of 13 cm. radius, approximately counterbalanced by a weight. The shaft of this sector runs in 4 rollers, each 4 cm. in diameter. Attached to it is another sector in the form of a frame which holds the chart paper, backed up by a ladder of threads. At equal intervals of time a sharp point is brought down on this by a clock, being displaced parallel to the axis of the sectors to give a time scale.

—— A very beautiful barometer by Nairne & Blunt, London, in the office of the Director. The cistern is adjustable, and there is a float in the mercury, with an ivory stem bearing a transverse mark. This stem runs in an ivory sleeve with another mark. It has a hygroscopic hygrometer of some sort at the top, which may still be functioning. Probably about 1790.

GREENWICH, ENGLAND. NATIONAL MARITIME MUSEUM.

(One of the most spacious and beautifully arranged of museums. There is a catalog, *Instruments of Navigation* (London, 1958), in which barometers occupy pages 87–89. Not all of the barometers listed are on public view, but they can be examined by arrangement. Many of the exhibits are associated with famous mariners or ships.)

B.1 Marine barometer by Gautrau of Rochefort. The card and the catalog say eighteenth-century; but there are scales in both inches and lines and in cm. and mm., so this is somewhat doubtful. Only the inch scale has a vernier. The construction of the cistern is hard to guess.

B.2 Marine barometer by "Assier-Perica" (not Pinca, as in the catalog). Narrow tube in cylindrical wooden case only 2.5 cm. in diameter.

B.3 Aneroid barometer, about 15 cm. in diameter, inscribed "Elliott Bros. 30 Strand London." This is distinguished by having a vernier, reading to 0.01 in. It is marked "Compensated for Temperature" but has a curved Fahrenheit thermometer. It has royal associations.

B.4 A miniature aneroid barometer in a chased gold case, only 7/8 inch in diameter.

B.14 A marine barometer in gimbals, inscribed "Berge, London late Ramsden." (Not in printed catalog.) Berge was in business in Piccadilly from about 1800 to 1830.

HAMPTON COURT PALACE, ENGLAND. STATE APARTMENTS.

(Several barometers by famous makers, of course magnificently decorated. Only one seems to belong in our list.)

—— A siphon barometer signed "Tompion London," with the readings taken in the open limb, which alone is visible. It appears to be graduated in actual inches (the scale being, of course, inverted), so that there may be a good-sized bulb at the top of the column, hidden in a gilt urn.

KASSEL, GERMANY. LANDESMUSEUM.

(This museum houses a magnificent collection of scientific—especially astronomical—instruments, formed by the Landgraves of Hesse. With one exception, the barometers are of no scientific interest.)

—— A wheel barometer in a magnificent ormolu case signed "Gaudron, Paris." Weather indications only, no graduations or figures. It has an enormous spirit thermometer with a spherical bulb about 3 cm. in diameter, its scale in no recognizable units. This may well be one of the oldest existing barometers. (*See p. 128.*)

LEYDEN, NETHERLANDS. RIJKSMUSEUM VOOR DE GESCHIEDENIS DER NATUURWETENSCHAPPEN.

(Probably the liveliest of the history of science museums, beautifully arranged and easily accessible to the public. A number of barometers of scientific interest were on display in June, 1961. In addition there are a large number of "Contraroleur" barometers (*see p. 90*), a typical Dutch design, some on display and more in the storeroom.)

Th20 An enormous diagonal barometer with the nearly horizontal part of the tube almost 3 meters long, with a slope of about 1:50. (*See p. 112.*)

Th27   A siphon barometer with upper and lower chambers about 8 mm. in bore, and a movable brass scale about 2 mm. x 30 mm. in section, bearing the inscription "W. M. Logeman Haarlem." This maker died about 1860.

Th30   The so-called "oil barometer" patented by A. H. Borgesius. This is from the Kammerlingh-Onnes Laboratory at Leyden University, and dates from about 1914. It is really a greatly improved form of Hooke's "marine barometer." (See p. 383.)

Th32   A combination of a cistern barometer and a sympiesometer, signed on the ivory scale "J. Sewill Maker to the Admiralty Liverpool & London." There was a Joseph Sewell in Liverpool at 15 Canning Place at least as early as 1835. The firm, now J. Sewill & Co., is still in business at 61 South Castle Street.

——   An astonishing folding barometer which actually folds, by virtue of a ground joint in the tube. It has a small pear-shaped cistern with a tube of about 3 mm. bore extending upward from this for about 6 cm., and ending in a glass stopcock. It is carried in a rectangular mahogany case, after being folded. At a guess, the date might be 1870.

LONDON, ENGLAND. SCIENCE MUSEUM.

(One of the very large collections, covering both science and engineering. The meteorology section has a printed catalog [1922], long out of print; the cards displayed with the exhibits are very informative and generally reliable. The inventory numbers show the year in which the item was acquired. The collection of barometers is of a high level of scientific interest; the proximity of the Victoria and Albert Museum has probably discouraged people from presenting barometers of purely decorative purpose.)

1876-793   Kreil's Barometrograph, designed in 1841, made by Dressler of Prague, and set up at Kew Observatory in 1845. (See p. 291.) The card displayed in 1961 referred to this barograph as being the precursor of Changeux, who is erroneously given the date 1870 instead of 1780.

1876-814   A famous and often-illustrated diagonal barometer made by Watkins & Smith, supposedly in 1753. It has a very elegant and interesting "perpetual calendar" in the center of its large mahogany frame; the barometer rises on the left, and on the right is a spirit thermometer in °F. Setting pointers are provided both on the thermometer and on the sloping part of the barometer tube, which is graduated below from 28 to 31 (expanded) inches and arbitrarily above from 0-20.

1893-133   A mountain barometer by Dollond. The cistern (says the card) "is of the usual flexible-base type with a supplementary siphon tube, connected to the cistern by a narrow neck which can be closed by means of a cock. . . . For use, the cock is opened and the mercury allowed to attain a fixed height in the siphon tube. . . . For transport the mercury is withdrawn from the siphon tube and the cock closed, after which the flexible base of the cistern is depressed, i.e., raised by means of the adjusting screw until the mercury completely fills the barometer tube." The scale is in inches, the vernier reading to 0.002 in.

1893-134   A mountain barometer by West, in a cylindrical wooden mount with a rotating brass sheath to cover the scale and thermometer for transport. The cistern has a leather bottom and a plunger operated by a screw. The vernier reads to 0.002 in. West flourished in London about 1820.

1893-143   A rather simple cistern barometer, interesting both because it is signed "Ramsden London," and because the index goes behind the tube as well as in front of it, though there is no slot for light. The front part is hinged and has a notch at the center-line of the tube.

This may be one of the earliest barometers with a vernier (*see p. 197*). It was in the storeroom in 1961.

1894-119 Admiral Milne's barograph, 1857. It resembles 1876-793 in principle, but is "home-made" in appearance. (*See p. 293.*)

1905-74 A. S. Davis' "Piesmic barometer," a patented air barometer (*see p. 383 and Fig. 15.8*).

1908-83 A marine barometer made by Thomas Jones, probably about 1820. It is in a mahogany frame, the wooden cistern with its leather bottom being covered by a brass cylinder. "The bore of the barometer tube (says the card) is contracted for the greater portion of its length to minimize 'pumping.'" It is interesting that the thermometer is attached to the inside of the door that covers the scale when the instrument is not in use. Unless it were read as soon as the door is opened, serious errors might occur.

1908-86 A marine barometer made by Newman about 1824, marked "N118." This has his first type of iron cistern (*see p. 157*). The scale and vernier read to 0.002 in., but the index, only in front of the mercury column, and suitable only for reading the edge of the meniscus, probably rendered this precision illusory. At the left of the scale are the following indications: "Capacities ⅟₃₆. Exp. [?] action +.025. Neu^t. 29.920 Tempt^{re}. 32°."

1918-14 Sympiesometer, made by Alexander Adie, and as described in his patent. Inscribed "Patent, Adie & Son, Edinburgh." (*See p. 378.*)

1921-321 The modification of the Kew-pattern barometer patented in 1919, in which the reading slide is moved by means of a micrometer screw with a graduated collar, the vernier being dispensed with. One division of the scale on the collar corresponds to 0.005 in. (*See p. 201.*)

1922-125 The recording barometer invented by W. H. Dines, a beautiful piece of instrument making inscribed "Dines' self-recording barometer J. Hicks. Maker. 8. 9 & 10. Hatton Garden London." (*See p. 295.*)

1924-50 A siphon barometer of the type suggested by Gay-Lussac in 1816, with the scales graduated on the tubes, as in a burette. The upper part of the shorter tube is almost filled by a glass rod wound with cotton or some similar thread, perhaps a cleaning device. This instrument was made by Negretti & Zambra, probably about 1875, and is graduated in inches. (*See p. 141.*)

1924-51 A barometer of the general form designed by De Luc (*see p. 137*), and with all the various accessories of his barometer, such as two thermometers, a cleaning swab, and a plumb-bob, but having heavy moldings at the top and bottom of the case, and brass scales, without verniers. The tap is ivory, as is the elaborate stopper for the short limb. Made by Nairne & Blunt, late in the eighteenth century.

1927-1910 A barometer by Sisson, London (*see pp. 198 and 222*). To quote the catalog: "A glass barometer tube dips into a mercury reservoir contained in the lower half of the wooden hemisphere which is situated in the base of the mahogany stand. The upper half of this hemisphere is detachable. . . . A small low power microscope, adjustable for height, is used to focus on the mercury meniscus inside the tube. . . . A fixed ivory scale is situated just above the reservoir; against this scale slides an ivory vernier. . . ." The lower end of this is chisel-shaped, and by ensuring that it is in contact with the mercury surface, the correction for change in level can be read from the lower scale. The main scale and vernier are of brass. This barometer, very interesting on several counts, is most probably by the younger Sisson, Jeremiah, and may be dated about 1775. It is not quite complete, and in 1961 was in the storeroom at the Museum.

1929-962 A "Thermobarometer" according to J. G. Ronketti, made by Carpenter & Westley. This is derived from Adie's "Sympiesometer," but has a thermometer with its bulb in the working volume of gas. (*See p. 379.*)

1939-301 A barograph made by M. Pillischer, Optician, 88 New Bond St., London, dated about 1850 on the card. This is based on a siphon barometer, with an ivory float suspended by a thin wire from a quadrant. From an opposite quadrant of greater radius depends a weight carrying a sharp point, which moves up and down, with the fluctuations of the barometer, near a paper-covered drum. Each hour a bar controlled by the clock which rotates the drum drives the point against the paper. Except during the moment when it is struck, the recording point can move as freely as the construction of the instrument allows. There seems to be no provision for temperature compensation.

1948-227 A beautiful barometer by Daniel Quare, similar to Bracknell 1 (*q.v.*) and Florence, Museo, 1135. The signature is "Invented & Made by Dan. Quare London."

1950-252 The open-scale barometer patented in 1892 by C. O. Bartrum, and in principle exactly like Descartes' two-liquid barometer, except for a more suitable choice of liquids (*see p. 87*). The magnification is 7.8, and there are scales in (expanded) inches and centimeters.

—— A very large barometer inscribed on an ivory plate near the top "Meteorological Society Standard Robert Carr Woods 47 Hatton Garden London." The tube is about 15 mm. in bore and the cistern, of heavy glass, about 20 cm. across. This instrument, lacking many parts, is in the storeroom. (*See p. 239.*)

MILAN, ITALY. LINCEO G. BECCARIA.

(The physics department of a famous and venerable high school, in a commodious new building.)

67 A wheel barometer with a beautiful dial 35 cm. in diameter, inscribed "Barometro Multiplicatore di Antonio Frascoli Milano 1866 N 2." The float is partly supported by a fine thread which passes over a pulley about 12 mm. in diameter on a shaft with a diameter of 2.5 mm. held by 4 anti-friction wheels each about 40 mm. across. The scale is from 725 to 775 mm., about 15 times magnified. It seems to be in good working order and complete except for the loss of a plumb-bob, and may be considered an exceptionally well-designed and constructed example of the wheel barometer, made just 200 years after Hooke invented it.

444 A portable siphon barometer of the same general shape as that of De Luc, but with a single movable scale and vernier, reading to 0.01 inch.

447 Cistern barometer signed "Marelli, Milano 1818" on the scale. This has an adjustable cistern and a float with an ivory stem projecting from the top. This stem carries a scratch which can be made to coincide with a mark on a cut-away ivory guide. (*See p. 209.*)

MUNICH, GERMANY. DEUTSCHES MUSEUM.

(The immense and fascinating museum of science and engineering which occupies nearly all of an island—Museumsinsel—in the River Isar. There is a good collection of meteorological instruments, especially barometers. Attached to the Museum is a very large and valuable library.)

12 A portable siphon barometer by Hiacint Vaccano who worked in Munich near the beginning of the nineteenth century. Worth noting only for the list of mountains written beside the scales. (*See p. 131.*)

13 A siphon barometer unfortunately with no signature or date. This has a

beautifully constructed movable scale, of glass through most of its length, attached at the bottom to a brass piece bearing a rack. Attached to this brass piece is an arm carrying a circle with a cross wire, and an eyepiece through which the cross wire and the mercury meniscus can be seen and the zero of the scale adjusted to this level. A rack and pinion moves a similar device for reading the level of the mercury in the longer tube. The tubes have a bore of about 6 mm.

14 A mysterious instrument having two cisterns and two tubes connected at the top, and thus sharing a vacuum space. One tube is larger than the other, which suggests experiments on capillarity such as those carried out at Turin in 1759 (see p. 187), especially as it looks like a laboratory item, the cisterns being closed with sealing wax and everything rather rough. But it has a list of mountains beside the scales! This was in the storeroom in 1961.

17 A siphon barometer with a scale movable by rack and pinion, signed "Vaccano München." This is all mounted on one half of an elliptical cylinder of wood, terminated by elliptical brass plates. For transport, the other half of the wooden cylinder is attached, completely enclosing the barometer. There is a steel tap near the bottom of the shorter limb. The scales are in lines without any reference to inches, the divisions numbered in tens and with a vernier reading to 0.1 line.

23 A movable-scale siphon barometer, chambers about 4½ mm. in bore, vertically above one another, with a constricted portion about 25 cm. long. The indexes could be used for reading either the top or the edge of the meniscus, and this could be seen against the light. In Paris inches and lines, with a vernier on the upper index reading to 0.1 line.

25 A portable barometer obviously intended for mountain use, bearing the inscription "Haas & Hurter, London N⁰

38." This is a cistern barometer intended to be adjusted by overflow. The scales in French inches and lines, with a vernier to 20ths of a line, and in English inches and 20ths, with the vernier apparently reading to 1/1200 inch! This instrument is mounted on its own mahogany tripod, and has a plumb-bob built inside it. The card says "Gefässbarometer nach Schiegg." Ulrich Schiegg (1752–1810) was a Benedictine monk who, after being a professor of mathematics in various places, ended his life in the Bavarian public service.

106b The operating portion of the great tower barometer. A large aneroid barometer operating a servo-system (see p. 432 and Fig. 16.28).

543 Labelled "Kontra Barometer." Stated on the card to be a Huygens barometer; i.e., Hooke's two-liquid barometer (see p. 88). Interesting in having a long spiral tube joining the two halves of the barometer, and a glass tap just below the widening in which the mercury and the dark-colored liquid are divided. Probably nineteenth-century.

2061a ⎱ Two recording barometers made
2061b ⎰ by Jean Krapp of Mannheim for the Societas Meteorologica Palatina about 1790. They differ mainly in that 2061a has an awkward rectangular chart sliding sideways beneath the clock face, while 2061b has an annular chart revolving round the clock, much as in the barometer of Cumming (see p. 289), but once a week. Each instrument is in an elaborate inlaid case and stands about 2 meters high.

2606 A siphon barometer signed "V^ct Liebherr à [sic] Landshut." A scale on each limb, in Paris inches and lines, verniers to 0.1 line. Probably not much later than 1800.

2702 A siphon barometer signed "P. Rath in München" and bearing the number "P.R.321." This is similar to some barometers made for C. M. Bauernfeind about 1857. It was in the storeroom in 1961, lacking half of its wooden case which, when complete, formed a nearly circular elliptical cylinder.

2915 A "bottle-barometer" of distinctly superior construction, inscribed "Carl Theodor Palatinus 1780," and therefore presumably one of those built for the *Societas Meteorologica Palatina* at Mannheim (*see p. 134*). It has silvered brass scales in Paris inches and lines, a vernier reading to 0.1 line, and a Réaumur thermometer.

3949 A traveling Fortin-type barometer signed "Rumpf in Gottingen 1823." The scale is in inches from 18 to 31. There is a wooden tripod, arranged so that when it is closed the whole barometer is protected (*see p. 161*). The tube must be at least 7 mm. in bore.

5581 A recording and transmitting aneroid barograph, ascribed on the card to M. Hipp of Neuchâtel. This records pointwise by the action of an electromagnet, and in addition there are a number of switches which could be used to transmit the value of the pressure to a distance with a precision of 1 or 2 mm. (*See p. 429.*)

6281 A diagonal barometer, inscribed "Magnum Barometrum Morlandinum" and also "Approbata in Accademia Hallae Magaborgicae." Pear-shaped "bottle," and a sliding door which can be raised in order to examine this. At about 24 in. up, the tube bends at about 45° to the right for about 4 in., and then to the left at a slope of about 1 in 3. The scale seems completely arbitrary, going up and down from the middle, where it is marked "Veränderlich," "Variable," and "Variabile." Interesting because of the double bend.

7244 Hipp's recording aneroid, 1871 (*see Fig. 16.26 and p. 429*).

7358 "Barometer nach Koeppen." This is a siphon barometer with two tubes of almost the same length, the open one being bent over and extended by a very narrow tube which goes right down to the bottom again. At the junction of the larger tubes there is a tube pointing downward and provided with a glass stopcock, presumably for ease in filling. The scale is made out of a steel measuring tape, which has corroded badly. The instrument is in a mahogany case, obviously intended for portability.

8935 A siphon barometer inscribed "verf von J. G. Greiner jun. in Berlin." A well-constructed instrument on sound instrument-making principles, with a brass scale having small microscopes at the upper and lower ends for reading the levels of the mercury. Graduated from 18 to 32 (probably Paris) inches, divided into 20ths with a vernier reading to 0.002 in. The wooden frame is cut away so that the light can come past the mercury columns. Attached to the scale is a nicely constructed thermometer in °R. I should date it about 1860.

22338 Twentieth-century cistern barometer with movable scale, made by Wilhelm Lambrecht in Göttingen. The end of the scale which dips into the mercury is peculiar (*see p. 221*). Instruments of this design were being offered for sale in 1954.

22585 A very large diagonal barometer inscribed "Schwabach 1. Mai 1774." Rather large tube (6 or 7 mm. bore) with the nearly horizontal portion more than a meter long, so that the "bottle" cistern would scarcely seem adequate. There is an extended piece of printed matter at both sides of the vertical portion of the tube, giving a "Historie des Barometers," and referring to Ramazzini as the author of the most considerable improvement in barometers—just after a reference to De Luc! The paper scale, which is partly missing, appears from the inscription to be graduated in Rhenish inches (*see p. 112*).

27712 A large recording barometer by G. Wanke of Osnabrück, designed by F.

C. G. Müller, Brandenburg, in 1874 (*see p. 329 and Fig. 11.30*).

31980   A siphon barometer by R. Fuess, Berlin. This has the scales engraved on the glass tubes, the readings of the two having to be subtracted. The tube is bent so that the two limbs are coaxial, and apart from the fact that it has a Bunten air trap (*see p. 000*) seems to be on the principles laid down by Gay-Lussac (*see p. 140*).

50730   An aneroid barograph which records in rectangular co-ordinates, signed "C. P. Goerz Berlin Nr. 3560." (*See p. 427.*)

54905   By the same maker as 8935 and very similar, but with a much bigger tube, 8 or 9 mm. in bore, and with larger motions and microscopes.

61214   A portable siphon barometer, unsigned, but very similar to no. 17 above. The scale, which seems to be in millimeters, is not numbered.

62225   A fixed-cistern barometer signed "J. J. Hughes Ratcliff London." Scales in inches, on ivory; the tube mounted in a turned wooden case surmounted by a glass-fronted box for the scales. This firm was in business from about 1817 to 1840.

——   A cistern barometer with an ivory cistern, apparently without means of adjustment, signed "Breitinger à Zuric." This has scales in English inches and tenths and in French inches and (surprisingly) tenths, with verniers reading to 0.01 inch. It has a thermometer in °F. and a self-contained plumb-bob.

MUNICH, GERMANY. STERNWARTE.

(The famous astronomical observatory, somewhat declined from its former state. The meteorological instruments developed by Lamont are no longer in evidence.)

——   A siphon barometer with a tube of uniform bore throughout, about 10 mm., and a Bunten air trap. Interesting because of the scale. The tube is mounted on a piece of mirror about 8 cm. wide, ruled at millimeter intervals from one side of the mirror to the other, with numbers every cm. The rulings are on the unsilvered side, so that they are reflected in the mirror to make a simple means of avoiding parallax.

OXFORD, ENGLAND. UNIVERSITY. HISTORY OF SCIENCE MUSEUM.

(A very fine collection, especially of small astronomical instruments. The barometers are of a high level of interest.)

33.25   A cistern barometer with a silvered scale signed by John Bird, London (died 1776). The interest is in the adjustable cistern, out of the top of which projects a short tube terminating in an inverted glass vial. In this is an ivory rod, presumably attached to a float, and also a vertical ivory plate with a horizontal scratch on it. There is reason to suspect that it has been modified (*see p. 207*).

218   A diagonal barometer in a mahogany frame with a silvered brass scale inscribed "Whitehurst Derby." The scale is graduated arbitrarily 0 to 60. Whitehurst left Derby in 1775, became an F.R.S.

——   A diagonal barometer and thermometer in a rectangular mahogany frame, with the barometer at the right, the diagonal sloping upward to the left. The thermometer, of which the tube is missing, has the old Royal Society scale of temperature, 0° being extremely hot and 95° extremely cold.

——   A large cistern barometer signed "Carpenter & Westley. 24 Regent St. London."* The tube is ¾ inch in bore

* Carpenter and Westley were in business at 24 Regent Street from about 1840 to 1924, but mainly as opticians in the twentieth century.

and the cistern is huge. It has a turret and a glass vial (now broken) with an ivory float and index, similar to no. 33.25 above, but larger.

—— A "folded barometer" of Amontons pattern (*see p. 143*) in a mahogany case, with alternate tubes of mercury and a red liquid. Inscribed "Torre fecit."

—— A diagonal barometer and thermometer on a large rectangular mahogany frame, enclosing a perpetual calendar. Signed "F. Watkins, London."

—— Two barometers, one in the storeroom, signed "Cary, London." Adjustable cisterns, each with a little turret emerging from the top, with a saw cut in the side of the turret. A screw plug seals the cistern for portability. (*See p. 206.*)

—— A "marine barometer" of Hooke's type, ascribed to John Patrick, probably dating from about 1700.

—— A siphon barometer of the Gay-Lussac type in a brass tube, inscribed "Bunten quai pelletier 30 Paris 1840." (*See p. 142.*)

—— A double-tube diagonal barometer signed "Charles Howarth. Fecit. Halifax." Scales on paper, 27.9 to 29.6 in. and 29.4 to 31.0 in. The tubes not over 3 mm. in bore.

PARIS, FRANCE. ACADÉMIE DES SCIENCES.
(At the offices of the Academy, quai de Conti.)

—— A two-liquid barometer mounted on a painted wooden panel, with the inscription "Fait par Jean Baptiste Prieur rue Ste Marguerite Faubourg St Antoine." Probably early eighteenth century, and the only remaining barometer that certainly belonged to the Académie Royale des Sciences and has descended to the present Academy.

PARIS, FRANCE. CONSERVATOIRE NATIONAL DES ARTS ET METIERS.
(One of the largest collections of scientific and technical apparatus, particularly well arranged from the standpoint of the serious student, though a little crowded. The collection of barometers is one of the most interesting that I have seen.)

1575   A peculiar siphon barometer inscribed "J. A. C. Charles Par J. H. Hassenfratz. Baromètre à Graduation Compensée." (J. A. C. Charles, 1746–1823, was a famous French physicist.) The tube is about 6 mm. in bore except for the short limb, which is about 15 mm. There are scales on each limb which are apparently inversely proportional in their divisions to the cross-sectional areas of the limbs. They go up and down from a *niveau* about 4 cm. from the upper end of the short limb. The idea is that the distance from the mark 24, say, on the upper scale to the mark 24 on the lower will be exactly 24 inches. What advantage this was supposed to confer is a mystery. The scale is elegantly divided on a strip of silvered brass about 10 cm. wide, and set in a wooden frame, but there are no verniers or indexes. The scale was "Divisé & Gravé par Lorichon, floreal an 13" (April 1805).

1578   A Hooke-type two-liquid barometer with a magnification of about 10. Chiefly interesting for the signature: "Baromètre composé par Mossy constructeur des instruments de phisique [*sic*] en verre de l' Academie des Sciences et de la société Roiale del'Medicine Quay Pelletier à la croix d' or à paris 1780."

1579   Folded barometer of Amontons' pattern. Inscribed "Baromètre Composé Par Mossy Approuvé de l'Academie Royalle des S^{ces}" and, lower down, "Paris 1768" (*see p. 143 and Fig. 7.3*).

1580 Conical barometer, about 1.8 meters long. Only the top of the column has a scale, and the magnification would seem to be about 5. Inscribed "Baromètre d' Amontons." (*See p. 119.*)

1582 A traveling cistern barometer in a turned wooden mounting, signed "Baromètre portatif par Mossy quai Pelletier à paris." The scale, graduated in inches and lines, has a vernier reading in twelfths of lines. The upper part of the case has a wooden semi-cylinder about 3 mm. thick which can be turned between ivory mounts to cover up the tube, scale, and thermometer. This may be one of the first barometers fitted with a vernier (*see p. 199*).

1583 Stated on the card to be by Mossy; but the scale is inscribed "Bodeur à Paris à [??] Quai de l'horloge (Année 1819)." This looks as if it were a frank imitation of no. 1582, the main differences being in the length of the scale and the slightly different shape of the housing of the cistern. The ivory mounts and the grain and finish of the wood are almost identical.

2628 A portable siphon barometer of the Gay-Lussac type by Bunten, inscribed "Bunten quai pelletier 30 Paris 1841." There are scales at top and bottom engraved on silver, and verniers reading to 0.1 mm., the reading slides being moved by pinions working in racks cut part way through the rather thin brass tube, which has a diameter of about 20 mm. near the bottom, less toward the top. There are slots at the back so that the meniscus can be seen against the light. Elegant workmanship.

4245 A well-constructed barometer on a mahogany frame, the silvered brass scale inscribed "Dollond London." Graduated in English inches, with a vernier to 0.01 in. The cistern is adjustable, the level presumably adjusted by overflow (*see p. 205*).

4249 Blondeau's iron siphon barometer, about 1780. (*See p. 158.*)

5411 A Fortin-type barometer of large size and excellent quality, with a tube probably 12 mm. in bore. There is a very beautifully divided scale on silver in mm. and half-mm., with a vernier reading to 0.01 mm. The cistern is of the usual form. This instrument is signed "Fastré aîné Quai des grands Augustins 63 à Paris (1849)."

8517 A cistern barometer inscribed "R. Bianchy Rue St. honoré No. 252." The cistern adjusted by overflow. An elegant scale on silver with a vernier reading to 1/12 line. Bianchy flourished about 1785.

8519 The cistern and glass tube of a Fortin-type barometer, attributed to Fortin himself.

8761 A barometer with two tubes dipping in the same cistern inscribed "Barometre de Megnié pour Mr Delavoisier De l'Academie Rle des Sciences No. 6. 1779." The idea was that it was unlikely that if they read the same they would both have a bad vacuum. This is no. 6 of a series; Lavoisier had eight made, and presented them to friends at various places (*see p. 250*). There are two more of these in the C.N.A.M., nos. 7658 and 9949.

8767 Conté's second weighing barometer, of steel, used on Napoleon's expedition to Egypt (*see p. 158*).

8917 Aneroid barometer after Goldschmid, signed "Hottinger & Cie. Zurich" (*see p. 418*).

8918 Similar to 8917 but very much smaller.

9136 ⎫ Three specimens of the air ba-
9524 ⎬ rometer of Hans and Hermary, an
10916 ⎭ ingenious combination of a liquid thermometer and an air thermometer. Looking at these will offer no indication of how they work. (*See p. 380.*)

12008   A movable-scale cistern barometer, stated to have been used by Dulong and Arago in their experiments on the gas laws. The scale and vernier are of highly unorthodox construction. (*See p. 221 and Fig. 9.15.*)

13164   A recording mercury barometer by Richard, of Paris, based on a very large siphon barometer, the tube quite 20 mm. inside diameter, with a float and a simple lever motion, the float being supported by a thin cord from a sector. There is no special provision for keeping the float away from the walls of the tube. The zero adjustment is performed by raising or lowering the entire siphon tube in relation to the mechanism by means of a screw with a large knob. Dates from before 1899.

14388[1] } Two instruments described on
14388[2] } the cards as "Baromètre à air système Montrichard" (*see p. 392*). They seem to be experimental models, crudely graduated, and the second seems incomplete.

16409   A barometer consisting of a large Bourdon tube with one end attached to the short arm of a steelyard, the other end of the tube being weighted. The pressure is measured by sliding a weight along the longer arm of the steelyard, at the end of which a pointer moves near a scale divided from +4 to −4 in 10ths.

17540   One of the first Vidie aneroid barometers, graduated from 63 to 80 cm. of mercury, in mm. The mechanism cannot be seen.

19948   A barometer with a very large interestingly shaped glass cistern, about 15 cm. in diameter, and a tube of about 6 mm. bore. It was apparently used as a fixed-cistern barometer. Scale in inches and lines and a vernier reading to 0.1 line. This instrument is signed "Mossy breveté du Roi quai Pelletier à Paris 1789," but the scale is inscribed "Divisé par Richer."

RICHMOND, ENGLAND. KEW OBSERVATORY.

(The famous old Observatory is still a working meteorological station, but the front hall is lined with glazed cabinets containing many historic instruments.)

152   Newman no. 2, one of those in which the scale moves and the vernier is fixed, almost exactly similar to Bracknell 4108B above. (*See p. 200.*)

154   A portable siphon barometer of very fine workmanship by "Thomas Jones, 62 Charing cross." This is of the Gay-Lussac type; its scales are in inches, 10ths, and 50ths, with verniers reading directly to 1/500 inch.

160   The copper cistern of Jordan's glycerine barometer, and the glass part of the tube, with its scale, looking rather like the water gauge of a steam boiler. (*See p. 370.*)

——   G. M. Whipple's modification of the barometer designed by Captain C. George, R.N., to be filled with mercury in the field, without boiling. (*See p. 171.*) It dates from 1881.

ROME, ITALY. OSSERVATORIO SU MONTE MARIO. MUSEO COPERNICANO.

(This museum, dedicated to the memory of Copernicus, contains a large number of astronomical instruments of great value, and a most interesting collection of meteorological instruments. In 1961 it was uncatalogued and almost unarranged, but I am informed that this situation will shortly be put right.)

126   A very large and well-made Fortin-type barometer which appears to have belonged to Father Secchi. It is by Secretan of Paris. It has a beautifully engraved scale in mm. and half-mm. on silver, the vernier reading to 0.02 mm. The tube is probably about 10 mm. in bore, the cistern about 40 mm., with a fine ivory point; the mercury quite clean and the

whole instrument in excellent condition in 1961.

—— Secchi's "Universal Meteorograph." A tremendous brass and steel instrument standing about 2½ meters high and comprising a barograph, a recording raingauge, a recording psychrometer, an anemograph and a recording wind vane. Everything is extremely sturdy and durable. The instrument received a grand prize at the Paris Universal Exhibition in 1867. (*See p. 304.*)

—— A barometer in a round mahogany case with half a Cardan mounting to hang it on a bracket on the wall. It is inscribed "Troughton London." Scale in English inches and tenths, vernier to 0.01 inch; also one in Paris inches and lines, vernier to 0.1 line. The reading index is obviously intended for the bottom of the meniscus; it could not be used for the top, because it is at the top of the slide. The meniscus cannot be illuminated from the back. The cistern seems to be adjustable, but there is no reference point for the zero of the scale. John and Edward Troughton were in business at 136 Fleet Street from about 1795 to 1825.

—— A peculiar barometer with a pear-shaped glass cistern and a tube about 4½ mm. in bore. Graduated in Paris inches from 25 to 30. It had two thermometers; one is missing, but the remaining one had scales in °F. and °R. This instrument, which is difficult to date, still had an excellent vacuum in 1961.

—— A barometer signed "Tecnomasio Milano." This is a siphon barometer with very large chambers; the lower one, which I could measure, is 28 mm. in bore. Movable scale with an index at the lower end by which the zero could be brought to the level of the mercury; the upper index and the scale both moved by micrometer screws. The scale from 715 to 830 mm., vernier to 0.1 mm. Probably intended to be read at the top of the meniscus. This might well have been a very accurate instrument, as the capillary correction

would be very small, but the workmanship leaves a little to be desired. It was given by Commendatore Castiello, Minister of Agriculture and Commerce, to the Meteorological Station Scarpellini "in Roma sul Campilogio." The firm Tecnomasio was established at Milan in 1863, originally to make physical instruments, but in the 1870's began to concentrate on electrical apparatus, and is now associated with Brown-Boveri.

—— An interesting and quite original barometer, having a bathtub-shaped iron cistern on a base with three leveling screws. From this base rises a post which supports the brass tube carrying the scale and vernier slide. The glass tube is loose inside this. The scale-tube ends in a long iron point, and the whole assembly can be moved up and down in relation to the cistern by means of a long screw, so as to set the point to the level of the mercury in the cistern. This instrument bears a strong resemblance to that described in 1872 by G. Uzielli. (*See p. 169.*)

SALEM, MASS., U.S.A. PEABODY MUSEUM.

(Primarily a maritime museum, with a very large collection of the instruments used in connection with navigation, by no means all on display.)

M.751 A marine barometer in an elaborate wooden frame, signed on the ivory scale "James Bassnett Maker Liverpool." The cistern has a brass cover. This maker first appears in the Liverpool directory in 1829 as Basnett, clockmaker. He doubled the "s" in 1841, and in 1855 the firm became James Bassnett & Son. This dates the barometer between 1841 and 1855. (I am indebted for this information to the Director of the Peabody Museum, Mr. M. V. Brewington.)

M.1723 A barometer stated on a paper label to have been manufactured by Alexander Marsh for the proprietors, John M. Merrick & Co., 7 Central Exchange, Worcester, Mass. On the scale is engraved

"Timby's Patent. Nov. 3.rd 1857." This is simply a "bottle barometer" (*see p. 133*) with a tap at the base of the "bottle" and a short length of rubber tubing below it, to allow for expansion of the mercury during transport. (*See p. 351.*) The tap is accessible from the back of the rectangular wooden case. The patent was U.S. Patent 18,560.

M.10485   A fixed-cistern barometer in a brass tubular case, with an iron cistern, and a marine mounting. The scale is in inches and 20ths, the vernier to 0.002 inch. Signed "Adie London No. 35" and dates from 1856.

TEDDINGTON, ENGLAND. NATIONAL PHYSICAL LABORATORY.

(The two barometers referred to below are in a glass case outside the door of the library at N.P.L.)

—— The Kew Standard barometer of 1854, somewhat modified. (*See p. 262.*)

—— The Kew Standard of 1858. Much like its predecessor except that the cistern is rectangular with two observing windows in the front of its wooden case, and that the barometer is hung from a wall bracket and held vertical by three screws passing through a ring mounted on a floor pedestal at the bottom.

UTRECHT, NETHERLANDS. UNIVERSITY MUSEUM.

(This beautifully arranged collection was catalogued [in Dutch] on large cards by the late Director, Dr. P. H. Van Cittert. Through the kindness of his widow Dr. J. G. Van Cittert-Eymers, the present director, I was able to obtain photocopies of the cards dealing with barometers. They are a model of how such things should be done, frequently embellished with diagrams showing internal structure. Most of these cards have been summarized below.)

G31   A cistern barometer with its own tripod, in a leather case, signed "J. Newman, 122 Regent Street, London" and bearing the serial number 65. The barometer hangs from the top of the tripod by a Cardan suspension; the tripod has telescopic legs. The tube is in a cylinder of mahogany, with a large opening in the front of the upper end so that the scale can be seen, and a narrower slit at the back for the light. A brass tube with corresponding slits can be turned to open or close these apertures. The scale is in English inches; the vernier can be pushed up and down by hand, and there is a screw for fine adjustment. On the brass cover of the cistern is engraved "Neutral point 30.006/ Capacities $\frac{1}{36}$/ Capillary Action 0.052/ Temperature 55°." The cistern is of Newman's later design (*see p. 148*) in two parts which can be rotated relative to one another, with the excellent addition of a little cup mounted on the partition and surrounding the end of the tube, to prevent air, which might come through the communicating holes when the instrument is turned right way up, getting into it. A thermometer has its bulb in the cistern.

G32   A cistern barometer by Dollond in the form of a walking stick. There are slits so that the meniscus can be seen against the light, as in G31. The cistern is iron with a porous wood top, and a stopper can be screwed up from below to close the end of the tube. There is a thermometer with its bulb in the cistern and also one near the scale. On an ivory plate is engraved "No. 262 / N.P.30.280/ Capa. $\frac{1}{57}$ / Temp. 73," but Van Cittert notes that this is no longer significant, as the glass tube has been renewed. The stick carries a 30-inch scale, so that it may also be used as a measuring rod. (Evidently Dollond's had used some of Newman's ideas, or vice versa; *see p. 240.*)

G33   A walking-stick barometer, also by Dollond and like G32 except that there is no measuring scale on the stick.

G34　Another walking-stick barometer by Dollond, but more primitive.

G35　A movable-scale siphon barometer by Greiner, dating from 1834. It had microscopes (of which one is lost) for setting the indexes.

G36　Two barometers with constant-level reservoirs according to Prins (*see p. 222*). These cisterns are made of wood.

G37　Two more cisterns for Prins-type barometers, one of wood and quite simple, the other an ingenious iron one in which the plane on which the mercury spreads out can be flooded from a second reservoir beneath, adjustable by means of a piston that can be screwed up and down. The surplus mercury drains back through a hole.

G38　A very elegant fixed-cistern marine barometer with a Cardan suspension, by A. Lipkens of Delft, dating from about 1820. It is remarkable for having a thermometer with a vernier reading to 0.1°C. It also has a remarkable amount of information engraved on two brass strips beside the tube. On the left: "Le pesanteur Specifique du mercure employé = 13.598 à la temp.^{re} 0°. Le moyen diam.^{r} intérieur du tube dans L'interval des indications extrêmes est 5.5 mm., d'ou resulte une dépression de 1.4 mm. Le diamètre intérieur de la Cuvette est de 71.5 mm." Then the scale in mm. from 800 to 600. Below this, a table of corrections for the change of level, and then the notice: "Tous les corrections relatives à la température, on peut avoir recours aux tables barométriques." On the right-hand strip: "On est dispensé de tenir Compte de la dépression, on y a eu éguard en déterminant la distance du plan tangent au 0 du vernier. Le point de l'eau bouillante a été divisées à 20° Centigrade." Under this the thermometer, and lower down still: "Le laiton se dilate de 0° à 100° de 0,^{m} 0018782 . . . ," and finally a table of corrections for the dilatation of 800 mm. of the barometer scale. As Van Cittert

remarks on the card, it is a very handsome piece of work. What a sailor would make of it is another matter.

G39　A cistern barometer with the scale on a wooden lath, supported by a wooden ring floating on the mercury in the cistern (*see p. 209*). The silvered brass scale carries three sets of divisions, in French, English, and Rhenish Inches. The weather signs are in Dutch, which leads one to assume that it was made in Holland.

G40　A "square" (Bernouilli) barometer by "L. Primaues, Amsteldam," dated about 1750 by Van Cittert. The horizontal portion of the tube is about 110 cm. long and the magnification about 15. There is a scale of Amsterdam inches on this, and on the vertical tube four scales: (1) The weight of the atmosphere in Amsterdam lbs. per square Rhenish foot; (2) Rhenish inches and 36ths: (3) A scale in 18ths of Rhenish inches going each way from "Veranderlijk;" (4) A scale in Amsterdam inches, from 28 to 31. There are also the usual weather signs, in Dutch.

G41　An air barometer, intended for marine use, devised in 1832 by S. Brouwer, a teacher at Groningen. There is a bulb with a thermometer sealed into it, which holds a sample of air. A mercury manometer measures the pressure, in excess of that exerted by the atmosphere, necessary to compress this sample to a standard volume, defined by an index. A piston, working in a cylinder, is used to exert the pressure. The atmospheric pressure is then derived from a calibration, with the aid of Boyle's and Gay-Lussac's laws.

VIENNA, AUSTRIA—ZENTRALANSTALT FÜR METEOROLOGIE UND GEODYNAMIK.

(The headquarters of the Austrian weather service. There has been an admirable tendency to preserve instruments of historical interest.)

——　A Redier barograph of the second type (*see p. 325*), not operative but in beautiful mechanical condition.

—— A barograph according to Kreil, excellently preserved (*see p. 291*).

—— The barometer described by Pozděna in 1911 (*see p. 281*), intact though not operative.

—— A Sprung weight barograph, of the type in which the rolling weight is driven by a screw, this being turned by friction wheels controlled by a magnetic clutch. It is said to have been in continuous use since 1890.

WASHINGTON, D.C., U.S.A. SMITHSONIAN INSTITUTION.

(The national museum of science and technology. As this is pre-eminently a museum for public education, few barometers are on display; but the remainder can easily be made accessible to the scholar.)

224,424   A barometer with an adjustable cistern, inscribed "Smithsonian Instn Jas Green," and in much bigger and more deeply engraved letters, "N.Y." This makes it likely that the instrument dates from about 1849 (*see p. 344*). The cistern has two rectangular glass windows between boxwood bushings, but no "ivory point"; it is made adjustable by a leather bag and a screw. In use the mercury surface was lined up with the top of the windows. (*See p. 346.*)

230,002   A "Holosteric Barometer—Compensated," made by Naudet & Co. Marked on the back of the case, "U.S. Signal Service—1101." There are several other rather similar aneroid barometers.

247,924   An aneroid, "U.S.G.[eological] S.[urvey] no. 60. James Green, New York."

247,925   Another, U.S.G.S. no. 181, which also bears the name "Jas Pitkin, London."

308,190   An aneroid barograph by Richard, having no less than 16 vacuum chambers and a large drum, the magnification about 3. A thermometer in °F. is attached.

313,693   An aneroid of which the scale covers an arc of only about 90° and is divided in inches and tenths. It is marked "Beaumont's Barometer, 175 Center Street, New York. Patented June 14, 1859." Beaumont's patent, no. 24,365, covered multiple elastic chambers and was primarily concerned with steam gauges.

314,525   Part of the recording mechanism of Foreman's barograph (*see p. 325*).

314,545   An aneroid of the Goldschmid pattern marked "Ship-Barometer, No. 116, Hottinger and Co., Zurich." It is 3½ in. in diameter and 2½ in. deep.

314,608   A rather elementary barometer with an iron cistern, patented in 1860 under no. 28,626 by L. Woodruff of Ann Arbor, Michigan, and made by Chas. Wilder of Peterboro, N.H. It is stated to have been used by Dr. John King of Cincinnati, Ohio.

314,625   A beautiful mountain barometer of the Fortin type by Ernst of Paris, the brass protecting tube being only about 18 mm. in diameter. It is stated to have been carried on expeditions organized by the Smithsonian Institution.

314,712   A specimen of the Tuch barometer (*see p. 350*), marked "J. Green, N.Y."

314,941   One of the large movable-scale barometers, No. 562, by Newman (*see p. 218*).

316,421   A simple barometer with a fiducial point and an adjustable cistern, made by the Central Scientific Co., and covered by U.S. Patents 1,353,482 and 1,632,084.

316,288 A barograph by Friez, ca. 1939, incorporating the means of temperature-compensation devised by W. G. Boettinger (U.S. Patent 2,165,744). This also has a damping mechanism consisting of two oil-filled cylinders and pistons at either end of an intermediate lever. It was made for the U.S. Signal Corps.

316,500 "Compensated Barograph No. 1, Marvin System, made by Schneider Bros., N.Y." This is the recording siphon barometer referred to on page 296.

316,941 Draper's barograph, complete except for the mercury (see p. 316). The printed chart remaining on the instrument has the heading "Lick Observatory, Mt. Hamilton, Calif."

317,417 ⎫ Three examples of Hough's
318,283 ⎬ barograph, in various states of
318,284 ⎭ completeness (see p. 324.)

319,469 A diagonal barometer and thermometer in a large rectangular mahogany frame, marked "Watkins, St Ja^s Street London." The frame bears a "perpetual calendar" beginning with 1753, apparently printed from the same plate as that on barometer 1876–814 in the Science Museum, London (see above). The styles of the mountings of these two barometers are so different that one is led to doubt whether the date 1753 is necessarily that of either instrument.

ZURICH, SWITZERLAND. ZENTRALANSTALT FÜR METEOROLOGIE.

(The headquarters of the national weather service.)

—— Four substantial movable-scale barometers of similar design, in honorable retirement, of a type used in the Swiss meteorological network shortly after it was set up in the 1860's. (See p. 220.)

—— A simple siphon barometer on a wood frame with brass scales, marked "Fuess, Berlin, no. 306." As described in Fuess' very early catalogs.

—— A weight barograph of Sprung's last type, made by Fuess, (see p. 315 and Fig. 11.19.) Still in operation and functioning very well. There is a spare tube for it, that has never been needed, in the storeroom.

Index

For exhibitions and museums look under the names of the cities concerned.

B

THE HISTORY OF THE BAROMETER
W. E. KNOWLES MIDDLETON

designer:	Edward D. King
typesetter:	Baltimore Type
typefaces:	Caledonia text, Perpetua display
printer:	Universal Lithographers, Inc.
paper:	Perkins & Squier F
binder:	Moore & Company
cover material:	Bancroft Arrestox C

Milton Keynes UK
Ingram Content Group UK Ltd.
UKHW011459250224
438379UK00002B/303